T0328224

BIOCHAR FROM BIOMASS AND WASTE

BIOCHAR FROM BIOMASS AND WASTE

Fundamentals and Applications

Edited by

YONG SIK OK

Korea Biochar Research Center, O-Jeong Eco-Resilience Institute (OJERI) & Division of Environmental Science and Ecological Engineering, Korea University, Seoul, Republic of Korea

DANIEL C.W. TSANG

Department of Civil and Environmental Engineering, The Hong Kong Polytechnic University, Hung Hom, Kowloon, Hong Kong, P.R. China

NANTHI BOLAN

Global Centre for Environmental Remediation (GCER), Faculty of Science, The University of Newcastle, Callaghan, NSW, Australia

J.M. NOVAK

U.S. Department of Agriculture, Agricultural Research Service, Coastal Plains Research Center, Florence, SC, United States

Elsevier
Radarweg 29, PO Box 211, 1000 AE Amsterdam, Netherlands
The Boulevard, Langford Lane, Kidlington, Oxford OX5 1GB, United Kingdom
50 Hampshire Street, 5th Floor, Cambridge, MA 02139, United States

British Library Cataloguing-in-Publication Data
A catalogue record for this book is available from the British Library

Library of Congress Cataloging-in-Publication Data
A catalog record for this book is available from the Library of Congress

ISBN: 978-0-12-811729-3

For Information on all Elsevier publications
visit our website at https://www.elsevier.com/books-and-journals

Publisher: Candice Janco
Acquisition Editor: Candice Janco
Editorial Project Manager: Hilary Carr
Production Project Manager: Prem Kumar Kaliamoorthi
Cover Designer: Greg Harris

Typeset by MPS Limited, Chennai, India

Contents

List of Contributors

Tahir Abbas Department of Environmental Sciences and Engineering, Government College University, Faisalabad, Pakistan

Nahida Akter Department of Chemistry, Bangladesh University of Engineering and Technology (BUET), Dhaka, Bangladesh

Md. Samrat Alam Department of Earth and Atmospheric Sciences, University of Alberta, Edmonton, AB, Canada

Daniel S. Alessi Department of Earth and Atmospheric Sciences, University of Alberta, Edmonton, AB, Canada

Arshad Ali Spatial Ecology Lab, School of Life Science, South China Normal University, Guangzhou, Guangdong, China

Shafaqat Ali Department of Environmental Sciences and Engineering, Government College University, Faisalabad, Pakistan

Md. Shafiul Azam Department of Chemistry, Bangladesh University of Engineering and Technology (BUET), Dhaka, Bangladesh

Arooj Bashir Department of Environmental Sciences and Engineering, Government College University, Faisalabad, Pakistan

Luke Beesley The James Hutton Institute, Environmental and Biochemical Sciences Group, Aberdeen, United Kingdom

Nanthi Bolan Global Centre for Environmental Remediation (GCER), Faculty of Science, The University of Newcastle, Callaghan, NSW, Australia

Zhaohui Cai Institute of Coastal Environmental Pollution Control, Key Laboratory of Marine Environment and Ecology, Ministry of Education, College of Environmental Science and Engineering, Ocean University of China, Qingdao, P.R. China

Liqiang Cui School of Environmental Science and Engineering, Yancheng Institute of Technology, Yancheng, China

Caleb E. Egene Department of Green Chemistry and Technology, Ghent University, Gent, Belgium

Ali El-Naggar Department of Soil Sciences, Faculty of Agriculture, Ain Shams University, Cairo, Egypt; Korea Biochar Research Center, O-Jeong Eco-Resilience Institute (OJERI) & Division of Environmental Science and Ecological Engineering, Korea University, Seoul, Republic of Korea

Su Yun Gladys Choo Department of Building, National University of Singapore, Singapore, Singapore

Sameera R. Gunatilake College of Chemical Sciences, Institute of Chemistry Ceylon, Rajagiriya, Sri Lanka

K. Hall Department of Soil, Water, and Climate, University of Minnesota, St. Paul, MN, United States

Noha E.E. Hassan Agricultural Extension and Rural Development Research Institute, Agricultural Researcher Center, Giza, Egypt

Barbora Hudcová Department of Environmental Geosciences, Faculty of Environmental Sciences, Czech University of Life Sciences Prague, Prague, Czech Republic

James A. Ippolito Department of Soil and Crop Sciences, Colorado State University, Fort Collins, CO, United States

Shih-Hao Jien Department of Soil and Water Conservation, National Pingtung University of Science and Technology, Pingtung, Taiwan

Mark G. Johnson U.S. Environmental Protection Agency, National Health and Environmental Effects Research Laboratory/ORD Western Ecology Division, Corvallis, OR, United States

Michael Komárek Department of Environmental Geosciences, Faculty of Environmental Sciences, Czech University of Life Sciences Prague, Prague, Czech Republic

Harn Wei Kua Department of Building, National University of Singapore, Singapore, Singapore

Eilhann E. Kwon Department of Environment and Energy, Sejong University, Seoul, Republic of Korea

Jechan Lee Department of Environmental and Safety Engineering, Ajou University, Suwon, Republic of Korea

Qiao Li School of Environmental and Biological Engineering, Nanjing University of Science and Technology, Nanjing, P.R. China

Xiaodian Li Department of Environmental Engineering, P.R. China Jiliang University, Hangzhou, Zhejiang, P.R. China

Bingjie Liu Institute of Coastal Environmental Pollution Control, Key Laboratory of Marine Environment and Ecology, Ministry of Education, College of Environmental Science and Engineering, Ocean University of China, Qingdao, P.R. China

Guocheng Liu College of Resource and Environment, Qingdao Agricultural University, Qingdao, P.R. China

Shou-Heng Liu Department of Environmental Engineering, National Cheng Kung University, Tainan, Taiwan

Jiewen Luo Shanghai Key Laboratory of Atmospheric Particle Pollution and Prevention (LAP3), Department of Environmental Science and Engineering, Fudan University, Shanghai, P.R. China; Shanghai Institute of Pollution Control and Ecological Security, Shanghai, P.R. China

Arosha Maqbool Department of Environmental Sciences and Engineering, Government College University, Faisalabad, Pakistan

E. Moore Department of Environmental Science, School of Earth and the Environment, Rowan University, Glassboro, NJ, United States

Nabeel Khan Niazi Institute of Soil and Environmental Sciences, University of Agriculture Faisalabad, Faisalabad, Pakistan; Southern Cross GeoScience, Southern Cross University, Lismore, NSW, Australia

J.M. Novak U.S. Department of Agriculture, Agricultural Research Service, Coastal Plains Research Center, Florence, SC, United States

Nadeeka L. Obadamudalige Global Centre for Environmental Remediation (GCER), Faculty of Science, The University of Newcastle, Callaghan, NSW, Australia

Yong Sik Ok Korea Biochar Research Center, O-Jeong Eco-Resilience Institute (OJERI) & Division of Environmental Science and Ecological Engineering, Korea University, Seoul, Republic of Korea

Chathuri Peiris College of Chemical Sciences, Institute of Chemistry Ceylon, Rajagiriya, Sri Lanka

Muhammad Zia ur Rehman Institute of Soil and Environmental Sciences, University of Agriculture, Faisalabad, Pakistan

Jörg Rinklebe University of Wuppertal, School of Architecture and Civil Engineering, Institute of Foundation Engineering, Water- and Waste-Management, Laboratory of Soil- and Groundwater-Management, Wuppertal, Germany

Muhammad Rizwan Department of Environmental Sciences and Engineering, Government College University, Faisalabad, Pakistan

Ajit K. Sarmah Department of Civil & Environmental Engineering, The Faculty of Engineering, The University of Auckland, Auckland, New Zealand

Sabry M. Shaheen Department of Soil and Water Sciences, University of Kafrelsheikh, Faculty of Agriculture, Kafr El-Sheikh, Egypt; University of Wuppertal, School of Architecture and Civil Engineering, Institute of Foundation Engineering, Water- and Waste-Management, Laboratory of Soil- and Groundwater-Management, Wuppertal, Germany

Hua Shang Shanghai Key Laboratory of Atmospheric Particle Pollution and Prevention (LAP3), Department of Environmental Science and Engineering, Fudan University, Shanghai, P.R. China; Shanghai Institute of Pollution Control and Ecological Security, Shanghai, P.R. China

K.A. Spokas US Department of Agriculture, Agricultural Research Service, Soil Water Management Research Unit, St. Paul, MN, United States

Filip M.G. Tack Department of Green Chemistry and Technology, Ghent University, Gent, Belgium

Rizwan Tareq Department of Materials and Metallurgical Engineering, Bangladesh University of Engineering and Technology (BUET), Dhaka, Bangladesh

Lukáš Trakal Department of Environmental Geosciences, Faculty of Environmental Sciences, Czech University of Life Sciences Prague, Prague, Czech Republic

Daniel C.W. Tsang Department of Civil and Environmental Engineering, The Hong Kong Polytechnic University, Hung Hom, Kowloon, Hong Kong, P.R. China

Meththika Vithanage Ecosphere Resilience Research Center, Faculty of Applied Sciences, University of Sri Jayewardenepura, Nugegoda, Sri Lanka

Martina Vítková Department of Environmental Geosciences, Faculty of Environmental Sciences, Czech University of Life Sciences Prague, Prague, Czech Republic

Hailong Wang School of Environment and Chemical Engineering, Foshan University, Foshan, Guangdong, P.R. China; Key Laboratory of Soil Contamination Bioremediation of Zhejiang Province, Zhejiang A & F University, Hangzhou, Zhejiang, P.R. China

Jianxu Wang State Key Laboratory of Environmental Geochemistry, Institute of Geochemistry, Chinese Academy of Sciences, Guiyang, P.R. China; University of Wuppertal, School of Architecture and Civil Engineering, Institute of Foundation Engineering, Water- and Waste-Management, Laboratory of Soil- and Groundwater-Management, Wuppertal, Germany

Xiaonan Wang Department of Chemical and Biomolecular Engineering, National University of Singapore, Singapore

Jayani J. Wewalwela Department of Agricultural Technology, Faculty of Technology, University of Colombo, Colombo, Sri Lanka

A. Williams Davidson Laboratory, Stevens Institute of Technology, Hoboken, NJ, United States

Xinni Xiong Department of Civil and Environmental Engineering, The Hong Kong Polytechnic University, Hung Hom, Kowloon, Hong Kong, P.R. China

Yilu Xu Global Centre for Environmental Remediation (GCER), Faculty of Science, The University of Newcastle, Callaghan, NSW, Australia

Yubo Yan School of Chemistry and Chemical Engineering, Huaiyin Normal University, Jiangsu, P.R. China

Xiaodong Yang Institute of Resources and Environment Science, Xinjiang University, Urumqi, Xinjiang, China

Siming You Division of Systems, Power & Energy, School of Engineering, University of Glasgow, Glasgow, United Kingdom

Iris K.M. Yu Department of Civil and Environmental Engineering, The Hong Kong Polytechnic University, Hung Hom, Kowloon, Hong Kong, P.R. China

Chenchen Zhang Institute of Coastal Environmental Pollution Control, Key Laboratory of Marine Environment and Ecology, Ministry of Education, College of Environmental Science and Engineering, Ocean University of China, Qingdao, P.R. China

Ming Zhang Department of Environmental Engineering, P.R. China Jiliang University, Hangzhou, Zhejiang, P.R. China

Shicheng Zhang Shanghai Key Laboratory of Atmospheric Particle Pollution and Prevention (LAP3), Department of Environmental Science and Engineering, Fudan University, Shanghai, P.R. China; Shanghai Institute of Pollution Control and Ecological Security, Shanghai, P.R. China

Weihua Zhang School of Environmental Science and Engineering, Sun Yat-sen University, Guangzhou, P.R. China

Cheng Zhao Department of Environmental Engineering, P.R. China Jiliang University, Hangzhou, Zhejiang, P.R. China

Hao Zheng Institute of Coastal Environmental Pollution Control, Key Laboratory of Marine Environment and Ecology, Ministry of Education, College of Environmental Science and Engineering, Ocean University of China, Qingdao, P.R. China

Shaojie Zhou Shanghai Key Laboratory of Atmospheric Particle Pollution and Prevention (LAP3), Department of Environmental Science and Engineering, Fudan University, Shanghai, P.R. China; Shanghai Institute of Pollution Control and Ecological Security, Shanghai, P.R. China

Xiangdong Zhu Shanghai Key Laboratory of Atmospheric Particle Pollution and Prevention (LAP3), Department of Environmental Science and Engineering, Fudan University, Shanghai, P.R. China; Shanghai Institute of Pollution Control and Ecological Security, Shanghai, P.R. China

PART I

BIOCHAR PRODUCTION

CHAPTER

1

Production and Formation of Biochar

Jechan Lee[1], *Ajit K. Sarmah*[2] *and Eilhann E. Kwon*[3]

[1]Department of Environmental and Safety Engineering, Ajou University, Suwon, Republic of Korea [2]Department of Civil & Environmental Engineering, The Faculty of Engineering, The University of Auckland, Auckland, New Zealand [3]Department of Environment and Energy, Sejong University, Seoul, Republic of Korea

1.1 INTRODUCTION

Feeding the world's burgeoning population in the coming decades is a daunting but achievable goal. Throughout the world, there has been extremely high demand for fossil fuels used to generate energy, which has led to current energy crises, especially in industrial and developed countries. The total energy consumption in the world is predicted to increase by approximately 65% by the year 2040 (EIA, 2016; BP, 2016). The depletion in availability of fossil fuels and deleterious environmental issues caused by using fossil fuels are some of the most urgent problems mankind faces today. The reduction of fossil fuel consumption and replacement with clean renewable sustainable energy resources would be an effective way to resolve these environmental problems in the coming years.

Biomass is derived from living or recently living organic materials. It includes lignocellulosic materials such as wood, plants, forestry residuals, and agricultural crop residuals and nonlignocellulosic materials such as animal manure and organic parts of municipal solid waste (MSW) (Demirbaş, 2001). Biomass grows relatively quickly compared to fossil fuels and is abundant on Earth and thus considered as a potential resource for sustainable energy production (Perlack et al., 2005). In addition, biomass is the only renewable carbon-based energy source that can be converted into carbonaceous gaseous, liquid, and solid fuels (Özbay et al., 2001).

Lignocellulosic biomass, such as plants and trees, is comprised typically of cellulose, hemicellulose, and lignin. The three components are cross-linked and chemically bonded by noncovalent forces. These components also have their own unique physical and

Biochar from Biomass and Waste
DOI: https://doi.org/10.1016/B978-0-12-811729-3.00001-7

chemical properties. Cellulose is a homopolymer of D-glucose, linked by β-1,4 glycosidic bonds that form long chains (Fengel and Wegener, 1984) with the molecular formula of $(C_6H_{10}O_5)_n$. Hemicellulose is a hetropolymer of glucose, mannose, xylan, xylose, arabinose, galactose, and sugar acids (Hendriks and Zeeman, 2009). Lignin is a complex organic polymer comprised of three phenyl-propane monomers such as coniferyl, trans-*p*-coumaryl, and sinapyl alcohols (Hendriks and Zeeman, 2009). The amorphous heteropolymer is oriented by a different degree of methoxylation of the monomers forming large molecular structures (Glasser and Sarkanen, 1989). Hemicellulose is thermally degraded relatively easily compared to other components (LeVan et al., 1990; Winandy, 1995). Lignin is most thermally stable among the components in biomass. Compositions of these three components are highly dependent on the type of biomass.

Although biomass is considered as a renewable energy resource, given the inferior properties of biomass (e.g., low heating values, high contents of moisture and volatile components, fibrous nature) (Pimchuai et al., 2010; Khan et al., 2009; Yip et al., 2010; Demirbas, 2004a), it is not ideal for fuel. In addition, seasonal variation and wide diversity of energy densities, chemical compositions, and physical shapes of different biomass feedstocks lead to discontinuous availability, inefficient transportation, handling, sizing, and storage of the feedstocks. Due to these limitations and challenges, treatment of biomass is required for it to be used as a renewable energy resource. A broad range of biomass treatment processes are available to enhance the combustion properties of biomass and convert it into gaseous or liquid fuels (Goyal et al., 2008b; Saxena et al., 2009). Among the various biomass treatment technologies, thermochemical processes are generally preferred because of advantages such as high conversion efficiency and short reaction time (Liu et al., 2013).

Recently, biochar, a thermochemically converted solid carbonaceous product produced from the pyrolysis process of biomass, has shown promise due to its captivating extensive applications such as reducing greenhouse gas emissions by means of carbon capture and storage (CCS) and increasing soil fertility in an environmentally friendly way. Most recently, biochar has been used in other nonsoil-related applications such as in biocomposite development (Das et al., 2016a,b; Poulose et al., 2018) and in green concrete production (Akhtar and Sarmah, 2018). The history of biochar is associated with the discovery of *Terra Preta di indio* (black earth) in the Amazon basin. Studies have shown that the soil improved crop productivity compared to other topical soils (Zech et al., 1990). Early research into soil interaction with biochar has shown that biochar is one of the key components of the distinct properties of certain soils found in the Amazonian basin (Glaser et al., 2001). These findings led to a number of investigations in various parts of the world with regard to biochar and its application as soil amendment (Lehmann, 2009; Libra et al., 2011; Amonette and Joseph, 2009). Thereafter, biochar was considered as a new tool for establishing environmental management and sustainable energy production beyond soil applications (Lee et al., 2017; Qian et al., 2015).

Biochar is defined by various authors by different terms such as charcoal, char, and agrichar produced from nonfossil-based carbonaceous organic materials such as biomass. However, most definitions of biochar are interrelated with regard to its production and applications (Ahmad et al., 2014). According to the International Biochar Initiative (IBI),

"Biochar is a solid material obtained from the thermochemical conversion of biomass in an oxygen-limited environment." (IBI, 2012). This definition of biochar has since become the mostcommon. It is critical to differentiate between terminologies such as charcoal and biochar. Charcoal is a carbon-rich solid material prepared by charring biomass, utilized as a solid fuel to produce energy. Biochar is an alternative to charcoal, employed for purposes such as CCS and soil conditioners (Lehmann and Joseph, 2015). Biochar is a solid byproduct produced with a dry carbonization process (e.g., pyrolysis). Biochar production processes have significant effects on the physicochemical properties of the biochar (Fuertes et al., 2010; Wiedner et al., 2013).

Hydrochar is another term, when char is made by hydrothermal carbonization (a process where biomass is heated at a temperature range of 200–300°C in the presence of water), and is comprised of two phases: liquid and solid (i.e., slurry) (Manyâ, 2012; Sohi et al., 2010; Funke and Ziegler, 2010). Note that the terms "biochar" and "hydrochar" are distinctively different and therefore hydrothermal carbonization of wet biomass resulting in "hydrochar" cannot be considered as "biochar."

The physicochemical properties of biochar are highly associated with the raw materials used, the type of process, and the process conditions. This chapter introduces how the different kinds of biochar can be made depending on the types of feedstock, production processes, and their operation conditions.

1.2 RAW MATERIALS OF BIOCHAR

From a production point of view, it is important to classify the feedstock of biochar and the production method used as well as its feasibility as it is highly dependent on the type of feedstock chosen (especially, dry or wet). The initial moisture content of biomass determines the classification of dry or wet feedstocks. Dry biomass such as some wood species and agricultural residues have low moisture content (lower than 30%) after harvested (Knežević, 2009). On the other hand, wet biomass with high moisture content (higher than 30%) typically refers to freshly harvested biomass like sewage sludge, algae, vegetable wastes, animal wastes, and so on (Knežević, 2009). Wet feedstocks need to be dried prior to processing, but biomass drying processes are energy intensive and thus reduce the economic efficiency of the process (Mani et al., 2006; Sokhansanj and Fenton, 2006).

Biomass can also be categorized as purpose-grown biomass (i.e., energy crops) and waste biomass (Lehmann et al., 2006). Compared to other crops, energy crops (e.g., switchgrass, miscanthus) need low maintenance and have high energy content. They also have low moisture content (<10%) upon harvesting, not requiring additional drying processes. Energy crops are primarily used by biorefinery industries for producing liquid fuels (Brosse et al., 2012). Waste biomass includes food wastes, sewage sludge, organic fraction of MSW, agroforestry waste, and animal manure (Brick and Lyutse, 2010). The employment of waste biomass in the production of biochar is practical because the waste feedstock does not compete with food and energy crops and for arable land. Nevertheless, there is no universal consensus on what the (ideal) composition of waste biomass is (Perlack and Turhollow, 2003), and overuse of waste biomass can interrupt the environmental lifecycle. Despite these negative factors, it is considered that employing waste biomass as a feedstock to produce biochar is not only ecofriendly but can also pave the way to effectively utilize waste streams (Lehmann and Joseph, 2015; Sohi et al., 2010).

1.3 PROCESSES FOR BIOCHAR PRODUCTION

Biochar is produced through thermochemical conversion of biomass. Thermochemical processes include pyrolysis, torrefaction (dry or wet), gasification, and hydrothermal processing. It is essential to choose the proper technique and the right operating conditions (e.g., temperature, vapor and solid residence time, heating rate, reaction environment) for biochar production.

Among the thermochemical processes, dry torrefaction and gasification are not suitable to make biochar. Dry torrefaction heats biomass feedstock at 200–300°C at ambient pressure under an inert environment for 30 minutes to a few hours (Rousset et al., 2012; Bach and Tran, 2015). During dry torrefaction, biomass feedstock loses only 10% of the energy contained in the feedstock in the form of gaseous products, thereby increasing its energy density (Bridgeman et al., 2008; Kambo and Dutta, 2015). Thus, dry torrefaction has been used as a pretreatment process of biomass prior to combustion (Bergman et al., 2005). Nevertheless, the solid product resulting from dry torrefaction of biomass still contains volatile organic compounds from the raw biomass. In addition, the torrefied biomass has physicochemical properties in between biomass and biochar. Therefore, torrefied biomass is technically not considered as "biochar." However, this is still debatable as literature is sparse on the use and comparison between "hydrochar" and "biochar" for a specific purpose. Therefore, as part of the discussion, we have also included information on hydrochar in this chapter.

Gasification is a partial combustion process of biomass at higher temperatures (600–1200°C) than pyrolysis with residence time of 10–20 seconds (McKendry, 2002; Brewer et al., 2009). The primary purpose of the gasification process is to convert carbonaceous feedstocks into a mixture of gases (H_2, CO, and CO_2), for example, synthesis gas (syngas) and producer gas (Bridgwater, 1995; Kirubakaran et al., 2009; Puig-Arnavat et al., 2010). Thus, technically, an ideal gasification process should not produce biochar. In actual practice, however, biochar is produced with very low yields (less than 10%) in gasification (Brewer et al., 2009). The biochar made from gasification of biomass contains a variety of inorganic elements and polycyclic aromatic hydrocarbons (PAHs) (Ippolito et al., 2012). PAHs are toxic byproducts produced at high temperatures, so using biochar containing such toxic compounds for environmental remediation can be problematic (Sivula et al., 2012; Laird et al., 2011). Furthermore, IBI states that harmful air-pollutant emissions and toxicant accumulation in the final biochar material must be avoided (IBI, 2013). Hence, in this chapter, biochar production via pyrolysis and hydrothermal carbonization is further considered. Table 1.1 lists the characteristics of different thermochemical processes and typical yields of the solid product achieved by each process.

TABLE 1.1 Classification of Different Thermochemical Pretreatments in Terms of Operating Conditions and Biochar Yield (Kambo and Dutta, 2015; Liu et al., 2015)

Process		Operation Temp. (°C)	Heating Rate (°C min^{-1})	Residence Time	Biochar Yield (wt.%)
Pyrolysis	Slow	300–800	5–7	>1 h	35–50
	Fast	400–600	300–800	0.5–10 s	15–35
	Flash	400–1000	~1000	<2 s	10–20
Dry torrefaction		200–300	10–15	30 min–4 h	60–80
Gasification		600–1200	50–100 (°C s^{-1})	10–20 s	<10
Hydrothermal carbonization		180–260	5–10	5 min–12 h	45–70

1.3.1 Pyrolysis

Biomass pyrolysis is the principal technique used to produce biochar. Pyrolysis is a process that thermally decomposes biomass by heating it at elevated temperatures under controlled inert conditions (i.e., very little O_2 or under an inert gas atmosphere such as N_2). During pyrolysis, the organic components in biomass are thermally decomposed to release a vapor phase while residual solid phase (i.e., biochar) remains. In the vapor phase, polar and high-molecular-weight compounds are cooled to produce liquid phase (i.e., bio-oil) while low-molecular-weight compounds (i.e., noncondensable gases such as H_2, CH_4, C_2H_2, CO, and CO_2) remain as the gaseous phase.

Pyrolysis can be primarily categorized as fast and slow pyrolysis according to its operating parameters, especially heating rate. Fast pyrolysis occurs at 400–600°C with heating rates higher than 300°C min^{-1} for short vapor residence time (0.5 to 10 seconds). Fast pyrolysis mainly yields bio-oil, but produces biochar with a yield of 15–30 wt.%. Slow pyrolysis is mostly preferred to produce biochar because it gives the highest solid product yield (35 to 50 wt.%) (Mohan et al., 2006). This process is carried out at temperatures from 300°C to 800°C with heating rates of 5–10°C min^{-1}. Vapor residence time in slow pyrolysis ranges from a few minutes to a few hours (Onay and Kockar, 2003).

Physical changes and chemical reactions occur simultaneously in pyrolysis, and strongly depend on the reactor conditions, operation parameters, and nature of the feedstock. Biochar yield and its properties are highly related to such variables. For example, a low pyrolysis temperature and a slow heating rate lead to high yield of solid product (Karaosmanoğlu et al., 1999; Onay, 2007). In contrast, a high pyrolysis temperature, fast heating rate, and long residence time have effects on the carbon percentage, surface area, and high heating value (HHV) of biochar (Ronsse et al., 2013). The content of fixed carbon in biochar usually increases with an increase in pyrolysis temperature. This can be explained as follows. In the early stage of pyrolysis (i.e., at low temperatures), volatile matter, such as CO, CO_2, C_xH_y, $C_xH_yO_z$, H_2O, HCN, and NH_3, are randomly released from the biomass feedstock. The random release of volatile matter decreases the yield of biochar (Becidan et al., 2007). A further increase in pyrolysis temperature decreases the release of carbon-rich compounds ($C_xH_yO_z$) whereas other volatile compounds (CO, CO_2, HCN, etc.) are continuously released, leading to increased content of fixed carbon in the remaining biochar. Apart from the biochar yield and fixed carbon content, texture features (e.g., surface area and pore size distribution) of biochar are influenced by pyrolysis temperature. For instance, increasing the pyrolysis temperature results in the release of more volatile matter from the biomass surface, which produces a biochar with greater surface area and more pores.

It has been found that in addition to pyrolysis temperature and heating rate, at high pressures, high contents of moisture (e.g., 40%–60%) increase biochar yield (Manyâ, 2012). This suggests that biomass with high moisture content may be preferred to produce biochar but could be energy intensive. The inherent chemical composition of biomass (i.e., composition of cellulose, hemicellulose, lignin, and inorganics) is another important parameter for biochar production. It was reported that biomass with a high lignin content such as spruce wood and pine wood begets biochar with a high yield and high content of fixed carbon (Antal and Grúnli, 2003). Inorganic species such as alkaline earth metals contained in biomass feedstock have catalytic effects on the decomposition of biomass and formation of char (Yaman, 2004). Biochar yield has been found to decrease when biomass feedstock is pretreated with hot water or acid (Vassilev et al., 2012; Uchimiya et al., 2011). The available literature shows that for pyrolysis of different biomass samples (e.g., herbaceous biomass, woody biomass, poultry litter, microalgae) conducted under hydrogen atmospheres, the increase in temperature decreases biochar yield but increases the content of PAHs (McBeath et al., 2015). Among the samples, herbaceous biomass-derived biochars have low carbon content with high ratios of PAHs to total organic carbon. However, woody biomass-derived biochars have high carbon content with low ratios of PAHs to total organic carbon. Furthermore, the surface area and pH of biochar are affected mainly by pyrolysis temperature, although the total organic carbon content, mineral concentration, ash content, and carbon sequestration capacity of biochar are predominantly influenced by the type of biomass feedstock (Zhao et al., 2013). Table 1.2 summarizes some of the available information on biochar properties as a function of temperature and type of feedstock. The data indicate that judicious selection of feedstock and pyrolysis temperature make it possible to tune the properties of biochar for specific applications.

In addition to biochar, two other pyrolytic products (gas and liquid) are distributed in a ratio of 25%–30% for a typical slow pyrolysis process. The gaseous and liquid products are syngas and bio-oil, respectively, both considered renewable energies. Therefore, pyrolysis is an environmental friendly and cost-effective biochar production process due to recirculating the gaseous products and recovering the liquid products to improve the efficiency of slow pyrolysis (Zhang et al., 2010). A study examined the economic tradeoff between production of bio-oil and biochar via pyrolysis of biomass (Yoder et al., 2011). In the study, quadratic production functions for bio-oil and biochar during the pyrolysis were estimated, and the relationship between product yield and pyrolysis temperature was established. The results were used to infer optimum process temperature for a price ratio of biochar to bio-oil under various economic conditions. The biochar yield and its properties are highly associated with the characteristics of the particular process and the nature of the feedstock.

TABLE 1.2 Characteristics of Biochars Produced by Pyrolysis of Different Biomass Feedstocks (Zhao et al., 2013; Liu et al., 2015)

Feedstock	Production Temp. (°C)	Yield (%)	Total C (%, Dry Basis)	Fixed C (%, Dry Basis)	Volatile Matter (%, Dry Basis)	Ash (%)	pH	BET-N_2 Surface Area, ($m^2\ g^{-1}$)	Cation Exchange Capacity (cmol kg^{-1})
Wheat straw	200	99.3	38.7	22.5	70.2	7.2	5.4	2.5	32.1
	300	52.5	59.8	53.2	31.3	14.7	8.7	3.5	87.2
	400	29.8	62.9	63.7	17.6	18.0	10.2	33.2	95.5
	500	26.8	68.9	72.1	11.1	16.2	10.2	182.0	146.0
Sawdust	500	28.3	75.8	72.0	17.5	9.9	10.5	203.0	41.7
Grass	500	27.8	62.1	59.2	18.9	20.8	10.2	3.3	84.0
Peanut shell	500	32.0	73.7	72.9	16.0	10.6	10.5	43.5	44.5
Pig manure	200	98.0	37.0	12.6	50.7	35.7	8.2	3.6	23.6
	300	57.5	39.1	34.7	27.4	37.2	9.7	4.3	49.0
	400	38.5	42.7	40.2	11.0	48.4	10.5	47.4	82.8
	500	35.8	45.3	19.2	10.7	69.6	10.8	42.4	132.0
Cow manure	500	57.2	43.7	14.7	17.2	67.5	10.2	21.9	149.0
Shrimp hull	500	33.4	52.1	18.9	26.6	53.8	10.3	13.3	389.0
Chlorella	500	40.2	39.3	17.4	29.3	52.6	10.8	2.8	562.0
Bone dregs	500	48.7	24.2	10.5	11.0	77.6	9.6	113.0	87.9
Waste paper	500	36.6	56.0	16.4	30.0	53.5	9.9	133.0	516.0
Wastewater sludge	500	45.9	26.6	20.6	15.8	61.9	8.8	71.6	168.0
Waterweeds	500	58.4	25.6	3.8	32.4	63.5	10.3	3.8	509.0

Reprinted with permission from: Liu, W.-J., Jiang, H., Yu, H.-Q., 2015. Development of biochar-based functional materials: toward a sustainable platform carbon material, Chem. Rev. 115 (22), 12251-12285. https://doi.org/10.1021/acs.chemrev.5b00195, Copyright (2015) American Chemical Society.

1.3.2 Hydrothermal Carbonization

Hydrothermal carbonization, also known as wet torrefaction, gives a high conversion efficiency, requires no predrying step of feedstock, and operates at relatively low temperatures compared to other thermochemical processes used to produce biochar. It is carried out in water at 180–260°C for 5 minutes to 6 hours (Mumme et al., 2011; Hoekman et al., 2013). The process operates under saturation vapor pressure of water varying with reaction temperature (i.e., subcritical water). In subcritical water, water is still liquid (under pressure) and acts as a nonpolar solvent. For example, water acts as methanol at ~200°C and as acetone at ~300°C (Kritzer and Dinjus, 2001). This means that organic fractions of raw materials of biochar are soluble in water at such conditions. Also, sub- and supercritical water have unique properties that can be used for destroying hazardous wastes such as polychlorinated dibenzofurans and polychlorinated biphenyls (Weber et al., 2002).

Hydrothermal carbonization is suitable for treating wet feedstock (Hitzl et al., 2015; Zhao et al., 2014) because it is not affected by the moisture content of feedstock and takes place in the presence of water. Therefore, it can eliminate the predrying step for feedstock, a step that is highly energy intensive, potentially reducing the costs of biochar production (Benavente et al., 2015). In general, hydrothermal carbonization forms three phase products: biochar (solid phase), bio-oil (liquid phase), and small fractions of gaseous products, mainly CO_2. The distribution of the three phases and properties of each product are highly associated with process operating conditions. For hydrothermal carbonization, temperature is the governing operation parameter (Kambo and Dutta, 2014). The yield of biochar from hydrothermal carbonization is about 40–70 wt.% (Yan et al., 2010). However, a newly prepared biochar is a two-phase slurry (i.e., liquid and solid), so a series of dewatering steps (e.g., mechanical compressing, filtering, thermal/solar drying) are required before it can be used as a fuel. In hydrothermal carbonization, the moisture content of biochar can be less than 50% just by mechanical compressing because a fraction of oxygen is removed by dehydratization and decarboxylation reactions (Kambo and Dutta, 2015). This reduces the amount of supplementary energy and time consumption for thermal/solar drying to further reduce the moisture content.

The primary purpose of thermochemical processes is to decompose the rigid polymeric structure of feedstock (e.g., lignin) into low-molecular-weight and small chemical compounds. The rate of decomposition is dependent on reaction conditions (e.g., temperature, time, reaction medium). In hydrothermal carbonization processes, the decomposition reaction is initiated by hydrolysis because it operates in subcritical water. It favors rapid depolymerization of feedstock into water-soluble products such as monomers and oligomers (Yu et al., 2008; Bobleter, 1994). In the case of lignocellulosic biomass used as feedstock, different inorganics can be present such as Mg, Ca, Na, K, Fe, P, S, etc. For a typical biomass combustion, these elements exist as their corresponding oxide forms (e.g., MgO, CaO, Na_2O, K_2O, Fe_2O_3, P_2O_5, and SO_3) in ash (Tortosa Masiã et al., 2007). These metal oxides cause corrosion, fouling, slagging, and klinker formation during the process (Baxter et al., 1998). However, acetic acid forms in a byproduct stream of the hydrothermal carbonization process, and the metals are leached out in the acid via acid solvation, thereby reducing ash content in the solid product (Liu and Balasubramanian, 2014; Lynam et al., 2011). Furthermore, hydrothermal carbonization suppresses the formation of dioxins and corrosive species by converting organic Cl into inorganic Cl when treating feedstock with high Cl content (Zhao et al., 2014).

For an industrial hydrothermal carbonization plant, a major challenge is the continuous supply of water. In addition, a massive amount of water is required for an industrial full-scale operation, making it economically nonviable. For instance, for the production of one ton of dry biochar from miscanthus via hydrothermal carbonization at 260°C, 12 tons of liquid water are required at a 50 wt.% biochar yield and a ratio of dry biomass to water of 1:6 in spite of its superior combustion properties (Kambo and Dutta, 2014). Therefore, recirculating water is critical for industrial hydrothermal carbonization processes. The recirculation of water also decreases the cost of treating wastewater and lowers consumption of external heat (i.e., high heat recovery efficiency) (Stemann et al., 2013).

1.4 MECHANISM OF THE FORMATION OF BIOCHAR

Even though understanding the mechanisms of biochar formation (e.g., physical conversions of feedstock and chemical reactions taking place) is critical to make biochar with the desired physicochemical properties, it has not fully become understood yet. Typically known chemical reactions occurring during thermochemical processes for biochar production include decomposition and depolymerization of polymeric components, aromatization, condensation, decarbonylation, decarboxylation, dehydration, demethoxylation, intermolecular derangement, and so on (Liu et al., 2015). Reaction temperature is one of the most important parameters governing the dominant reaction(s), but it is difficult to maintain a uniform temperature profile in a real system. Therefore, in pyrolysis and hydrothermal carbonization, various chemical reactions take place simultaneously and/or successively, resulting in a complex mixture of reactants, intermediates, and products in different phases. The following sections focus on the various mechanisms of biochar and hydrochar formation.

1.4.1 Formation of Biochar Via Pyrolysis

The mechanism of pyrolysis of biomass to produce biochar is comprised of pyrolysis mechanisms of cellulose, hemicellulose, and lignin (Goyal et al., 2008a) because each component has different thermal characteristics. For example, cellulose decomposes at 240–350°C while hemicellulose breaks down at 200–260°C. Lignin starts to decompose between 280°C and 500°C (Babu, 2008).

Pyrolysis of cellulose can be characterized by decreasing the degree of polymerization (Shen et al., 2011), which can occur in decomposition and charring (Shen et al., 2010b) and in fast volatilization with forming levoglucosan (Shen and Gu, 2009), an important intermediate step in the pyrolysis of cellulose. In cellulose pyrolysis, depolymerization of cellulose into oligosaccharides first occurs. The formation of D-glucopyranose by cleaving glucosidic bonds occurs as follows. The D-glucopyranose undergoes an intramolecular rearrangement to produce levoglucosan (Li et al., 2001). Levoglucosenone can be produced by dehydration of levoglucosan, followed by further proceeding through aromatization, decarboxylation, dehydration, and intermolecular condensation to produce a complex solid network (i.e., biochar). In other pathways, levoglucosan undergoes a series of dehydration and rearrangement reactions forming hydroxymethylfurfural (HMF) which is an intermediate in the formation of syngas and bio-oil and a series of aromatization, polymerization, and intramolecular condensation reactions that make biochar (Mettler et al., 2012; Ronsse et al., 2012; Lin et al., 2009; Luo et al., 2004; Zhang et al., 2012; Patwardhan et al., 2011b; Banyasz et al., 2001).

Pyrolysis of hemicellulose begins with depolymerization of hemicellulose into oligosaccharides followed by cleaving glycosidic linkages in xylan chains and rearrangement of oligosaccharides to form 1,4-anhydro-D-xylopyranose (Shen et al., 2010a). This intermediate step is needed to form biochar through aromatization, decarboxylation, dehydration, and intramolecular condensation, similar to the case of cellulose pyrolysis (Peters, 2011; Peng and Wu, 2010; Huang et al., 2012; Patwardhan et al., 2011a; Sefain et al., 1985).

The mechanism of lignin pyrolysis is much more complex than that of cellulose and hemicellulose because the structure of lignin is more complex than cellulose and hemicellulose. One of the most important dominant pathways in lignin pyrolysis is a free radical reaction (Kosa et al., 2011; Cho et al., 2012; Mu et al., 2013). The cleavage of β-O-4 linkage in lignin generates free radicals. This is the initial step for the free radical reaction (Chu et al., 2013; Ben and Ragauskas, 2011). The free radicals capture protons of species with weak O − H or C − H bonds (e.g., phenyl group), leading to formation of decomposed products like creosol and vanillin. They are passed onto other species, causing chain propagation. When the reaction goes further and two radicals collide, stable compounds are formed, terminating the free radical chain reactions. However, observing the free radicals generated during pyrolysis is difficult; hence, the exact mechanism of the pyrolysis of lignin is not known and is a challenging task.

Apart from cellulose, hemicellulose, and lignin, inorganic species are also contained in biomass, and have significant effects on biochar production. Highly mobile inorganics, such as K and Cl, can easily vaporize at low temperatures in pyrolysis. Mg and Ca vaporize at relatively high temperatures and are covalently and ionically bound with organic compounds (Bourke et al., 2007). N, P, and S are covalently bound with complicated organic molecules in cells in plants (Marschner, 2012). Some inorganic species, especially alkaline earth metal elements (e.g., Ca, Mg, and K), can catalyze the biochar production process (Yaman, 2004). For instance, some chemical compounds containing K expedite the secondary cracking of the volatile organic compounds produced during the pyrolysis, leading to the production of more gaseous products such as CO, CO_2, CH_4, C_2H_4, and H_2 (Bridgwater, 2012). This results in the cracking of biochar. For pyrolysis of complex organic networks (e.g., lignocellulosic biomass), however, catalytic effects from the inherent organic species are not significant enough to make biochar with the desired properties. The development of effective catalysts for the production of biochar may be important not only to be able to engineer biochar with certain desired surface functionality and pore structure but also to reduce the process operation cost by moderating reaction conditions (decrease in reaction temperature and feedstock residence time, etc.).

1.4.2 Formation of Biochar Via Hydrothermal Carbonization

In hydrothermal carbonization processes, decomposition and depolymerization of cellulose, hemicellulose, and lignin occur at lower temperatures than pyrolysis (Yan et al., 2009). Cellulose and hemicellulose begin to decompose at 160−180°C while lignin does not decompose until critical point of water (Bobleter, 1994). The degradation mechanism of lignocellulosic biomass in hydrothermal carbonization is similar to that in pyrolysis. Nevertheless, decomposition of feedstock is initiated by hydrolysis because of the presence of subcritical water in the process. This leads to breakage of ester and ether bonds in monomeric sugars by the addition of a water molecule (Bobleter, 1994), which reduces the activation energy of decomposition of polymeric compounds in biomass (Libra et al., 2011). In addition, cellulose and hemicellulose are driven off from feedstock during hydrothermal carbonization. As a result, a product with high lignin content remains (i.e., char). Considering that HHV of lignin ($\sim$26 MJ kg^{-1}) is higher than other biomass constituents

($\sim$17–18 MJ kg^{-1}) (Demirbaş, 2005) and lignin is the most thermally stable component in biomass, higher energy densification and mass loss could be observed with a biomass feedstock with a lower ratio of lignin to hemicellulose for thermochemical processing of various biomass feedstocks (Yan et al., 2009; Demirbas, 2004b).

HMF is also produced during hydrothermal carbonization by decomposition of hemicellulose under aqueous conditions (Funke and Ziegler, 2010). HHV of HMF ($\sim$22 MJ kg^{-1}) is higher than that of cellulose and hemicellulose (Verevkin et al., 2009). The HMF formation in the aqueous phase highly depends on reaction temperature, residence time, and the amount of acid used in the process (Salak Asghari and Yoshida, 2006). Additionally, hydrochar made through hydrothermal carbonization often shows a slightly porous structure that provides improved adsorption capacity (Regmi et al., 2012). Deposition of HMF in the biochar surface can increase the overall energy density of the biochar. This can explain the higher energy density of the biochar obtained through hydrothermal carbonization than that of the biochar obtained through pyrolysis. The addition of acids and salts to reduce pH into hydrothermal carbonization processes could be considered as a means to catalyze decomposition of cellulose and hemicellulose (Lynam et al., 2011, 2012). Acids and salts facilitate solubilization and removal of organic species from polymeric structure. The addition of acetic acid increases the concentration of HMF in the pores of biochar, thereby increasing the HHV of the biochar (de Souza et al., 2012). However, too great an amount of acid in the process rehydrates HMF into formic acid and levulinic acid (Nilges et al., 2012), decreasing the heating value of the biochar.

1.5 CONCLUSIONS

In this chapter, the definition, classification, and properties of, production processes, and reaction chemistry of biochar were discussed. However, biochar production from a broader range of biomass feedstocks (e.g., organic waste, industrial wastes) requires additional research. Biochar may be a versatile tool in a wide range of environmental and industrial areas because the morphological features, physicochemical composition, and functionalities can be modified. The modification and potential applications of biochar will be discussed in other parts of this book.

References

Ahmad, M., Rajapaksha, A.U., Lim, J.E., Zhang, M., Bolan, N., Mohan, D., et al., 2014. Biochar as a sorbent for contaminant management in soil and water: a review. Chemosphere 99, 19–33.

Akhtar, A., Sarmah, A.K., 2018. Novel biochar-concrete composites: manufacturing, characterization and evaluation of the mechanical properties. Sci. Total Environ. 616–617, 408–416.

Amonette, J.E., Joseph, S., 2009. Characteristics of biochar: microchemical properties. In: Lehmann, J., Joseph, S. (Eds.), Biochar for Environmental Management: Science and Technology. Earthscan, London, UK.

Antal, M.J., Grúnli, M., 2003. The art, science, and technology of charcoal production. Ind. Eng. Chem. Res. 42, 1619–1640.

Babu, B.V., 2008. Biomass pyrolysis: a state-of-the-art review. Biofuels, Bioprod. Biorefin. 2, 393–414.

Bach, Q.-V., Tran, K.-Q., 2015. Dry and wet torrefaction of woody biomass—a comparative study on combustion kinetics. Energy Procedia 75, 150–155.

Banyasz, J.L., Li, S., Lyons-Hart, J., Shafer, K.H., 2001. Gas evolution and the mechanism of cellulose pyrolysis. Fuel 80, 1757–1763.

Baxter, L.L., Miles, T.R., Miles, T.R., Jenkins, B.M., Milne, T., Dayton, D., et al., 1998. The behavior of inorganic material in biomass-fired power boilers: field and laboratory experiences. Fuel Process. Technol. 54, 47–78.

Becidan, M., Skreiberg, Ø., Hustad, J.E., 2007. NO_x and N_2O precursors (NH_3 and HCN) in pyrolysis of biomass residues. Energy Fuels 21, 1173–1180.

Ben, H., Ragauskas, A.J., 2011. NMR characterization of pyrolysis oils from Kraft lignin. Energy Fuels 25, 2322–2332.

Benavente, V., Calabuig, E., Fullana, A., 2015. Upgrading of moist agro-industrial wastes by hydrothermal carbonization. J. Anal. Appl. Pyrolysis 113, 89–98.

Bergman, P.C.A., Boersma, A.R., Zwart, R.W.R., Kiel, J.H.A., 2005. Torrefaction for biomass co-firing in existing coal-fired power stations. Energy Centre of Netherlands, Report No. ECN-C-05–013.

Bobleter, O., 1994. Hydrothermal degradation of polymers derived from plants. Prog. Polym. Sci. 19, 797–841.

Bourke, J., Manley-Harris, M., Fushimi, C., Dowaki, K., Nunoura, T., Antal, M.J., 2007. Do all carbonized charcoals have the same chemical structure? A model of the chemical structure of carbonized charcoal. Ind. Eng. Chem. Res. 46, 5954–5967.

BP, 2016. Statistical Review of World Energy. BP Global.

Brewer, C.E., Schmidt-Rohr, K., Satrio, J.A., Brown, R.C., 2009. Characterization of biochar from fast pyrolysis and gasification systems. Environ. Prog. Sustain. Energy 28, 386–396.

Brick, S. & Lyutse, S. 2010. Biochar: Assessing the Promise and Risks to Guide US Policy. Natural Resource Defence Council, USA.

Bridgeman, T.G., Jones, J.M., Shield, I., Williams, P.T., 2008. Torrefaction of reed canary grass, wheat straw and willow to enhance solid fuel qualities and combustion properties. Fuel 87, 844–856.

Bridgwater, A.V., 1995. The technical and economic feasibility of biomass gasification for power generation. Fuel 74, 631–653.

Bridgwater, A.V., 2012. Review of fast pyrolysis of biomass and product upgrading. Biomass Bioenergy 38, 68–94.

Brosse, N., Dufour, A., Meng, X., Sun, Q., Ragauskas, A., 2012. Miscanthus: a fast-growing crop for biofuels and chemicals production. Biofuels Bioprod. Biorefin. 6, 580–598.

Cho, J., Chu, S., Dauenhauer, P.J., Huber, G.W., 2012. Kinetics and reaction chemistry for slow pyrolysis of enzymatic hydrolysis lignin and organosolv extracted lignin derived from maplewood. Green Chem. 14, 428–439.

Chu, S., Subrahmanyam, A.V., Huber, G.W., 2013. The pyrolysis chemistry of a β-O-4 type oligomeric lignin model compound. Green Chem. 15, 125–136.

Das, O., Bhattacharyya, D., Sarmah, A.K., 2016a. Sustainable eco–composites obtained from waste derived biochar: a consideration in performance properties, production costs, and environmental impact. J. Cleaner Prod. 129, 159–168.

Das, O., Sarmah, A.K., Bhattacharyya, D., 2016b. Biocomposites from waste derived biochars: mechanical, thermal, chemical, and morphological properties. Waste Manage. 49, 560–570.

Demirbaş, A., 2001. Biomass resource facilities and biomass conversion processing for fuels and chemicals. Energy Convers. Manage. 42, 1357–1378.

Demirbas, A., 2004a. Combustion characteristics of different biomass fuels. Prog. Energy Combust. Sci. 30, 219–230.

Demirbas, A., 2004b. Effects of temperature and particle size on bio-char yield from pyrolysis of agricultural residues. J. Anal. Appl. Pyrolysis 72, 243–248.

Demirbaş, A., 2005. Estimating of structural composition of wood and nonwood biomass samples. Energy Sources 27, 761–767.

De Souza, R.L., Yu, H., Rataboul, F., Essayem, N., 2012. 5-Hydroxymethylfurfural (5-HMF) production from hexoses: limits of heterogeneous catalysis in hydrothermal conditions and potential of concentrated aqueous organic acids as reactive solvent system. Challenges 3, 212–232.

EIA, 2016. International Energy Outlook. US Energy Information Administration.

Fengel, D., Wegener, G., 1984. Wood: Chemistry, Ultrastructure, Reactions. Walter de Gruyter.

Fuertes, A.B., Arbestain, M.C., Sevilla, M., Maciã-Agullõ, J.A., Fiol, S., Lõpez, R., et al., 2010. Chemical and structural properties of carbonaceous products obtained by pyrolysis and hydrothermal carbonisation of corn stover. Soil Res. 48, 618–626.

Funke, A., Ziegler, F., 2010. Hydrothermal carbonization of biomass: a summary and discussion of chemical mechanisms for process engineering. Biofuels, Bioprod. Biorefin. 4, 160–177.

Glaser, B., Haumaier, L., Guggenberger, G., Zech, W., 2001. The 'Terra Preta' phenomenon: a model for sustainable agriculture in the humid tropics. Naturwissenschaften 88, 37–41.

Glasser, W.G., Sarkanen, S., 1989. Lignin: Properties and Materials. ACS Publications.

Goyal, H.B., Saxena, R.C., Seal, D., 2008a. Thermochemical Conversion of Biomass to Liquids and Gaseous Fuels. CRC Press, Taylor & Francis Group.

Goyal, H.B., Seal, D., Saxena, R.C., 2008b. Bio-fuels from thermochemical conversion of renewable resources: a review. Renew. Sustain. Energy Rev. 12, 504–517.

Hendriks, A.T.W.M., Zeeman, G., 2009. Pretreatments to enhance the digestibility of lignocellulosic biomass. Bioresour. Technol. 100, 10–18.

Hitzl, M., Corma, A., Pomares, F., Renz, M., 2015. The hydrothermal carbonization (HTC) plant as a decentral biorefinery for wet biomass. Catal. Today 257, 154–159.

Hoekman, S.K., Broch, A., Robbins, C., Zielinska, B., Felix, L., 2013. Hydrothermal carbonization (HTC) of selected woody and herbaceous biomass feedstocks. Biomass Convers. Biorefin. 3, 113–126.

Huang, J., Liu, C., Tong, H., Li, W., Wu, D., 2012. Theoretical studies on pyrolysis mechanism of xylopyranose. Comput. Theor. Chem. 1001, 44–50.

IBI, 2012. Standardized product definition and product testing guidelines for biochar that is used in soil. International Biochar Initiative (IBI) biochar standards.

IBI, 2013. Pyrolysis and gasification of biosolids to produce biochar. IBI White Paper.

Ippolito, J.A., Laird, D.A., Busscher, W.J., 2012. Environmental benefits of biochar. J. Environ. Qual. 41, 967–972.

Kambo, H., Dutta, A., 2014 Hydrothermal carbonization (HTC): an innovative process for the conversion of low quality lignocellulosic biomass to hydrochar for replacing coal. In: Proceedings of the 9th Annual Green Energy Conference (IGEC-IX), 2014 Tianjin, China, pp. 25–28.

Kambo, H.S., Dutta, A., 2015. A comparative review of biochar and hydrochar in terms of production, physicochemical properties and applications. Renew. Sustain. Energy Rev. 45, 359–378.

Karaosmanoğlu, F., Tetik, E., Gøllþ, E., 1999. Biofuel production using slow pyrolysis of the straw and stalk of the rapeseed plant. Fuel Process. Technol. 59, 1–12.

Khan, A.A., De Jong, W., Jansens, P.J., Spliethoff, H., 2009. Biomass combustion in fluidized bed boilers: potential problems and remedies. Fuel Process. Technol. 90, 21–50.

Kirubakaran, V., Sivaramakrishnan, V., Nalini, R., Sekar, T., Premalatha, M., Subramanian, P., 2009. A review on gasification of biomass. Renew. Sustain. Energy Rev. 13, 179–186.

Knežević, D. 2009. Hydrothermal Conversion of Biomass (Ph.D. thesis). University of Twente.

Kosa, M., Ben, H., Theliander, H., Ragauskas, A.J., 2011. Pyrolysis oils from CO_2 precipitated Kraft lignin. Green Chem. 13, 3196–3202.

Kritzer, P., Dinjus, E., 2001. An assessment of supercritical water oxidation (SCWO): existing problems, possible solutions and new reactor concepts. Chem. Eng. J 83, 207–214.

Laird, D., Rogovska, N., Garcia-Perez, M., Collins, H., Streubel, J. & Smith, M., 2011. Pyrolysis and biochar–opportunities for distributed production and soil quality enhancement. Sustainable alternative fuel feedstock opportunities, challenges and roadmaps for six US regions. In: Proceedings of the Sustainable Feedstocks for Advanced Biofuels Workshop, 2011. SWCS publisher Atlanta, GA, pp. 257–281.

Lee, J., Kim, K.-H., Kwon, E.E., 2017. Biochar as a catalyst. Renew. Sustain. Energy Rev. 77, 70–79.

Lehmann, J., 2009. Terra Preta Nova – Where to from here? In: Woods, W.I., Teixeira, W.G., Lehmann, J., Steiner, C., Winklerprins, A., Rebellato, L. (Eds.), Amazonian Dark Earths: Wim Sombroek's Vision. Springer, Netherlands, Dordrecht.

Lehmann, J., Gaunt, J., Rondon, M., 2006. Bio-char sequestration in terrestrial ecosystems—a review. Mitigat. Adapt. Strat. Global Change 11, 395–419.

Lehmann, J., Joseph, S., 2015. Biochar for Environmental Management: Science, Technology and Implementation. Routledge.

Levan, S., Ross, R.J., Winandy, J.E., 1990. Effects of fire retardant chemicals on the bending properties of wood at elevated temperatures. US Department of Agriculture, Forest Service, Forest Products Laboratory Madison, WI, USA.

Li, S., Lyons-Hart, J., Banyasz, J., Shafer, K., 2001. Real-time evolved gas analysis by FTIR method: an experimental study of cellulose pyrolysis. Fuel 80, 1809–1817.

Libra, J.A., Ro, K.S., Kammann, C., Funke, A., Berge, N.D., Neubauer, Y., et al., 2011. Hydrothermal carbonization of biomass residuals: a comparative review of the chemistry, processes and applications of wet and dry pyrolysis. Biofuels 2, 71–106.

Lin, Y.-C., Cho, J., Tompsett, G.A., Westmoreland, P.R., Huber, G.W., 2009. Kinetics and mechanism of cellulose pyrolysis. J. Phys. Chem. C 113, 20097–20107.
Liu, W.-J., Jiang, H., Yu, H.-Q., 2015. Development of biochar-based functional materials: toward a sustainable platform carbon material. Chem. Rev. 115, 12251–12285.
Liu, Z., Balasubramanian, R., 2014. Upgrading of waste biomass by hydrothermal carbonization (HTC) and low temperature pyrolysis (LTP): a comparative evaluation. Appl. Energy 114, 857–864.
Liu, Z., Quek, A., Kent Hoekman, S., Balasubramanian, R., 2013. Production of solid biochar fuel from waste biomass by hydrothermal carbonization. Fuel 103, 943–949.
Luo, Z., Wang, S., Liao, Y., Cen, K., 2004. Mechanism study of cellulose rapid pyrolysis. Ind. Eng. Chem. Res. 43, 5605–5610.
Lynam, J.G., Coronella, C.J., Yan, W., Reza, M.T., Vasquez, V.R., 2011. Acetic acid and lithium chloride effects on hydrothermal carbonization of lignocellulosic biomass. Bioresour. Technol. 102, 6192–6199.
Lynam, J.G., Toufiq Reza, M., Vasquez, V.R., Coronella, C.J., 2012. Effect of salt addition on hydrothermal carbonization of lignocellulosic biomass. Fuel 99, 271–273.
Mani, S., Sokhansanj, S., Bi, X., Turhollow, A., 2006. Economics of producing fuel pellets from biomass. Appl. Eng. Agric. 22, 421–426.
Manyâ, J.J., 2012. Pyrolysis for biochar purposes: a review to establish current knowledge gaps and research needs. Environ. Sci. Technol. 46, 7939–7954.
Marschner, P. (Ed.), 2012. Mineral Nutrition of Higher Plants. Elsevier.
Mcbeath, A.V., Wurster, C.M., Bird, M.I., 2015. Influence of feedstock properties and pyrolysis conditions on biochar carbon stability as determined by hydrogen pyrolysis. Biomass Bioenergy 73, 155–173.
Mckendry, P., 2002. Energy production from biomass (part 1): overview of biomass. Bioresour. Technol. 83, 37–46.
Mettler, M.S., Paulsen, A.D., Vlachos, D.G., Dauenhauer, P.J., 2012. Pyrolytic conversion of cellulose to fuels: levoglucosan deoxygenation via elimination and cyclization within molten biomass. Energy Environ. Sci. 5, 7864–7868.
Mohan, D., Pittman, C.U., Steele, P.H., 2006. Pyrolysis of wood/biomass for bio-oil: a critical review. Energy Fuels 20, 848–889.
Mu, W., Ben, H., Ragauskas, A., Deng, Y., 2013. Lignin pyrolysis components and upgrading—technology review. BioEnergy Res. 6, 1183–1204.
Mumme, J., Eckervogt, L., Pielert, J., Diakitë, M., Rupp, F., Kern, J., 2011. Hydrothermal carbonization of anaerobically digested maize silage. Bioresour. Technol. 102, 9255–9260.
Nilges, P., Dos Santos, T.R., Harnisch, F., Schroder, U., 2012. Electrochemistry for biofuel generation: electrochemical conversion of levulinic acid to octane. Energy Environ. Sci. 5, 5231–5235.
Onay, O., 2007. Influence of pyrolysis temperature and heating rate on the production of bio-oil and char from safflower seed by pyrolysis, using a well-swept fixed-bed reactor. Fuel Process. Technol. 88, 523–531.
Onay, O., Kockar, O.M., 2003. Slow, fast and flash pyrolysis of rapeseed. Renew. Energy 28, 2417–2433.
Özbay, N., PþTþN, A.E., Uzun, B.B., PþTþN, E., 2001. Biocrude from biomass: pyrolysis of cottonseed cake. Renew. Energy 24, 615–625.
Patwardhan, P.R., Brown, R.C., Shanks, B.H., 2011a. Product distribution from the fast pyrolysis of hemicellulose. ChemSusChem 4, 636–643.
Patwardhan, P.R., Dalluge, D.L., Shanks, B.H., Brown, R.C., 2011b. Distinguishing primary and secondary reactions of cellulose pyrolysis. Bioresour. Technol. 102, 5265–5269.
Peng, Y., Wu, S., 2010. The structural and thermal characteristics of wheat straw hemicellulose. J. Anal. Appl. Pyrolysis 88, 134–139.
Perlack, R.D., Turhollow, A.F., 2003. Feedstock cost analysis of corn stover residues for further processing. Energy 28, 1395–1403.
Perlack, R.D., Wright, L.L., Turhollow, A.F., Graham, R.L., Stokes, B.J., Erbach, D.C., 2005. Biomass as feedstock for a bioenergy and bioproducts industry: the technical feasibility of a billion-ton annual supply. DTIC Document.
Peters, B., 2011. Prediction of pyrolysis of pistachio shells based on its components hemicellulose, cellulose and lignin. Fuel Process. Technol. 92, 1993–1998.
Pimchuai, A., Dutta, A., Basu, P., 2010. Torrefaction of agriculture residue to enhance combustible properties. Energy Fuels 24, 4638–4645.

Poulose, A.M., Elnour, A.Y., Anis, A., Shaikh, H., Al-Zahrani, S.M., George, J., et al., 2018. Date palm biochar-polymer composites: an investigation of electrical, mechanical, thermal and rheological characteristics. Sci. Total Environ. 619–620, 311–318.

Puig-Arnavat, M., Bruno, J.C., Coronas, A., 2010. Review and analysis of biomass gasification models. Renew. Sustain. Energy Rev. 14, 2841–2851.

Qian, K., Kumar, A., Zhang, H., Bellmer, D., Huhnke, R., 2015. Recent advances in utilization of biochar. Renew. Sustain. Energy Rev. 42, 1055–1064.

Regmi, P., Garcia Moscoso, J.L., Kumar, S., Cao, X., Mao, J., Schafran, G., 2012. Removal of copper and cadmium from aqueous solution using switchgrass biochar produced via hydrothermal carbonization process. J. Environ. Manage. 109, 61–69.

Ronsse, F., Bai, X., Prins, W., Brown, R.C., 2012. Secondary reactions of levoglucosan and char in the fast pyrolysis of cellulose. Environ. Prog. Sustain. Energy 31, 256–260.

Ronsse, F., Van Hecke, S., Dickinson, D., Prins, W., 2013. Production and characterization of slow pyrolysis biochar: influence of feedstock type and pyrolysis conditions. GCB Bioenergy 5, 104–115.

Rousset, P., Macedo, L., Commandrë, J.M., Moreira, A., 2012. Biomass torrefaction under different oxygen concentrations and its effect on the composition of the solid by-product. J. Anal. Appl. Pyrolysis 96, 86–91.

Salak Asghari, F., Yoshida, H., 2006. Acid-catalyzed production of 5-hydroxymethyl furfural from D-fructose in subcritical water. Ind. Eng. Chem. Res. 45, 2163–2173.

Saxena, R.C., Adhikari, D.K., Goyal, H.B., 2009. Biomass-based energy fuel through biochemical routes: a review. Renew. Sustain. Energy Rev. 13, 167–178.

Sefain, M.Z., El-Kalyoubi, S.F., Shukry, N., 1985. Thermal behavior of holo- and hemicellulose obtained from rice straw and bagasse. J. Polym. Sci.: Polym. Chem. Ed. 23, 1569–1577.

Shen, D., Xiao, R., Gu, S., Luo, K., 2011. The pyrolytic behavior of cellulose in lignocellulosic biomass: a review. RSC Adv. 1, 1641–1660.

Shen, D.K., Gu, S., 2009. The mechanism for thermal decomposition of cellulose and its main products. Bioresour. Technol. 100, 6496–6504.

Shen, D.K., Gu, S., Bridgwater, A.V., 2010a. Study on the pyrolytic behaviour of xylan-based hemicellulose using TG–FTIR and Py–GC–FTIR. J. Anal. Appl. Pyrolysis 87, 199–206.

Shen, D.K., Gu, S., Bridgwater, A.V., 2010b. The thermal performance of the polysaccharides extracted from hardwood: cellulose and hemicellulose. Carbohydr. Polym. 82, 39–45.

Sivula, L., Oikari, A., Rintala, J., 2012. Toxicity of waste gasification bottom ash leachate. Waste Manag. 32, 1171–1178.

Sohi, S.P., Krull, E., Lopez-Capel, E., Bol, R., 2010. A review of biochar and its use and function in soil. Adv. Agron. 105, 47–82.

Sokhansanj, S., Fenton, J., 2006. Cost Benefit of Biomass Supply and Pre-Processing. BIOCAP Canada Foundation.

Stemann, J., Erlach, B., Ziegler, F., 2013. Hydrothermal carbonisation of empty palm oil fruit bunches: laboratory trials, plant simulation, carbon avoidance, and economic feasibility. Waste Biomass Valorizat. 4, 441–454.

Tortosa Masiã, A.A., Buhre, B.J.P., Gupta, R.P., Wall, T.F., 2007. Characterising ash of biomass and waste. Fuel Process Technol. 88, 1071–1081.

Uchimiya, M., Wartelle, L.H., Klasson, K.T., Fortier, C.A., Lima, I.M., 2011. Influence of pyrolysis temperature on biochar property and function as a heavy metal sorbent in soil. J. Agric. Food. Chem. 59, 2501–2510.

Vassilev, S.V., Baxter, D., Andersen, L.K., Vassileva, C.G., Morgan, T.J., 2012. An overview of the organic and inorganic phase composition of biomass. Fuel 94, 1–33.

Verevkin, S.P., Emel'yanenko, V.N., Stepurko, E.N., Ralys, R.V., Zaitsau, D.H., Stark, A., 2009. Biomass-derived platform chemicals: thermodynamic studies on the conversion of 5-hydroxymethylfurfural into bulk intermediates. Ind. Eng. Chem. Res. 48, 10087–10093.

Weber, R., Yoshida, S., Miwa, K., 2002. PCB destruction in subcritical and supercritical water evaluation of PCDF formation and initial steps of degradation mechanisms. Environ. Sci. Technol. 36, 1839–1844.

Wiedner, K., Rumpel, C., Steiner, C., Pozzi, A., Maas, R., Glaser, B., 2013. Chemical evaluation of chars produced by thermochemical conversion (gasification, pyrolysis and hydrothermal carbonization) of agro-industrial biomass on a commercial scale. Biomass Bioenergy 59, 264–278.

Winandy, J.E., 1995. Effects of fire retardant treatments after 18 months of exposure at 150 F (66 C), US Department of Agriculture, Forest Service, Forest Products Laboratory.

Yaman, S., 2004. Pyrolysis of biomass to produce fuels and chemical feedstocks. Energy Convers. Manage. 45, 651–671.

Yan, W., Acharjee, T.C., Coronella, C.J., VãSquez, V.R., 2009. Thermal pretreatment of lignocellulosic biomass. Environ. Prog. Sustain. Energy 28, 435–440.
Yan, W., Hastings, J.T., Acharjee, T.C., Coronella, C.J., VãSquez, V.R., 2010. Mass and energy balances of wet torrefaction of lignocellulosic biomass. Energy Fuels 24, 4738–4742.
Yip, K., Tian, F., Hayashi, J.-I., Wu, H., 2010. Effect of alkali and alkaline earth metallic species on biochar reactivity and syngas compositions during steam gasification. Energy Fuels 24, 173–181.
Yoder, J., Galinato, S., Granatstein, D., Garcia-Pëerez, M., 2011. Economic tradeoff between biochar and bio-oil production via pyrolysis. Biomass Bioenergy 35, 1851–1862.
Yu, Y., Lou, X., Wu, H., 2008. Some recent advances in hydrolysis of biomass in hot-compressed water and its comparisons with other hydrolysis methods. Energy Fuels 22, 46–60.
Zech, W., Haumaier, L., Reinhold, H., 1990. Ecological aspects of soil organic matter in tropical land use. Humic Substances in Soil and Crop Sciences: Selected Readings 187–202.
Zhang, L., Xu, C., Champagne, P., 2010. Overview of recent advances in thermo-chemical conversion of biomass. Energy Convers. Manage. 51, 969–982.
Zhang, X., Yang, W., Blasiak, W., 2012. Thermal decomposition mechanism of levoglucosan during cellulose pyrolysis. J. Anal. Appl. Pyrolysis 96, 110–119.
Zhao, L., Cao, X., Mašek, O., Zimmerman, A., 2013. Heterogeneity of biochar properties as a function of feedstock sources and production temperatures. J. Hazard. Mater. 256, 1–9.
Zhao, P., Shen, Y., Ge, S., Chen, Z., Yoshikawa, K., 2014. Clean solid biofuel production from high moisture content waste biomass employing hydrothermal treatment. Appl. Energy 131, 345–367.

PART II

BIOCHAR CHARACTERIZATION

CHAPTER

2

Physical Characteristics of Biochars and Their Effects on Soil Physical Properties

Shih-Hao Jien

Department of Soil and Water Conservation, National Pingtung University of Science and Technology, Pingtung, Taiwan

2.1 INTRODUCTION

The physical characteristics of biochars contribute to their function for soil quality. Their physical characteristics can be both directly and indirectly related to the way in which they affect soil systems. According to Brady and Weil (2008), soils each have their own distinct physical properties depending on the nature of mineral and organic matter, their relative amounts, and the way in which minerals and organic matter are associated with the soils. When biochar is incorporated into soils, its contribution to the physical nature of the system may be significant, with effects on soil structure, porosity, consistency, and tilth through changing the bulk density, aggregate changing, pore-size distribution, etc. Many studies have shown that biochar is a useful resource and can improve the physicochemical properties of soil, effectively maintaining soil organic matter (SOM) levels, increasing fertilizer use efficiency, and increasing crop production, particularly for long-term cultivated soils in subtropical and tropical regions (Chan et al., 2007, 2008; Deenik et al., 2011; Van Zwieten et al., 2010). Furthermore, the application of biochar to soils might be a practical method to aid in the long-term maintenance of soil organic carbon contents and soil fertility. The application of biochar to soils can also maintain SOM levels and soil aggregation stability (Kimetu and Lehmann, 2010; Tejada and Gonzalez, 2007; Trompowsky et al., 2005) because biochar is characterized by recalcitrant C from microbial degradation and by a charged surface with organic functional groups. Reducing

Biochar from Biomass and Waste
DOI: https://doi.org/10.1016/B978-0-12-811729-3.00002-9

soil-erosion potential, maintaining SOM, and improving soil aggregative stability are critical processes. Previous studies have demonstrated the importance of SOM to the physiochemical properties of soil (Materechera, 2009; Wuddivira et al., 2009) and erosion susceptibility (Auerswald et al., 2003; Tejada and Gonzalez, 2007). Many studies have reported on the use of biochar as an amendment for crop production and for improving the chemical properties of highly weathered tropical soils (Iswaran et al., 1980; Liang et al., 2006). However, few studies have investigated the effects of biochar on soil physical properties and soil erodibility (Atkinson et al., 2010; Hammes and Schmid, 2009).

This chapter mainly focuses on the physical properties of biochars, relating how their surface characteristics influence soil quality, particularly in soil physical properties.

2.2 BIOCHAR STRUCTURE AND MICROSTRUCTURE

2.2.1 Surface Properties of Biochars

Surface properties, including surface area, charge density, pore structure, and distribution, are important characteristics as they influence the essential functions for retention capacity of water, nutrients, and microbial activity. The physical features of biochars are important in soil processes. Lehmann and Joseph (2009) showed that operating parameters, including processing heating rate, highest treatment temperature (HTT), pressure, reaction residence time, reaction vessel (orientation, dimensions, stirring regime, catalysts, etc.), pretreatment (drying, comminution, chemical activation, etc.), the flow rate of ancillary inputs, all influence the resultant physical properties of biochar made with any biomass feedstock.

Based on our unpublished data (Fig. 2.1), the hydrophilic capacity of biochar also seems to be influenced by HTT. The contact angles of the water on the different biochars (RHB: rice husk biochar; WB: wood biochar) were observed in our experiment. The results showed that the hydrophilic property (contact angle≧90 degrees) increased in the biochars pyrolized with temperatures higher than 400°C regardless of biochar type. We assumed that the HTT increases more structured regular spacing between the planes results. Interplanar distances also decrease with the increased ordering and organization of molecules, all of which result in larger surface areas per volume. Higher specific surface area and more micropores resulting from higher pyrolization temperatures might be the critical factor to retaining water in biochars. Just like soil systems, the limited capacity of sandy soil to store water and plant nutrients is partly related to the relatively small surface area of its soil particles (Troeh and Thompson, 2005). We will discuss porosity and pore distribution in Section 2.2.2.

In order to identify the changing functional groups during the pyrolization process of biomass, semiquantitative analysis of functional groups of the biochars were determined by CPMAS ^{13}C NMR (solid-state ^{13}C cross-polarization magic-angle nuclear magnetic resonance). Spectra showed that more aromatic-C proportion was found in the biochar with higher pyrolized temperature (≥400°C) (Fig. 2.2). In theory, aromatic-C seems to be much more hydrophobic than alkyl-C/*O*-alkyl-C, which implies that biochars pyrolyzed by higher temperatures (≥400°C) should also be more hydrophobic. However, the hydrophilic capacity observed from our contact angle experiment showed contrary results. We

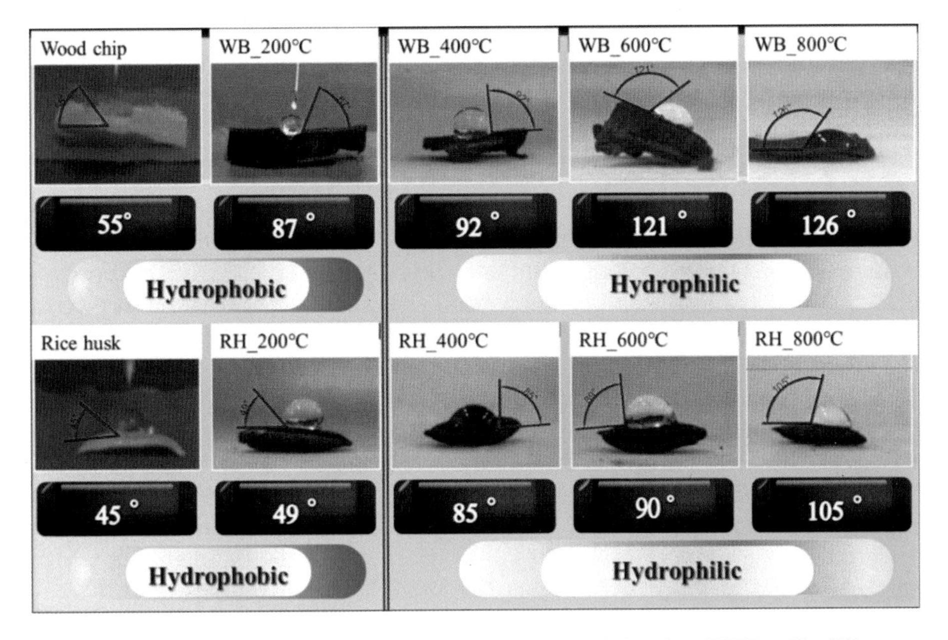

FIGURE 2.1 Contact angle of wood biochar (WB) and rice husk biochar (RHB) with different pyrolized temperatures.

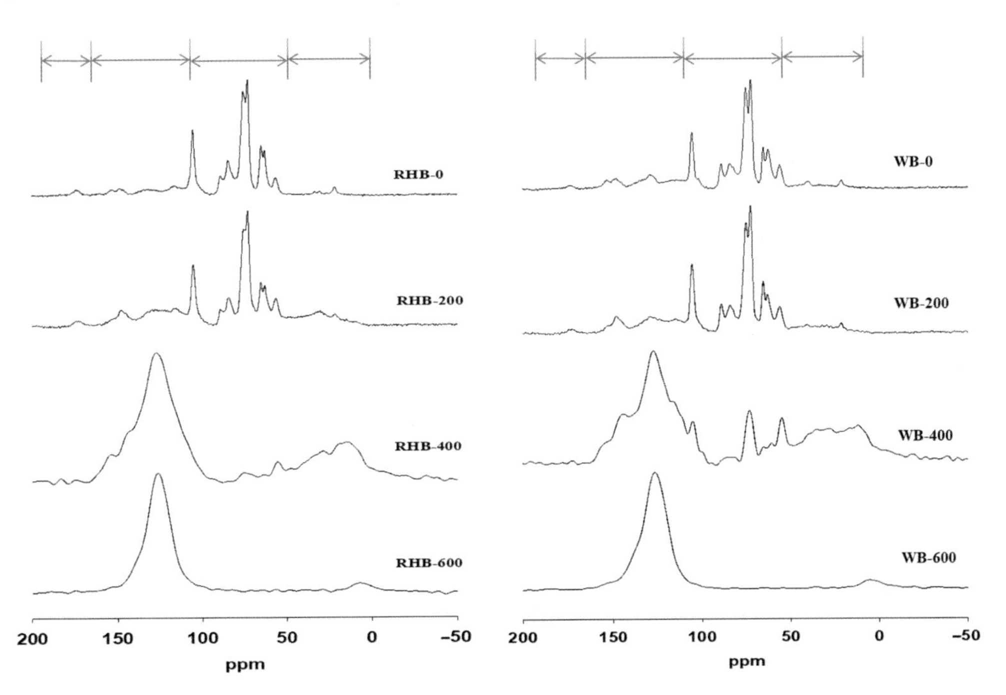

FIGURE 2.2 Solid-state ^{13}C cross-polarization magic-angle nuclear magnetic resonance spectra for wood biochar (WB) and rice husk biochar (RHB) with different pyrolized temperatures.

therefore assumed that the hydrophilic capacity of the biochars was determined by integrated parameters. Porous characteristics might be the most critical parameter to determine the hydrophilic capacity of the biochars.

2.2.2 Pore Distribution and Surface Area of Biochars

To realize the surface morphology of biochars, the biochar samples were viewed by optical microscopy with reflected light and then scanning electron microscopy (SEM) (Hitachi, S-3000N, Japan) to identify microscale structure. A back-scattered electron image representing the mean atomic abundance in a back-and-white image was observed from the surface of the samples coated by Au. The mineral phases of the sample were identified using SEM and energy-dispersive spectroscopy (EDS) (Horiba, EMAX-ENERGY EX-200, Japan), with 15 kV and 180 Pa for the acceleration voltage and beam current, respectively, in a vacuum of 25 Pa with an Au coating.

Fig. 2.2 indicates SEM observations of RHB and WB, which were pyrolized from different temperatures. More pores with diameter $\leqq 10\,\mu m$ were found in the biochars with higher pyrolization temperature (600°C) than lower temperature (400°C) (Fig. 2.2). Pyrolysis processing of biomass enlarges the crystallites and makes them more ordered, which increases with HTT. Lua et al. (2004) showed that increasing the pyrolysis temperature from 250 to 500°C increases the BET surface area due to the increasing evolution of volatiles from pistachio-nut shells, resulting in enhanced pore development in biochars.

From our unpublished data, the lignosulfonate (LS), which was a byproduct of the process of neutral sulphite semichemical pulping (NSSC) for pulp manufacturing, was used to produce biochar under different temperatures (Table 2.1). In this study, slow pyrolization was carried out, with heat rates as follows: 110°C, 30 min, 5°C min^{-1} (the first step), 300°C, 30 min, 5°C min^{-1} (the second step), and 400–1000°C, 30 min, 5°C min^{-1} (the final step). We used mercury porosimeter to determine the surface porous characteristics of the lignin biochar. The results showed that porosity (%) and average pore diameter (nm) obviously decreased with pyrolized temperature, and total pore area ($m^2\,g^{-1}$) significantly increased with temperature. However, due to the instrumental limits, the pores we determined in this experiment belonged to macropores (>50 nm) based on Rouquerol et al. (1999), who indicated that the total pore volume of the biochar should be divided

TABLE 2.1 Porous Characteristic of Lignin With Different Pyrolization Temperatures

Carbons	Porosity (%)	Average Pore Diameter (nm)	Total Pore Area ($m^2\,g^{-1}$)
CH400	78.2	5206	4.64
CH600	77.7	1833	7.30
CH800	76.1	2101	11.0
CH1000	72.7	435.8	30.3

CH: carbon washing by DI water.

into micropores (pores of internal diameter less than 2 nm), mesopores (pores of internal width between 2 nm and 50 nm), and macropores (pores of internal width greater than 50 nm).

Troeh and Thompson (2005) noted that macropores are relevant to vital soil functions such as aeration and hydrology. Lehmann and Joseph (2009) also indicated that macropores are also relevant to the root extension through soil and as habitats for a vast variety of soil microbes. Although micropore surface areas are significantly larger than macropore surface areas in biochars, macropore volumes can be larger than micropore volumes. It is possible that these broader volumes could result in greater functionality in soils than narrow surface areas. Jien and Wang (2013) and Jien et al. (2015) also showed that these macropores could protect added organic materials (compost) from decomposing by microbial. Wildman and Derbyshire (1991) suggested that as anticipated from the regular size and arrangement of plant cells in most biomass from which biochars are derived, the macropore size distribution is comprised of discrete groups of pores sizes rather than a continuum. Additionally, another consideration is the type of microbial communities that utilize soil pores as a preferred habitat. Microbial cells typically range in size from 0.5 to 5 μm, and consist predominantly of bacteria, fungi, actinomycetes, and lichens. The macropores present in biochars (Fig. 2.3) might therefore provide suitable dimensions for clusters of microorganisms to inhabit.

Regarding mesopores and micorpores, Lehmann and Joseph (2009) compiled some of the data available in the literature to demonstrate the relationship between micropore volume and total surface area of biochars. This provides evidence that pore sizes distributed in the micropore range make the greatest contribution to total surface area. The development of microporosity with higher temperatures and longer reaction residence

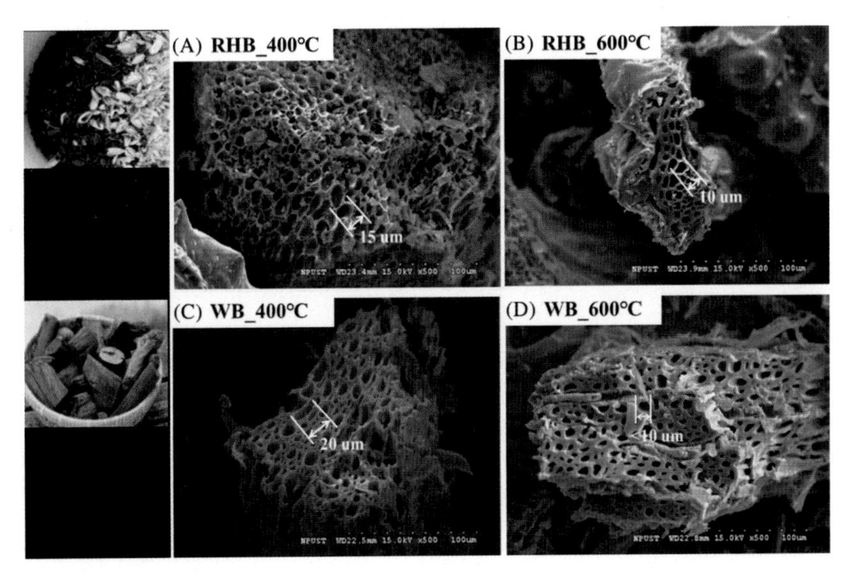

FIGURE 2.3 Surface structure characteristics of wood biochar (WB) and rice husk biochar (RHB) with different pyrolized temperatures using scanning electron microscopy (SEM).

times has been demonstrated by several research groups. Elevated temperatures provide the activation energies and longer retentions allow the time for the reactions to reach completion, leading to greater degrees of order in the structures. The surface area of biochars generally increases with increasing HTT until it reaches the temperature at which deformation occurs, resulting in subsequent decreases in surface area. A typical example is provided by Brown et al. (2006), who produced biochar from pine in a laboratory oven purged with N_2 at a range of final temperatures varying from 450 to 1000°C, and heating rates varying from 30°C h^{-1} to 1000°C h^{-1}. Brown et al. (2006) found that independent of heating rate, maximum surface area, as measured by BET (N_2), was realized at a final temperature of 750°C. At the lowest HTT (i.e., 450°C), all of the surface areas were found to be less than 10 m^2 g^{-1}, while those produced at intermediate temperatures of 600 to 750°C had a surface area of approximately 400 m^2 g^{-1} (Brown et al., 2006).

2.3 SOIL PHYSICAL PROPERTIES OF BIOCHAR-AMENDED SOILS

We have published some articles on the impacts of the biochar application on soil physical properties (Jien and Wang, 2013; Hseu et al., 2014; Jien et al., 2015, 2017; Lee et al., 2018). For highly weathered soil, Jien and Wang (2013) noted that after incubation of 105 d, the Bd of the biochar-amended soils significantly decreased from 1.42 to 1.15 Mg m^{-3}, and the rate of decease increased with the biochar application rate. Other than changes in the Bd, the biochar-amended soils exhibited significantly higher total porosities (50%) than the unamended controls (41%) after 105 d incubation. Fig. 2.6F further shows a variation of porosity during the incubation duration; the control and 2.5% biochar-amended soil presented unobvious changes throughout the duration, and a gradual decrease in porosity appeared in the 5% biochar-amended soil. The MWD of soil aggregation was consistently higher for the biochar-amended soils than the control after incubation of 21 d; however, significant differences between the amended soils and the control were found after incubation of 84 d. Furthermore, applying biochar to the soil caused a significant increase in the saturated hydraulic conductivity (K_{sat}). At the end of the incubation, the K_{sat} values of the amended soils were twice as high as the control soils, although there were great variances at the beginning of the incubation, especially for the 5% biochar-amended soil. After 21 d after biochar was incorporated into the soil, the K_{sat} stabilized gradually and kept higher consistency for the biochar-amended soils to the end of the incubation.

We further extended our study into field scale in a highly weathered soil (Typic Paleudult), and some physical properties after 1 year were estimated to evaluate the variation after biochar application, such as aggregative stability, water retention ability, and porosity (Figs. 2.4–2.6). Fig. 2.4 indicates that formation of soil aggregates after WB application, obvious coarse aggregates (>0.25 mm) were found in the treatments of WB_2%, compost_1% + WB_2% and PAM 50 ppm. The most distinct aggregation was found in the treatment of compost_1% + WB_2%, and the coarsest aggregate was about 40 mm.

In our earlier-mentioned field experiment, the water characteristic curve was also carried out for the treatments of the control, wood biochar_4% (WB_4%), compost_1% + wood biochar_4% (CWB_4%), and PAM 50 ppm (PAM). Fig. 2.5 indicates that water retention capacity was the best (>60%) in the treatment of CWB_4% compared with the

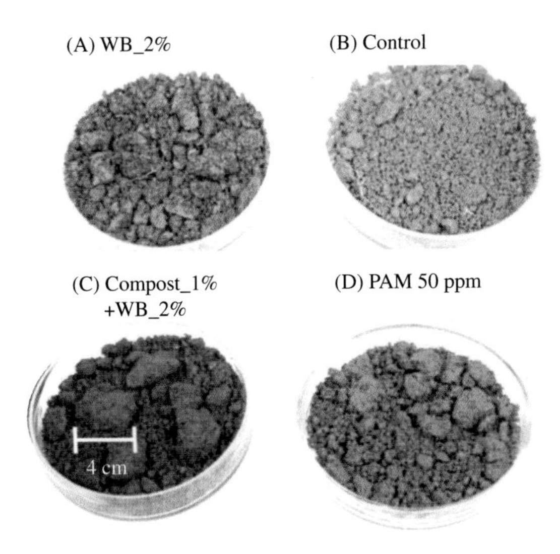

FIGURE 2.4 Formation of soil aggregates in different treatments after 1-year field experiment: (A) wood biochar_2%; (B) control; (C) compost_1% + wood biochar_2%; (D) polyacrylamide_50 ppm.

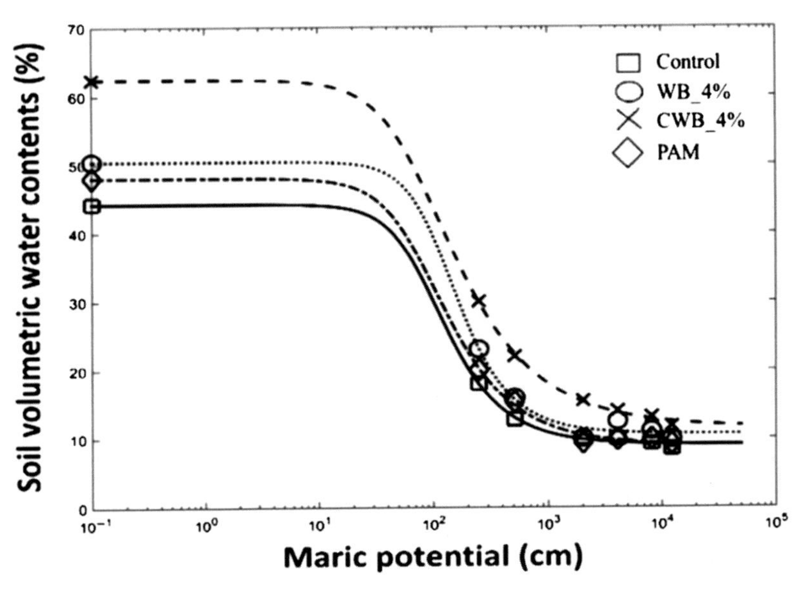

FIGURE 2.5 Water characteristic curve of the soils with different treatments after 1-year field experiment.

control (~43%). Fig. 2.5 also shows that macropore (≧50 μm) proportion of the amended soil increased by 3%–10%, particularly in the treatment of CWB_4%.

After our results for aggregation and water retention capacity, we further tried to look into microscale of the soil peds to evaluate the changing of the soil structure especially for porous structure between the control and the biochar-amended soil by X-ray computed tomography (CT).

The CT technique permits nondestructive investigation of the soil pore system and results in 3D maps of X-ray attenuation (De Gryze et al., 2006). The attenuation expressed in values of the Hounsfield unit (HU) correlates with local composition of solid, water,

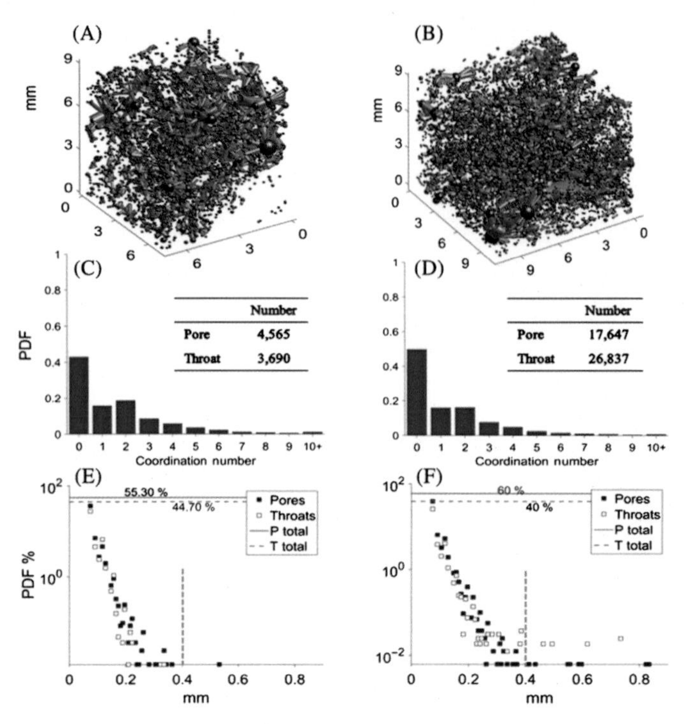

FIGURE 2.6 The spatial distribution of pores with different sizes by using high-resolution X-ray CT: pore and throat distribution of (A) the control and (B) wood biochar_4%; coordination number and frequency of (C) the control and (D) wood biochar_4%; pore size distribution and frequency of (E) the control and (F) wood biochar_4%.

and air phases and allows the small-scale 2D and 3D determination of soil physical properties (Schrader et al., 2007). For estimating 3D soil bulk density distributions, the X-ray attenuation further requires information on water content and solid particle density at a representative local scale (Rogasik et al., 2003). The spatial resolution of medical CT scanners is currently in a range of about 500 μm, which is low compared to about 200 nm of nano-CT; however, medical CT facilitates scanning of larger (101 to 102 cm) samples, while scanning at higher spatial resolution is limited to smaller (10^{-1} to 100 cm) samples (Cnudde et al., 2006). For the ultrahigh-resolution 3D X-ray microscope (Xradia 520 Versa, Zeiss) we used in this study, the spatial resolution is about 100 μm.

Based on our X-ray microscope results, the numbers of pores and throats (connection channels between pores) obviously increased by 3 to 7 times compared with the control (Fig. 2.6A–D). Fig. 2.6E and F further shows that the increased soil pores and throats in this study were <0.4 mm in diameter. There were about 0.5%–1% increases of pores and throats with ≥0.4 mm in diameter in this study.

2.3.1 Effects of Biochars on CO_2 Emission

Based on our previous study (Jien et al., 2015), we conducted a short-term (70 days) incubation experiment to assess the effects of biochar application on the decomposition of added bagasse compost in three rural soils with different pH values and textures. Two rice hull biochars, produced through slow pyrolization at 400°C (RHB-400) and 700°C

(RHB-700), with application rates of 1%, 2%, and 4% (w/w), were separately incorporated into soils with and without compost (1% (w/w) application rate). Experimental results indicated that C mineralization rapidly increased at the beginning in all treatments, particularly in those involving 2% and 4% biochar. The biochar addition increased C mineralization by 7.9%–48% in the compost-amended soils after 70 days incubation while the fractions of mineralized C to applied C significantly decreased. Moreover, the estimated maximum of C mineralization in soils treated with both compost and biochar was obviously lower than expected and were calculated by a double exponential model (two pool model). Based on the micromorphological observation, added compost was wrapped in the soil aggregates formed after biochar application and then protected from decomposing by microbes. Coapplication of compost with biochar may be more efficient to stabilize and sequester C than individual application in the studied soils, especially for biochar produced at high pyrolization temperature.

To determine the interactions among the compost, biochar, and soil particles, microstructures were observed using a polarized microscope (Leica DM EP, TX, USA). To illustrate, the microscope images of the thin section of a clayey soil and a sandy soil treated with 1% compost and 2% biochars are shown in Fig. 2.7. After 70 days,

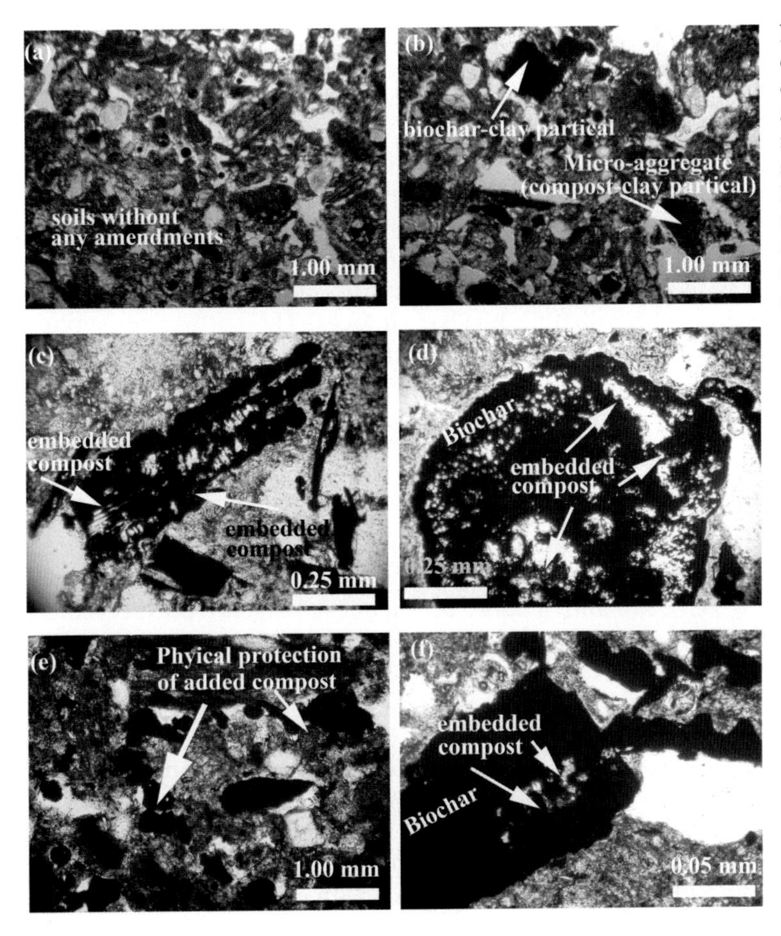

FIGURE 2.7 Microstructural observations in the biochar- and compost-amended soils (Lo soil and Sp soil) using a polarized microscope: (A) unamended Lo soil; (B, C) the treatment of 2% RHB-400 + 1% compost in Lo soil, plain polarized light (PPL); (D)–(F) the treatment of 2% RHB-700 + 1% compost in Sp soil with PPL.

macroaggregates formed during the mutual interaction among the soil particles, biochar, and compost (Fig. 2.7A and B). Microstructure changed from single-spaced porphyric (unamended treatment) to single-spaced equal enaulic (biochar treatment) based on the micromorpholigical description guidelines. Fig. 2.7C and F further shows that added compost (brown color) was obviously embedded or adsorbed into the micropores and surface of the biochar.

Furthermore, in this study, the biochar produced at lower pyrolization temperature seemed to induce more cumulative CO_2 emissions in the biochar-amended soils than the biochar produced at a higher temperature. This may be attributable to a greater proportion of recalcitrant C and lower ROC content in RHB-700 than in RHB-400 (Qayyum et al., 2014; Zimmerman et al., 2011; Keith et al., 2011). We assume that coapplication of biochars with composts may be a better way to stabilize SOM and sequestrate carbon in the soils than individual application, especially for a biochar produced with higher temperatures. To clarify the interaction among biochar, compost, and soil, a microscale observation was carried out by polarized microscope. From our microstructure observation (Fig. 2.7), a mechanism of SOM stabilization by biochar addition could be deduced as follows: soil structure was changed and some macroaggregate were formed after biochar incorporation (Fig. 2.7A–B), which was also shown in our previous studies (Jien and Wang, 2013; Hseu et al., 2014). The formation of the new aggregates wrapped the biochar and compost in the aggregates, and thus may prevent rapid decomposing by microbes (Fig. 2.7C–F). Accordingly, the decreases in unstable carbon pool may also result from the sorption of compost-derived carbon onto the biochar, either within the biochar pores (Fig. 2.7C–F) or onto the external biochar surfaces (Fig. 2.7E). Cornelissen et al. (2005) and Sobek et al. (2009) reported that biochars exhibit extremely high adsorption affinity for organic matter and may suppress organic C mineralization. In addition, Kasozi et al. (2010) reported that the organic matter sorption onto biochar surfaces is kinetically limited by slow diffusion into the subnanometer-sized pores dominating biochar surfaces. The various organomineral interactions lead to aggregations of clay particles and organic materials, which stabilizes both soil structure and the carbon compounds within the aggregates.

2.3.2 Nutrients Retention of Biochar-Amended Soils

Biochar application in degraded soil could also be useful to prevent nutrients loss due to erosion or runoff. In our previous study (Lee et al., 2018), we applied a wood biochar made of driftwood (mainly zelkova) and a kind of chemical reagent, polyacrylamide (PAM) to a slopeland (comprised of Ultisol) to determine the best method to prevent soil mass and nutrient lost from erosion. The heavy annual rainfall ($\sim 2500\ mm\ y^{-1}$) in the study area exposes the soil to intensive leaching and soil erosion. During the experiment, applications of most of the amendments significantly decreased the runoff under natural rainfall conditions ($P < 0.05$). The coapplication of WB and green waste compost (GWC) was the most effective at decreasing the runoff (by 16.7%). SOM can strengthen the soil aggregates with relatively large soil pores, which increases water infiltration rate and hydraulic conductivity (Lado et al., 2004a,b; Tejada and Gonzalez, 2007). The biochar-involved treatments showed that biochar application, either with or without compost, was most likely to decrease water runoff and improve water conservation (Beck et al., 2011; Sadeghi et al., 2016), and indicated a similar or higher efficiency compared to the PAM application. In general, less runoff water leads to less soil erosion. In comparison with the control, the amount of soil loss was significantly lower for the PAM, GWC, and WB_4% + compost_1% (WBC4) treatments; however, no significant difference was found among them. This result revealed that the application of compost, as a natural treatment, may protect soil from water erosion with the same efficiency as PAM application. Previous studies showed that biochar treatment had the same protective efficiency under ordinary rainfall events (Hseu et al., 2014; Jien and Wang, 2013; Jien et al., 2015, 2017; Li et al., 2017). Our previous study revealed that the driftwood biochar application into the same soil used in this study may improve soil physical properties including bulk density, saturated soil hydraulic conductivity, and mean weight diameter of soil aggregates during an incubation experiment (Jien and Wang, 2013). However, soil loss was significantly higher when WB4 treatment was applied, compared to the control. The biochar was found floating in the runoff water during severe precipitation events and thus caused observable loss from the subplots (unquantified), even though the biochar had been well mixed with the topsoil. Furthermore, stronger splash erosion and sheet erosion were observed in the WB2- and WB4-applied subplots than in the other subplots. This may be explained by the stronger floating and hydrophobicity of biochar than of soil particulates and compost. The negative effect of biochar on soil erosion was not found with the coapplication treatments with compost (WBC2 and WBC4). Doan et al. (2015) also reported that under natural rainfall, coapplication of biochar with vermicompost resulted in the lowest soil loss compared with individual application of other organic matter or biochar. For best results, the biochar should be coapplied to a soil with other organic matter resources to protect the soil from water erosion under intensive precipitation events.

All treatments decreased the total IN loss through runoff, except for the WB2 and GWC treatments (Table 2.2). The N loss decreased from 3.23 to 0.99 g m^{-2} via the WB4 treatment compared with the control. Compost application resulted in the highest P loss amount, 12.2 mg m^{-2}, through runoff. Only the PAM and WB2 treatments significantly decreased the P loss level. On the other hand, the compost increased soil nutrients to a

TABLE 2.2 Chemical Properties of Biochar-Amended Soil at the End of Incubation Time (After 105 d)

Treatment (%, Biochar Application Rate)	Bd (Mg m^{-3})	Porosity (%)	K_{sat} (cm h^{-1})	MWD (cm)	SER (g m^{-2} h^{-1})
0	1.42 ± 0.10a	41.0 ± 4.24a	16.7 ± 2.83a	45.6 ± 1.34a	1458 ± 50.0a
2.5	1.15 ± 0.05b	51.0 ± 1.41b	30.0 ± 0.28b	48.2 ± 0.49ab	730 ± 94.6b
5	1.08 ± 0.02b	52.0 ± 1.41b	33.1 ± 0.39c	50.0 ± 0.57b	532 ± 106c

Values followed by the same letter within a column are not significantly different at $P < 0.05$ level based on Duncan's test.
SOC: soil organic carbon; CEC: cation exchange capacity; BS: base saturation percentage; Bd: bulk density; MBC: microbial biomass carbon; Po: porosity; K_{sat}: saturated hydraulic conductivity; MWD: mean weight diameter of soil aggregates; SER: soil-erosion rate.

TABLE 2.3 Runoff Water, Soil Loss, and Runoff Water IN Content in the 1-Year Experiment

Treatments	W_{runoff} (mm)	Soil Loss (kg m^{-2})	$NH_4^+-N_{runoff}$ (g m^{-2})	$NO_3^--N_{runoff}$ (g m^{-2})	P_{runoff} (mg m^{-2})
Control	382 (5.8)^{a}	15.5 (0.04)^{bc}	1.41 (0.03)^{a}	1.82 (0.30)^{a}	6.81 (0.64)^{b}
PAM	358 (9.8)^{bc}	11.3 (3.98)^{d}	0.74 (0.57)^{bc}	1.01 (0.90)^{bc}	4.44 (0.04)^{d}
B2	362 (4.8)^{b}	17.5 (1.14)^{ab}	1.03 (0.06)^{ab}	1.44 (0.31)^{ab}	5.19 (0.32)^{c}
B4	345 (3.2)^{cd}	19.6 (1.15)^{a}	0.43 (0.19)^{c}	0.56 (0.12)^{c}	5.27 (1.18)^{bc}
GWC	334 (6.3)^{de}	12.5 (0.57)^{d}	1.04 (0.57)^{ab}	1.41 (0.24)^{ab}	12.2 (0.20)^{a}
WBC2	318 (2.7)^{e}	14.0 (0.89)^{cd}	0.84 (0.29)^{bc}	1.09 (0.18)^{bc}	7.27 (0.37)^{b}
WBC4	319 (2.2)^{e}	11.6 (1.98)^{d}	0.84 (0.10)^{bc}	1.04 (0.00)^{bc}	6.93 (0.56)^{b}

Values shown are means (standard deviations). Individual letters after means indicate no significant difference between the corresponding treatments ($P < 0.05$).
B2 and B4: biochar application of 2% and 4%, respectively; Com: compost application of 1%; CB2 and CB4: coapplication of compost (1%) and biochar (2% and 4%, respectively).

higher level, thereby increasing their leaching via runoff. Compost itself can be a source of nutrients, which can directly transport along with or release nutrients into the runoff water. Therefore, while GWC treatment significantly decreased both runoff water and soil loss, the inorganic N loss via runoff remained statistically the same and P loss was even higher than that of control. Biochar, however, normally contains less N and P than compost. Thus, biochar applied to soils can adsorb more nutrients, retain more soil water, and consequently decrease nutrient losses from runoff, as shown in Table 2.3. The coapplication of biochar and compost resulted in less IN and P loss through runoff compared with the control, whereas the sole application of compost had no significant effect on the IN loss and even increased the P loss during the experiment. A similar result of coapplication of biochar and vermicompost was also revealed by Doan et al. (2015) (Fig. 2.8).

2.4 FUTURE RESEARCH

Studies regarding biochar application in pot and/or field scale have recently been published to evaluate the effects on soil physiochemical properties and crop production. Application rate and production process (HTT and retention time) of biochars are purpose specific. Different porous characteristics, electric charge, and charge density through different produced conditions may be controlled or adjusted to adapt to specific purposes, such as for pollutant adsorbents, soil amendments, prevention of plant diseases and insect pests, etc. In future, the aging effects on surface behaviors of biochars in soils and a long-term field experiment could further determine new directions to evaluate biochar functions in soil systems.

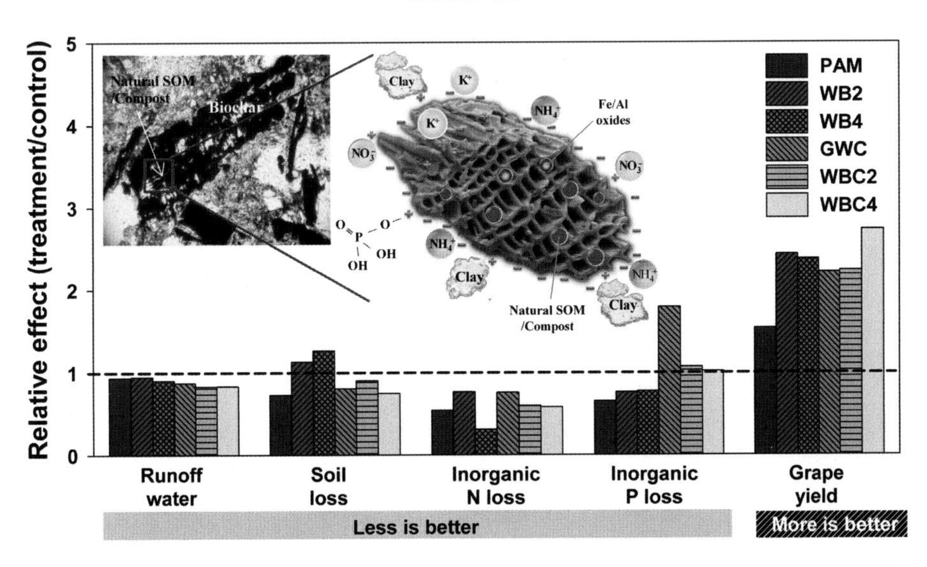

FIGURE 2.8 The relative effect (treatment/control) for retention of water, soil, and nutrients for biochar-amended soil (1-year field experiment).

References

Atkinson, C.J., Fitzgerald, J.D., Hipps, N.A., 2010. Potential mechanisms for achieving agricultural benefits from biochar application to temperature soils: a review. Plant Soil 337, 1–18.

Auerswald, K., Kainz, M., Fiener, P., 2003. Soil erosion potential of organic versus conventional farming evaluated by USEL modeling of cropping statistics for agricultural districts in Bavaria. Soil Use Manage. 19, 305–311.

Beck, D.A., Johnson, G.R., Spolek, G.A., 2011. Amending greenroof soil with biochar to affect runoff water quantity and quality. Environ. Pollut. 159, 2111–2118.

Brady, N.C., Weil, R.R., 2008. An Introduction to the Nature and Properties of Soils, 14th ed. Prentice Hall, Upper Saddle River, NJ.

Brown, R.A., Kercher, A.K., Nguyen, T.H., Nagle, D.C., Ball, W.P., 2006. Production and characterization of synthetic wood chars for use as surrogates for natural sorbents. Org. Geochem. 37, 321–333.

Chan, K.Y., Van Zwieten, L., Meszaros, I., Downie, A., Joseph, S., 2007. Agronomic values of green waste biochar as a soil amendment. Aust. J. Soil Res. 45, 629–634.

Chan, K.Y., Van Zwieten, L., Meszaros, I., Dowie, A., Joseph, S., 2008. Using poultry litter biochars as soil amendments. Aust. J. Soil Res. 46, 437–444.

Cnudde, V., Masschaele, B., Dierick, M., Vlassenbroeck, J., Van Hoorebeke, L., Jacobs, P., 2006. Recent progress in X-ray CT as a geosciences tool. Appl. Geochem. 21, 826–832.

Cornelissen, G., Gustafsson, Ö., Bucheli, T.D., Jonker, M.T.O., Koelmans, A.A., Noort, P.C.M., 2005. Extensive sorption of organic compounds to black carbon, coal, and kerogen in sediments and soils: mechanisms and consequences for distribution, bioaccumulation, and biodegradation. Environ. Sci. Technol 39, 6881–6895.

Deenik, J.L., Diarra, A., Uehara, G., Campbell, S., Sumiyoshi, Y., Antal, M.J., 2011. Charcoal ash and volatile matter effects on soil properties and plant growth in an acid Ultisol. Soil Sci. 176, 336–345.

De Gryze, S., Jassogne, L., Six, J., Bossuyt, H., Wevers, M., Merckx, R., 2006. Pore structure changes during decomposition of fresh residue: X-ray tomography analyses. Geoderma 134, 82–96.

Doan, T.T., Henry-des-Tureaux, T., Rumpel, R., Janeau, J.L., Jouquet, P., 2015. Impact of compost, vermicompost and biochar on soil fertility, maize yield and soil erosion in Northern Vietnam: a three year mesocosm experiment. Sci. Total Environ. 514, 147–154.

Hammes, K., Schmid, M.W.I., 2009. Changes in biochar in soils. In: Lehmann, M., Joseph, S. (Eds.), Biochar for Environmental Management Science and Technology. Earthscan, London, pp. 169–182.

Hseu, Z.Y., Jien, S.H., Chien, W.H., Liou, R.C., 2014. Impacts of biochar on physical properties and erosion potential of a mudstone slopeland soil. Sci. World J. Available from: https://doi.org/10.1155/2014/602197.

Iswaran, V., Jauhri, K.S., Sen, A., 1980. Effect of charcoal, coal and peat on the yield of moog, soybean and pea. Soil Biol. Biochem. 12, 191–192.

Jien, S.H., Wang, C.S., 2013. Effects of biochar on soil properties and erosion potential in a highly weathered soil. Catena 110, 225–233.

Jien, S.H., Wang, C.C., Lee, C.H., Lee, T.Y., 2015. Stabilization of organic matter by biochar application in compost-amended soils with contrasting pH values and textures. Sustainability 7, 13317–13333.

Jien, S.H., Chen, W.C., Ok, Y.S., Awad, Y.M., Liao, C.S., 2017. Short-term biochar application induced variations in C and N mineralization in a compost-amended tropical soil. Environ. Sci. Pollut. Res. Int. Available from: https://doi.org/10.1007/s11356-017-9234-8.

Kasozi, G.N., Zimmerman, A.R., Nkedi-Kizza, P., Gao, B., 2010. Catechol and humic acid sorption onto a range of laboratory-produced black carbons (biochars). Environ. Sci. Technol 44, 6189–6195.

Keith, A., Singh, B., Singh, B.P., 2011. Interactive priming of biochar and labile organic matter mineralization in a smectite-rich soil. Environ. Sci. Technol. 45, 9611–9618.

Kimetu, J.M., Lehmann, J., 2010. Stability and stabilisation of biochar and green manure in soil with different organic carbon contents. Aust. J. Soil Res. 48, 577–585.

Lado, M., Paz, A., Ben-Hur, M., 2004a. Organic matter and aggregate size interactions in infiltration, seal formation, and soil loss. Soil Sci. Soc. Am. J. 68, 935–942.

Lado, M., Paz, A., Ben-Hur, M., 2004b. Organic matter and aggregate-size interactions in saturated hydraulic conductivity. Soil Sci. Soc. Am. J. 68, 234–242.

Lee, C.H., Wang, C.C., Lin, H.H., Lee, S.S., Tsang, D.C., Jien, S.H., et al., 2018. In-situ biochar application conserves nutrients while simultaneously mitigating runoff and erosion of an Fe-oxide-enriched tropical soil. Sci. Total Environ 619–620, 665–671.

Lehmann, J., Joseph, S.D., 2009. Biochar for Environmental Management. Science and Technology, Earthscan, London.

Li, Z.G., Gu, C.M., Zhang, R.H., Ibrahim, M., Zhang, G.S., Wang, L., et al., 2017. The benefic effect induced by biochar on soil erosion and nutrient loss of slopping land under natural rainfall conditions in central China. Agric. Water Manage 185, 145–150.

Liang, B., Lehmann, J., Solomon, D., Kinyangi, J., Grossman, J., O'Neill, B., et al., 2006. Black carbon increases cation exchange capacity in soils. Soil Sci. Soc. Am. J. 70, 1719–1730.

Lua, A.C., Yang, T., Guo, J., 2004. Effects of pyrolysis conditions on the properties of activated carbons prepared from pistachio-nut shells. J. Anal. Appl. Pyrolysis 72, 279–287.

Materechera, S.A., 2009. Aggregation in a surface layer of a hardsetting and crusting soil as influenced by the application of amendments and grass mulch in a South African semi-arid environment. Soil Tillage Res. 105, 251–259.

Qayyum, M.F., Steffens, D., Reisenauer, H.P., Schubert, S., 2014. Biochars influence differential distribution and chemical composition of soil organic matter. Plant Soil Environ. 60, 337–343.

Rogasik, H., Onasch, I., Brunotte, J., Jegou, D., Wendroth, O., 2003. Assessment of soil structure using X-ray computed tomography. Geo. Soc. London Publ. 215, 151–165.

Rouquerol, F., Rouquerol, I., Sing, K., 1999. Adsorption by Powders and Porous Solids. Academic Press, London, UK.

Sadeghi, S.H., Hazbavi, Z., Harchegani, M.K., 2016. Controllability of runoff and soil loss from small plots treated by vinasse-produced biochar. Sci. Total Environ. 541, 483–490.

Schrader, S., Rogasik, H., Onasch, I., Jegou, D., 2007. Assessment of soil structural differentiation around earthworm burrows by means of X-ray computed tomography and scanning electron microscopy. Geoderma 137, 378–387.

Sobek, A., Stamm, N., Bucheli, T.D., 2009. Sorption of phenyl urea herbicides to black carbon. Environ. Sci. Technol. 43, 8147–8152.

Tejada, M., Gonzalez, J.L., 2007. Influence of organic amendments on soil structure and soil loss under simulated rain. Soil Tillage Res 93, 197–205.

Troeh, F.R., Thompson, L.M., 2005. Soils and Soil Fertility. Blackwell Publishing, Iowa, US.

Trompowsky, P.M., Benites, V.D.M., Madari, B.E., Pimenta, A.S., Hockaday, W.C., Hatcher, P.G., 2005. Characterization of humic like substances obtained by chemical oxidation of eucalyptus charcoal. Org. Geochem. 36, 1480–1489.

Van Zwieten, L., Kimber, S., Morris, S., Chan, K.Y., Downie, A., Rust, J., et al., 2010. Effects of biochar from slow pyrolysis of papermill waste on agronomic performance and soil fertility. Plant Soil 327, 235–246.

Wildman, J., Derbyshire, F., 1991. Origins and functions of macroporosity in activated carbons from coal and wood precursors. Fuel 70, 655–661.

Wuddivira, M.N., Stone, R.J., Ekwue, E.I., 2009. Structure stability of humid tropical soils as influenced by manure incorporation and incubation duration. Soil Sci. Soc. Am. J. 73, 1353–1360.

Zimmerman, A.R., Gao, B., Ahn, M.Y., 2011. Positive and negative carbon mineralization priming effects among a variety of biochar-amended soils. Soil Biol. Biochem. 43, 1169–1179.

CHAPTER

3

Elemental and Spectroscopic Characterization of Low-Temperature (350°C) Lignocellulosic- and Manure-Based Designer Biochars and Their Use as Soil Amendments

J.M. Novak[1] *and Mark G. Johnson*[2]

[1]U.S. Department of Agriculture, Agricultural Research Service, Coastal Plains Research Center, Florence, SC, United States [2]U.S. Environmental Protection Agency, National Health and Environmental Effects Research Laboratory/ORD Western Ecology Division, Corvallis, OR, United States

DISCLAIMER

The information in this document has been funded in part by the US Environmental Protection Agency. It has been subjected to review by the National Health and Environmental Effects Research Laboratory's Western Ecology Division and approved for publication. Approval does not signify that the contents reflect the views of the Agency, nor does mention of trade names or commercial products constitute endorsement or recommendation for use. USDA is an equal opportunity provider and employer.

3.1 INTRODUCTION

Biochar has gained world attention as a potential amendment in the agronomic and environmental sectors. Biochars are produced from a variety of organic substances under

Biochar from Biomass and Waste
DOI: https://doi.org/10.1016/B978-0-12-811729-3.00003-0

a wide range of pyrolysis conditions (Lehmann and Joseph, 2015). Pyrolysis conditions such as temperature and redox environments play an important role in the carbonization processes where raw parent feedstocks are converted into syngas, bio-oil, and biochar (Sohi et al., 2009; Boateng et al., 2015). Biochars can be designed with specific characteristics by manipulating variables such as feedstock selection, blends, activating agents, and pyrolysis temperatures. Thus, these variables are important to consider since they determine the biochar chemical makeup, which ultimately controls its interaction with organic chemicals, modification of soil pH, and release of nutrients into soil (Lehmann and Joseph, 2015). Therefore, understanding the elemental and spectroscopic properties of biochars is critical to explain the reaction mechanisms in soils and their ability to reactant with pollutants.

3.2 BIOCHAR DEFINITION

For this chapter, biochar is defined as an amendment for soil application. Biochar is distinguished from other thermally produced C-enriched material (e.g., black carbon, charcoal, soot) when it is used or produced for this purpose (Lehmann and Joseph, 2015). Others will argue, however, that biochar is not a sole product, but rather a continuum of black carbon forms, each of which possess a set of unique chemical properties (Jones et al., 1997; Spokas, 2010). Black carbon is an umbrella term used to define many forms of pyrogenic carbonaceous material produced through thermochemical processing of organic substances (Lehmann and Joseph, 2015). Nonetheless, biochar has been defined and a list of general characteristics provided by the International Biochar Initiative (IBI, 2017).

3.3 BIOCHAR FEEDSTOCKS

Biochar production technology using pyrolysis has been previously reviewed (Bridgewater, 2003; Laird et al., 2009; Sohi et al., 2009; Boateng et al., 2015). Within these reviews, the authors describe how different types of organic-based feedstocks can be pyrolyzed to produce biochars including lignocellulosic materials (e.g., wood chips, crop residues); manures; industrial byproducts (e.g., wood debris, flooring); and municipal wastes (e.g., cardboard, biosolids). For this chapter, two lignocellulosic-based and two manure-based feedstocks were collected and processed for pyrolysis (Fig. 3.1). Here, these four uncharred feedstocks are referred to as "parent feedstocks," They were later pyrolyzed using a low temperature (350°C) to produce biochar. The parent feedstocks were also blended together prior to pyrolysis to produce designer biochars (Novak et al., 2009b; Novak and Busscher, 2012).

Both the Loblolly pine chip (*Pinus taeda*) and switchgrass (*Panicum virgatum*) parent feedstocks were obtained from the Coastal Plain region of South Carolina (SC), USA. Pine chips were recovered from logging piles remaining after tree-thinning operations while switchgrass was harvested from plots related to a previous biofuel experiment (Fig. 3.1). The parent poultry-litter and swine-solid feedstocks were obtained from animal production facilities in South Carolina (SC) and North Carolina (NC), USA, respectively. Poultry

Loblolly pine chip (Berkeley Co.,

Switchgrass (Darlington Co., SC)

Poultry litter (Orangeburg Co., SC)

Swine solids (Sampson Co., NC)

FIGURE 3.1 Feedstock types and collection location. *Source: J.M. Novak.*

litter was collected from a facility that used softwood shavings as bedding material (Cantrell et al., 2012). Swine solids were obtained through chemical precipitation and subsequent physical separation from liquid effluent using a rotary press separator (Fig. 3.1; Vanotti et al., 2009). These four parent feedstocks and subsequent blends were pyrolyzed at 350°C under production conditions as outlined in Cantrell et al. (2012) and Novak et al. (2012). The biochars were then analyzed for their proximate (ash, volatile C, and fixed C) and ultimate (carbon, hydrogen, oxygen, nitrogen, and sulfur) composition by Hazen Research Inc., (Golden, CO, USA) using ASTM D 3172 and 3176 standard methods, respectively (ASTM, 2006). The biochars were acid digested using EPA method 3051 (EPA, 2007), and the inorganic composition measured using inductively coupled plasma-mass spectrometry (ICP-MS) (Novak et al., 2009b). Scanning electron microscopy (SEM) was used to obtain the surface images of biochars produced from parent feedstocks by the Microscopy and NanoImaging facility at Iowa State University, Ames, Iowa (USA) using methods as outlined at http://microscopy.biotech.iastate.edu/services. Biochar structural characteristics were determined using ^{13}C nuclear magnetic resonance (NMR) and Fourier transformed infrared (FTIR) spectroscopy described in Novak et al. (2009b) and Cantrell et al. (2012). The ^{13}C NMR peak spectral assignments and %C distribution were determined as described by Wang et al. (2007).

3.4 BIOCHAR PRODUCTS

Pyrolysis of organic feedstocks will produce three products including a liquid (bio-oil), a syngas, and a solid (biochar). Bio-oil is created during pyrolysis because natural polymeric constituents (i.e., lignin, cellulose, fats) are broken down into volatile gasses containing O- and H- containing forms. These gasses can be later recondensed along the pyrolysis continuum into bio-oil or recycled to facilitate energy required for upstream feedstock drying and carbonization. Carbonization furthers the removal of polar functional groups and rearrangement of ring/linear-shaped organic structures into polycondensed aromatic sheets in the resultant biochar (Chen et al., 2008; Keiluweit et al., 2010).

The pyrolysis process is conducted under very low oxygen concentrations, at pyrolysis temperatures ranging from 300 to 1100°C, using variable residence times (seconds to hours; Manyà, 2012). Low oxygen levels are necessary for carbonization to occur while minimizing CO_2 and NO_x production (Antal and Grønli, 2003). The pyrolysis process has different modes of processing including slow/fast pyrolysis, flash pyrolysis, and gasification. Slow pyrolysis, such as with traditional kilns, exposes feedstocks in a batch thermal mode for days to a week. Fast pyrolysis uses a traditional method of advancing the feedstock through a retort/oven for several minutes using a continuous feed system. In the flash pyrolysis procedure, feedstocks are exposed to a burst of thermal energy, usually from 1 to 5 seconds. Finally, gasification exposes the feedstock to high temperatures up to 1100°C to maximize conversion into syngas materials. Final product recovery under these modes includes bio-oil, biochar, and gases (Manyà, 2012; Boateng et al., 2015). In general, flash pyrolysis yields 60% biochar and 40% bio-oil and syngas, while fast pyrolysis yields higher bio-oil recoveries with minimal biochar leftovers. Slow pyrolysis produces almost equal yields of bio-oil, biochar, and syngas (Laird et al., 2009). Biochars produced for agronomic purposes prefer a slow pyrolysis mode to maximize biochar yields (Song and Guo, 2012).

3.5 GENERAL CHARACTERISTICS OF BIOCHARS

Biochar has variable chemical and physical properties due to differences in parent feedstock selection, pyrolysis temperature, and their pyrolysis production modes. The literature has shown that biochars produced from animal manure feedstocks tend to have lower C, O, and H content than lignocellulosic-based biochar feedstocks. Lower C, H, and O contents in manure-based biochars are related to the feedstock containing more volatile OC compounds that are lost during the dry and carbonization phases of pyrolysis (Lehmann and Joseph, 2015). Additionally, manure-based biochars with higher pH values contain higher ash contents, and their ash fraction contains more elements suitable for plant nutrients (Chan et al., 2008; Cantrell et al., 2012; Ippolito et al., 2015). The manure-based biochars have a calcareous nature because the ash contains significant amounts of Ca, K, and Mg (Sistani and Novak, 2006; Novak et al., 2009a). Thus, manure-based biochars can act as weak-liming agents for acidic soils (Yuan and Xu, 2011; Hass et al., 2012; Ameloot et al., 2015).

On the other hand, lignocellulosic-based biochars tend to have higher fixed C content that represents the C material resistant to loss during pyrolysis (Domingues et al., 2017). Therefore, biochar yields are generally higher when the biochar has a higher fixed C content. Additionally, the fixed C content of biochars is considered a surrogate for their longevity in soil since fixed C is comprised of aromatic structures not readily mineralized (Novak et al., 2009b; Crombie et al., 2013; Domingues et al., 2017). If the agronomic goal is to add more-stable C forms to soil, then a biochar such as wood-based with a high fixed C percentage is desirable (Novak et al., 2009b; Domingues et al., 2017). Biochars produced from lignocellulosic materials, in comparison to manure-based feedstocks, contain less nutrients, and have a lower $CaCO_{3\text{-eq}}$ (Camps-Arbestain et al., 2015) due to their lower ash contents (Novak et al., 2012; Jindo et al., 2014). Biochars produced from lignocellulosic-based feedstocks have favorable characteristics that can improve soil OC contents, but also can be a suitable blending matrix with other feedstock types (Novak et al., 2014b) or as a carrier for inorganic fertilizers (Hagemann et al., 2017; Wisnubroto et al., 2017). However, in general, manure-based biochars have additional characteristics that make them relatively more effective as a soil fertility amendment compared to lignocellulosic-based biochars.

3.6 LOW-TEMPERATURE PYROLYZED DESIGNER BIOCHARS

In this chapter, we also report on elemental and spectroscopic properties for biochars produced at low pyrolysis temperatures (350°C) from four parent feedstocks and through feedstock blending. Biochar characterization results from single-parent feedstock are common in the biochar literature. The parent feedstocks are usually of a local origin, a waste product, or material available in large qualtities.

Biochars can also be produced by blending parent feedstocks at specific weight ratios prior to pyrolysis (Novak and Busscher, 2012). We highlight the properties of blends because the parent feedstocks often have special characteristics that can be utilized for targeted soil deficiencies. As an example, biochar produced from poultry litter contains high concentrations of total and water soluble P (Novak et al., 2014a). Here, the goal was to manipulate the P content of the poultry- litter feedstock by blending in specific weight ratios with a lignocellulosic-based feedstock containing much lower P concentrations. After soil addition, the blends have been shown to deliver sufficient plant-available P needed for producing a corn crop in a SC sandy soil (Novak et al., 2014a). The blends were made at a plant:manure ratio of 90:10; 80:20, and terminated with a 50:50 blend as an upper threshold for fertilizer testing. Incubation experiments revealed that the 80:20 blend delivered sufficient soil Mehlich 3 extractable P levels needed for the corn crop. Thus, the goal was achieved whereby a biochar was designed with an intended service and a critical plant nutrient (P) was not overapplied. Of course, to design biochars as nutrient delivery systems, a biochar's chemical and elemental composition of both the biochar and the feedstock sources used to produce it must be determined beforehand.

3.6.1 Ultimate, Proximate, and Inorganic Composition

The ultimate, proximate and inorganic composition of biochars produced from the lignocellulosic- and manure-based parent feedstocks and through blends are shown in Tables 3.1 and 3.2, respectively.

Similar to what is reported in the literature, the lignocellulosic-based biochars have lower volatile C, fixed C, ash, and C, N, and S content than the manure-based biochars (Table 3.1). Manure-based biochars have higher ash and N and S content due to the

TABLE 3.1 Physical and Chemical Composition of Biochars Pyrolyzed at 350°C From Feedstocks and Blends (Means of n = 3; odw = Oven Dry Weight)

Feedstock	Volatile C (%)	Fixed C (%)	Ash (%)	C (%)	H (%)	N (%)	O (%)	S (%)
Pine chip (PC)	40.4	57.8	1.8	78.7	4.9	0.4	14.2	0.0
Switchgrass (SG)	41.4	55.4	3.2	75.5	4.6	0.5	16.2	0.0
Poultry litter (PL)	36.2	31.8	32.	51.5	3.6	4.1	6.9	0.9
Swine solids (SS)	37.6	27.4	35	51	3.7	5.9	3.2	1.2
PC:PL 50:50	30.6	50.9	18.5	63.7	3.8	3.4	10.2	0.4
PC:PL 80:20	33.7	58.9	7.3	75.6	4.6	1.3	10.8	0.2
PC:PL 90:10	37.2	58.5	4.4	78.1	4.8	0.9	11.7	0.1
SG:SS 80:20	33.5	52.7	13.7	69.2	4.1	2.5	10	0.4
SG:SS 90:10	38.5	52.4	9.6	74.5	4.9	2.0	8.6	0.2

Source: J.M. Novak.

presence of salts (Novak et al., 2014b) and N (Gul and Whalen, 2016) excreted from animals. In contrast, the O and H contents are higher for the lignocellulosic- than the manure-based biochars (Table 3.1). This is due to the presence of −OH groups associated with the cellulose, hemicellulose, and lignin components of the plants structural makeup (Jacobsen and Wyman, 2000).

The van Krevelen diagram (van Krevelen, 1950) is a convenient method to compare the impact of drying and carbonization on the structural and functional group properties of biochars. Using the ultimate results from biochar (Table 3.1), the van Krevelen diagram gives both the atomic H/C and O/C ratios (Fig. 3.2). These plots commonly show that certain biochars will cluster in regions depicted in the van Krevelen diagram.

The unpyrolyzed parent lignocellulosic and manure feedstocks were also analyzed for their proximate and ultimate analysis (data not reported). From these results, their atomic ratios were calculated and plotted (noted as 0°C in Fig. 3.2) in the diagram to compare with subsequent properties of their respective biochars. The unpyrolyzed parent feedstocks have higher O/C and H/C ratios compared to their ensuing biochars, and thus occur in a different region of the van Krevelen diagram.

At 350°C pyrolysis temperature, all biochars in the van Krevelen diagram cluster around the same region. In fact, biochars produced from the parent feedstocks cluster very close to their respective blends (Fig. 3.2). The pattern in the van Krevelen diagram is consistent with the loss of H- and O-containing compounds during the drying and carbonization phases of pyrolysis. Biochars produced at higher temperatures (>350°C) and soot, in contrast, all have very low H/C and O/C atomic ratios because of the severe loss of -O and -H molecules from carbonization (Fig. 3.2).

Among the four parent feedstocks, biochar produced from poultry litter had the highest concentrations of 16 of the 18 elements assayed (Table 3.2). Poultry litter biochar can serve as a good source of plant macro- (i.e., Ca, Mg, K, P) and micronutrients (i.e., Co, Cu, Mn, Zn) compared to biochar prepared from lignocellulosic-based parent feedstocks. As

TABLE 3.2 Inorganic Elemental Composition of Biochars Pyrolyzed at 350°C From Parent Feedstocks and Blends (mg kg^{-1}; n = 1, Oven-Dry Weight Basis)[a]

Feedstock	Al	As	Ca	Co	Cr	Cu	Fe	K	Mg
Pine chip (PC)	230	0.0	3047	0.1	0.4	12	345	1752	1039
Switchgrass (SG)	24	0.1	2130	0.1	2.8	29	396	1947	1709
Poultry litter (PL)	226	18	21963	2.3	3.4	250	1841	30040	6370
Swine solids (SS)	45	0.0	2475	2.4	0.1	2300	48	2202	1838
PC:PL 50:50	300	18	21709	1.1	2.3	163	1237	41418	7063
PC:PL 80:20	216	6.7	6704	0.6	2.3	64	1573	9345	2936
PC:PL 90:10	184	4.2	4791	0.4	2.1	34	790	5790	1846
SG:SS 80:20	291	0.4	13538	1.3	6.7	661	1515	4927	9971
SG:SS 90:10	145	0.2	5535	0.7	5.5	311	816	4995	5751

	Mn	Na	Ni	P	Pb	Si	Ti	V	Zn
Pine chip (PC)	74	2.1	0.7	207	1.8	295	4.4	0.1	39
Switchgrass (SG)	38	43	27	384	2.1	528	1.9	0.1	33
Poultry litter (PL)	462	7050	6.5	15937	0.5	732	13	2.1	957
Swine solids (SS)	101	246	1.9	495	2.4	698	3.7	0.1	2974
PC:PL 50:50	496	7600	6.2	13777	2.2	414	12	2.3	573
PC:PL 80:20	189	2697	2.6	4832	1.1	308	6.1	0.6	230
PC:PL 90:10	122	1678	2.3	3490	2.0	176	3.8	0.4	126
SG:SS 80:20	587	1203	10	14831	2.2	925	29	2.9	1531
SG:SS 90:10	246	572	5.3	8254	1.2	639	15	1.5	732

[a]*Digested using EPA method 3051 (conc.* HNO_3*) and quantified using ICP-MS.*
Source: J.M. Novak.

described earlier, the general trend is that lignocellulosic-based biochars contain lower plant macro- and micronutrient concentrations compared to manure-based biochars (Table 3.2). Nevertheless, lignocellulosic-based biochars do contain critical plant nutrients, but biochars produced from manures are weak liming agents and offer a richer and higher concentration of plant nutrients. Thus, manure-based biochars are a more effective amendment for improving soil fertility (Chan et al., 2008; Song and Guo, 2012).

As shown in Table 3.2, biochars produced from the parent feedstocks are concentrated in major plant nutrients including Ca, K, and Mg, and especially total P. Biochars derived from these parent feedstocks when used as a soil amendment should be carefully considered, especially for P-enriched manure-based biochars applied to sandy soils. If overapplied, P dissolved from the biochar can leach through sandy soils potentially creating groundwater quality issues. Thus, four parent feedstocks were blended at specific weight

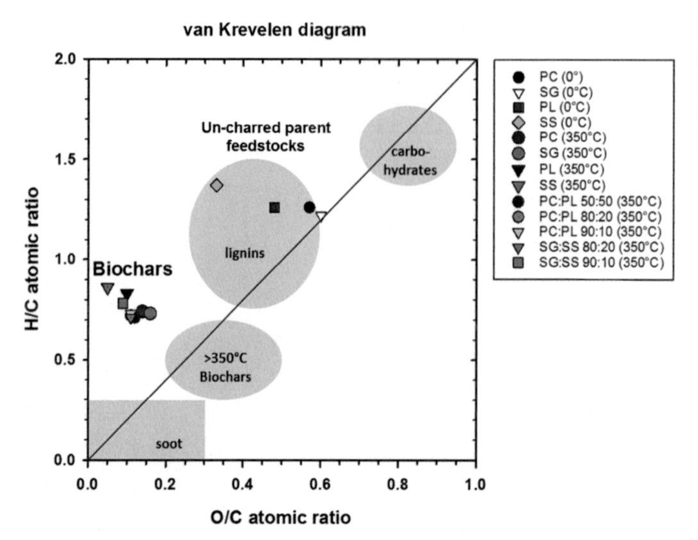

FIGURE 3.2 Regional plots of elemental composition from the parent feedstock and low-temperature biochars shown on the van Krevelen diagram. *Source: J.M. Novak and M.G. Johnson.*

ratios. Blending the P-enriched manure-based parent feedstocks with low P-containing lignocellulosic-based feedstocks resulted in dramatic TP declines (Table 3.2). In fact, the TP concentration in 100% poultry-litter biochar declined almost five-fold when blended at a 90:10 ratio using pine chips. Following suit, the high plant micronutrient concentrations of Cu and Zn were also reduced by blending with lignocellulosic-based feedstocks. Blending of feedstocks prior to pyrolysis offers the flexibility to design the biochar to enhance specific nutrient concentrations in soils, or reduce their concentration to be in balance with plant nutritional requirements.

3.6.2 Spectroscopic Characteristics

3.6.2.1 SEM Images

SEM offers the ability to visually inspect the biochar surface morphology for detection of pores, fissures, and other surface materials/deposits. Thus, collecting SEM images is a common tool for understanding biochar surface properties. Images were collected of the lignocellulosic-based biochars produced from pine chips (Fig. 3.3A and B) and from switchgrass (Fig. 3.3C and D) at a gross-scale (200 μm) and a finer-scale (50 to 20 μm) resolution. The gross-scale SEM images of the biochars revealed large differences in the parent feedstock structural properties. The SEM images collected at 200-μm scale show that remnants of the structural materials in the pine-chip biochar occur as fibers (Fig. 3.3A). A close-up image reveals that the pine-chip biochar surface is marginally coated with small particles, and has a few both large (200 μm) and smaller diameter pores (Fig. 3.3B, <5 μm). The fibrous nature of this pine-based biochar is similar to the SEM images reported by Cetin et al. (2005).

On the other hand, SEM images of switchgrass biochar shows remnants of the individual cell walls inside the grass fiber (Fig. 3.3C; 200 μm). Images taken at a closer scale (50 μm) of the switchgrass biochar show that the vesicles in the cell walls (i.e., Golgi complex) have ruptured and created pores that are about 10 to 50 μm in diameter (Fig. 3.3D).

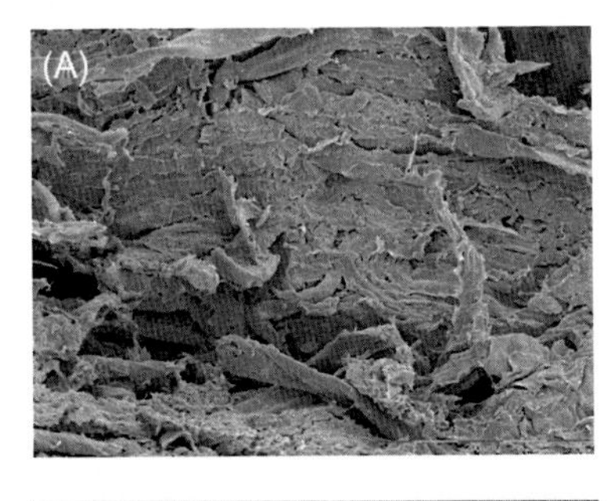

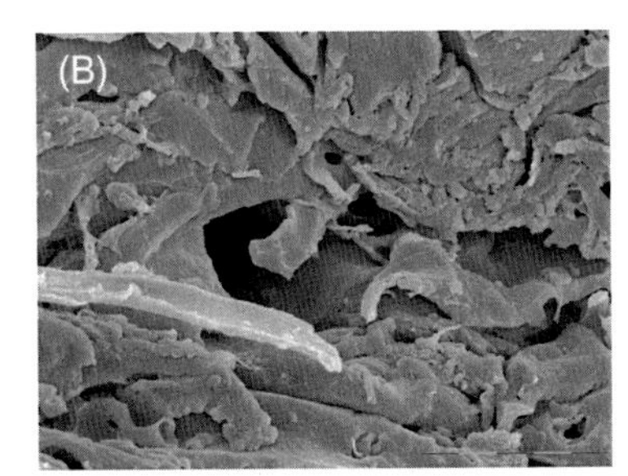

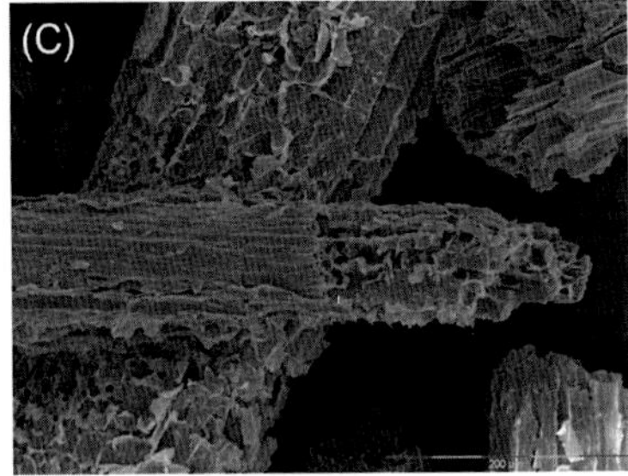

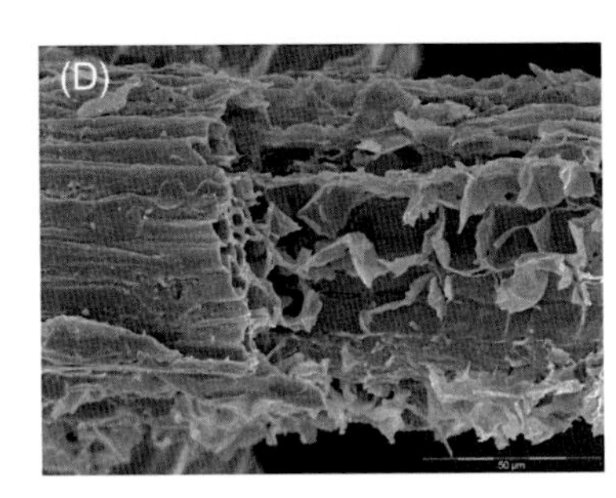

FIGURE 3.3 SEM images of biochars produced from parent pine-chip feedstock (A and B) and parent switchgrass feedstock (C and D). *Source: J.M. Novak.*

Similarly, pores in switchgrass biochars were also observed in the SEM images reported by Pilon and Lavoie (2011) and Bayan (2015). These pores were speculated to be important mechanisms at improving total water retention by trapping H_2O in a sandy soil (Novak and Watts, 2013). Switchgrass is also a popular feedstock for biochar production and subsequent usage as a soil amendment since it has been reported to reduce nutrient leaching (Ippolito et al., 2012) and improve soil microbial/enzymatic activity (Ducey et al., 2013; Kelly et al., 2015).

The SEM images of the two manure-based biochars revealed a contrasting surface morphology compared to the lignocellulosic-based biochar. At a gross scale (200 μm), both SEM images of biochar produced from poultry litter (Fig. 3.4A) and swine solids (Fig. 3.4C) reveal a surface containing pores, void of spaces, and extensively covered with irregular-shaped particles. Closer images of both biochars at 10 to 20 μm scale reveal the coverage extent and shapes of these particles (Fig. 3.4B and D). Similarly, Spokas et al. (2014) reported that surface examination of poultry-litter biochar using SEM showed coatings of small irregular shaped particles. Spokas et al. (2014) further scanned their surfaces using electron dispersive spectroscopy and found that the coatings were comprised of inorganic salts (e.g., Ca, Mg, P). For these two manure-based biochars shown here, the particles are probably salts excreted from animals during manure production. This would explain the rich mineral composition in the manure-based biochars (Table 3.2). Joseph et al. (2010) also reported the presence of nanoscale particles on the surfaces of manure-based biochars. Particle accumulation was theorized to be due to ions from the dissolved salt being attracted to sites involved with redox reactions and due to electrostatic interactions with electron dense regions (Joseph et al., 2010). Spokas et al. (2014) noted that some of the salts were easily displaced with deionized H_2O, and thus salt coatings on animal-manure based biochars add a fertility benefit.

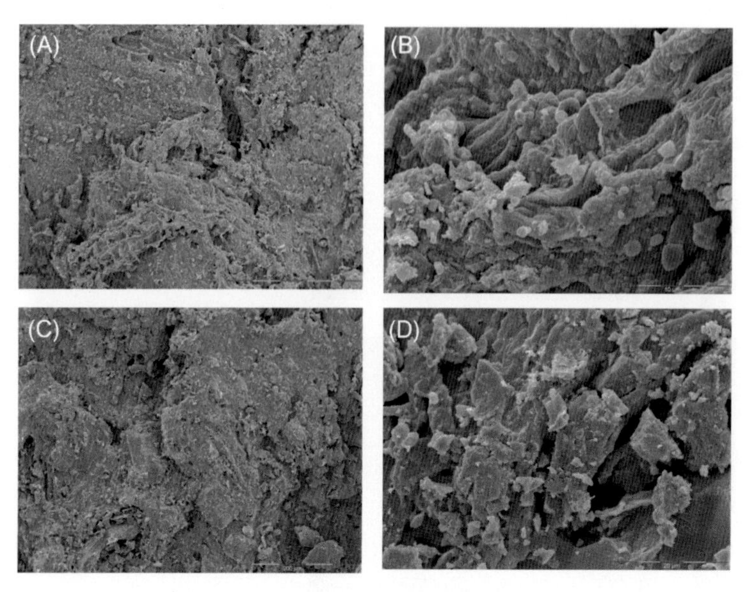

FIGURE 3.4 SEM images of biochars produced from parent poultry-litter feedstock (A and B) and parent swine-solid feedstock (C and D). *Source: J.M. Novak.*

3.6.2.2 Structural and Functional Group Properties of Biochars Revealed With ^{13}C NMR and FTIR Spectroscopy

Spectroscopic methods using ^{13}C NMR spectroscopy can be used to determine the structural and functional group makeup of biochars (Baldock and Smernik, 2002; Brewer et al., 2009; Cheng et al., 2010). In NMR analysis, biochar is subject to a strong magnetic field and radiofrequency pulse over a period. The ^{13}C nuclear active isotope among all the ^{14}C atoms absorbs radiofrequency within a narrow band and produces a resonance frequency due to shielding of nearby electrons (Smernik, 2017). A detectable magnetic spectrum is obtained and differences in signal intensity plotted along an *x*-axis. These are chemical shifts using units of parts per million (ppm; Pavia et al., 1979). Thus, interpretation of ^{13}C NMR spectra is conducted by assignment of chemical shift regions in the spectrum to specific organic structures, such as C in unsubstituted aliphatic compounds (0 to 50 ppm) and in core structural features such as aromatic-C (109–145 ppm). The area under each chemical shift region is determined and reported as a percentage C distribution.

Biochars are also commonly analyzed using FTIR spectroscopy because the method is nondestructive, spectra are obtained in a short time period, and the spectral technology is not as difficult to understand as NMR (Johnston, 2017). FTIR is a vibrational spectroscopy method that provides a means of characterizing the constituents of organic and inorganic samples (Parikh et al., 2014). In FTIR spectroscopy, the samples can be presented to the instrument in a variety of ways including transmission (where the IR beam passes through the sample); diffuse reflectance, or DRIFT (where the incident IR beam is sorbed by the sample and is then re-emitted diffusely and gathered by the instrument); or via attenuated total reflectance, or ATR (where the sample is placed in contact with an optically dense crystal [e.g., a diamond], and the IR beam is passed through it into the sample and absorbance information is reflected into the crystal and then by the detector). The biochar sample is subject to IR radiation and bonds in the matrix will absorb energy at specific wavelengths within the mid-IR (4000 to 400 wavenumbers), resulting in molecular vibrations. The molecules are excited to a higher energy state, and depending on the bond type and mass of the bonded atoms, will cause the bond to vibrate in one of several modes (Pavia et al., 1979; Parikh et al., 2014). Biochar samples are usually scanned with spectral wavelengths between 4000 and 550 wavenumbers (cm^{-1}). Absorption wavelengths of light are assigned to specific core structural and functional groups in biochar and to mineral/impurities (Johnston, 2017). In most biochar characterization studies, FTIR and NMR spectroscopy are used because when the results are combined, the biochar core structures and presence of specific functional groups can be distinguished.

The ^{13}C NMR spectral data for the biochars produced from the parent lignocellulosic, parent manure, and blends are presented in Figs. 3.5–3.8. The ^{13}C NMR spectral assignments and percent C distribution in the biochars is presented in Table 3.3.

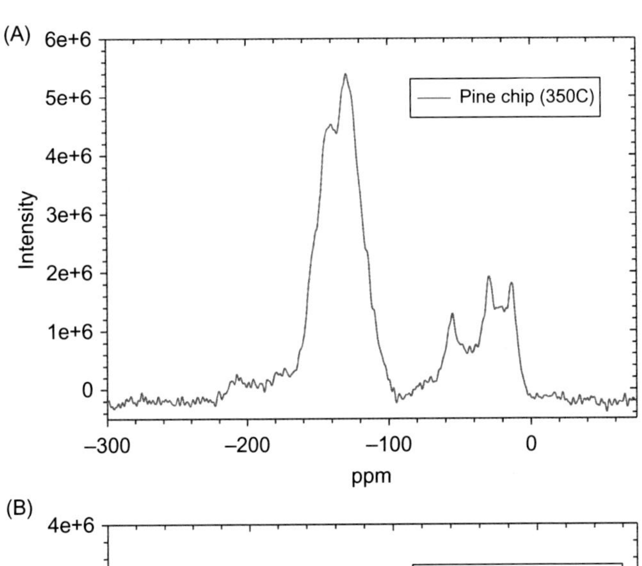

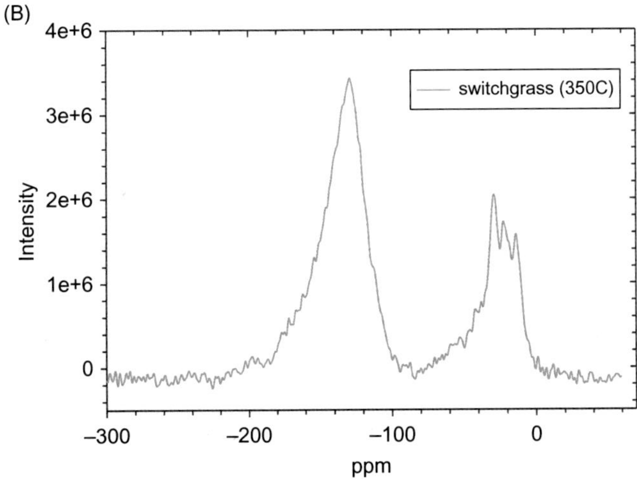

FIGURE 3.5 ^{13}C NMR spectra from biochars produced from parent pine-chip feedstock (A) and parent switchgrass feedstock (B). *Source: J.M. Novak.*

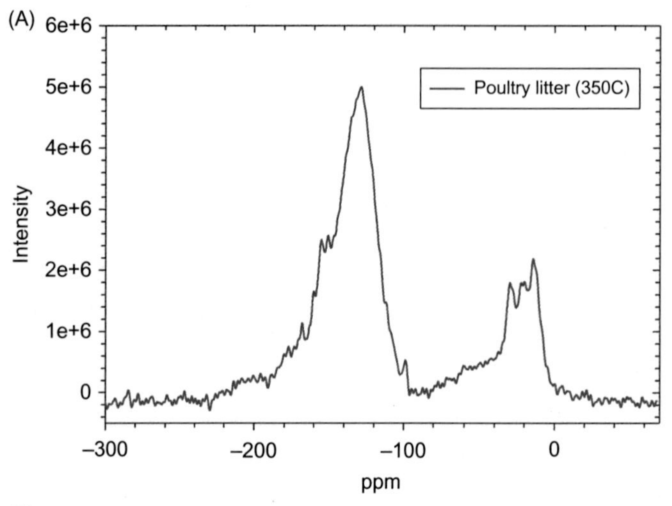

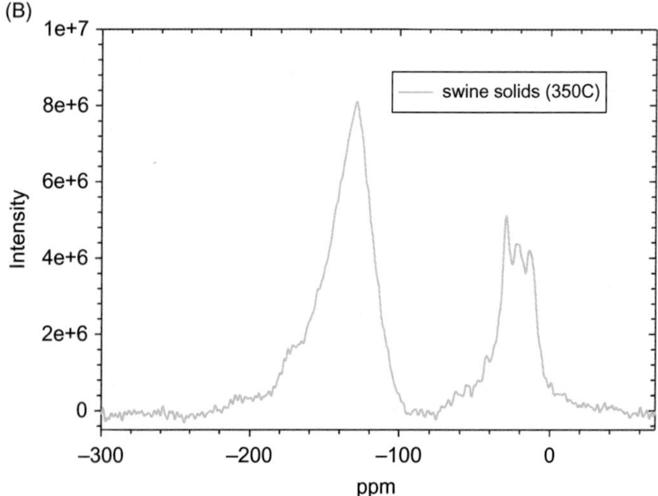

FIGURE 3.6 ^{13}C NMR spectra from biochars produced from parent poultry-litter feedstock (A) and parent swine-solid feedstock (B). *Source: J.M. Novak.*

Biochars produced from the parent pine-chip and switchgrass feedstocks exhibit two groups of distinct ^{13}C NMR spectral peaks between 0 to 60 ppm and 100 to 190 ppm (Fig. 3.5A and B). These two groups of peaks are assigned to C in unsubstituted aliphatic and n-alkyl compounds, and C in aromatic structures along with C in phenolic and carboxylic acid functional groups. Biochar pyrolyzed from the parent switchgrass has more C as aliphatic compounds (alkanes, fatty acids), but then has less aromatic C-containing structures as compared to biochar from the parent pine chip (Table 3.3). The parent pine-chip spectra exhibit a distinguishing singlet peak at 58 ppm due to proteinaceous and methoxy (-OCH_3) compounds (Fig. 3.5A) whereas this peak is lacking in the parent switchgrass spectra (Fig. 3.5B).

Biochars produced from the parent poultry-litter and swine-solid feedstocks also show a near similar ^{13}C NMR spectral pattern (Fig 3.6A and B) as the lignocellulosic-based

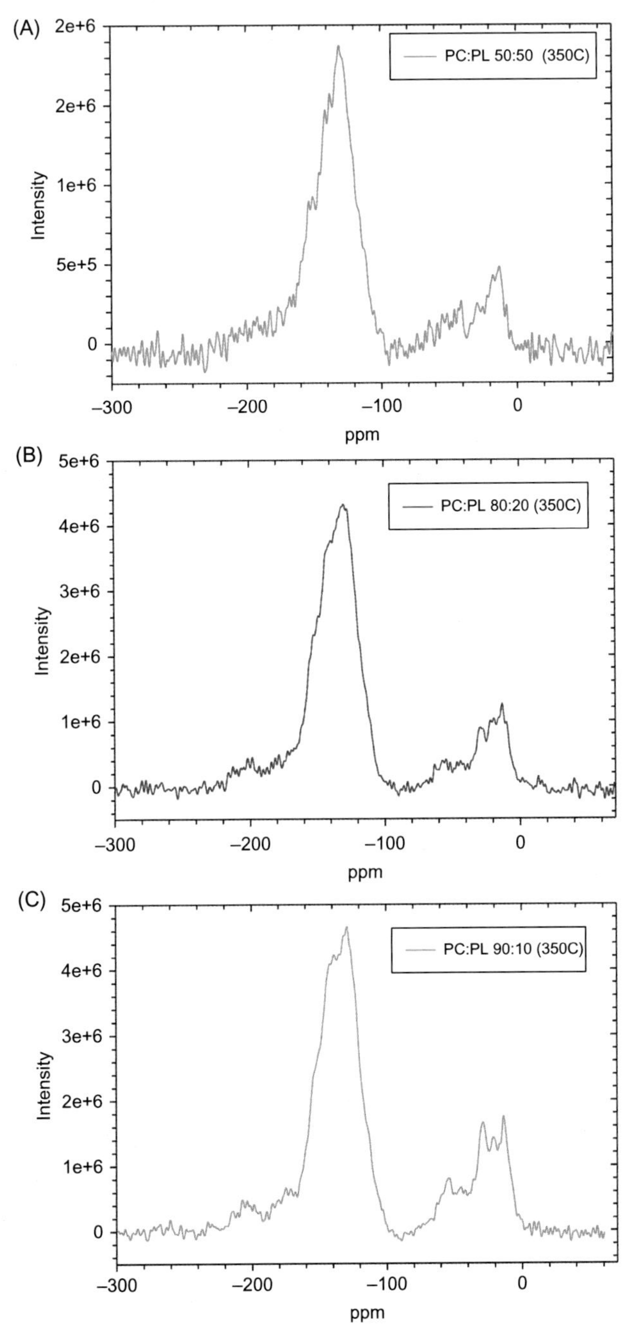

FIGURE 3.7 ^{13}C NMR spectra from biochars produced from parent pine-chip and poultry-litter feedstocks at (A) 50:50, (B) 80:20, and (C) 90:10 weight ratios. *Source: J.M. Novak.*

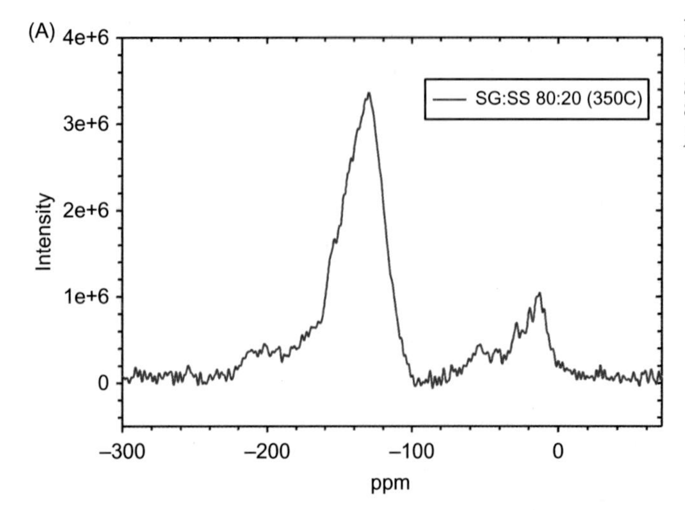

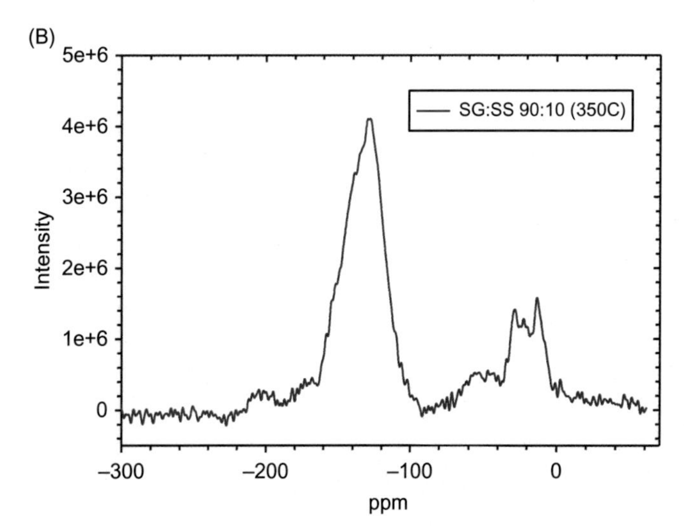

FIGURE 3.8 ^{13}C NMR spectra from biochars produced from parent switchgrass and swine-solid feedstocks at (A) 80:20 and (B) 90:10 weight ratios. *Source: J.M. Novak.*

biochars. While both manure-based parent feedstocks have similar C as aromatic groups, biochar from poultry litter has more C distributed as fatty acids and carboxylic acid compounds (Table 3.3), biochars produced from parent swine solids have more C as proteins, cellulose, and esters and amides (Table 3.3). In both spectra from the manure-based biochars, there is a lack of the methoxy peak at about 58 ppm; in contrast, this same peak occurs as a prominent singlet crest in the parent pine-chip biochar spectra.

Making biochars from blends of pine-chip and poultry-litter feedstocks had a pronounced impact on their structural and functional group composition (Fig. 3.7A–C). For example, pyrolyzing blends of the pine chip: poultry litter at the highest blend ratio (50:50), there is more C in five different spectral regions, followed by a dramatic decline in aromatic C compounds (Table 3.3). As the ratio of pine-chip feedstock increases in the blend with poultry litter, there is a two to six-fold increase among the 0 to 50, 109–145,

TABLE 3.3 Peak Spectral Areas From ^{13}C NMR Spectroscopic Characterization of Biochars Pyrolyzed at 350°C From Parent Feedstocks and Blends (Spectral Interpretation After Wang et al., 2007)

	0–50	50–61	61–96	96–109	109–145	145–163	163–190	190–220 ppm
Feedstock	% C distribution							
Pine chip (PC)	20	4	1	2	55	15	2	1
Switchgrass (SG)	28	2	0	2	49	13	5	0
Poultry litter (PL)	21	2	2	3	49	15	7	2
Swine solids (SS)	16	4	4	3	47	16	5	5
PC:PL 50:50	8	16	18	6	25	3	21	3
PC:PL 80:20	15	2	0	1	55	17	6	4
PC:PL 90:10	20	3	1	2	49	16	6	4
SG:SS 80:20	15	2	2	2	49	16	8	6
SG:SS 90:10	21	3	2	3	53	13	4	5

Spectral Interpretation

ppm	Chemical structures	Examples
0–50	Unsubstituted aliphatic	Alkanes, fatty acids
50–61	N-alkyl compounds	Amino acids, proteins
61–96	Carbohydrates	cellulose
96–109	Anomeric C	Monosaccharides
109–145	Aromatic C	Benzene-like compounds
145–163	Phenolic C	Phenol, nitrophenol
163–190	Carboxylic C	Acetic acid
190–220	Ketonic C = O	Esters and amides

Source: J.M. Novak.

and 145 to 163 ppm regions suggesting inclusion of more alkanes, fatty acids, aromatic, and phenolic-like structures, respectively (Table 3.3). Essentially, inclusion of more pine-chip feedstock into the poultry-litter blend transfers more lignocellulosic characteristic through the addition of cellulose, hemicelluloses, and lignin compounds. Blending the switchgrass with swine solids produces biochars with relatively minor structural differences (Fig. 3.8A and B). Adding 10% more by weight of swine solids to the switchgrass feedstock blend increased the C amount as unsubstituted aliphatic and aromatic compounds, followed by a decrease in C as phenolic and carboxylic acid groups (Table 3.3).

The FTIR spectra of the parent poultry-litter and pine-chip biochars along with the three blends are presented in Fig. 3.9. The seven spectra were presented in one plot to allow for comparison of spectral attributes of each biochar.

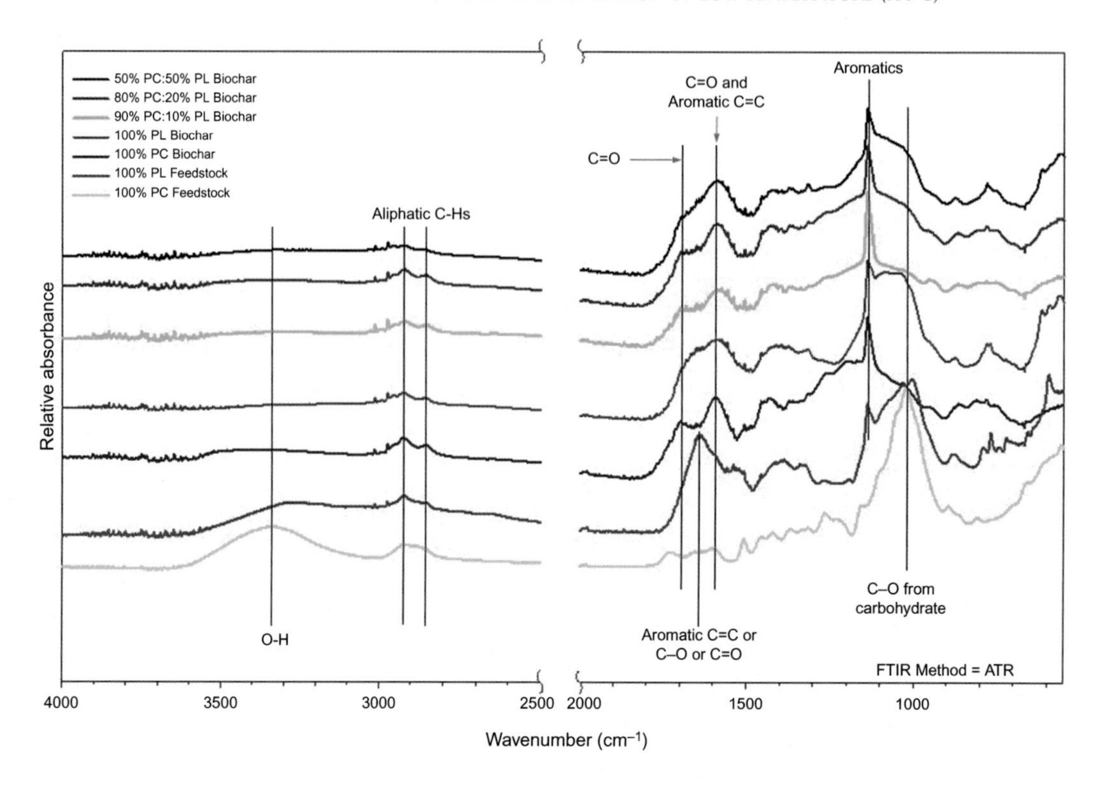

FIGURE 3.9 FTIR spectra of uncharred parent feedstocks and from biochars produced from pine-chip and poultry-litter feedstocks and their three blends at 50:50, 80:20, and 90:10 weight ratios. *Source: M. G. Johnson.*

The uncharred pine-chip and poultry-litter feedstocks show pronounced C–O stretching peaks from carbohydrate and carbohydrate-like compounds in their FTIR spectra (Fig. 3.9). There are also very strong C = C, C–O, and C = O spectral peaks due to aromatic compounds in the uncharred poultry-litter spectral scan. This is probably the result of lignin-like material from the poultry-litter bedding. This same peak intensity is not as prominent in the uncharred pine-chip spectra. Wood shavings, chopped straw, cardboard and paper wastes, which are rich in cellulose, are commonly used for poultry-litter bedding (Bolan et al., 2004). Moving along to the biochars, the parent pine-chip biochar (noted as 100% PC) shows minimal intensity from carbohydrates C–O groups, but does show prominent peaks due to C = C and C = O groups associated with COOH functional groups and aromatics (Fig. 3.9). Also prominent in the spectra are the two peaks near 2800 cm^{-1}, due to aliphatic C–H stretching. This is consistent with the 20%C distributed in the aliphatic region of the pine-chip biochar NMR spectra (Table 3.3). The parent poultry-litter biochar (noted as 100% PL) has distinct spectral peaks from C–O from carbohydrates, but a decrease in C = O from aromatic C groups. The decrease in aromatic signal intensity in the poultry-litter biochar is consistent with a noted decrease in aromatic %C (Table 3.3). Biochars from the pine-chip: poultry-litter (PC:PL) blends show an increase in the carbohydrate peak intensity and a small decrease in the aromatic signal when more poultry litter is added to the blend (Fig. 3.9). This is consistent with the %C distribution upward trend in the 50 to 61 and 61 to 96 ppm region and in the 109 to 145 ppm region of the NMR spectra (Table 3.3). Adding more poultry litter to the pine-chip blend thus shifts the structural components in the biochar from being aromatic to having more protein, cellulose, and COOH functional groups.

The FTIR spectra of the parent switchgrass and swine-solid biochars along with the two blends are presented in Fig. 3.10. The six spectra were presented in one plot to allow for comparison of spectral attributes of each biochar and their blends.

The uncharred parent switchgrass and swine solid (noted as 100% SG and SS feedstock, respectively) show strong signal intensity from carbohydrates C–O groups and medium intensity peaks due to C = C and C = O groups associated with COOH functional groups and aromatics (Fig. 3.10). The parent switchgrass biochar (noted as 100% SG biochar) has lost the C–O spectral peaks from carbohydrates, but exhibits a sharp C = O from aromatic C groups. The increase in aromatic signal intensity in the switchgrass spectra is probably due to carbonization of aliphatic/carbohydrate material into compounds possessing aromatic character. Similarly, biochar from the swine solids (noted as 100% SS biochar) shows

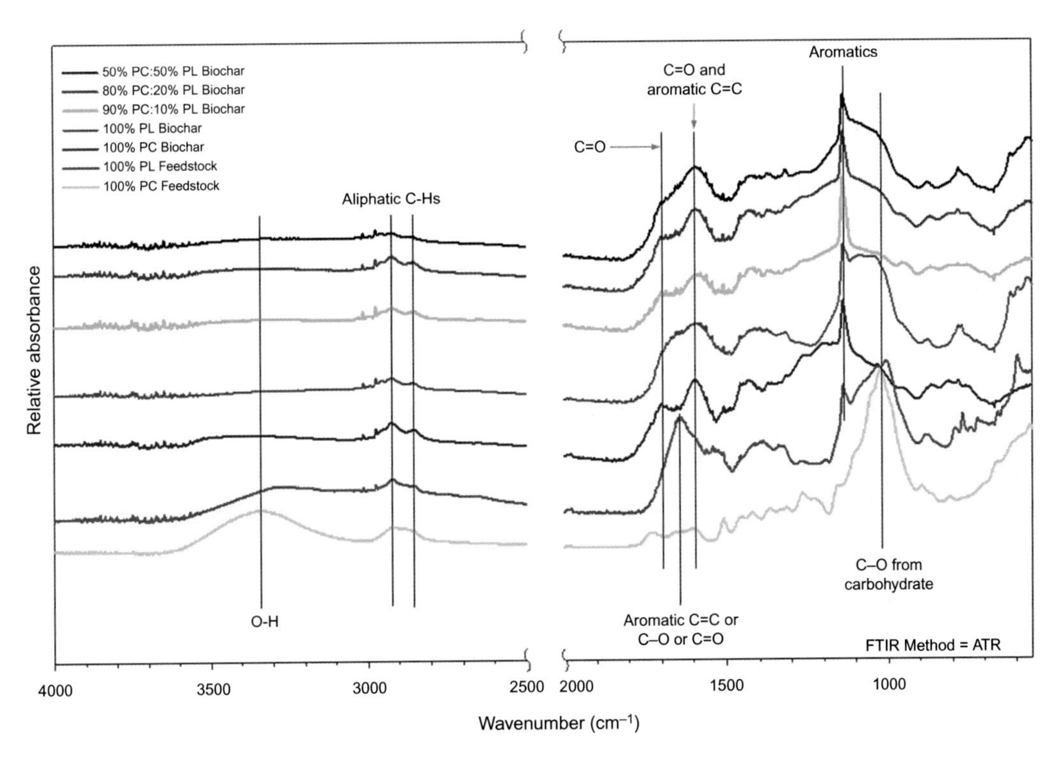

FIGURE 3.10 FTIR spectra from the uncharred switchgrass and swine solid, and biochars produced from the switchgrass and swine-solid feedstocks and their two blends at 80:20 and 90:10 weight ratios. *Source: M. G. Johnson.*

an increase in aromatic signals (Fig. 3.10). The two switchgrass and swine-solid blends (80:20 and 90:10 SG: SS biochars) show the loss of carbohydrate peak intensity in the spectra, but the two aromatic signals remain relatively strong. Additionally, there is a decline in spectral intensity from aliphatic C-Hs in the two SG:SS blends compared to biochar produced from their respective 100% feedstocks. Blending more swine solids into the switchgrass pool resulted in a decline in FTIR spectral behavior of aliphatic compound, which is consistent with the drop of 21% to 15% C in the 0 to 50 ppm region of the NMR peak spectral area (Table 3.3).

3.7 COMPARISON OF LOW VERSUS HIGH TEMPERATURE-PRODUCED BIOCHARS AS A SOIL AMENDMENT

The primary objective of this chapter was to characterize the chemical composition and spectral properties of low-temperature (350°C) biochars and discuss their suitability as soil amendments. A balanced evaluation of biochars performance as a soil amendment should also involve a description of chemical properties on high temperature-produced (>350°C) biochars. We have reported that pyrolysis of feedstocks at low temperatures (300 to 350°C) results in biochars that contain more C=O and C−H functional groups that can serve as cation-exchange sites after oxidation in soil (Glaser et al., 2002). Our spectroscopic results presented in this chapter are consistent with the results presented by Glaser et al. (2002). In general, biochars produced at these lower pyrolysis temperatures have a more diversified organic structural makeup because there is less loss of volatile compounds. The presence of more labile compounds in biochars such as proteins, fatty acids, and carbohydrates will stimulate microbial activity (Lehmann et al., 2011; Rutigliano et al., 2014), and lead to some complex shifts in community composition and associations (Nielson et al., 2014). Thus, low temperature-produced biochars are potentially more amendable as soil amendments for improving soil microbial properties and increasing soil cation-exchange capacities.

Biochars produced at high pyrolysis temperature (400 to 800°C) have declines in their O/C and H/C atomic ratio because of the loss of volatile compounds and a gain in polycondensed aromatic structures (Spokas, 2010; Novak et al., 2012). Carbonization reactions increase the biochars surface area by released gases creating fissures and void spaces in the condensed aromatic sheets (Chia et al., 2015). High-temperature biochars also have higher amounts of C in aromatic structures (Novak and Busscher, 2012). A dominance of aromatic C in the biochars structural composition makes them more recalcitrant to soil microbial oxidation (Schmidt and Noack, 2000; Glaser et al., 2002), so they would be a more appropriate selection for increasing long-term C storage. High-temperature biochar has additional characteristics that make them important amendments in the agronomic and environmental sectors because of their ability to modify soil C sequestration (Laird, 2008), soil fertility characteristics (Novak et al., 2009a; Jeffery et al., 2011), physical properties (Novak and Busscher, 2012), and heavy-metal binding potentials (Agrafioti et al., 2014; Janus et al., 2015; Puga et al., 2015). Additionally, biochars with high surface areas or high aromatic character are regarded as ideal amendments for remediating mine impacted soils/spoils (Beesley et al., 2011; Xie et al., 2015).

3.8 CONCLUSIONS

Biochars can be produced from a variety of lignocellulosic and manure feedstocks utilizing a range of pyrolysis temperatures. To further expand biochar usage as an amendment, designer biochars can be created through blending feedstocks, residence time, and temperature selection. Effective use of biochars requires examination of their elemental and spectroscopic properties because these characteristics control their interaction with target materials and their performance as an amendment or sorbent for contaminants. Low pyrolysis temperature biochars have unique properties that make them suitable as soil amendments for fertility improvement, nutrient retention, and stimulation of microbial activity. These soil improvements are further enhanced when a specific parent feedstock is selected for biochar production.

References

Agrafioti, E., Kalderis, D., Diamadopoulos, E., 2014. Arsenic and chromium removal from water using biochars derived from rice husk, organic solid wastes and sewage sludge. J. Environ. Manage. 133, 309–314.

Ameloot, N., Sleutel, S., Das, K.C., Kanagaratnam, J., De Neve, S., 2015. Biochar amendment to soils with contrasting organic matter level; effect on N mineralization and biological soil properties. GCB Bioenergy 7, 135–144.

Antal, M.J., Grønli, M., 2003. The art, science, and technology of charcoal production. Indust. Eng. Chem. Res. 42, 1619–1640.

ASTM, 2006. Petroleum Products, Lubricants, and Fossil Fuels: Gaseous Fuels, Coal, and Coke. ASTM International, West Conshohochen, PA, USA.

Baldock, J.A., Smernik, R.J., 2002. Chemical composition and bioavailability of thermally altered pinus resinosa (Red Pine) wood. Org. Geochem. 33, 1093–1109.

Bayan, M.R., 2015. Adsorption of methylene blue by biochar produced through torrification and slow pyrolysis of switchgrass. Res. J. Phar. Biol. Chem. Sci. 6 (4), 8–17.

Beesley, L., Moreno-Jimenez, E., Gomez-Eyles, J.L., Harris, E., Robinson, B., Sizmur, T., 2011. A review of biochars' potential role in the remediation, revegetation and restoration of contaminated soils. Environ. Pollut. 159, 3269–3282.

Boateng, A.A., Garcia-Perez, M., Masek, M., Brown, R., del Campo, B., 2015. Biochar production technology. In: Lehmann, J., Joseph, S. (Eds.), Biochar for Environmental Management: Science, Technology and Implementation, second ed. Earthscan, Routledge Publ., London, UK, pp. 64–87.

Bolan, N.S., Adriano, D.C., Mahimairaja, S., 2004. Distribution and bioavailability of trace elements in livestock and poultry manure byproducts. Crit. Rev. Environ. Sci. Tech. 34, 291–338.

Brewer, C.E., Schmidt-Rohr, K., Satrio, J.A., Brown, R.C., 2009. Characterization of biochar from fast pyrolysis and gasification systems. Environ. Prog. Sust. Energy 28 (3), 386–396.

Bridgewater, A.V., 2003. Renewable fuels and chemicals by thermal processing of biomass. Chem. Eng. J. 91, 87–102.

Camps-Arbestain, M., Amonette, J.E., Singh, B., Wang, T., Schmidt, H.P., 2015. A biochar classification system and associated test method. In: Lehmann, J., Joseph, S. (Eds.), Biochar for Environmental Management: Science, Technology and Implementation, second ed. Earthscan, Routledge Publ., London, UK, pp. 165–193.

Cantrell, K.B., Hunt, P.G., Uchimiya, M., Novak, J.M., Ro, K.S., 2012. Impact of pyrolysis temperature and manure source on physicochemical characteristics of biochar. Biores. Technol. 107, 419–428.

Cetin, E., Gupta, R., Moghtaderi, B., 2005. Effects of pyrolysis pressure and heating rate on radiate pine char structure and apparent gasification reactivity. Fuel 84, 1328–1334.

Chan, K.Y., Van Zwieten, L., Meszarnos, I.A., Downie, A., Joseph, S., 2008. Using poultry litter biochars as soil amendments. Aust. J. Soil Res. 46, 437–444.

Chen, B., Zhou, D., Zhu, L., 2008. Transitional adsorption and partition of nonpolar and polar aromatic contaminants by biochars of pine needles with different pyrolytic temperatures. Environ. Sci. Technol. 42, 5137–5143.

Cheng, H.N., Wartelle, L.H., Klasson, K.T., Edwards, J.C., 2010. Solid-state NMR and ESR studies of activated carbons produced from pecan shells. Carbon. N. Y. 48, 2455–2469.

Chia, C.H., Downie, A., Munroe, P., 2015. Characterization of biochar: Physical and structural properties. In: Lehmann, J., Joseph, S. (Eds.), Biochar for Environmental Management: Science, Technology and Implementation, 2nd ed. Earthscan, Routledge Publ., London, UK, pp. 89–137.

Crombie, K., Masek, O., Sohi, S.P., Brownsort, P., Cross, A., 2013. The effects of pyrolysis conditions on biochar stability as determined by three methods. GCB Bioenergy 5, 122–131.

Domingues, R.R., Trugilho, P.F., Silva, C.A., de Melo, I.C., Melo, L.C.A., Magriotis, Z.M., et al., 2017. Properties of biochar derived from wood and high-nutrient biomass with the aim of agronomic and environmental benefits. PLoS. One 12 (5), e0176884.

Ducey, T.F., Ippolito, J.A., Cantrell, K.B., Novak, J.M., Lentz, R.D., 2013. Addition of activated switchgrass biochar to an aridic subsoil increases microbial nitrogen cycling gene abundance. App. Soil Ecol. 65, 65–72.

EPA, 2007. EPA method 3050. Microwave assisted acid digestion of sediments, sludges, soils, and oils. Available online at https: www.epa.gov/sites/production/files/2015-12/documents/3051a.pdf.

Glaser, B., Lehmann, J., Zech, W., 2002. Ameliorating physical and chemical properties of highly weathered soils in the tropics with charcoal—a review. Biol. Fertil. Soils 35, 219–230.

Gul, S., Whalen, J.K., 2016. Biochemical cycling of nitrogen and phosphorus in biochar-amended soils. Soil. Biol. Biochem. 103, 1–15.

Hagemann, N., Kammann, C.I., Schmidt, H.P., Kappler, A., Behrens, S., 2017. Nitrate capture and slow release in biochar amended compost and soils. PLoS. One 12 (2), e0171214.

Hass, A., Gonzalez, J.M., Lima, I.M., Godwin, H.W., Halvorson, J.J., Boyer, D.G., 2012. Chicken manure biochar as liming and nutrient source for acid Appalachian soil. J. Environ. Qual. 41, 1096–1106.

International Biochar Initiative (IBI). 2017. Standardized product definition and product testing guidelines for biochar that is used in soil. Available at: http://www.biochar-international.org/characterizationstandard.

Ippolito, J.A., Novak, J.M., Busscher, W.J., Ahmedna, M., Rehrah, D., Watts, D.W., 2012. Switchgrass biochar affects two Aridisols. J. Environ. Qual. 41 (4), 1123–1130.

Ippolito, J.A., Spokas, K.A., Novak, J.M., Lentz, R.D., Cantrell, K.B., 2015. Biochar elemental composition and factors influencing nutrient retention. In: Lehmann, J., Joseph, S. (Eds.), Biochar for Environmental Management: Science, Technology and Implementation, 2nd ed. Earthscan, Routledge Publ., London, UK, pp. 139–164.

Jacobsen, S.E., Wyman, C.E., 2000. Cellulose and hemicelluloses hydrolysis models for application to current and novel pretreatment processes. Appl. Biochem. Biotech. 84–86, 81–96.

Janus, A., Pelfrêne, A., Heymans, S., Deboffe, C., Douay, F., Waterlot, C., 2015. Elaboration, characteristics and advantages of biochars for the management of contaminated soils with a specific overview on Miscanthus biochars. J. Environ. Manage. 162, 275–289.

Jeffery, S., Verheijen, F.A., van der Velde, M., Bastos, A.C., 2011. A quantitative review of the effects of biochar application to soils on crop productivity using meta-analysis. Agric. Ecosyst. Environ. 144, 175–187.

Jindo, K., Mizumoto, H., Sawada, Y., Sanchez-Monedero, M.A., Sonoki, T., 2014. Physical and chemical characteristics of biochars derived from different agricultural residues. Biogeosciences 11, 6613–6621.

Johnston, C.T., 2017. Biochar analysis by Fourier-transform infra-red spectroscopy. In: Singh, B., Camps-Arbestain, M., Lehmann, J. (Eds.), Biochar: A Guide to Analytical Methods. CRC Press, Boca Raton, FL, USA, pp. 199–213.

Jones, T.P., Chaloner, W.G., Kuhlbusch, T.A.G., 1997. Proposed Biogeochemical and Chemical Based Terminology for Fire-Altered Plant Matter. Springer-Verlag, Berlin, Germany.

Joseph, S.D., Camps-Arbestain, M., Lin, Y., Munroe, P., chia, C.H., Hook, J., et al., 2010. An investigation into the reactions of biochar in soil. Aust. J. Soil Res. 48, 501–515.

Keiluweit, M., Nico, P.S., Johnson, M.G., Kleber, M., 2010. Dynamic molecular structure of plant biomass-derived black carbon (biochar). Environ. Sci. Technol. 44, 1247–1253.

Kelly, C., Calderon, F.C., Acosta-Martinez, V., Maysoon, M., Benjamin, J.G., Rutherford, D., et al., 2015. Switchgrass biochar effects on plant biomass and microbial dynamics in two soils from different regions. Pedosphere 25 (3), 329–342.

Laird, D.A., 2008. The charcoal vision: A win-win-win scenario for simultaneously producing bioenergy, permanently sequestering carbon, while improving soil and water quality. Agron. J. 100, 178–181.

Laird, D.A., Brown, R.C., Amonette, J.E., Lehmann, J., 2009. Review of the pyrolysis platform for coproducing bio-oil and biochar. Biofuels, Bioprod. Bioref. 3, 547–562.

Lehmann, J., Joseph, S., 2015. Biochar for Environmental: Science, Technology and Implementation, 2nd ed. Earthscan, London.

Lehmann, J., Rillig, M.C., Theis, J., Masiello, C.A., Hockaday, W.C., Crowley, D., 2011. Biochar effects on soil biota – a review. Soil. Biol. Biochem. 43, 1812–1836.

Manyà, J.J., 2012. Pyrolysis for biochar purposes: A review to establish current knowledge gaps and research needs. Environ. Sci. Tech. 46, 7939–7954.

Nielson, S., Minchin, T., Kimber, S., van Zwieten, L., Gilbert, J., Munroe, P., et al., 2014. Comparative analysis of the microbial communities in agricultural soil amended with enhanced biochars or traditional fertilizers. Agric. Ecosyst. Environ. 191, 73–82.

Novak, J.M., Busscher, W.J., 2012. Selection and use of designer biochars to improve characteristics of southeastern USA coastal plain soils. In: Lee, J.W. (Ed.), Advanced Biofuels and Bioproducts., Vol. 1. Springer, NY, USA, pp. 69–97.

Novak, J.M., Watts, D.W., 2013. Augmenting soil water storage using uncharred switchgrass and pyrolyzed biochar. Soil Use Manage. 29, 98–104.

Novak, J.M., Busscher, W.J., Laird, D.L., Ahmedna, M., Watts, D.W., Niandou, M.A.S., 2009a. Impact of biochar on fertility of a southeastern coastal plain soil. Soil Sci. 174, 105–112.

Novak, J.M., Lima, I., Xing, B., Gaskin, J.W., Steiner, C., Das, K.C., et al., 2009b. Characterization of designer biochar produced at different temperatures and their effects on a loamy sand. Ann. Environ. Sci. 3, 195–206.

Novak, J.M., Cantrell, K.B., Watts, D.W., 2012. Compositional and thermal evaluations of lignocellulosic and poultry litter chars via high and low temperature pyrolysis. Bioenerg. Res 6, 114–130.

Novak, J.M., Cantrell, K.B., Watts, D.W., Busscher, W.J., Johnson, M.G., 2014a. Designing relevant biochars as soil amendments using lignocellulosic-based and manure-based feedstocks. J. Soils Sed. 14, 330–343.

Novak, J.M., Spokas, K.A., Cantrell, K.B., Ro, K.S., Watts, D.W., Glaz, B., et al., 2014b. Effects of biochars and hydrochars produced from lignocellulosic and animal manure fertility of a Mollisol and Entisol. Soil Use Manage. 30, 175–181.

Parikh, S.J., Margenot, A.J., Mukome, F.M.D., Calderon, F., Goyne, K.W., 2014. Soil chemical insights provided through vibrational spectroscopy. Adv. Agron. 126, 1–148.

Pavia, Dl. G.M., Lampman, Kriz Jr, G.S., 1979. Introduction toSpectroscopy: A Guide for Students of Organic Chemistry. Sanders College Publ., Philadelphia, PA, USA.

Pilon, G., Lavoie, J.M., 2011. Characterization of switchgrass char produced in torrification and pyrolysis conditions. BioResources 6 (4), 4824–4839.

Puga, A.P., Abreu, C.A., Melo, L.C.A., Beesley, L., 2015. Biochar application to a contaminated soil reduces the availability and plant uptake of zinc, lead, and cadmium. J. Environ. Manage. 159, 86–93.

Rutigliano, F.A., Romano, M., Marzaioli, R., Baglivo, I., Baronti, S., Miglietta, F., et al., 2014. Effect of biochar addition on soil microbial community in a wheat crop. Eur. J. Soil Biol. 60, 9–15.

Schmidt, M.W., Noack, A.G., 2000. Black carbon in soils and sediments: analysis, distribution, implications, and current challenges. Global. Biogeochem. Cycles 14, 777–793.

Sistani, K.R., Novak, J.M., 2006. Trace metal accumulation, movement and remediation in soils receiving animal manure. In: Prasad, M.N., Sajwan, K.S., Naidu, K.S. (Eds.), Trace Elements in the Environment, Biogeochemistry, Biotechnology, and Bioremediation. CRC Press, Boca Raton, FL, pp. 689–706.

Smernik, R.J., 2017. Analysis of biochars by 13C nuclear magnetic resonance spectroscopy. In: Singh, B., Camps-Arbestain, M., Lehmann, J. (Eds.), Biochar: A Guide to Analytical Methods. CRC Press, Boca Raton, FL, USA, pp. 151–161.

Sohi, S., E. Loez-Capel, E. Krull, and R. Bol. 2009. Biochar's roles in soil and climate change: A review of research needs. CSIRO Land and Water Science Report. 05/09. Glen Osmond, Australia.

Song, W., Guo, M., 2012. Quality variations of poultry litter biochar generated at different pyrolysis temperatures. J. Anal. App. Pyrol. 94, 138–145.

Spokas, K.A., 2010. Review of the stability of biochar in soils: Predictability of O:C molar ratios. Carbon Manage. 1, 289–303.

Spokas, K.A., Novak, J.M., Masiello, C.A., Johnson, M.G., Colosky, E.C., Ippolito, J.A., et al., 2014. Physical disintegration of biochar: an overlooked process. Environ. Sci. Technol. Lett. 1, 326–332.

van Krevelen, D.W., 1950. Graphical-statistical method for the study of structure and reaction of coal. Fuel 29, 269–284.

Vanotti, M.B., Szogi, A.A., Millner, P.D., Loughrin, J.H., 2009. Development of a second-generation environmentally superior technology for treatment of swine manure in the USA. Bioresour. Technol. 100, 5406–5416.
Wang, X., Cook, R., Tao, S., Xing, B., 2007. Sorption of organic contaminants by biopolymers: Role of polarity, structure and domain spatial arrangement. Chemosphere 66, 1476–1484.
Wisnubroto, E.I., Utomo, W.H., Soelistyari, H.T., 2017. Biochar as carrier for nitrogen plant nutrients: the release of nitrogen from biochar enriched with ammonium sulfate and nitrate acid. Inter. J. App. Eng. Res. 12 (6), 1035–1042.
Xie, T., Reddy, K.R., Wang, C., Yargicoglu, E., Spokas, K.A., 2015. Characteristics and application of biochar for environmental remediation: a review. Crit. Rev. Environ. Sci. Technol. 45, 939–969.
Yuan, J.H., Xu, R.K., 2011. The amelioration effects of low temperature biochar generated from nine crop residues on acidic Ultisols. Soil Use Manage. 27, 110–115.

Further Reading

Bolan, N.S., Szogi, A.A., Chuasavathi, T., Seshadri, B., Rothrock, M.J., Panneerselvam, P., 2010. Uses and management of poultry litter. Worlds Poultry Sci. J. 66 (4), 673–698.
Kleber, M.W., Hockaddy, Nico, P.S., 2015. Characteristics of biochar: macro-molecular properties. In: Lehmann, J., Joseph, S. (Eds.), Biochar for Environmental Management: Science, Technology and Implementation, 2nd ed. Earthscan, Routledge Publ., London, UK, pp. 111–137.

CHAPTER

4

Modeling the Surface Chemistry of Biochars

Md. Samrat Alam and Daniel S. Alessi

Department of Earth and Atmospheric Sciences, University of Alberta, Edmonton, AB, Canada

4.1 INTRODUCTION

Biochar, a carbon-rich solid, can be produced through reductive thermal processing and/or pyrolysis of biomass derived from a variety of feedstocks (Cao et al., 2009; Cao and Harris, 2010; Dong et al., 2011; Ahmad et al., 2014; Alam et al., 2016). It is also a major byproduct of forest fires globally, and has profound influences on nutrient cycling, such as soil nitrogen (N) dynamics, in the underlying soils and sediments (DeLuca et al., 2006; Thomas and Gale, 2015). Biochars contain a range of surface functional groups activated by thermal alteration of the parent organic materials (Ahmad et al., 2014). Time, heat, pressure, and physical characteristics of biomass are important parameters to consider during the commercial production of biochar. In particular, pyrolysis temperature plays an important role in the density and nature of functional groups of the resulting biochar (Uchimiya et al., 2012; Vithanage et al., 2015). Variability in biochar quality and surface chemistry can be controlled if these parameters are well constrained and understood during the production process (Woolf et al., 2010; Ahmad et al., 2014).

Over the past decade, biochar has received increasing worldwide attention as a sorbent to remediate contaminated soils and waters (Cao et al., 2009; Cao and Harris, 2010; Mohan et al., 2011; Choppala et al., 2012; Alam et al., 2016; Thompson et al., 2016; Alam et al., 2018a); to improve soil fertility by increased retention of nutrients (Uchimiya et al., 2012; Laird and Rogovska, 2015; Lehmann et al., 2015); and for its role as an important sink for atmospheric carbon dioxide (CO_2) (Woolf et al., 2010). It is cost effective compared to many sorbents, including activated carbon (AC), graphene oxides, wheat straw, activated sludge, bagasse fly ash, and blast furnace waste and coals that have been used to remove metals from aqueous media, Biochar has also been shown to have among the highest

Biochar from Biomass and Waste
DOI: https://doi.org/10.1016/B978-0-12-811729-3.00004-2

sorption capacity for contaminant removal from aqueous solution (Chen and Lin, 2001, Mohan and Pittman, 2006; Mohan et al., 2011; Choppala et al., 2012; Teixidó et al., 2013; Hadjittofi and Pashalidis, 2015; Thompson et al., 2016). Although promising for numerous applications, quantifying biochar surface chemistry is challenging because it is comprised of a complex mixture of carbonaceous and inorganic phases that vary widely by source feedstock and production conditions (Ahmad et al., 2014; Inyang et al., 2016).

Determining biochar surface chemistry and reactivity is not only important for engineered remediation solutions, but also for understanding its behavior toward metals in nature. An understanding of specific metal–biochar adsorption reactions is needed in order to develop mechanistic and predictive geochemical models of metal adsorption to biochars. Two metal-adsorption modeling approaches are common: empirical models and surface complexation models (SCMs) (Bethke and Brady, 2000; Koretsky, 2000). Empirical models, which do not often consider the formation of discrete surface complexes, may not be reliable because they cannot account for changes to factors such as pH, ionic strength, and sorbent-to-sorbate ratio in dynamic environmental systems (Bethke and Brady, 2000; Koretsky, 2000). On the other hand, surface complexation modeling is a quantitative approach, is premised on considering the full aqueous speciation of the metal(s) of interest, and is grounded on balanced chemical reactions. Therefore, it can account for changes in water chemistries including pH and adsorbent:adsorbate ratios (Beveridge and Murray, 1980; Manceau and Charlet, 1994; Bethke and Brady, 2000; Koretsky, 2000; Borrok et al., 2005; Alessi and Fein, 2010).

Not only is an understanding of biochar surface chemistry needed to determine how it will sequester metals, but knowledge of metal speciation in solution is critical to developing predictive models of metal adsorption. Indeed, the fate, mobility, and bioavailability of trace metals in the natural environment is strongly controlled by their speciation (Tack and Verloo, 1995; Van Leeuwen et al., 2005; Reeder et al., 2006). A range of processes, including adsorption of aqueous ions and complexes onto environmental surfaces (i.e., bacterial, organic and mineral surfaces), precipitation, dissolution of minerals, and oxidation-reduction reactions, control the fate, mobility, distribution, and bioavailability of trace metals (Sparks, 2005). Developing predictive geochemical models of the mobility of trace metals in the environment by sorbents such as biochar then depends on accurate quantification of the biochar surface functional groups (sites), their protonation constants, and their reactivity toward the specific metal species that develop in aqueous solution as a function of factors including metal concentration, ionic strength, and solution pH (Koretsky, 2000; Borrok et al., 2005; Alessi and Fein, 2010). In this chapter, we focus on discussing the development of predictive SCMs of metal adsorption to biochars, a needed step to advance the application of biochars to applications such as water treatments and soil and sediment amendments.

4.2 SURFACE COMPLEXATION MODELING

To accurately predict the fate and transport of heavy metals by biochars, it is critical to know the identity and density of surface functional groups (sites) and their reactivity to protons (i.e., to changes in pH) and metals (Koretsky, 2000; Borrok et al., 2005; Alessi and Fein, 2010). It is well established that bacteria, clay, HFO, silica, and biochar can adsorb metals to their negatively charged functional groups (Beveridge and Murray, 1980; Manceau and Charlet, 1994; Manning and Goldberg, 1997; Uchimiya et al., 2012; Alam et al., 2018b), and that they play important roles in the distribution and mobility of metals in many geologic systems (Karthikeyan and Elliott, 1999; Marmier et al., 1999; Borrok et al., 2005; Alessi and Fein, 2010; Tan et al., 2011; Alam et al., 2018a).

As noted, metal-adsorption data are modeled using empirical models and SCMs. Studies using empirical formulas to model adsorption data may not be able to make reliable predictions, because they do not consider factors such as metal competition for attachment to surface functional groups and surface heterogeneity of sorbents in dynamic environmental systems (Bethke and Brady, 2000; Koretsky, 2000). For example, the distribution coefficient (K_d) is a type of empirical model that defines the partitioning of a metal between aqueous solution and the bulk geologic solids as a simple linear ratio. Though K_d models are an easy way to model heavy metal distribution, molecular-scale information on binding mechanisms is not considered. Furthermore, the K_d model assumes an unlimited number of available surface sites for metal sorption, which, for example, ignores the metal saturation of the sorbent surface (Bethke and Brady, 2000; Koretsky, 2000). Commonly used adsorption modeling approaches such as the Freundlich and Langmuir (K_F, K_L) isotherms expand on the K_d approach by allowing for decreasing metal-adsorption capacity with increasing saturation of the solids with metal; however, metal speciation in solution or on the solid sorbent is not considered. The K_L approach also does not consider a sorbent's surface heterogeneity and competition of metals for sorption at the biochar surface sites (Bethke and Brady, 2000; Koretsky, 2000; Borrok et al., 2005). Generally the constants calculated using empirical models are not valid across changes in metal concentration, pH, competing ions, and temperature (Bethke and Brady, 2000; Koretsky, 2000; Borrok et al., 2005). If these factors change, new experiments must be conducted in the laboratory to calculate a new empirical constant (K_d K_F, or K_L) that is valid for the new system conditions (Bethke and Brady, 2000; Koretsky, 2000; Borrok et al., 2005). Therefore, the application of empirical partitioning constants can be problematic for dynamic environmental systems.

Unlike empirical adsorption modeling approaches, the SCM approach can predict the acid/base and ion/metal-binding behaviors onto a variety of environmental surfaces (Fein et al., 1997; Karthikeyan and Elliott, 1999; Marmier et al., 1999; Bethke and Brady, 2000; Koretsky, 2000; Borrok et al., 2005; Fein et al., 2005; Alessi and Fein, 2010; Tan et al., 2011). Using the SCM approach, the overall distribution of metals in a system can be predicted from stability constants of sorbent-metal surface complexes (Fein et al., 1997; Bethke and Brady, 2000; Borrok et al., 2005). The SCM approach can also account for heterogeneity of the sorbents, electrostatic effects due to change in the surface charge of the sorbent, changes in pH, ionic strength, and metal speciation, and competition of the metal of interest with other metals in solution for adsorption onto the solid surfaces (Bethke and Brady, 2000; Koretsky, 2000; Borrok et al., 2005).

Fig. 4.1 illustrates the workflow commonly used to develop an SCM for metal binding to sorbents such as biochars. The first step in developing an SCM is to conduct potentiometric titrations of the sorbent (biochar in this case). For example, in the most general sense, protonation/deprotonation and cation metal-binding reactions of the functional groups present on the biochar surface can then be described as:

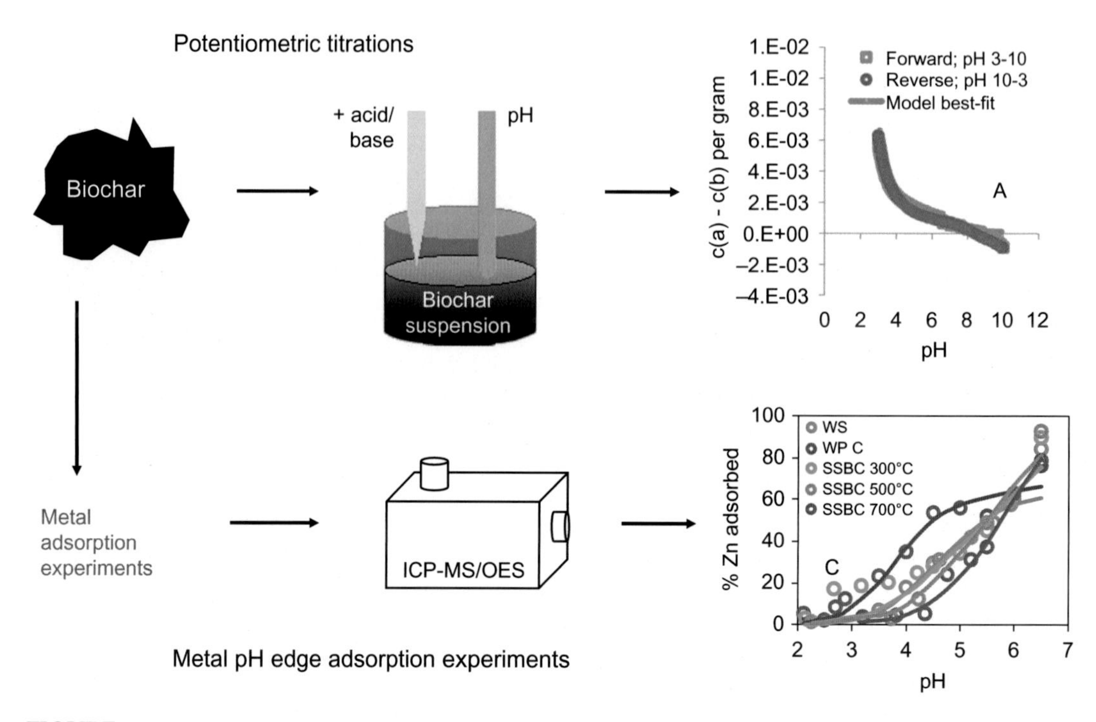

FIGURE 4.1 Experimental approaches to develop a surface complexation model. To determine proton-binding constants (K_a) and corresponding site concentrations, potentiometric titrations are first performed on biochar to determine its proton reactivity across a range of environmentally relevant pH values (top). pH-edge metal adsorption experiments are then conducted, and in consort with the protonation model, used to calculate metal-binding constants (K_M) for each biochar surface site (bottom).

$$R - A(H)^0 \leftrightarrow H^+ + R - A^- \tag{4.1}$$

$$R - A(H)^0 + M^{++} \leftrightarrow R - A(M)^+ + H^+ \tag{4.2}$$

where R represents the biochar macromolecule to which the functional group type A is attached and M^{++} represents metals that biochar can sorb.

The acidity constant, K_a, for reaction (1) can be calculated as:

$$K_a = \frac{[R - A^-]a_{H^+}}{[R - AH^0]} \tag{4.3}$$

where a_{H^+} represents the activity of protons in the bulk solution and $[R{-}A^-]$ and $[R{-}AH^0]$ represent the concentration of deprotonated and protonated sites at equilibrium, respectively. Protonation data collected from potentiometric titrations are commonly modeled using software such as FITEQL (Westall, 1982) or ProtoFit (Turner and Fein, 2006), and the resulting protonation model can predict how biochar surface charge develops as a function of solution pH. The primary surface functional groups of biochar are thought to be carboxyl, phenolic, and hydroxyl moieties (Uchimiya et al., 2011a), and similar to many environmental materials, the surface charge of biochar is increasingly negative with increasing pH. However, the identity of specific surface functional groups can only be constrained by analyzing samples with methods including Raman, X-ray absorption, X-ray photoelectron, and IR spectroscopies (discussed in Section 4.3).

Following titrations, metal-adsorption experiments must be performed to quantify the affinity of individual metal species to the set of proton-active functional groups described in Eq. (4.1). pH-edge adsorption experiments are typically employed to develop a model that can predict metal binding as a function of changing biochar surface chemistry and aqueous metal speciation, although metal isotherm-type data can also be modeled. The metal-binding equilibrium constant for metal M^{++}, K_M, for Eq. (4.2) can be written as:

$$K_{M^{++}} = \frac{[R - AM^{+}]a_{H^{+}}}{[R - AH^{0}]a_{M^{++}}} \tag{4.4}$$

where $[R{-}AM^{+}]$ represents the concentration of biochar functional group A that is complexed with M^{++}, and $a_{M^{++}}$ is the activity of M^{++} in solution after equilibrium is attained. In principle, the resulting set of protonation (K_a) and metal-binding (K_M) constants can be used in consort with aqueous speciation constants for the metal(s) of interest to predict the adsorption behavior of the metal(s) over a wide range of metal and sorbent concentrations and pH. If solution ionic strength is of concern, electrostatic SCMs can be used, which include an electrostatic term in the mass action expression (e.g., in Eqs. 4.3 and 4.4). A thorough discussion of the use of electrostatic SCM approaches can be found in articles including Davis and Kent (1990).

4.3 SPECTROSCOPIC AND CALORIMETRIC APPROACHES

While SCMs can model the adsorption of metal ions onto sorbents across variable chemical conditions, the SCM approach does not provide information about the surface chemistry of biochars nor the coordination of adsorbed ions. To fill this gap, spectroscopic and calorimetric approaches can be employed. For example, Raman spectroscopy can give information about the relative order of the graphene sheets of carbonaceous materials (Ferrari and Robertson, 2000) that comprise an increasing fraction of the carbon in biochars with increasing pyrolysis temperature (Uchimiya et al., 2011b). While the surface functional group concentration (per unit mass biochar) declines markedly with increasing pyrolysis temperature, biochar surface area and the presence of delocalized π-electrons increases dramatically, and these π-electrons are thought to influence the adsorption of charged species (Rodríguez-Vila et al., 2018) and the redox transformations of certain metals (Kappler et al., 2014; Rajapaksha et al., 2017). Fourier transform infrared (FTIR) spectroscopy is another commonly used technique to characterize the surface chemistry of biochar (Calvelo Pereira et al., 2015). Besides identifying the stretching bands (e.g., $C=C$, $C=O$, $C{-}O$) associated with major groups of biochar surface functional groups, FTIR can identify $Si{-}O{-}Si$ modes indicative of the presence of silica, which is particularly important for biochars derived from source materials such as grasses (Silber et al., 2010). However, caution must be used in interpreting FTIR data as a proxy for quantitative assignment of surface functional groups. Definitive characterization of surface functional groups (and even calculation of K_a values) using IR spectroscopy should be performed by analyzing a set of biochar samples as a function of pH, and then using changes in the stretching bands to determine if specific functional groups are indeed proton-active (see examples in Leone et al., 2007; Alessi et al., 2014).

Synchrotron-based extended X-ray absorption fine structure (EXAFS) studies are a powerful tool that can provide insights into metal-surface complexes formed, at an atomic level, and this information is key for developing rigorous predictive models by the SCM approach (Scheidegger et al., 1997; Elzinga and Reeder, 2002; Tan et al., 2011; Betts et al., 2013). Numerous authors have used EXAFS to investigate the coordination of metals with biochar surface functional groups; for example: Ippolito et al. (2012) to separate out Cu sorption versus precipitation onto a pecan-shell biochar; Ahmad et al. (2014) to determine the speciation of Pb in a biochar-amended shooting range soil; and Alam et al. (2018a) to determine the coordination of Ni and Zn bound to wheat-straw biochar. In the latter paper, the authors used the EXAFS shell-by-shell fitting results to determine reactions representing the metal-surface coordinations used in developing a nonelectrostatic surface complexation model (Fig. 4.2). In this way, the generalized surface complexation reactions

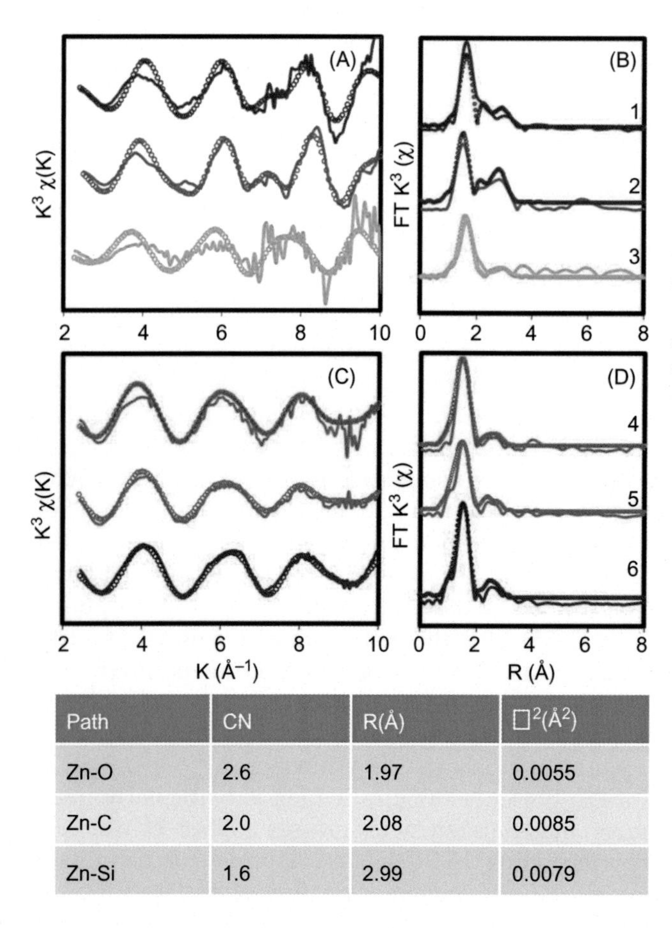

Path	CN	R(Å)	□^{2}(Å^2)
Zn-O	2.6	1.97	0.0055
Zn-C	2.0	2.08	0.0085
Zn-Si	1.6	2.99	0.0079

FIGURE 4.2 EXAFS and corresponding Fourier-transformed data for Ni (A, B) and Zn (C, D) adsorption to wheat-straw biochar. EXAFS data fitting revealed interatomic distances consistent with the presence of carboxyl and silanol groups (table), information that was subsequently used to inform the development of the corresponding surface complexation model. Source: *Modified from Alam, M.S., Lewis, D. G., Chen, N., Konhauser, K.O., Alessi, D.S., 2018b. Thermodynamic analysis of nickel (II) and zinc (II) adsorption to biochar. Environ. Sci. Technol. 52 (11), 6246–6255.*

(e.g., Eq. 4.1) were definitively assigned specific surface functional groups, or the SCM modified to reflect insights gained from EXAFS analyses.

Biochars commonly contain a considerable fraction of inorganic minerals (Ahmad et al., 2014), some of which may be reactive toward charged species. X-ray diffraction (XRD) coupled with sequential extractions and digestions can be used to characterize the mineralogy of biochars and the relative recalcitrance of the various solid phases that comprise the heterogeneous mixture of materials that comprise them (von Gunten et al., 2017). Quartz (SiO_2), calcite ($CaCO_3$), and other carbonate minerals are among the most commonly found minerals in biochars (Li et al., 2016; Yang et al., 2016; Alam et al., 2018a). The presence of carbonates in biochars provides alkalinity, which can lead to the precipitation of cations at elevated pH (e.g., Shen et al., 2018a). Furthermore, silanol functional groups on SiO_2 are proton-active, and as such can lead to the adsorption of charged species. Thus, if SiO_2 comprises a considerable fraction of a biochar, its potential role in metal adsorption should be investigated using the spectroscopic techniques described earlier.

While spectroscopic approaches give information about the coordination of metal ions at the biochar surface, they do not provide information to determine the thermodynamic driving forces of adsorption, which may involve processes such as bond formation, dehydration, enthalpy, and entropy of the metal-surface reaction(s) (Gorman-Lewis et al., 2006 Gorman-Lewis, 2014; Du et al., 2016). Isothermal titration calorimetry (ITC) provides direct measurements of the enthalpy of reactions (ΔH) (Gorman-Lewis et al., 2006; Gorman-Lewis, 2014). Formation constants calculated from surface complexation modeling can be used to calculate Gibbs energy (ΔG_r), where R is the rate constant and T is the absolute temperature, according to Eq. (4.5). The entropy of reaction (ΔS_r) can then be calculated using Eq. (4.6). Combining calorimetric data and surface complexation modeling yields the data needed to determine the driving forces for reactions (see, e.g., Gorman-Lewis et al., 2006; Gorman-Lewis, 2014).

$$\Delta G = 2.3026 \times R \times T \times \log K \tag{4.5}$$

$$\Delta G = \Delta H - T\Delta S \tag{4.6}$$

Combinations of SCM, spectroscopy, and ITC are rigorous approaches to understanding the molecular-scale mechanisms of metals adsorption onto various types of biochars. Overall, these combined approaches promise to provide a better understanding of molecular-scale mechanisms of metals adsorption onto the surface of biochars, and will hopefully lead to models of adsorption that are predictive and may uncover universal metal-adsorption behaviors.

4.4 STATE OF BIOCHAR SURFACE CHEMISTRY MODELING

Despite the need for theoretical models grounded in thermodynamics and informed by spectroscopy, almost all recent studies of metal uptake by biochars have relied on empirical models, without understanding the underlying metal-adsorption reactions at the biochar surface (Mohan and Pittman, 2006; Cao et al., 2009; Cao and Harris, 2010; Mohan et al., 2011; Choppala et al., 2012; Zhang et al., 2013; Hadjittofi and Pashalidis, 2015;

Zhou et al., 2017). To address this gap, a handful of recent studies have focused on developing a molecular-level understanding of heavy-metal adsorption and reduction by biochars, in order to develop predictive and mechanistic models of metal binding to biochar surfaces. Examples include Zhang and Luo (2014), who used the SCM approach to model copper (Cu) adsorption to biochar; Vithanage et al. (2015) who used SCM to model antimony (Sb) adsorption; and Alam et al. (2018a) who modeled the sorption of Se(VI) and Cd(II) to biochar and its mixture with agricultural soils; however, these authors did not directly measure the surface coordination of adsorbed ions at the biochar surface, and thus the metal-surface coordinations they invoke are speculative. More recently, Alam et al. (2018b) used two thermodynamic approaches—surface complexation modeling (SCM) and isothermal titration calorimetry (ITC), further supported by synchrotron-based EXAFS and FTIR spectroscopy—to develop predictive and mechanistic models of competitive Ni and Zn binding to biochars.

In addition to its use in water treatment, there is an increasing drive to amend soils and sediments with biochar, to control metal mobility. Long-term risks to ecosystems and humans can be caused by high concentrations of potentially toxic metals in soils, which are often persistent and difficult to remove once they accumulate in soils (Gray et al., 2006; Querol et al., 2006; Liu et al., 2009; Jiang et al., 2012). Inorganic minerals and organic materials including zeolite, lime, and red-mud and chicken manure have been proposed and studied as soil amendments to immobilize these metals in soils. However, their uses have been costly and have not produced sufficiently positive results (Gray et al., 2006; Querol et al., 2006; Liu et al., 2009). In contrast, recent studies have showed that biochar is highly effective at increasing soil fertility and immobilizing trace metals because of its microporous structure, charged functional groups, and cation exchange capacity (CEC) (Chen and Lin, 2001; Uchimiya et al., 2012; Rees et al., 2014; Laird and Rogovska, 2015; Lehmann et al., 2015). In order to understand how biochar amendment will influence soil chemistry, it is critical to determine the underlying reactions that occur between metals and biochar in complex, multicomponent systems. This knowledge forms the basis for developing a flexible model that can predict the metal mobility in biochar-amended agricultural soils. Multisorbent systems including soils, mineral assemblages, and microbes have been well studied (Dzombak and Morel, 1990; Fein et al., 1997; Cox et al., 1999; Fein, 2006; Komárek et al., 2015; Flynn et al., 2017), but such studies on systems containing biochar are only emerging.

Within the SCM approach, there are two commonly used methods to model metal distribution in multicomponent systems or complex assemblages of sorbents: the component additivity (CA) approach and the general composite (GC) approach (Davis et al., 1998). The CA approach can predict the adsorption of metals onto multicomponent sorbents by invoking the proton- and metal-surface complex stability constants for each individual sorbent in the mixture (Davis et al., 1998; Fowle and Fein, 1999; Pagnanelli et al., 2006). In contrast, the GC approach investigates the reactivity of a multicomponent sorbent by assigning generalized sites to the mixture, rather than groups specific to individual sorbents (Davis et al., 1998; Fowle and Fein, 1999; Pagnanelli et al., 2006). Such an approach does not identify specific metal-surface functional group pairs but allows for metal adsorption data to be considered within the flexible and predictive framework of the SCM theory described in Section 4.2. Previous studies showed that the CA approach can successfully predict metal adsorption to mixtures of pure minerals and bacteria (Davis et al., 1998; Fowle and Fein, 1999; Pagnanelli et al., 2006; Alessi and Fein, 2010). The CA approach may be a useful first step in modeling metal adsorption in complex mixtures containing biochar, such as soil (e.g., Alam et al., 2018a).

Besides its ability to adsorb metals, biochar can donate, accept, and transfer electrons via either biotic or abiotic pathways in their surrounding environments (Keiluweit et al., 2010; Kappler et al., 2014; Yu et al., 2015). Biochar also has significant redox capacity and can function as a reducing agent (Joseph et al., 2010; Kappler et al., 2014), although the reduction of redox-sensitive metals by biochar is not well studied. To enhance the use of biochar as a technology to treat redox-sensitive metals, it is important to understand the kinetics of redox reactions involving biochar across a range of solution pH. For example, Fe^{II}-bearing minerals and Fe^{II} adsorbed to clay minerals can provide a pool of reduction capacity whereby redox-sensitive contaminant metals can be reduced and immobilized via abiotic reduction (Eary and Rai, 1988; Patterson et al., 1997; Fredrickson et al., 2000). Additionally, Cossio (2017) and Rajapaksha et al. (2018) showed preliminary direct spectroscopic evidence of Cr(VI) reduction to Cr(III) by biochars using synchrotron based X-ray absorption spectroscopy (XAS). Further study is needed to better constrain the kinetics of redox reactions involving biochars across a range of solution pH, and to determine the chemistry of the products of contaminant-metal reduction. Metal adsorption at surface functional groups is often a precursor to redox reactions, and thus the SCM approach may be a valuable tool in studying the metal-redox transformations induced by biochars.

Composites of Fe(II)-bearing solid phases and biochars have also been shown to be effective in adsorbing and reducing redox-active metals from aqueous solution, because they are inexpensive and possess high surface area (He et al., 2005; Jung et al., 2007). However, mineral nanoparticles such as magnetite start to aggregates at neutral pH, resulting in dramatically lower adsorption and reduction capacity of redox-sensitive metals (Peterson et al., 1997). Composites of biochar with redox-active minerals such as zero-valent iron and magnetite could have tremendous potential to remove redox-active metals from aqueous solution at a range of pH because the biochar component can enhance the dispersity of the mineral particles and avoid agglomeration and passivation of the minerals under near-neutral and basic pH. Indeed, biochar has been used as a substrate to support zero valent iron (ZVI) and magnetite to enhance the sorption of Cr(VI). However, the underlying Cr(VI) reduction processes were not well studied (Su et al., 2016; Mandal et al., 2017), and further research on the surface chemistry, including the SCM approach, is needed to assess the efficacy of these composite materials.

4.5 OUTLOOK

The future application of biochar for water treatment, soil remediation, and agricultural purposes depends on an understanding of how its surface behaves across wide changes in solution chemistry. Studies that investigate specific mechanisms of metal adsorption, and models that can predict the behavior of metals toward biochar in dynamic environmental conditions, are increasingly needed. The challenges in developing such models are clear: they are more data-intensive, require a greater user knowledge of surface chemistry, and often involve time-consuming and potentially costly spectroscopic analyses. Despite these challenges, selecting the appropriate biochar for a particular application depends on a thorough, mechanistic understanding of how it is likely to behave in the system of interest.

Considerable challenges remain in applying the SCM approach to predict the removal of contaminants from solution. For example, biochar porosity creates micro- and nano-sized zones in which solution chemistry is considerably different than bulk solution, and where equilibrium may not be rapidly achieved. The SCM approach assumes that metal adsorption on biochar surface functional groups reaches equilibrium, and so if a considerable fraction of the total metal is held in micro- or nanopores, the SCM approach may fail. Furthermore, the effects of aging on the overall chemistry of biochar is only beginning to be understood via accelerated aging studies (e.g., Shen et al., 2018b). Changing surface chemistry through time will impact the way in which a biochar interacts with charged species in solution. Finally, the role of nonspecific metal adsorption is challenging to quantify and difficult to model in an SCM framework. For example, Rodríguez-Vila et al. (2018), Cu and Zn adsorption to biochar produced at relatively high pyrolysis temperatures may be controlled by attraction to delocalized π-electrons and not proton-active surface functional groups. Only through continued detailed characterization of biochar surface chemistry will a comprehensive understanding of metal–biochar interactions emerge, which can lead to a predictive modeling framework useful in biochar applications.

References

Ahmad, M., Rajapaksha, A.U., Lim, J.E., Zhang, M., Bolan, N., Mohan, D., et al., 2014. Biochar as a sorbent for contaminant management in soil and water: a review. Chemosphere 99, 19–33.

Alam, M.S., Cossio, M., Robinson, L., Kenney, J.P.L., Wang, X., Konhauser, K.O., et al., 2016. Removal of organic acids from water using biochar and petroleum coke. Environ. Technol. Innovat. 6, 141–151.

Alam, M.S., Swaren, L., Gunten, K.V., Cossio, M., Robbins, L.J., Flynn, S.L., et al., 2018a. Application of surface complexation modeling to trace metals uptake by biochar-amended agricultural soils. Appl. Geochem. 88, 103–112.

Alam, M.S., Lewis, D.G., Chen, N., Konhauser, K.O., Alessi, D.S., 2018b. Thermodynamic analysis of nickel (II) and zinc (II) adsorption to biochar. Environ. Sci. Technol. 52 (11), 6246–6255.

Alessi, D.S., Fein, J.B., 2010. Cadmium adsorption to mixtures of soil components: testing the component additivity approach. Chem. Geol. 270 (1), 186–195.

Alessi, D.S., Lezama-Pacheco, J.S., Stubbs, J.E., Janousch, M., Bargar, J.R., Persson, P., et al., 2014. The product of microbial uranium reduction includes multiple species with U(IV)-phosphate coordination. Geochim. Cosmochim. Acta 131, 115–127.

Bethke, C.M., Brady, P.V., 2000. How the Kd approach undermines ground water cleanup. Ground Water 38 (3), 435–443.

Betts, A.R., Chen, N., Hamilton, J.G., Peak, D., 2013. Rates and mechanisms of Zn^{2+} adsorption on a meat and bonemeal biochar. Environ. Sci. Technol. 47, 14350–14357.

Beveridge, T., Murray, R., 1980. Sites of metal deposition in the cell wall of Bacillus subtilis. J. Bacteriol. 141 (2), 876–887.

Borrok, D., Turner, B.F., Fein, J.B., 2005. A universal surface complexation framework for modeling proton binding onto bacterial surfaces in geologic settings. Am. J. Sci. 305 (6–8), 826–853.

Calvelo Pereira, R., Camps Arbestian, M., Vazquez Sueiro, M., Maciá-Agulló, J.A., 2015. Assessment of the surface chemistry of wood-derived biochars using wet chemistry, Fourier transform infrared spectroscopy, and X-ray photoelectron spectroscopy. Soil Res. 53 (7), 753–762.

Cao, X., Harris, W., 2010. Properties of dairy-manure-derived biochar pertinent to its potential use in remediation. Bioresour. Technol. 101, 5222–5228.

Cao, X.D., Ma, L.N., Gao, B., Harris, W., 2009. Dairy-manure derived biochar effectively sorbs lead and atrazine. Environ. Sci. Technol. 43, 3285–3291.

Chen, J.P., Lin, M., 2001. Equilibrium and kinetic of metal ion adsorption onto a commercial H-type granular activated carbon: experimental and modelling studies. Water Res. 35 (10), 2385–2394.

Choppala, G., Bolan, N., Megharaj, M., Chen, Z., Naidu, R., 2012. The influence of biochar and black carbon on reduction and bioavailability of chromate in soils. J. Environ. Qual. 41 (4), 1175–1184.

Cossio, M., 2017. Mechanisms of the reductive immobilization of hexavalent chromium by Wheat Straw biochar. MSc Thesis, University of Alberta.

Cox, J.S., Smith, D.S., Warren, L.A., Ferris, F.G., 1999. Characterizing heterogeneous bacterial surface functional groups using discrete affinity spectra for proton binding. Environ. Sci. Technol. 33, 4514–4521.

Davis, J.A., Kent, D., 1990. Surface complexation modeling in aqueous geochemistry. Rev. Mineral. Geochem. 23 (1), 177–260.

Davis, J.A., Coston, J.A., Kent, D.B., Fuller, C.C., 1998. Application of the surface complexation concept to complex mineral assemblages. Environ. Sci. Technol. 32, 2820–2828.

DeLuca, T.H., MacKenzie, M.D., Gundale, M.J., Holben, W.E., 2006. Wildfire-produced charcoal directly influences nitrogen cycling in ponderosa pine forests. Soil Sci. Soc. Am. J. 70, 448–453.

Dong, X., Ma, L.Q., Li, Y., 2011. Characteristics and mechanisms of hexavalent chromium removal by biochar from sugar beet tailing. J. Hazard. Mater. 190 (1), 909–915.

Du, H., Chen, W., Cai, P., Rong, X., Dai, K., Peacock, C.L., et al., 2016. Cd(II) sorption on montmorillonite-humic acid-bacteria composites. Sci. Rep. 6, 19499.

Dzombak, D.A., Morel, F.M.M., 1990. Surface Complexation Modeling: Hydrous Ferric Oxide. John Wiley & Sons, New York, NY, 393 pp.

Eary, L.E., Rai, D., 1988. Chromate removal from aqueous wastes by reduction with ferrous ion. Environ. Sci. Technol. 22 (8), 972–977.

Elzinga, E.J., Reeder, R.J., 2002. X-ray absorption spectroscopy study of Cu(II) and Zn(II) adsorption complexes at the calcite surface: implications for site-specific metal incorporation preferences during calcite crystal growth. Geochim. Cosmochim. Acta 66, 3943–3954.

Fein, J.B., 2006. Thermodynamic modeling of metal adsorption onto bacterial cell walls: current challenges. Adv. Agron. 90, 179–202.

Fein, J.B., Daughney, C.J., Yee, N., Davis, T.A., 1997. A chemical equilibrium model for metal adsorption onto bacterial surfaces. Geochim. Cosmochim. Acta 61, 3319–3328.

Fein, J.B., Boily, J-.F., Yee, N., Gorman-Lewis, D., Turner, B.F., 2005. Potentiometric titrations of *Bacillus subtilis* cells to low pH and a comparison of modeling approaches. Geochim. Cosmochim. Acta 69, 1123–1132.

Ferrari, A.C., Robertson, J., 2000. Interpretation of Raman spectra of disordered and amorphous carbon. Phys. Rev. B 61, 14095–14107.

Flynn, S.L., Gao, Q., Robbins, L.J., Warchola, T., Weston, J.N.J., Alam, M.S., et al., 2017. Measurements of bacterial mat metal binding capacity in alkaline and carbonate-rich systems. Chem. Geol. 451, 17–24.

Fowle, D.A., Fein, J.B., 1999. Competitive adsorption of metal cations onto two gram positive bacteria: testing the chemical equilibrium model. Geochim. Cosmochim. Acta 63, 3059–3067.

Fredrickson, J.K., Zachara, J.M., Kennedy, D.W., Duff, M.C., Gorby, Y.A., Shu-mei, W.L., et al., 2000. Reduction of U(VI) in goethite (α-FeOOH) suspensions by a dissimilatory metal-reducing bacterium. Geochim. Cosmochim. Acta 64 (18), 3085–3098.

Gorman-Lewis, D., 2014. Enthalpies and entropies of Cd and Zn adsorption onto Bacillus licheniformis and enthalpies and entropies of Zn adsorption onto Bacillus subtilis from isothermal titration calorimetry and surface complexation modeling. Geomicrbiol. J. 31, 383–359.

Gorman-Lewis, D., Fein, J.B., Jensen, M.P., 2006. Enthalpies and entropies of proton and cadmium adsorption onto Bacillus subtilis bacterial cells from calorimetric measurements. Geochim. Cosmochim. Acta 70, 4862–4873.

Gray, C.W., Dunham, S.J., Dennis, P.G., Zhao, F.J., McGrath, S.P., 2006. Field evaluation of in situ remediation of a heavy metal contaminated soil using lime and red-mud. Environ. Pollut. 142, 530–539.

von Gunten, K., Alam, M.S., Hubmann, M., Ok, Y.S., Konhauser, K.O., Alessi, D.S., 2017. Modified sequential extraction for biochar and petroleum coke: metal release potential and its environmental implications. Bioresour. Technol. 236, 106–110.

Hadjittofi, L., Pashalidis, I., 2015. Uranium sorption from aqueous solutions by activated biochar fibres investigated by FTIR spectroscopy and batch experiments. J. Radioanal. Nucl. Chem. 304 (2), 897–904.

He, Z.L., Yang, X.E., Stoffella, P.J., 2005. Trace elements in agroecosystems and impacts on the environment. J. Trace Elem. Med. Biol. 19 (2 – 3), 125–140.

Inyang, M.I., Gao, B., Yao, Y., Xue, Y., Zimmerman, A., Mosa, A., et al., 2016. A review of biochar as a low-cost adsorbent for aqueous heavy metal removal. Crit. Rev. Environ. Sci. 46 (4), 406–433.

Ippolito, J.A., Strawn, D.G., Scheckel, K.G., Novak, J.M., Ahmedena, M., Niandou, M.A.S., 2012. Macroscopic and molecular investigations of copper sorption by a steam-activated biochar. J. Environ. Qual. 41, 150–156.

Jiang, J., Xu, R., Jiang, T., Li, Z., 2012. Immobilization of Cu (II), Pb (II) and Cd (II) by the addition of rice straw derived biochar to a simulated polluted Ultisol. J. Hazard. Mater. 229–230, 145–150.

Joseph, S., Camps-Arbestain, M., Lin, Y., Munroe, P., Chia, C., Hook, J., et al., 2010. An investigation into the reactions of biochar in soil. Soil Res. 48 (7), 501–515.

Jung, Y., Choi, J., Lee, W., 2007. Spectroscopic investigation of magnetite surface for the reduction of hexavalent chromium. Chemosphere 68, 1968–1975.

Kappler, A., Wuestner, M.L., Ruecker, A., Harter, J., Halama, M., Behrens, S., 2014. Biochar as an electron shuttle between bacteria and Fe (III) minerals. Environ. Sci. Technol. Lett. 1 (8), 339–344.

Karthikeyan, K., Elliott, H.A., 1999. Surface complexation modeling of copper sorption by hydrous oxides of iron and aluminum. J. Colloid Interface Sci. 220 (1), 88–95.

Keiluweit, M., Nico, P.S., Johnson, M.G., Kleber, M., 2010. Dynamic molecular structure of plant biomass-derived black carbon (biochar). Environ. Sci. Technol. 44, 1247–1253.

Komárek, M., Koretsky, C.M., Stephen, K.J., Alessi, D.S., Chrastný, V., 2015. Competitive adsorption of Cd(II), Cr (VI), and Pb(II) onto nanomaghemite: a spectroscopic and modeling approach. Environ. Sci. Technol. 49 (21), 12841–12859.

Koretsky, C., 2000. The significance of surface complexation reactions in hydrologic systems: a geochemist's perspective. J. Hydrol. 230 (3), 127–171.

Laird, D., Rogovska, N., 2015. Biochar effects on nutrient leaching. In: Lehmann, J., Joseph, S. (Eds.), Biochar for Environmental Management: Science, Technology and Implementation. Routledge, UK, pp. 521–542.

Lehmann, J., Kuzyakov, Y., Pan, G., Ok, Y.S., 2015. Biochars and the plant-soil interface. Plant Soil 395, 1–5.

Leone, L., Ferri, D., Manfredi, C., Persson, P., Shchukarev, A., Sjöberg, S., et al., 2007. Modeling the acid-base properties of bacterial surfaces: a combined spectroscopic and potentiometric study of the Gram-positive bacterium *Bacillus subtilis*. Environ. Sci. Technol. 41, 6465–6471.

Li, F., Shen, K., Long, K., Wen, J., Xie, X., Zeng, X., et al., 2016. Preparation and characterization of biochars from *Eichornia crassipes* for cadmium removal in aqueous solutions. PLoS One 11 (2), e0148132.

Liu, L.N., Chen, H.S., Cai, P., Lianga, W., Huang, Q.Y., 2009. Immobilization and phytotoxicity of Cd in contaminated soil amended with chicken manure compost. J. Hazard. Mater. 163, 563–567.

Manceau, A., Charlet, L., 1994. The mechanism of selenate adsorption on goethite and hydrous ferric oxide. J. Colloid Interface Sci. 168 (1), 87–93.

Mandal, S., Sarkar, B., Bolan, N., Ok, Y.S., Naidu, R., 2017. Enhancement of chromate reduction in soils by surface modified biochar. J. Environ. Manage. 186 (2), 277–284.

Manning, B.A., Goldberg, S., 1997. Adsorption and stability of arsenic (III) at the clay mineral-water interface. Environ. Sci. Technol. 31 (7), 2005–2011.

Marmier, N., Delisée, A., Fromage, F., 1999. Surface complexation modeling of Yb (III) and Cs (I) sorption on silica. J. Colloid Interface Sci. 212 (2), 228–233.

Mohan, D., Pittman, C.U., 2006. Activated carbons and low cost adsorbents for remediation of tri-and hexavalent chromium from water. J. Hazard. Mater. 137 (2), 762–811.

Mohan, D., Rajput, S., Singh, V.K., Steele, P.H., Pittman, C.U., 2011. Modeling and evaluation of chromium remediation from water using low cost bio-char, a green adsorbent. J. Hazard. Mater. 188 (1), 319–333.

Pagnanelli, F., Bornoroni, L., Moscardini, E., Toro, L., 2006. Non-electrostatic surface complexation models for protons and lead(II) sorption onto single minerals and their mixture. Chemosphere 63, 1063–1073.

Patterson, R.R., Fendorf, S., Fendorf, M., 1997. Reduction of hexavalent chromium by amorphous iron sulfide. Environ. Sci. Technol. 31, 2039–2044.

Peterson, M.L., White, A.F., Brown, G.E., Parks, G.A., 1997. Surface passivation of magnetite by reaction with aqueous Cr(VI): XAFS and TEM results. Environ. Sci. Technol. 31, 1573–1576.

Querol, X., Alastuey, A., Moreno, N., Alvarez-Ayuso, E., Garcia-Sanchez, A., Cama, J., et al., 2006. Immobilization of heavy metals in polluted soils by the addition of zeolitic material synthesized from coal fly ash. Chemosphere 62, 171–180.

Rajapaksha, A.U., Alam, M.S., Chen, N., Alessi, D.S., Igalavithana, A.D., Tsang, D.C.W., et al., 2018. Removal of hexavalent chromium in aqueous solutions using biochar: chemical and spectroscopic investigations. Sci. Total Environ. 625, 1567–1573.

Reeder, R.J., Schoonen, M.A.A., Lanzirotti, A., 2006. Metal speciation and its role in bioaccessibility and bioavailability. Rev. Mineral. Geochem. 64, 59–113.

Rees, F., Simonnot, M.O., Morel, J.L., 2014. Short-term effects of biochar on soil heavy metal mobility are controlled by intra-particle diffusion and soil pH increase. Eur. J. Soil Sci. 65, 149–161.

Rodríguez-Vila, A., Selwyn-Smith, H., Enunwa, L., Smail, I., Covelo, E.F., Sizmur, T., 2018. Predicting Cu and Zn sorption capacity of biochar from feedstock C/N ratio and pyrolysis temperature. Environ. Sci. Pollut. Res. Int. 25 (8), 7730–7739.

Scheidegger, A.M., Lamble, G.M., Sparks, D.L., 1997. Spectroscopic evidence for the formation of mixed-cation hydroxide phases upon metal sorption on clays and aluminum oxides. J. Colloid Interface Sci. 186, 118–128.

Shen, Z., Zhang, Y., Jin, F., Alessi, D.S., Zhang, Y., Wang, F., et al., 2018a. Comparison of nickel adsorption on biochars produced from softwood and Miscanthus straw. Environ. Sci. Pollut. Res. 25 (15), 14626–14635.

Shen, Z., Hou, D., Zhao, B., Xu, W., Ok, Y.S., Bolan, N.S., et al., 2018b. Stability of heavy metals in soil washing residue with and without biochar addition under accelerated ageing. Sci. Total Environ. 619–620, 185–193.

Silber, A., Levkovitch, I., Graber, E.R., 2010. pH-dependent mineral release and surface properties of cornstraw biochar: agronomic implications. Environ. Sci. Technol. 44 (24), 9318–9323.

Sparks, D.L., 2005. Toxic metals in the environment: the role of surfaces. Elements 1 (4), 193–197.

Su, H., Fang, Z., Tsang, P.E., Zheng, L., Cheng, W., Fang, J., et al., 2016. Remediation of hexavalent chromium contaminated soil by biochar-supported zero-valent iron nanoparticles. J. Hazard. Mater. 318, 533–540.

Tack, F., Verloo, M.G., 1995. Chemical speciation and fractionation in soil and sediment heavy metal analysis: a review. Int. J. Environ. Anal. Chem. 59 (2–4), 225–238.

Tan, X., Hu, J., Montavon, G., Wang, X., 2011. Sorption speciation of nickel (II) onto Ca-montmorillonite: batch, EXAFS techniques and modeling. Dalton Trans. 40 (41), 10953–10960.

Teixido, M., Hurtado, C., Pignatello, J.J., Beltrán, J.L., Granados, M., Peccia, J., 2013. Predicting contaminant adsorption in black carbon (biochar)-amended soil for the veterinary antimicrobial sulfamethazine. Environ. Sci. Technol. 47 (12), 6197–6205.

Thomas, S.C., Gale, N., 2015. Biochar and forest restoration: a review and meta-analysis of tree growth responses. New Forest. 6 (6), 601–614.

Thompson, K.A., Shimabuku, K.K., Kearns, J.P., Knappe, D.R.U., Summer, S., Cook, S.M., 2016. Environmental comparison of biochar and activated carbon for tertiary wastewater treatment. Environ. Sci. Technol. 50, 11253–11262.

Turner, B.F., Fein, J.B., 2006. Protofit: a program for determining surface protonation constants from titration data. Comput. Geosci. 32 (9), 1344–1356.

Uchimiya, M., Klasson, K.T., Wartelle, L.H., Lima, I.M., 2011a. Influence of soil properties on heavy metal sequestration by biochar amendment: 1. Copper sorption isotherms and the release of cations. Chemosphere 82, 1431–1437.

Uchimiya, M., Wartelle, L.H., Klasson, K.T., Fortier, C.A., Lima, I.M., 2011b. Influence of pyrolysis temperature on biochar property and function as a heavy metal sorbent in soil. J. Agric. Food Chem. 59 (6), 2501–2510.

Uchimiya, M., Bannon, D.I., Wartelle, L.H., 2012. Retention of heavy metals by carboxyl functional groups of biochars in small arms range soil. J. Agric. Food Chem. 60 (7), 1798–1809.

Van Leeuwen, H.P., Town, R.M., Buffle, J., Cleven, R.F., Davison, W., Puy, J., et al., 2005. Dynamic speciation analysis and bioavailability of metals in aquatic systems. Environ. Sci. Technol. 39 (22), 8545–8556.

Vithanage, M., Rajapaksha, A.U., Ahmad, M., Uchimiya, M., Dou, X., Alessi, D.S., et al., 2015. Mechanisms of antimony adsorption onto soybean stover-derived biochar in aqueous solutions. J. Environ. Manage. 151, 443–449.

Westall, J.C., 1982. FITEQL, a computer program for determination of chemical equilibrium constants from experimental data, Version 2.0. Department of Chemistry, Oregon State University, Corvallis, OR. Report 82-02.

Woolf, D., Amonette, J.E., Street-Perrott, F.A., Lehmann, J., Joseph, S., 2010. Sustainable biochar to mitigate global climate change. Nat. Commun. 1, 56.

Yang, F., Zhao, L., Gao, B., Xu, X.Y., Cao, X.D., 2016. The interfacial behavior between biochar and soil minerals and its effect on biochar stability. Environ. Sci. Technol. 50, 2264–2271.

Yu, L., Yuan, Y., Tang, J., Wang, Y., Zhou, S., 2015. Biochar as an electron shuttle for reductive dechlorination of pentachlorophenol by *Geobacter sulfurreducens*. Sci. Rep. 5, 16221.

Zhang, Y., Luo, W., 2014. Adsorptive removal of heavy metal from acidic wastewater with biochar produced from anaerobically digested residues: kinetics and surface complexation modeling. BioResources 9 (2), 2484–2499.

Zhang, Z., Cao, X., Liang, P., Liu, Y., 2013. Adsorption of uranium from aqueous solution using biochar produced by hydrothermal carbonization. J. Radioanal. Nucl. Chem. 295, 1201–1208.

Zhou, B., Wang, Z., Shen, D., Shen, F., Wu, C., Xiao, R., 2017. Low cost earthworm manure-derived carbon material for the adsorption of Cu^{2+} from aqueous solution: impact of pyrolysis temperature. Ecol. Eng. 98, 189–195.

PART III

APPLICATIONS

CHAPTER

5

Biochar for Mine-land Reclamation

James A. Ippolito[1], *Liqiang Cui*[2], *J.M. Novak*[3] *and Mark G. Johnson*[4]

[1]Department of Soil and Crop Sciences, Colorado State University, Fort Collins, CO, United States [2]School of Environmental Science and Engineering, Yancheng Institute of Technology, Yancheng, China [3]U.S. Department of Agriculture, Agricultural Research Service, Coastal Plains Research Center, Florence, SC, United States [4]U.S. Environmental Protection Agency, National Health and Environmental Effects Research Laboratory/ORD Western Ecology Division, Corvallis, OR, United States

DISCLAIMER

The information in this document has been funded in part by the US Environmental Protection Agency. It has been subjected to review by the National Health and Environmental Effects Research Laboratory's Western Ecology Division and approved for publication. Approval does not signify that the contents reflect the views of the Agency, nor does mention of trade names or commercial products constitute endorsement or recommendation for use.

5.1 INTRODUCTION

Abandoned mine lands are those lands, waters, and surrounding watersheds where extraction, beneficiation, or processing of ores and minerals has occurred (US EPA, 2017a). Globally, abandoned mines number in the hundreds of thousands. Examples include:

- The United Kingdom, with more than 2000 abandoned mines (SRK Consulting, 2017);
- Italy, with approximately 3000 abandoned mines (Italy Europe 24, 2015);

Biochar from Biomass and Waste
DOI: https://doi.org/10.1016/B978-0-12-811729-3.00005-4

- France, with ~4000, Germany 100 s, Poland 1000–1500, Turkey > 700, and Korea >1000 abandoned mines (International Commission on Mine Closure, International Society for Rock Mechanics, 2008);
- Abandoned mines in Australia number >50,000 (Mining Technology, 2015);
- Canada, with ~13,000 abandoned mines (Mackasey, 2000);
- The United States is estimated to contain ~500,000 abandoned mines (BLM, 2016).

Furthermore, if the wastes contain sulfide ores, then oxidation reactions generate acid mine drainage, a large environmental concern because it leads to significant environmental impacts globally (Coetzee et al., 2010). In the western United States alone, approximately 33,000 mines sites generate acidity (Mittal, 2011), leading to increases in heavy metal bioavailability, degradation of mine-affected soils and surface waters (Ippolito et al., 2017), or contamination of shallow groundwaters (Ramontja et al., 2011). Thus, it is important that management strategies are available for reducing the continued environmental impact in mine lands with subsequent successful improvements in environmental quality.

Biochar may play a role in improving mine-land remediation by positively affecting changes in mine-land soil conditions. Biochar application has been shown to increase mine soil, tailings, and waste rock pH, cation-exchange capacity, water-holding capacity, organic matter content, nitrate concentration, biological *N*-fixation rates, and phosphatase and dehydrogenase activity (Fellet et al., 2011; Hanauer et al., 2012; Kelly et al., 2014; Kim et al., 2014; Reverchon et al., 2015). Biochar application to mine-land materials has also been shown to increase vegetative yield and cover (Aspen Center for Environmental Studies, 2011; Rodríguez-Vila et al., 2015). Reduction in bioavailable heavy metal concentration is speculated to produce a positive change in soil condition and improvement in plant growth. The following sections illustrate the important role biochar plays in sorbing, sequestering, precipitating, and ultimately reducing bioavailable metals in contaminated solutions, soils, and mine-land situations.

5.1.1 Cadmium

Exposure to Cd can lead to a variety of health effects, such as flu-like symptoms (short-term exposure to high Cd levels), kidney, bone, lung disease, and cancer (long-term exposure to low-level Cd levels) (US DOL-OSHA, 1993). Thus, reducing Cd exposure/ingestion by humans and animals is an important health concern. Biochars have been shown to play a role in reducing heavy Cd bioavailability in a variety of environmental settings.

Sekulić et al. (2018) showed that crushed apricot seeds, pretreated with phosphoric acid prior to pyrolysis, could almost linearly remove Cd from the aqueous phase as associated with increasing solution pH from 2 to 7. The authors attributed Cd removal to adsorption mechanisms up to ~pH 7 and Cd-hydroxide precipitation above pH 7. Karunanayake et al. (2018) used six different biochars (i.e., douglas fir biochar, magnetized douglas fir and switchgrass biochars, two different commercial biochars, pine wood biochar) to remove Cd from solution over a pH range of 2–7. Cadmium adsorption increased with increasing biochar surface area and, similar to the findings of Sekulić et al. (2018), sorption increased with increasing pH. Karunanayake et al. (2018) suggested that as solution pH increases, deprotonation of biochar surface carboxylic acid and other acidic hydroxyl groups leads to lower net positive charge, thus enhancing Cd sorption at negative biochar sites. Deng et al. (2017) used activated rice straw biochar to determine removal capabilities on solution-borne Cd. The authors reported that Cd sorption results at low pH values were similar to those reported by Karunanayake et al. (2018) and to Sekulić et al. (2018). Aslam et al. (2017) compared a low- and high-temperature biochar (350°C and 650°C, respectively) for removal of Cd from solution, and found that the lower temperature biochar was more effective at removal. It was speculated that this reduction was due to cation-exchange capacity and functional groups present. Shen et al. (2017a) used algae-derived biochar to effectively sorb and remove Cd from aqueous solution, with biochar sorbing up to 217 mg Cd g^{-1}. The authors attributed sorption to electrostatic interactions, ion exchange, and surface complexation.

In China, Cd accumulation in rice grown in Cd-contaminated soil from mining/smelting operations has been documented (Hu et al., 2016). As a remediation example, Cui et al. (2011) added wheat straw biochar at 0, 10, 20, and 40 Mg ha^{-1} to a rice paddy soil containing ~22 mg kg^{-1} total Cd. The amended biochar raised soil pH, decreased bioavailable soil Cd content by 5.5%–52%, and lowered rice Cd content by 17%–62%. Bian et al. (2013) found a similar response when adding wheat straw-derived biochar to rice paddies. Qi et al. (2018) showed that within soils containing greater sorption capacity, biochar made from wood shavings or chicken litter had no effect on bioavailable Cd concentrations. Under acidic conditions, however, the authors observed a decrease in soluble Cd content, attributing the reduction to soil pH increases. Yin et al. (2017) added rice straw biochar to a contaminated paddy soil, showing porewater Cd concentration reductions, which was likely due to immobilization via surface complexation and precipitation of insoluble Cd mineral phases. Ouyang et al. (2017) added increasing corn straw biochar applications (0%–5% by weight) to a soil containing elevated Cd concentrations from long-term use of Cd-containing phosphate fertilizers. Cd was found to be immobilized rather than transported offsite, likely due to Cd immobilization by biochar. In their study, a 3% application rate caused the greatest Cd reductions. Qian et al. (2017) added rice straw biochar to a soil slurry and a hydroponic system containing 5.6 mg Cd L^{-1}. Root growth was inhibited by the presence of Cd by 20%–50%, but when biochar was added root inhibition was only 4%–25%. Biochar was speculated by the authors to cause the positive root growth in the presence of Cd.

5.1.2 Copper

Long-term Cu exposure can irritate mucous membranes, cause headaches, dizziness, nausea, and diarrhea; drinking water containing excess Cu can lead to nausea, vomiting, diarrhea, and if taken intentionally, can cause liver and kidney damage, and even death (ATSDR, 2004). Since the greatest chance of ingesting excess Cu is via drinking water, it is important to reduce Cu release from piping, industrial applications, or from acid mine drainage. Again, biochar has been proven to remove Cu from various sources.

Biochar use for removing Cu from solution has been well studied. Ippolito et al. (2012) used a KOH-steam-activated pecan shell biochar to sorb Cu from solution. The authors noted that biochar could retain >42,000 mg Cu kg^{-1} biochar, and that under acidic conditions, Cu sorbed onto biochar as humic-like substances. In contrast, while under alkaline conditions, Cu oxide and carbonate precipitates were the dominate sequestration mechanisms. Deng et al. (2017) used activated rice straw biochar to remove Cu from solution over varying pH values. The authors showed that Cu removal rate increased with increasing pH up to ~pH of 5, likely due to deprotonation of functional groups leading to greater Cu removal. Furthermore, Cu removal capacity equaled between 36,000 and 47,000 mg Cu kg^{-1} biochar, within a range similar to Ippolito et al. (2012). Likewise, Batool et al. (2017) showed that farmyard manure biochar and poultry litter biochar could remove ~44,000 mg Cu kg^{-1} biochar. Tran et al. (2017) created biochar from several types of mono- or polysaccharides to remove Cu from solution. Similar to the previous findings, the authors showed that Cu removal was highly pH dependent and could sorb a maximum of ~57,000 mg Cu kg^{-1} biochar. He et al. (2017) created a microscale bagasse biochar/polysulfone membrane to remove Cu from water over varying pH values. Copper sorption was maximized at pH > 4.5 with a sorption capacity of ~14,000 mg Cu kg^{-1} membrane. The authors attributed the sorption to the membrane surface charge becoming negative above pH 4.4, along with biochar functional groups becoming deprotonated. Zhou et al. (2017) used Fe- or Zn-doped sawdust biochar to effectively remove Cu from solution, suggesting that hydrophobic and hydrophilic sites on the biochar caused a reduction in solution Cu content.

Biochar use for removing Cu from soils has been studied less. Cárdenas-Aguiar et al. (2017) spiked a soil with 1000 mg Cu kg^{-1}, then added 10% biochar derived from urban waste or biochar + compost. The authors then attempted to grow various plant species (i.e., mustard, cress, ryegrass). Mustard and cress growth was severely affected by Cu-contaminated soil alone. However, when biochar was added alone, there was positive mustard growth. Similarly, biochar + compost had a positive effect on cress productivity. Cárdenas-Aguiar et al. (2017) attributed these positive plant growth responses to biochar and/or compost immobilizing Cu. Rodríguez-Vila et al. (2017) amended Cu-contaminated mine soil with 0%, 20%, 40%, 80%, and 100% of a 95:5 part mixture of technosol:holm oak wood biochar. As with the biochar-Cu-solution removal findings above, the authors showed a decrease in Cu availability with increasing soil pH. They also observed an increase in plant growth using higher mixture application rates. Meier et al. (2017) added chicken manure or oat hull biochar to a Cu-contaminated soil at 0%, 1%, and 5% (w/w). Biochars increased soil pH, decreased Cu availability, and increased Cu associated with the organic and residual soil fraction. The authors also showed that microbial respiration increased, suggesting that biochars improved their habitat. Ippolito et al. (2017) added lodgepole pine or tamarisk-derived biochars (0%, 5%, 10%, and 15% by weight) to four mine-land soils containing elevated Cu concentrations. As with Meier et al. (2017), the authors observed a significant increase in pH and a decrease in Cu availability, with increasing Cu concentrations in the oxyhydroxide and carbonate soil phases.

However, the above biochar responses are not always observed. Ippolito (unpublished) utilized soils spiked with 0, 250, 500, and 1000 mg Cu kg^{-1} from a previous study (Ippolito et al., 2011), and then added increasing amounts (0%, 0.5%, 1%, and 2% by weight) of hardwood biochar. Their concern was to observe the effects on alfalfa growth and Cu uptake, and potential reductions in soil extractable Cu content (Fig. 5.1). Although increasing Cu content significantly decreased plant yield ($P < .001$), increasing the biochar application rate had no effect ($P = .293$) on improving plant growth. Increasing the biochar application rate decreased plant Cu concentrations ($P = .003$) but had no effect on available

FIGURE 5.1 Four replicates (front to back) of, from left to right in each image, increasing Cu concentrations of 0, 250, 500, and 1000 mg kg^{-1}. Fast pyrolysis hardwood biochar was added to all pots at (A) 0%, (B) 0.5%, (C) 1.0%, and (D) 2% by weight and then alfalfa was grown over a period of approximately 30 days. *Photos courtesy of Jim Ippolito.*

soil Cu concentrations ($P = .348$). This is an example of employing a biochar that has not been vetted for soil Cu removal. Thus, some caution exists in terms of using any type of biochar for remediation purposes.

5.1.3 Lead

Widespread extraction, processing, and use of Pb has resulted in extensive environmental contamination, human exposure, and significant health problems globally (World Health Organization, 2017). Issues with long-term human Pb exposure include distribution to the brain, liver, kidneys, and bones; Pb release from bones during pregnancy, which subsequently leads to developing fetus exposure; it is particularly harmful to young children; and there is no known level of Pb exposure that is considered safe (World Health Organization, 2017). Short-term Pb exposure can lead to abdominal pain, constipation, headaches, irritability, loss of appetite or memory, and weakness (Centers for Disease Control and Prevention, 2017).

As with other heavy metals described above, scientific reports have established that biochar can sorb Pb. Aslam et al. (2017) compared biochars (350°C and 650°C, respectively) for Pb removal from solution. The authors found approximately 70%–95% Pb removal from solution after shaking with the low-temperature biochar, compared to only 39%–60% removal with the high-temperature biochar over the same shaking period. The authors suggested that Pb was bound to a greater extent with low-temperature biochars due to Pb attraction to oxygen-containing functional groups or precipitation with different compounds present on biochar surfaces. Deng et al. (2017) used activated rice straw biochar to remove solution-borne Pb, noting that Pb removal may have been mainly due to functional groups present on the biochar. Lee et al. (2017) suggested that Pb sorption onto palm oil sludge biochar was due to cation-exchange mechanisms. He et al. (2017) used a bagasse biochar to remove Pb from water over varying pH values. As with their Cu observations, Pb sorption was maximized at $pH > 4.5$, with sorption attributed to negative surface charge development above pH 4.4 and functional groups becoming deprotonated. Karunanayake et al. (2018) found a similar Pb–pH response with two douglas fir biochars. Sekulić et al. (2018) showed that crushed apricot seed biochar could sorb approximately 179 mg Pb g^{-1}. The authors speculated that solution Pb removal occurred via a number of pathways, including a quick surface sorption phenomenon, then followed by a slow sorption via intraparticle diffusion, and finally maximum sorption was reached since intraparticle diffusion reached a minimum. Shen et al. (2017b) utilized four biochars (i.e., hardwood, wheat straw pellets, rice husk, soft wood pellets) to remove Pb from solution, followed by sequential extraction to identify Pb phases present. Lead was sorbed onto the biochars predominantly as potentially bioavailable phases. This Pb phase could possibly be released back into the environment.

Biochar has been added to various Pb-contaminated soils with varying degrees of success in terms of reducing bioavailable Pb concentrations. Fellet et al. (2011) added increasing orchard pruning biochar (0%, 1%, 5%, and 10%, w/w) to a Pb-contaminated mine tailing. The authors observed a significant reduction in bioavailable Pb concentrations with increasing biochar application rate. Jain et al. (2017) added lemongrass biochar to acidic coal field overburden material in order to study the effects of herb growth and Pb uptake. The results suggested that biochar applied at 15% or 20% (by weight) reduced Pb translocation from roots to aboveground plant tissue by 81%, and aboveground plant Pb accumulation was reduced by ~94% as compared to controls. Kelly et al. (2014) added pine wood biochar to acidic mine-land soils at rates of 0%, 10%, 20%, and 30% (v/v). Biochar application increased soil pH and decreased solution Pb leachate concentrations by 63% in one mine soil, but did not affect Pb leachate concentrations from another mine soil. Biochar produced from tobacco was used to successfully increase soil cation-exchange capacity, reduce soil bioavailable Pb concentrations, increase Chinese cabbage yield, and decrease plant Pb uptake in heavily contaminated Pb soils from China (Lahori et al., 2017). Moreno-Barriga et al. (2017) added 1% and 2% hog waste biochar and $CaCO_3$ waste to a Pb-contaminated, acidic mine tailing to assess growth effects of proso millet. Biochar rate increased soil pH, reduced Pb in a relatively bioavailable form, and increased Pb in a more recalcitrant form. However, the highest biochar application rate negatively affected plant growth likely by limiting other nutrients essential for plant growth. This is another example of prevetting biochar before its field use. Thus, as with the Cu observations by Ippolito (above; unpublished), a need for caution exists when utilizing biochars for remediation purposes.

5.1.4 Zinc

Most Zn enters the environment as a result of mining activities; purifying of Zn, Pb, and Cd ores; steel production; coal burning; and waste-burning or reutilization (e.g., biosolids land application) (Agency for Toxic Substances and Disease Registry, 2005). Only exposure to high Zn doses has toxic effects, and long-term high-dose Zn supplementation interferes with human Cu uptake; thus Zn toxic effects typically induce human Cu deficiency symptoms (Plum et al., 2010). However, at the cellular level, excess endogenous Zn can act on several molecular regulators of programmed cell death, which sometimes occurs in the brain (Plum et al., 2010). Areas that have been mined for other elements (e.g., Pb, Au, Ag) typically produces waste mine tailings with elevated Zn concentrations (i.e., Zn is not released during the mining/smelting process); these are specific locations where excess Zn may become toxic to microorganisms and plants.

Recent biochar studies have specifically targeted Zn issues in soils. Kumar et al. (2018) spiked a sandy soil with 880 mg Zn kg^{-1}, aged the soil for 60 days, applied 1%, 3%, and 5% (by weight) using either grain husk or cattle manure biochar, and then grew Ficus over a 180 day period. In general, leaf Zn content decreased with increasing biochar rate, and pot leachate Zn concentrations were lower than the control, by a factor of $\sim 1-2$. The authors also showed that biochar acted as a Zn-sink via chemical extraction of separated biochar particles. Xu et al. (2017) added 0%, 0.5%, 1%, and 2% (w/w) wine lees (i.e., dead yeast and other particulates remaining after the wine-making process) biochar to a topsoil containing elevated Zn concentrations (1580 mg kg^{-1}) contaminated from nearby industrial processes; rice was grown throughout the study. The percentage of soil-exchangeable Zn decreased with increasing biochar application rate, over various rice growth stages, and rice yield increased at the two highest biochar application rates as compared to the control. The authors attributed positive results to biochar increasing soil properties such as pH and cation-exchange capacity, as well as containing functional groups that sorb Zn and reduce bioavailability.

The addition of 1% or 2% hog waste biochar to a heavily contaminated Zn-containing mine tailing (>1000 mg total Zn kg^{-1}) caused Zn to shift from exchangeable to Fe/Mn oxide-bound phases (Moreno-Barriga et al., 2017). Shifts in metal phases were attributed to an increase in mine-tailing pH associated with biochar application. In addition, changes in pH and shifts in Zn soil fractions were similar to that found by Ippolito et al. (2017). Lahori et al. (2017) added 1% tobacco biochar to a Zn-contaminated soil (225 mg total Zn kg^{-1}; 29 mg plant-available Zn kg^{-1}) and observed a significant reduction in plant-available Zn concentration as compared to control soil. The addition of a liming source along with biochar caused further reductions in plant-available Zn, with results likely explained by shifts in Zn form toward oxide-bound phases (e.g., see Moreno-Barriga et al., 2017; Ippolito et al., 2017). Fellet et al. (2011) added increasing orchard pruning biochar (0%, 1%, 5%, 10% weight:weight) to a Zn-contaminated mine tailing ($>11{,}500$ mg total Zn kg^{-1}). As with their Pb findings, the authors observed a significant reduction in Zn bioavailability with increasing biochar application rate, likely attributed to shifts in soil Zn pools as observed by others.

Although most studies have shown significant reductions in soil Zn availability associated with biochar application, others have shown no change or significant negative effects on soil Zn concentrations. Peltz and Harley (2016) provided a case study for biochar use in the Upper Animas Mining District of southwestern Colorado, USA. Findings from a 2-year trial, with mine-affected soils receiving 30% biochar (by volume), showed soil leachate concentrations with little change in total Zn. Kelly et al. (2014) added pine wood biochar (0%, 10%, 20%, 30% volume:volume) to a heavily contaminated mine soil (>2700 mg total Zn kg^{-1}) and repeated leaching the soil over a 2-month period. Results showed that biochar application actually caused significant increases in leachate Zn concentrations near the end of the study period in one of two spoil materials investigated.

5.1.5 Recent Case Study—Biochar Use in Multielement-Contaminated Mine Waste

Utilizing mine spoils with mixed heavy metal contamination, Ippolito et al. (2017) recently reported that increasing lodgepole pine or tamarisk biochar application (0%, 5%, 10%, and 15% by weight) to several acid generating mine-land soils could increase soil pH and decrease metal bioavailability. Furthermore, the authors showed via a sequential extraction procedure that biochar addition caused shifts in soil metal pools toward Cd, Cu, Pb, and Zn-associated carbonate and oxyhydroxide phases. Similar results were found in a recent parallel study by Ippolito using mine-land soils amended with increasing switchgrass biochar (unpublished). Switchgrass biochar (pH = 9.4) was added at 0%, 5%, 10%, and 15% (by weight) to four different acidic mine-land soils from Creede and Leadville, Colorado, USA, and from near Coeur d'Alene, Idaho, USA. Increasing biochar application rate increased soil pH (Fig. 5.2A), and concomitantly decreased soil Cu, Pb, and Zn bioavailability (Fig. 5.2B–D, respectively). Following switchgrass biochar application, soil phases shifted toward carbonate bound and organically complexed heavy metal phases and were likely responsible for the reduction in heavy metal bioavailability.

5.1.6 Recent Case Study—Biochar Use in Cd- and Zn-Contaminated Paddy Soil

Cui (unpublished) recently studied biochar application effect on a heavy metal-contaminated soil in China. Specifically, wheat straw biochar was applied to a Cd and Zn-contaminated paddy soil (total Cd and Zn = 22.6 and 186 mg kg^{-1}, respectively) at 0%, 5%, and 15% by weight, with soil mixtures shaken in 0.01 M $CaCl_2$ solution over a 240 day incubation period. Samples were destructively sampled at various time intervals, and mixture pH and 0.01 M $CaCl_2$ extractable Cd and Zn concentrations were determined. After 240 days, the solid phase was allowed to air dry, and then subject to the European Community Bureau of Reference (BCR) sequential extraction procedure according to Ure et al. (1993); additional BCR extraction information can be found in Ippolito et al. (2017).

Increasing biochar application rate increased soil-biochar mixture pH over the 240-day study (Fig. 5.3A), and based on previously published work (above), it was anticipated that heavy metal concentrations would decrease. Indeed, the author observed significant decreases in bioavailable Cd and Zn concentrations with increasing biochar application rate at all sampling times (Fig. 5.3B and C). Bioavailable Cd and Zn concentrations

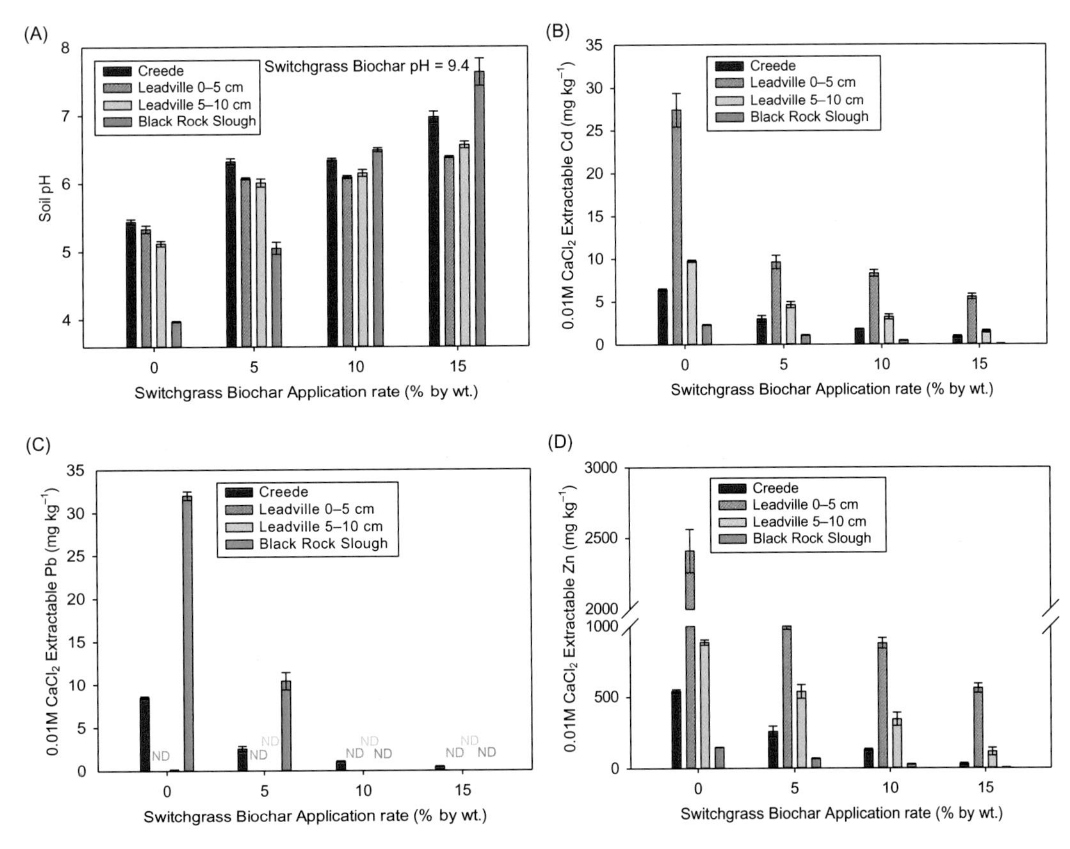

FIGURE 5.2 The effect of increasing switchgrass biochar rate on (A) pH and 0.01 M $CaCl_2$ extractable (B) Cd, (C) Pb, and (D) Zn concentrations in four mine-land soils. ND = nondetectable.

decreased by 53%–97% and 66%–98%, respectively, as compared to the control. Based on the BCR extraction procedure, Cui (unpublished) showed that Cd and Zn phases shifted toward carbonate and Fe/Mn oxyhydroxide phases. As shown and suggested by others (e.g., Ippolito et al., 2017; Ahmad et al., 2016; Park et al., 2011), these phases were likely responsible for the reduction in heavy metal bioavailability. Furthermore, these findings support the contention that biochars have the potential to be used to improve heavy metal-contaminated soils.

5.1.7 Recent Case Study—Designing Biochar Production and Use for Mine-Spoil Remediation

The Formosa mine site was established in Southwest Oregon, USA for silver, gold, and copper extraction from sulfide-containing rock deposits of the Rouge Formation. Mine spoils were discarded on the surface and ensuing oxidation of the sulfide ores caused their extreme acidification (Fig. 5.4A; $pH_{H2O} = 2.44$). Additionally, the spoils contained

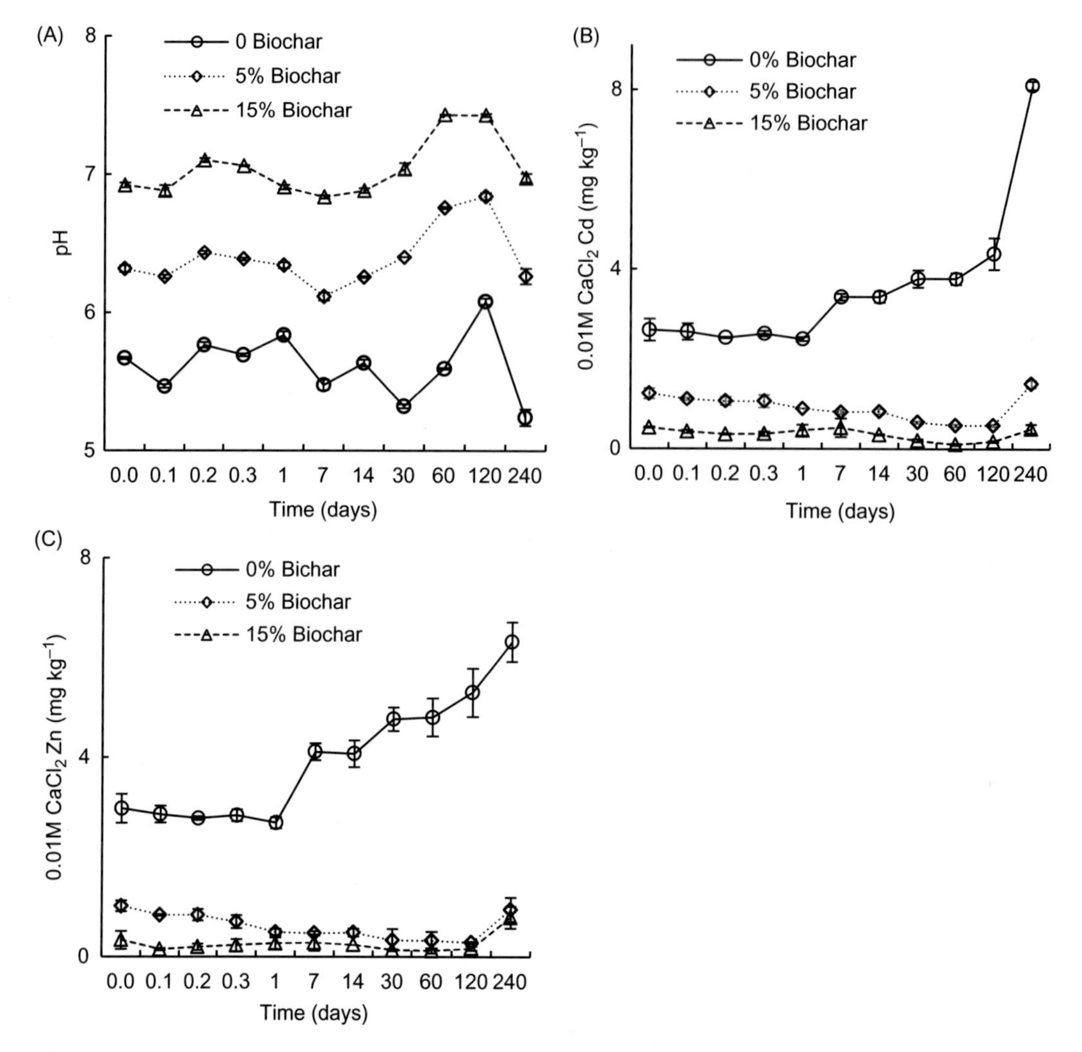

FIGURE 5.3 The effect of increasing wheat straw biochar application rate (0%, 5%, and 15% by weight) on Cd- and Zn-contaminated soil (A) pH and 0.01 M $CaCl_2$ extractable (i.e., bioavailable) (B) Cd and (C) Zn concentrations over a 240 day shaking/incubation period.

unextracted heavy metals such as Cu and Zn. Mine-spoil metal analysis revealed that the total, salt, and Mehlich-3 extractable Cu and Zn concentrations ranged between 609–651, 64–78, and 61–68 mg kg^{-1}, respectively. These spoil conditions have limited site revegetation and have probably stressed soil health characteristics (i.e., soil enzymatic and microbial activity). Thus, a reclamation plan should involve neutralizing the acid conditions, followed by reducing plant-available Cu and Zn concentrations. Additionally, restoring mine-spoil health characteristics could be facilitated by adding biochar, compost, and/or manure that increases spoil organic carbon (OC) content and enhances microbial growth conditions.

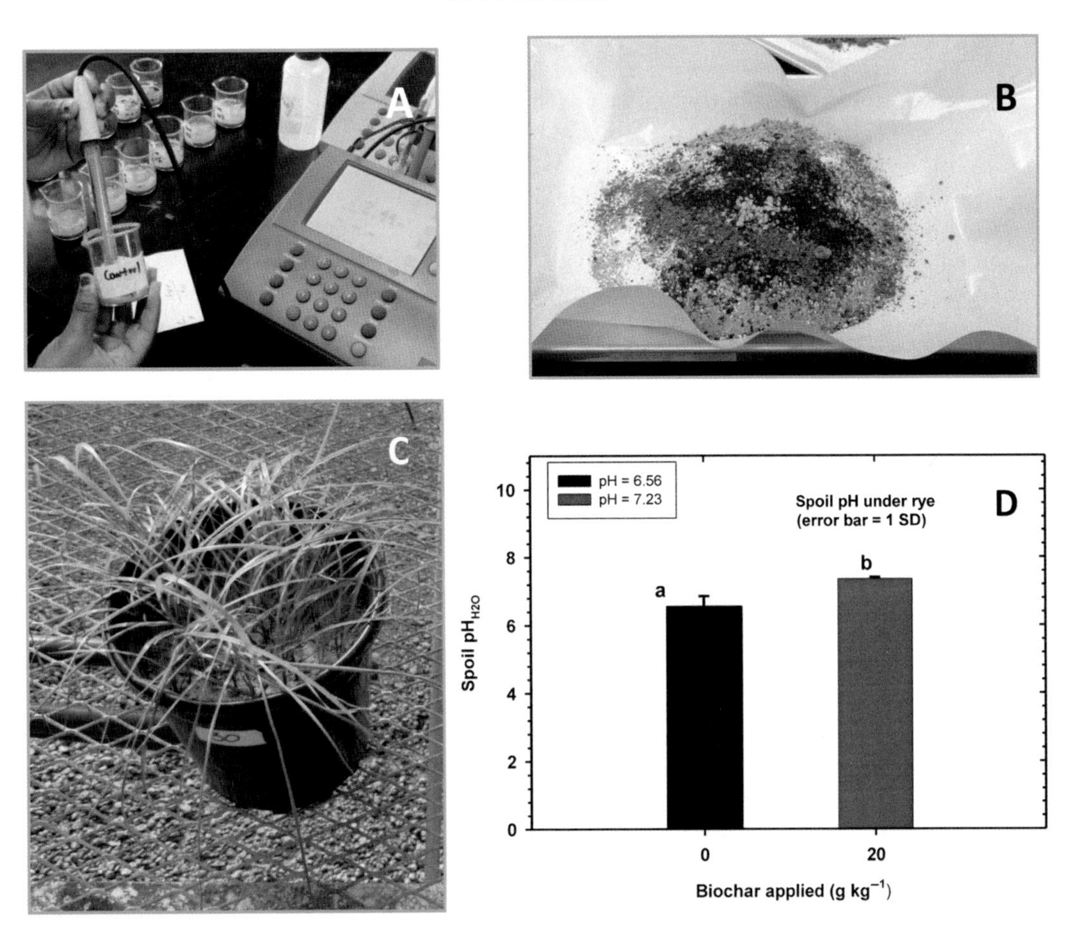

FIGURE 5.4 (A) pH of the Formosa mine spoil (located in southwest Oregon, USA). (B) Mixing 2% dairy manure biochar, 1% lime, and 0.5% poultry litter into the Formosa mine spoil. (C) Ryegrass grown in the amended Formosa mine spoil. (D) Average mine spoil pH values after the 100 day ryegrass experiment (the blue line in Fig. 5.4D signifies the upper pH threshold for cereal rye production). *Photos courtesy of Jeff Novak.*

A designer biochar was selected for addition to the Formosa mine spoil to assist with neutralizing the acidic pH and sequester both Cu and Zn concentrations. A manure feedstock and high pyrolysis temperature was selected since biochars produced from manures, at high temperature, are aklaline (Novak et al., 2009, 2012) and have an ash composition that can bind metals (Ehsan et al., 2014; Ippolito et al., 2017). Here, a dairy manure feedstock gasified at 800°C produced a designer biochar with a high ash content 52% (w/w) and an alkaline pH (10.2 in H_2O). These biochar characteristics have been implicated as key biochar traits for metal sequestration and as a lime source (Lehmann and Joseph, 2015).

The dairy manure biochar's ability to remediate the mine spoil was tested in a greenhouse experiment growing cereal rye (*Secale cereale*) as a test crop. Dairy manure biochar at 0 and 20 g kg^{-1} along with 10 g kg^{-1} of lime and 5 g kg^{-1} of fresh poultry litter manure were mixed into pots containing the Formosa mine spoil (Fig. 5.4B). The lime was added in case the acid-neutralization capability of this biochar was insufficient to effectively neutralize the Formosa mine-spoil acidic potential. Additionally, fresh poultry manure was added as an OC stimulant for microbial activity. Deionized H_2O was added to each mixture to bring the spoil moisture content to 10% (w/w). Each mine spoil was treated with 100 kg N ha^{-1} as a supplemental N source. Plant phosphorus (P) and potassium (K) nutritional requirements were assumed to be supplied by the dairy manure biochar because it contained 57 and 13 g kg^{-1} (by weight), respectively.

Rye was allowed to grow for 100 days (Fig. 5.4C). On the 72, 84, and 99 days of incubation, the rye was manually trimmed using scissors to a 6 cm height above the mine-spoil surface. The rye clippings were placed in beakers, oven-dried overnight at 60°C, and then weighed. Rye stubble (plant part below 6-cm trim height) was removed on day 100, dried and weighted in a similar fashion. Rye roots were separated from mine spoils by washing with deionized H_2O. Roots were placed in a beaker, dried in a similar manner and weighed. All rye material (cuttings, stubble, and roots) was later digested in concentrated HNO_3 and 30% H_2O_2. The digestate was filtered, and analyzed for Cu and Zn using ICP. A subsample of the mine spoil was collected for pH measurement in H_2O.

The greenhouse incubation revealed good rye growth after treatment without biochar (e.g., with lime alone; 0 g/kg biochar treatment) and with 20 g/kg biochar (Table 5.1). However, only the first plant cutting mean rye aboveground biomass yield was significantly different when compared to the control (Table 5.1). Although ryegrass grew well in the control (lime alone, 0 biochar), the benefits of adding the biochar can be observed in terms of plant metal uptake. Rye Cu uptake was reduced only in the third cutting and the root material. In contrast, biochar significantly reduced rye Zn uptake during all cuttings (and in roots) except at the first cutting. These results illustrate that the lime and biochar were capable of reducing plant uptake of Cu and Zn, although the results were mixed.

Mine-spoil mean pH values measured at the experiment end are shown in Fig. 5.4D. The blue line on Fig. 5.4D represents a soil pH of ~6.2, which is adequate for cereal growth. In both treatments, the pH at the end of the experiment exceeds this optimum pH range. The control treatment has a pH of 6.56 (spoil + lime alone) while the spoil treated with lime and biochar had a pH of 7.23. Thus, the amount of lime used in this experiment could be slightly reduced to < 10 g kg^{-1}. The results do show that there was a significant difference in the mean pH value between the control (0% biochar) and treated mine spoil. Raising the mine spoil pH to > 7 suggests that lime amounts can be reduced when used with this calcareous dairy manure biochar. It is important that the amount of lime and designer biochar addition be determined in preliminary experiments to ensure that the spoil pH is in a range that does not reduce the uptake of critical plant nutrients such as P and trace metals (i.e., B, etc.). One may also wish to consider the utility of lime and biochar when determining and neutralizing acid-generating potential in mine spoils containing sulfide-bearing mineral phases.

These results suggest that lime was important for raising the mine-spoil pH and the dairy manure biochar showed promise in minimizing both Cu and Zn uptake by rye. It is speculated that the reduced plant uptake of Cu and Zn may be through sequestration by components in the biochar ash, or by metal sorption or precipitation reactions associated with pH shifts as mentioned previously.

TABLE 5.1 Mean Rye Aboveground Biomass, Plant Cu and Zn Uptake at Different Cutting Days, and Cu and Zn in Roots at End of Study (Standard Deviation in Parentheses; odw = Oven Dry Weight; Data Not Published)[a]

	Biochar Treatment (g kg^{-1})	
Aboveground Biomass at Day of Cutting:	0	20
	Oven Dry Weight (g)	
1st cutting (d 72)	0.62(0.07)[a]	0.73 (0.06)[b]
2nd cutting (d 84)	1.23 (0.13)[a]	1.24 (0.17)[a]
3rd cutting (d 99)	2.15 (0.05)[a]	2.05 (0.16)[a]
Stubble (d 100)	5.73 (0.16)[a]	5.75 (0.48)[a]
Cu uptake in:	0	20
	mg kg^{-1}	
1st cutting (d 72)	26.7 (2.0)[a]	25.4 (1.5)[a]
2nd cutting (d 84)	25.4 (6.7)[a]	28.3 (8.5)[a]
3rd cutting (d 99)	22.3 (8.4)[a]	12.8 (0.9)[b]
Stubble (d 100)	10.2 (1.0)[a]	10.2 (2.4)[a]
Roots	205 (44)[a]	164 (14)[b]
Zn uptake in:	0	20
	mg kg^{-1}	
1st cutting (d 72)	79.5 (7.1)[a]	77.6 (6.1)[a]
2nd cutting (d 84)	87.9 (3.2)[a]	72.7 (6.0)[b]
3rd cutting (d 99)	73.6 (3.6)[a]	42.8 (3.6)[b]
Stubble (d 100)	73.7 (4.9)[a]	53.9 (4.2)[b]
Roots	202 (45)[a]	154 (16)[b]

[a]*Means compared between columns followed by a different letter are significantly different using a* t-test *at* $P < 0.05$ *level of significance.*

5.2 CONCLUSIONS

Over half a million abandoned mines exist globally. Many of these abandoned mine sites are capable of generating acidity, increasing metal solubility, and subsequently degrading environmental quality. Based on the current literature, biochar has been proven to play a role in alleviating acidity and heavy metal contamination through a number of reactions. Biochar reactions include the material acting as a liming source and raising soil pH. This allows for bioavailable metal concentrations to be reduced via precipitation, oxide, hydroxide, carbonate, and organic phase interactions. Understanding the on-site, heavy metal contamination conditions the initial metal forms present and the proper biochar application required for metal transformations is of paramount importance. Furthermore, understanding the potential for biochar-induced chemical reactions to occur, or lack thereof, prior to biochar land application, may save time and money during mine-land reclamation. Several examples were presented whereby biochar use did not affect soil heavy metal reductions, suggesting that caution exists in terms of haphazardly using any type of biochar for remediation purposes.

References

Agency for Toxic Substances and Disease Registry, 2004. Public health statement for copper. Available at https://www.atsdr.cdc.gov/phs/phs.asp?id = 204&tid = 37 (accessed 15.09.17.).

Agency for Toxic Substances and Disease Registry, 2005. Public health statement: zinc. Available at https://www.atsdr.cdc.gov/ToxProfiles/tp60-c1-b.pdf (accessed 15.09.17.).

Ahmad, M., Lee, S.S., Lee, S.E., Al-Wabel, M.I., Tsang, D.C.W., Ok, Y.S., 2016. Biochar-induced changes in soil properties affected by immobilization/mobilization of metals/metalloids in contaminated soils. J. Soils Sediments 17, 717–730.

Aslam, Z., Khalid, M., Naveed, M., Shahid, M., Aon, M., 2017. Evaluation of green waste and popular twigs biochar produced at low and high pyrolytic temperature for efficient removal of metals from water. Water Air Soil Pollut. 228, 432.

Aspen Center for Environmental Studies, 2011. Hope Mine Biochar Project. July 2010–August 2011, Aspen, Colorado, USA. Available at: https://www.aspennature.org/restore/forest-ecosystem-health/hope-mine-biochar-project (accessed 15.09.17.).

Batool, S., Idrees, M., Hussain, Q., Kong, J., 2017. Adsorption of copper (II) by using derived-farmyard and poultry manure biochars: efficiency and mechanism. Chem. Phys. Lett. 689, 190–198.

Bian, R., Chen, D., Liu, X., Cui, L., Li, L., Pan, G., et al., 2013. Biochar soil amendment as a solution to prevent Cd-tainted rice from China: results from a cross-site field experiment. Ecol. Eng. 58, 378–383.

BLM, 2016. Abandoned mine lands portal: extent of the problem. Available at: http://www.abandonedmines.gov/extent_of_the_problem (accessed 15.09.17.).

Cárdenas-Aguiar, E., Gascó, G., Paz-Ferreiro, J., Méndez, A., 2017. The effect of biochar and compost from urban organic waste on plant biomass and properties of an artificially copper polluted soil. Int. Biodeteri. Biodegrad. 124, 223–232.

Centers for Disease Control and Prevention, 2017. Health problems caused by lead. Available at https://www.cdc.gov/niosh/topics/lead/health.html (accessed 15.09.17.).

Coetzee, H., Hobbs, P.J., Burgess, J.E., Thomas, A., Keet, M., 2010. Mine water management in the Witwatersrand gold fields with special emphasis on acid mine drainage. Report to the Inter-Ministerial Committee on Acid Mine Drainage. Available at: https://www.gov.za/sites/default/files/ACIDReport_0.pdf (accessed 15.09.17.).

Cui, L., Li, L., Zhang, A., Pan, G., Bao, D., Chang, A., 2011. Biochar amendment greatly reduces rice Cd uptake in a contaminated paddy soil: a two-year field experiment. Bioresources 6, 2605–2618.

Deng, J., Liu, Y., Liu, S., Zeng, G., Tan, X., Huang, B., et al., 2017. Competitive adsorption of Pb(II), Cd(II) and Cu (II) onto chitosan-pyromellitic dianhydride modified biochar. J. Colloid Interface Sci. 506, 355–364.

Ehsan, M., Barakat, M.A., Husein, D.Z., Ismail, S.M., 2014. Immobilization of Ni and Cd in soil by biochar derived from unfertilized dates. Water Air Soil Pollut. 225, 2123–2133.

Fellet, G., Marchiol, L., Delle Vedove, G., Peressotti, A., 2011. Application of biochar on mine tailings: effects and perspectives for land reclamation. Chemosphere 83, 1262–1267.

Hanauer, T., Jung, S., Felix-Henningsen, P., Schnell, S., Steffens, D., 2012. Suitability of inorganic and organic amendments for in situ immobilization of Cd, Cu, and Zn in a strongly contaminated Kastanozem of the Mashavera valley, SE Georgia. I. Effect of amendments on metal mobility and microbial activity in soil. J. Plant Nutr. Soil Sci. 175, 708–720.

He, J., Song, Y., Chen, P., 2017. Development of a novel biochar/PSF mixed matrix membrane and study of key parameters in treatment of copper and lead contaminated water. Chemosphere 186, 1033–1045.

Hu, Y., Cheng, H., Tao, S., 2016. The challenges and solutions for cadmium-contaminated rice in China: a critical review. Environ. Int. 92–93, 515–532.

International Commission on Mine Closure, and International Society for Rock Mechanics, 2008. Mine closure and post-mining management international state-of-the-art. https://doi.org/10.13140/2.1.3267.8407.

Ippolito, J.A., Berry, C.M., Strawn, D.G., Novak, J.M., Levine, J., Harley, A., 2017. Biochars reduce mine land soil bioavailable metals. J. Environ. Qual. 46, 411–419.

Ippolito, J.A., Ducey, T.F., Tarkalson, D.D., 2011. Interactive effects of copper on alfalfa growth, soil copper, and soil bacteria. J. Agric. Sci. 3, 138–148.

Ippolito, J.A., Strawn, D.G., Scheckel, K.G., Novak, J.M., Ahmedna, M., Niandou, M.A.S., 2012. Macroscopic and molecular investigations of copper sorption by a steam-activated biochar. J. Environ. Qual. 41, 1150–1156.

Italy Europe 24, 2015. Mines to museums: 3,000 abandoned mining sites await a new frontier in Italian tourism. Available at: http://www.italy24. ilsole24ore.com/art/arts-and-leisure/2015-10-05/mines-to-museums-3000-abandoned-mining-sites-await-new-frontier-italian-tourism-165011.php?uuid = ACCA9OAB (accessed 15.09.17.).

Jain, S.A., Singh, P., Khare, D., Chanda, D., Mishra, K., Shanker, T.K., 2017. Toxicity assessment of *Bacopa monnieri* L. grown in biochar amended extremely acidic coal mine spoils. Ecol. Eng. 108, 211–219.

Karunanayake, A.G., Todd, O.A., Crowley, M., Ricchetti, L., Pittman Jr., C.U., et al., 2018. Lead and cadmium remediation using magnetized and nonmagnitized biochar from douglas fir. Chem. Eng. J. 331, 480–491.

Kelly, C.N., Peltz, C.D., Stanton, M., Rutherford, D.W., Rostad, C.E., 2014. Biochar application to hardrock mine tailings: soil quality, microbial activity, and toxic element sorption. Appl. Geochem. 43, 35–48.

Kim, M.S., Min, H.G., Koo, N., Park, J., Lee, S.H., Bak, G.I., et al., 2014. The effectiveness of spent coffee grounds and its biochar on the amelioration of heavy metals-contaminated water and soil using chemical and biological assessments. J. Environ. Manage. 146, 124–130.

Kumar, A., Tsechansky, L., Lew, B., Raveh, E., Frenkel, O., Graber, E.R., 2018. Biochar alleviates phytotoxicity in Ficus elastic grown in Zn-contaminated soil. Sci. Total Environ. 618, 188–198.

Lahori, A.H., Zhang, Z., Guo, Z., Li, R., Mahar, A., Awasthi, M.K., et al., 2017. Beneficial effects of tobacco biochar combined with mineral additives on (im)mobilization and (bio)availability of Pb, Cd, Cu and Zn from Pb/Zn smelter contaminated soils. Ecotoxicol. Environ. Saf. 145, 528–538.

Lee, X.J., Lee, L.Y., Hiew, B.Y.Z., Gan, S., Thangalazhy-Gopakumar, S., Ng, H.K., 2017. Multistage optimizations of slow pyrolysis synthesis of biochar from palm oil sludge for adsorption of lead. Bioresour. Technol. 245, 944–953.

Lehmann, J., Joseph, S., 2015. Biochar for environmental management: an introduction. In: Lehmann, J., Joseph, S. (Eds.), Biochar for Environmental Management: Science, Technology and Implementation, second ed. Earthscan, Routledge Publ., London, UK, pp. 1–13.

Mackasey, W.O., 2000. Abandoned mines in Canada. Available at: https://miningwatch.ca/sites/default/files/mackasey_abandoned_mines.pdf (accessed 15.09.17.).

Meier, S., Curaqueo, G., Khan, N., Bolan, N., Rillig, J., Vidal, C., et al., 2017. Effects of biochar on copper immobilization and soil microbial communities in a metal-contaminated soil. J. Soil Sediments 17, 1237–1250.

Mining Technology, 2015. Managing Australia's 50,000 abandoned mines. Available at: http://www.mining-technology.com/features/featuremanaging-australias-50000-abandoned-mines-4545378/ (accessed 15.09.17.).

Mittal, A.K., 2011. Abandoned Mines: Information on the number of hardrock mines, cost of clean up, and value of financial assurances (Testimony before the subcommittee on energy and mineral resources, Committee on Natural Resources, House of Representatives No. GAO-11-834T). United States Accountability Office. Available at: http://www.gao.gov/new.items/d11834t.pdf (verified 15 November, 2017).

Moreno-Barriga, F., Faz, A., Acosta, J.A., Soriano-Disla, M., Martínez-Martínez, S., Zornoza, R., 2017. Use of Piptatherum miliaceum for the phytomanagement of biochar amended Technosols derived from pyritic tailings to enhance soil aggregation and reduce metal(loid) mobility. Geoderma 307, 159–171.

Novak, J.M., Busscher, W.J., Laird, D.L., Ahmedna, M., Watts, D.W., Niandou, M.A.S., 2009. Impact of biochar on fertility of a southeastern coastal plain soil. Soil Sci. 174, 105–112.

Novak, J.M., Cantrell, K.B., Watts, D.W., 2012. Compositional and thermal evaluations of lignocellulosic and poultry litter chars via high and low temperature pyrolysis. BioEnergy Res 6, 114–130.

Ouyang, W., Huang, W.J., Hao, X., Tysklind, M., Haglund, P., Hao, F.H., 2017. Watershed soil Cd loss after long-term agricultural practice and biochar amendment under four rainfall levels. Water Res. 122, 692–700.

Park, J.H., Choppala, G.K., Bolan, N.S., Chung, J.W., Chuasavathi, T., 2011. Biochar reduces the bioavailability and phytotoxicity of heavy metals. Plant Soil 348, 439–451.

Peltz, C.D., Harley, A., 2016. Biochar application for abandoned mine land reclamation. In Gao, M., Z. He, and S.M. Uchimiya (Eds.), Agricultural and Environmental Applications of Biochar: Advances and Barriers. pp. 325–339.

Plum, L.M., Rink, L., Haase, H., 2010. The essential toxin: impact of zinc on human health. Int. J. Environ. Res. Public Health 7, 1342–1365.

Qi, F., Lamb, D., Naidu, R., Bolan, N.S., Yan, Y., Ok, Y.S., et al., 2018. Cadmium solubility and bioavailability in soils amended with acidic and neutral biochar. Sci. Total Environ. 610–611, 1457–1466.

Qian, L., Chen, B., Han, L., Yan, J., Zhang, W., Su, A., et al., 2017. Effect of culturing temperatures on cadmium phytotoxicity alleviation by biochar. Envrion. Sci. Pollut. Res. 24, 23843–23849.

Ramontja, T., Eberle, D., Coetzee, H., Schwarz, R., Juch, A., 2011. Critical challenges of acid mine drainage in South Africa's Witwatersrand gold mines and Mpumalanga coal fields and possible research areas for collaboration between South African and German researchers and expert teams. In: Merkel, B., Schipek, M. (Eds.), The New Uranium Mining Boom. Springer Geology. Springer, Berlin, Heidelberg, Germany.

Reverchon, F., Yang, H., Ho, T.Y., Yan, G., Wang, J., Xu, Z., et al., 2015. A preliminary assessment of the potential of using an acacia-biochar system for spent mine site rehabilitation. Environ. Sci. Pollut. Res. 22, 2138–2144.

Rodríguez-Vila, A., Covelo, E.F., Forján, R., Asensio, V., 2015. Recovering a copper mine soil using organic amendments and phytomanagement with *Brassica juncea* L. J. Environ. Manage. 147, 73–80.

Rodríguez-Vila, A., Forján, R., Guedes, R.S., Covelo, E.F., 2017. Nutrient phytoavailability in a mine soil amended with technosol and biochar and vegetated with *Brassica juncea*. J. Soils Sediments 17, 1653–1661.

Sekulić, M.T., Pap, S., Stojanović, Z., Radonić, J., Knudsen, T.S., 2018. Efficient removal of priority, hazardous priority and emerging pollutants with Prunuc armeniaca functionalized biochar from aqueous wastes: experimental optimization and modeling. Sci. Total Environm. 613–614, 736–750.

Shen, Y., Li, H., Zhu, W., Ho, S.-H., Yuan, W., Chen, J., et al., 2017a. Microalgal-biochar immobilized complex: a novel efficient biosorbent for cadmium removal from aqueous solution. Bioresour. Technol. 244, 1031–1038.

Shen, Z., Zhang, Y., Jin, F., McMillan, O., Al-Tabbaa, A., 2017b. Qualitative and quantitative characterization of adsorption mechanisms of lead on four biochars. Sci. Total Environ. 609, 1401–1410.

SRK Consulting, 2017. The United Kingdom has thousands of abandoned metal mines. Available at: http://www.srk.co.uk/en/newsletter/focus-waste-geochemistry/united-kingdom-has-thousands-abandoned-metal-mines (accessed 15.09.17.).

Tran, H.N., Lee, C.K., Vu, M.T., Chao, H.P., 2017. Removal of copper, lead, methylene green 5, and acid red 1 by saccharide-derived spherical biochar prepared at low calcination temperatures: adsorption kinetics, isotherms, and thermodynamics. Water Air Soil Pollut. 228, 401.

Ure, A.M., Quevauviller, P., Muntau, H., Griepink, B., 1993. Speciation of heavy metals in soils and sediments. Int. J. Environ. Anal. Chem. 51, 135–151.

U.S. Department of Labor – Occupational Safety and Health Administration, 1993. Occupational exposure to cadmium, Section 5 – V. Health Effects. Available at: https://www.osha.gov/pls/oshaweb/owadisp.show_document?p_table = PREAMBLES&p_id = 819 (accessed 15.09.17.).

U.S. EPA, 2017a. Abandoned mine lands. Available at https://www.epa.gov/superfund/abandoned-mine-lands (accessed 15.09.17.).

World Health Organization, 2017. Lead poisoning and health. Available at http://www.who.int/mediacentre/factsheets/fs379/en/ (accessed 15.09.17.).

Xu, M.H., Xia, J., Wu, G., Yang, X., Zhang, H., Peng, X., et al., 2017. Shifts in the relative abundance of bacteria after wine-lees-derived biochar intervention in multi metal-contaminated paddy soil. Sci. Total Environ. 599–600, 1297–1307.

Yin, D., Wang, X., Peng, B., Tan, C., Ma, L.Q., 2017. Effect of biochar and Fe-biochar on Cd and As mobility and transfer in soil-rice system. Chemosphere 186, 928–937.

Zhou, Y., Liu, X., Xiang, Y., Wang, P., Zhang, J., Zhang, F., et al., 2017. Modification of biochar derived from sawdust and its application in removal of tetracycline and copper from aqueous solution: adsorption mechanism and modelling. Bioresour. Technol. 245, 266–273.

Further Reading

U.S. EPA, 2017b. Learn about lead. Available at https://www.epa.gov/lead/learn-about-lead (accessed 15.09.17.).

CHAPTER

6

Potential of Biochar for Managing Metal Contaminated Areas, in Synergy With Phytomanagement or Other Management Options

Filip M.G. Tack and Caleb E. Egene

Department of Green Chemistry and Technology, Ghent University, Gent, Belgium

6.1 INTRODUCTION

Due to human activities including agriculture and industry, extended areas of land in many places of the world have been subjected to increased input of potentially toxic elements through diffuse contamination. Typical examples are metal smelter-affected areas, where decades of untreated exhaust fumes have led to a significant input of metals and to soil metal concentrations that are significantly increased over normal baseline concentrations. Currently, diffuse contamination is ongoing in the context of the arsenic crisis in West Bengal and Bangladesh. Groundwater, in many cases naturally containing elevated concentrations of arsenic, is used to irrigate agricultural land, including rice fields. This causes a gradual increase in soil arsenic concentrations over significant areas (Khan et al., 2010). Soils may also naturally contain elevated concentrations of trace metals, e.g., nickel in serpentine soils (Brady et al., 2005), or the white limestones in Jamaica with cadmium concentrations as high as 400 mg kg^{-1} (Lalor, 2008; Lalor et al., 1998).

All these cases of diffuse contamination are characterized by the presence of trace metals and metalloids in excess of normal baseline concentrations, but typically still at relatively moderate levels. These levels are not of an extent to cause acute issues such as significant impairment of plant growth or development of biota. These levels, however, cause

Biochar from Biomass and Waste
DOI: https://doi.org/10.1016/B978-0-12-811729-3.00006-6

ecosystem stresses, have the potential to affect the development of plants and biota, and ultimately lead to an overall increased exposure of humans to the elements concerned. While many trace elements evidently are toxic at high concentrations, the chronic hazardous effects resulting from long-term exposure of elements such as cadmium and arsenic have become very clear through incidents with worldwide visibility, such as the occurrence of itae-itae in Japan related to chronic cadmium exposure in the 1950s (Nogawa, 1981), and the recent arsenic crisis in Bangladesh and West Bengal (Alam et al., 2002; Chowdhury, 2004; Tondel et al., 1999).

Metals at moderate contamination levels are of significant concern, despite the apparent absence of visible immediate deleterious effects. Because the contamination is moderate but extended over significant areas, effective cleanup using conventional engineering techniques is not an option because of the excessive costs associated with removing relatively low metal concentration from very large volumes of soil. This human-caused contamination, therefore, is to be considered as irreversible damage to these soils and constitutes an irreversible loss of clean land. The only option apart from taking these lands out of use is to continue using them, but in a risk-based approach, i.e., ensuring that their use does not entail a risk to the general population or other negative effects. As such, there is a need to more actively manage such contaminated lands. This is conceptualized in the idea of risk-based land management (RBLM) (Ferguson et al., 1998).

Metals and metalloids, as chemical elements, are bound to remain in the soil or move to other compartments. They cannot be destroyed nor be broken down. Therefore, management options of metal-contaminated soil will focus on either reducing the potential mobility of the metals present in the soil, or on controlling the flow of metals, e.g., deliberately enhancing the mobility toward produced and harvested biomass to promote a gradual decline of total and potentially mobile/bioavailable metal contents in the soil (phytoextraction), or aiming for immobilization to reduce metal leaching and uptake in crops or for an overall reduction in the cycling of these elements in ecosystems developing on these soils (phytostabilization) (Ferguson et al., 1998; Naidu et al., 2015).

Biochar has been promoted as a soil additive to mitigate climate change through carbon sequestration, while simultaneously improving soil fertility and enhancing crop production (Kookana et al., 2011; Lehmann and Joseph, 2009). In addition, biochar has a number of properties that may render it suitable for improving lands that contain levels of metals and metalloids above baseline concentrations, and in particular, above safe levels, when applied to land. First, it constitutes a very stable fraction of organic matter, with stability of several hundred to thousand years as opposed to natural soil organic matter that may mineralize within years (Lehmann et al., 2009). As such, application of biochar contributes to the overall beneficial effects of organic matter in soils for many years. The material typically is characterized by an elevated pH, and thus has liming properties (Ahmad et al., 2014), which is generally favorable to decrease the mobility of trace metals in soils. In relation to its highly porous structure and presence of functional groups, biochar exhibits different mechanisms of interaction with inorganic contaminants, which, depending on the element and properties of the soil and biochar, may result in either immobilization or increased solubility of different elements.

Although it appears promising to apply biochar as a soil amendment with long-term stability for managing moderately metal-contaminated soils, large-scale application is currently not yet ongoing (O'Connor et al., 2018). This is due, in part, to the idea being relatively new. In addition, biochar currently is not yet sufficiently available and affordable for the large-scale use required for this type of application. Effects of biochar on soil and, in particular, its interactions with metals are intensely being explored as evidenced by the vast literature that is emerging.

6.2 METALS AND METALLOIDS IN SOIL

Many metals and metalloids are natural components of any soil. Part of the trace element contents in the soil originates from the parent materials out of which the soil was formed. The geochemical background content of soils is determined by the natural processes of weathering and soil formation, natural aerial inputs of volcanic and other dusts, and sedimentation (Matschullat et al., 2000; Reimann and Garrett, 2005). As mankind significantly influenced the natural biogeochemical cycling of metals (Nriagu, 1979; Nriagu and Pacyna, 1988), current ambient concentrations of trace elements, especially in densely populated areas, may partly have been determined by anthropogenic activities (Kabata-Pendias et al., 1992).

Anthropogenic sources of trace metals and metalloids in soils include motor vehicle exhaust gas emissions, disposal of wastes, and industrial emissions. Use of agrochemicals such as fertilizers and pesticides, and soil amendments such as lime, composts, biosolids, sewage sludge, and livestock manure, can be a significant source of metal input into agricultural soils. Other potential sources of contamination include irrigation with contaminated water, or contamination from nearby industries through surface runoff, and atmospheric deposition (Adriano, 2001).

Trace elements in soils are distributed over different soil compartments (Fig. 6.1). Complex interactions govern the relations between metals in the solid phase and the soil solution, strongly depending on the specific chemical forms by which trace metals and metalloids are incorporated into the solid phase (Tack, 2010). These include adsorbed, complexed, or occluded pools. Trace elements sorbed or complexed to surfaces of soil components including clay minerals, hydrated oxides of iron and manganese, or organic matter can easily be released into the soil solution and accordingly will be part of the most bioavailable fractions of the elements present in the soil. When occluded within soil solids, they are, at least temporarily immobilized unless the phase concerned is dissolved or destroyed, which may be a matter of weeks, months, or a few years. Trace elements structurally incorporated in soil minerals are not likely to become available as they are only released upon weathering of the minerals, which typically is a very slow process where significant changes are only noticed over an extended period of time (Augusto et al., 2000; Bain and Langan, 1995).

Metals and metalloids that have accumulated in a soil overall tend to be rather strongly sorbed by the soil components. Cadmium, for example, might typically be present in soils at concentration levels up to 0.5 mg kg^{-1}, whereas cadmium levels 0.01 up to 5 μg represent typical soil-solution concentrations. Accordingly, depending on actual concentrations in solid and liquid phase, moisture content and bulk density, soluble cadmium typically

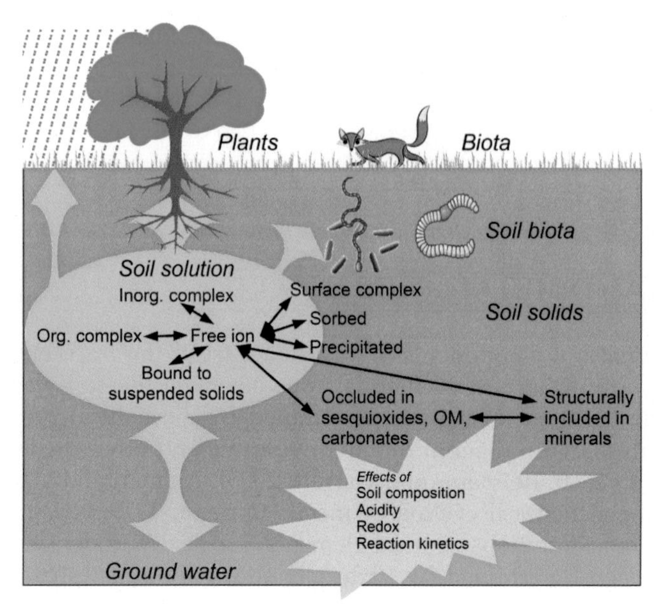

FIGURE 6.1 Conceptual representation of various pools of trace elements in the soil. Source: *Drawn from Cottenie, A., Verloo, M., 1984. Analytical diagnosis of soil pollution with heavy metals. Fresenius. J. Anal. Chem. 317, 389–393; Bernhard, M., Brinckman, F.E., Sadler, P.J., 1986. The Importance of Chemical "Speciation" in Environmental Processes. Springer-Verlag, Berlin; and Tack, F.M.G., 2010. Trace elements: general soil chemistry, principles and processes, In: Trace Elements in Soils. Wiley-Blackwell, Chichester, UK, pp. 9–37. This figure is licensed by F.M.G. Tack under a Creative Commons Attribution 4.0 International License.*

represents much less than 1/1000 of the total cadmium in the solid phase. Still, these tiny fractions in solution can be of significant environmental concern (Tack, 2010).

A limited number of trace elements are essential for biota to complete their lifecycle. Any metal or metalloid, however, whether essential or not, may become toxic to biota if certain levels of exposure are exceeded. Trace metals and metalloid exhibit acute toxicity when an organism is exposed to a high dose. Very commonly, however, hazardous effects of long-term exposure to relatively low concentrations will manifest only slowly after a long period. Accordingly, issues from moderate soil contamination with metals and metalloids may not easily be recognized and therefore are typically slow in being picked up by regulation and policy.

6.3 BIOCHAR AS A SOIL AMENDMENT FOR RISK-BASED LAND MANAGEMENT

Available methods for in situ or ex situ soil remediation can be technically challenging and complex, and may be prohibitively expensive. Accordingly, they generally are not applied widely in many countries (Naidu et al., 2015) or are limited to hotspots with highly significant contamination. Whereas most legislation with respect to soil contamination is still largely based on the concept of reducing total contaminant levels, two shifts in policy have characterized a more pragmatic approach to the management of contaminated soils: (1) a recognition that remediation of legacy contaminated soils should be based on risk management, and (2) a recognition that risk is determined by the dose-response relationship for each chemical compound (Naidu et al., 2015). RBLM strives for a beneficial and sustainable use of contaminated land accounting for the current and proposed land

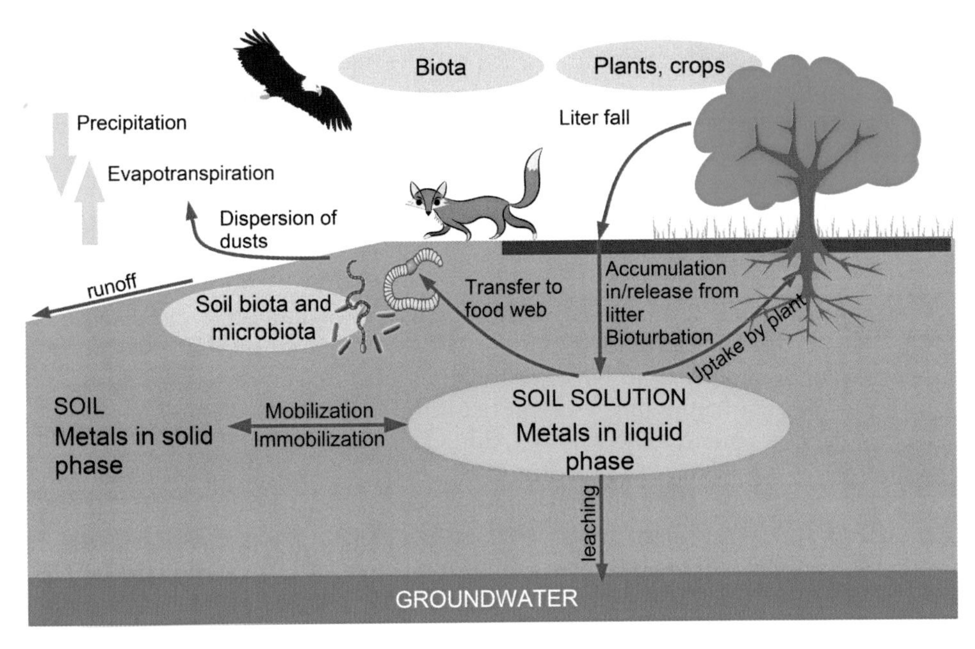

FIGURE 6.2 Conceptual migration pathways of metal from a contaminated substrate into the ecosystem. Source: *This figure is licensed by F.M.G. Tack under a Creative Commons Attribution 4.0 International License.*

use, expectations of the community and government, and available resources (Ferguson et al., 1998; Naidu et al., 2015).

Metals and metalloids present in the contaminated soil can be subject to a range of processes of solubilization, mobilization, and transport, and can move to different ecosystem compartments (Fig. 6.2). Concentrations in the soil solution and the intensity by which these soluble concentrations are maintained by buffering mechanisms of the solid phase determine the extent by which potentially toxic metals and metalloids may be mobilized, be transported to other ecosystem compartments or areas, and become bioavailable towards biota. "Bioavailability" is a term that refers to how much of a compound is available to a living organism (Naidu et al., 2011). It describes the extent and rate of absorption of a xenobiotic from the exposure site into an organism (Chen et al., 1995). Uptake of metals and metalloids by plants and biota directly living in the soil constitutes first steps in the transfer of these elements into the food web. Trace elements in vegetation may return to the soil through the shedding of leaves in the case of trees, and upon death and decay of the plant. Especially on sloped and barren surfaces, runoff, erosion, and movement of sediment from the soil can be a major transport mechanism for trace elements from fields (Korentajer et al., 1993). Metals may also be removed by leaching, where metals move along with the soil solution. RBLM aims to control the potential hazard of metals and metalloids in contaminated land either through reducing their mobilization and transport, or by steering and controlling the pathways and their associated risk, e.g., in phytoextraction.

Biochar as a soil amendment is distinguished from other carbonaceous products such as activated carbon and charcoal because it provides numerous agricultural and environmental benefits. Verheijen et al. (2010) captured these benefits in their definition of biochar as "biomass that has been pyrolyzed in a zero or low oxygen environment, applied to soil at specific site that is expected to sustainably sequester carbon and concurrently improve soil functions under current and future management, while avoiding short- and long-term detrimental effects to the wider environment as well as human and animal health." In recent years, biochar has gained important recognition for its multifunctionality that includes carbon sequestration and soil fertility enhancement (Laird et al., 2010), bioenergy production (Field et al., 2013), and environmental remediation (Oliveira et al., 2017). Several recent studies have provided evidence of the ability of biochar to remove organic and inorganic contaminants from water (Ahmad et al., 2014; Inyang and Dickenson, 2015), to reduce transfer of contaminants from agricultural soil to crops (Xu et al., 2012), and to stabilize and restore contaminated soils (Ahmad et al., 2014; Beesley et al., 2011; Fellet et al., 2011; O'Connor et al., 2018; Wang et al., 2018; Zhang et al., 2013) and sediments (Wang et al., 2018).

As biochar is a relatively new product, few commercial pyrolysis facilities exist that can supply it. Therefore, the price and availability of biochar varies widely. For the UK, a cost between USD 222 and 584 per tonne was assessed for biochar produced, delivered, and spread on fields (Shackley et al., 2011). A 2014 report by the International Biochar Initiative (Jirka and Tomlinson, 2014) indicated a global average price for pure biochar of USD 2650 USD per tonne, with on average a range from USD 90 in the Philippines to USD 5060 in the UK. The authors noted that in the compilation of these prices, it was not always clear whether the prices quoted were retail or wholesale/producer prices. In 2016, between USD 51 and 771 per tonne was mentioned as a cost for the production of biochar, with an outlier of USD 5668 reported for the UK (Ahmed et al., 2016). Granatstein et al. (2009) anticipated that future prices could go as low as USD 200 per tonne when the energy produced during pyrolysis is utilized and renewable energy credits are included. This is still significantly more than the price for agricultural lime (USD 25–50 per tonne), for example, or compost (USD 20–50 per tonne) (Fornes et al., 2015). However, the long-term effectiveness of biochars may compensate for the higher price compared to conventional soil amendments.

It can be expected that availability and applications of biochar will increase, as it will take a significant place in the bio-based economy (Vanholme et al., 2013). The bio-based economy is moving away from relying on limited fossil resources toward using renewable biomass for obtaining commodity chemical, renewable fuel, and energy resources (Ronsse et al., 2014; Vanholme et al., 2013) (Fig. 6.3). Pyrolysis may become a key process in the conversion of biomass to bio-oil and char. Bio-oil has the potential to substitute petroleum for the production of commodity chemicals including ethylene, benzene, methanol, styrene, and ethylene oxide, among others, which in turn are the starting materials used to manufacture polymers and high-value specialty chemicals, including adhesives, agrochemicals, cleaning materials, and elastomers. There is ongoing research into the production of transportation fuels from these pyrolysis oils that can replace conventional petroleum-based fuels.

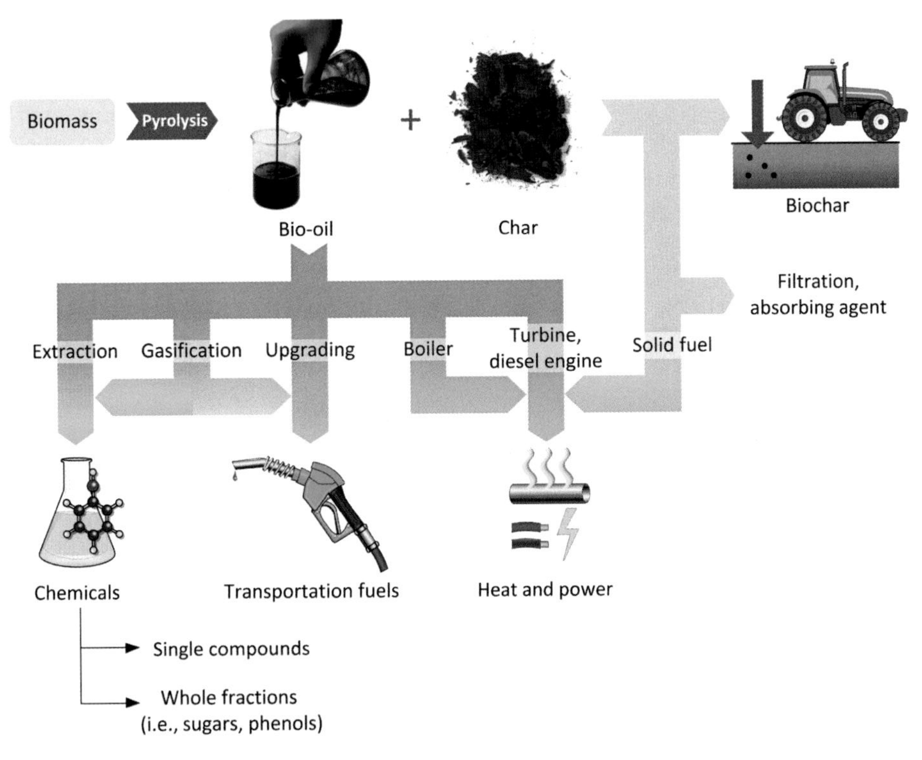

FIGURE 6.3 Biochar as a product in thermochemical approaches to convert biomass to renewable fuel, chemical, and energy resources (Ronsse et al., 2014). Source: *Reproduced with permission from the author.*

6.4 PROPERTIES OF BIOCHAR IN RELATION TO TRACE ELEMENT SORPTION

The potency of a given biochar for metal sorption depends on a combination of conditions including biochar source material, pyrolysis temperature, surface area, polarity, sorption temperature, and target contaminant (Ahmad et al., 2014). Generally, the goal would be to be able to design biochars that are tailored to remediate a specific soil issue (Novak et al., 2009). Two parameters that have received the most attention by researchers due to their significant influence on the metal-sorption capacity of biochar are the feedstock used and the pyrolysis temperature.

The type of feedstock used has a major influence on the physical and chemical properties of the biochar and thus its performance as a soil amendment. For example, Novak et al. (2014) found differences in the characteristics and elemental composition between manure-based and lignocellulosic biomass-based biochars. The manure feedstocks were found to be more alkaline and enriched in N and P. The high level of P is particularly significant as it can form insoluble phosphate complexes with metals and has been linked to the removal of lead, copper, zinc, and cadmium in contaminated systems (Bolan and

Duraisamy, 2003; Cao et al., 2003). Pyrolysis of lignocellulose-based feedstocks such as grasses and hardwood produces some oxygen-containing compounds including substituted and unsubstituted catechols, anhydrosugars, diols, and other compounds that are effective in chelating copper, zinc, and chromium (Ahmad et al., 2014). However, Kasozi et al. (2010) showed that changes in the sorption capacity of these organic compounds in biochars were more dependent on pyrolysis temperature than on biomass type.

In general, chars pyrolyzed at higher temperatures have higher carbon content, pH, and surface area (Ahmad et al., 2014). These are favored properties if the objective is to produce biochar for metal-adsorption applications. The higher carbon content and pH are a result of the more extensive removal of volatile matter in the original feedstock (Keiluweit et al., 2010; Uchimiya et al., 2012). During the pyrolysis process, the carbon and ashes are better able to withstand the high temperatures and are thus found in higher percentages in the biochar structure, compared to oxygen and hydrogen, which are easily volatilized. The higher ash content is partly responsible for the higher pH value, due to the increased presence of basic cations.

While extreme pyrolysis conditions increase the carbon content of biochar, it also decreases the oxygen content and surface hydrophilicity (Mohan et al., 2014). Chun et al. (2004) showed that wheat residue pyrolyzed between 500–700 °C had much higher oxygen-containing functional groups (e.g., –COOH, –OH, and –CHO) on the biochar surface than those produced at lower temperatures. This has implications for metal immobilization as oxygen-containing functional groups on biochar surfaces are important binding sites for metal sorption. An investigation by Uchimiya et al. (2012) revealed a direct correlation between the metal (especially lead and copper) stabilization ability of biochar with the amount of oxygen functional groups.

In summary, pyrolysis temperature could have contrasting effects on biochar properties, leading to contrasting effects on metal sorption (Li et al., 2017). This suggest that there is no single ideal pyrolysis temperature for metal adsorption. Each case should be tested and examined individually.

6.5 EFFECTS OF ADDING BIOCHAR TO SOIL

The beneficial effects of biochar mimic observations of black carbon (BC) in natural soils. BC exists in many soils as a result of wildfires (Certini, 2005; Sohi et al., 2009). Its origin can be traced back to the ancient Amerindian communities in the Amazon region where dark-colored soils, created by burying of charcoal over hundreds of years, was linked to sustained fertility. Studies in this region found that BC contributes significantly to the soil cation-exchange capacity (CEC), leading to a higher CEC per unit organic C compared to BC poor soils (Liang et al., 2006). Increases in pH, electrical conductivity, organic carbon, total nitrogen, available phosphorous, and exchangeable bases have also been reported (Lu et al., 2015; Nigussie et al., 2012).

When biochar is applied as a soil amendment, it will influence the soil environment in various ways. The relative contribution of any of the influencing factors is difficult to generalize because of differences in properties of the soil and biochar, weather and climate conditions, soil management and land use (Verheijen et al., 2010). Biochar properties are highly dependent on pyrolysis conditions and feedstock, and accordingly exhibit wide ranges between biochars. For example, CEC can range from very low to 400 $cmol_c\,kg^{-1}$, and C:N ratio may be anything between 7 and 500 or more. The pH, in contrast, is less variable, where most biochars exhibit a neutral to alkaline pH (Verheijen et al., 2010). Verheijen et al (2010) performed a meta-analysis on publications reporting on the effects of biochar on soil properties and plant growth. They found that overall, biochar has a small positive effect on plant growth, with greater effects seen on more acidic soils, whereas studies on calcareous soils did not reveal overall significant effects.

A key feature of biochar compared to other soil amendments is its long-term stability. Biochar is believed to reside in the soil for periods of hundreds to thousands of years, i.e., 10–1000 times longer than typical soil organic matter (Lehmann et al., 2009; Verheijen et al., 2010). Long-term decay of biochar is effectuated by a range of processes, including abiotic degradation, biotic decomposition, erosion, leaching and eluviation, and bioturbation (Lehmann et al., 2009). The surface area of fresh biochars, which is hydrophobic and has a relatively low surface charge, is initially changed rapidly mainly through abiotic processes (Lehmann et al., 2009). Hydrolysis and oxidation gives rise to the formation of carboxylate and phenolate groups that provide negative charge (Cheng et al., 2006; Lehmann et al., 2009). This abiotic oxidation is expected to facilitate the microbial metabolization of the very recalcitrant hydrophobic condensed aromatic structures of the biochar. Particles break down physically, but the specific dominant mechanisms of the particle size reduction are not well identified. Freezing and thawing are definitely important factors in cold regions, and also swelling and shrinking in vertisols can be factors. Particle size reduction contributes to the stabilization of biochars. Stability mechanisms include the inherent recalcitrance of the material, protection by aggregation, and interactions with other soil components. Biochar stability is promoted because it tends to be included in soil aggregates where it is less exposed to microorganisms. It is believed that the oxidation of the outer layer actually protects the inner regions of a biochar particle from access by microorganisms and associated enzymes. The outer layer subsequently becomes coated with mineral particles and soil organic matter (Lehmann et al., 2009).

Biochar can effectuate changes in physical properties of the soil such as bulk density, texture, structure, and pore size distribution. As such, it will have effects on the water-holding capacity, permeability for water and air, dynamics of swelling and shrinking, and workability and soil fragmentation in tillage (Downie et al., 2009). Biochar has a bulk density that is much below that of mineral soil, and therefore will tend to cause a decrease of the bulk density of the soil. Biochar increases the overall net soil surface area and accordingly also tends to improve soil water retention and soil aeration, especially in fine-textured soils (Verheijen et al., 2010). The dark color of biochar may contribute in reducing the soil albedo. A reduction of 37% was reported on a study site in Ghana (Oguntunde et al., 2008). Dark soils absorb more solar energy, and thus may facilitate soil warming in the spring. This effect, however, can be countered if the biochar increases the water retention of the soil and therefore also the specific heat of the soil (Verheijen et al., 2010).

Most biochars are neutral to alkaline, and thus provide a liming effect to the soil. The liming effect usually appears as the most important mechanism in increasing plant productivity after biochar application. Biochars with high liming capacity may provide great benefit to soils that require liming, and could be applied more frequently at lower rates, replacing conventional lime (Verheijen et al., 2010).

Biochars may exhibit a range of CECs, from very low to 400 $cmol_c\,kg^{-1}$ (Verheijen et al., 2010). Biochar from woody material is generally lower in CEC than biochar from nonwoody materials such as leaves or tree bark (Kookana et al., 2011). The CEC of biochar changes after incorporation into the soil. This is mainly caused by oxidation of the surface area, which results in the creation of additional oxygen-containing functional groups, mainly carboxylic and phenolic functional groups (Cheng et al., 2008). When added to soil biochar will thus enhance nutrient cycling and protect against leaching loss (Sohi et al., 2010).

Metals in bioavailable forms can adversely affect the activities of soil microbes by reducing their populations and changing the community structure and diversity (Renella et al., 2005). A reliable measure of the effect of such changes is the reduction in microbial enzyme activities in contaminated soils (Chen et al., 2012; Li et al., 2009). Cui et al. (2013) found that there was a correlation between the concentration of exchangeable heavy metals, soil enzyme activity, and pH.

Biochar promotes soil microbial activity by providing an appropriate habitat for soil microorganisms through its porous structures (Fischer and Glaser, 2012). Several studies reported a significant increase of microbial biomass, activity and growth rates, and changes in the composition of the microbial communities (Birk et al., 2009; Steiner et al., 2004). Cui et al. (2013) and Yang et al. (2016) monitored soil enzyme activities and showed that biochar amendments can improve the health of microbial communities in contaminated soils. This is likely a result of the alleviated toxicity of soil contaminants to microbes due to the biochar-mediated immobilization of metals and other pollutants and the consequent reduction in their bioavailability (Seneviratne et al., 2017; Zhu et al., 2017).

Biochar's distinctive physical and chemical properties give it the ability to interact with metals and potentially immobilize them, either by precipitation or increased sorption. The highly porous structure, active functional groups, and generally high pH and CEC of biochar have been identified as the key determinants of its metal immobilization capacity (Beesley et al., 2011; Zhang et al., 2013). These properties of biochar have encouraged widespread interest in its applications for soil remediation while its high C content and biochemically recalcitrant nature suggest a considerable potential to enhance the long-term carbon pool (Lehmann et al., 2006). Chemically, biochars possess oxygen-containing surface functional groups such as carbonyl, hydroxyl, carboxyl, quinones, and phenolic groups, which are formed as a result of biotic and abiotic oxidation (Cheng et al., 2006). The formation of these negatively charged functional groups and adsorption sites on the internal and external surfaces of biochar could influence its CEC, thereby enhancing the capacity of biochar-amended soils to form complexes with metal ions that are expected to be more stable than those formed by other forms of organic matter (Namgay et al., 2010).

Researchers have demonstrated the effectiveness of biochar in reducing high concentrations of trace metals in multielement-polluted soils as in the case of soluble cadmium and zinc (Beesley et al., 2010; Egene et al., 2018); enhanced immobilization of copper, nickel, and cadmium by complex formation with available ligands (Uchimiya et al., 2010); and significant reduction in extractable cadmium and lead over a 3-year period (Bian et al., 2014).

Studies have demonstrated the ability of biochar to reduce plant uptake of metals from soils in both tropical and temperate climates, but the extent of the effects depends on site-specific conditions and thus varies greatly between reports (O'Connor et al., 2018). The immobilization of bioavailable metals in soils resulting from biochar application is often identified as the main mechanism responsible for the reduced metal concentrations in plants (Herath et al., 2015; Zhang et al., 2017), and is in many cases dominantly related to the pH effect of the added biochar, although the effect of increased sorption to soil solids also remains important. In some cases, the reduction in plant metal concentrations may be attributed to a simple dilution effect as a result of increased plant biomass, where a similar uptake of elements is distributed in a larger biomass (Houben et al., 2013a; Park et al., 2011).

The biochar-mediated reduction in plant metal uptake has been demonstrated for different metals and a variety of plants. For example, Bian et al. (2014) observed a significant decrease in uptake of cadmium and lead by rice plants after application of wheat straw biochar over a 3-year period. Also, Herath et al. (2015) reported a 93%–97% decrease in the bioaccumulation of chromium, nickel, and manganese in tomato plants grown in 5% biochar-amended soil compared to the unamended soil. In a field study, Moreno-Jiménez et al. (2016) found decreased concentrations of cadmium, lead, and arsenic in barley grain grown in unfertilized soil amended with holm-oak biochar. The impact of biochar on metal uptake is largely controlled by the application rate. Studies have shown that increasing the amount of biochar applied to metal-contaminated soils generally results in lower plant uptake.

The feedstock quality determines the inherent metal concentration of the biochar. Biochars produced using natural feedstock (e.g., manure, plant residue) typically have low metal contents and as such their application to soils carries no significant risk for plant uptake. This was demonstrated by Lucchini et al. (2014a) who studied the effect of repeated biochar application (at rates of 25 and 50 metric ton ha^{-1}) on metal uptake of barley and beans over two cropping cycles. They found no significant effects in the total metal concentrations in the soil and plants, although small changes in soil metal distribution were observed. However, care must be taken when using biochar derived from anthropogenic wastes (e.g., biosolids) and contaminated feedstock (e.g., chemically contaminated plant residue). Biochar produced from waste wood materials containing high levels of copper introduced copper exceeding the metal immobilization capacity of the soil and biochar, resulting in phytotoxic metal concentrations (Lucchini et al., 2014b).

6.6 MANAGEMENT OPTIONS

Appropriate soil and plant management can help reduce the risk posed by metals in contaminated soils. These approaches have been referred to as in situ gentle remediation (Gupta et al., 1999; Onwubuya et al., 2009). The type of metal and its concentration as well as the soil properties are the key factors that determine the efficiency of the management approach used. The following section discusses some management approaches that may be suitable for diffusely metal-contaminated soils.

6.6.1 Biochar Amendment in Combination With Phytomanagement

Phytomanagement involves the manipulation of soil–plant systems to affect the fluxes of trace metals in the environment, with the goal of remediating contaminated soils, recovering valuable metals, or increasing micronutrient concentrations in crops (Robinson et al., 2009). Phytomanagement comprises a range of technologies that utilize specialized plants to extract, stabilize, degrade, or volatilize contaminants. The most applicable technologies for diffusely metal-contaminated soils are phytoextraction and phytostabilization. In phytoextraction, plants are used to move soil contaminants to the harvestable plant tissues for further treatment. This technique depends heavily on the ability of hyperaccumulator plants to extract high levels of a specific element of up to 1%–5% of their dry weight (Baker and Brooks, 1989). Some mechanisms used by these plants to withstand metal toxicity include chelation, compartmentalization in plant tissues, biotransformation, and cellular repair mechanism (Salt et al., 1998). However, the effectiveness of hyperaccumulator plants in phytoextraction is limited by their slow growth, low biomass production, and high metal selectivity (Tack and Meers, 2010). Furthermore, calculations reveal that it would take an unacceptable amount of time to remediate soils that are moderately or highly contaminated (Conesa et al., 2006). These limitations have steered the shift from phytoextraction toward phytostabilization. There are big opportunities to add value to phytoremediation by integrating it with other environmental agenda such as biomass energy, biodiversity, watershed management and protection from erosion, carbon sequestration, soil quality, and soil health (Dickinson et al., 2009).

In phytostabilization, plants are used to reduce metal mobility and to prevent their migration to groundwater (leaching) and agricultural land (wind transport), thus preventing their entry into the food chain (Tack and Meers, 2010). Trees are particularly recommended for phytostabilization purposes due to their extensive root systems and high transpiration capacity (Pulford and Watson, 2003). The mechanism of immobilization may be through absorption and accumulation by roots, adsorption on to roots, or precipitation within the rhizosphere (Fitz and Wenzel, 2002). Moreover, the presence of vegetation improves the chemical and biological properties of the contaminated soil by increasing the organic matter content, nutrient levels, CEC, and biological activity (Arienzo et al., 2004). This ultimately accelerates the development of a viable nutrient cycle and self-sustaining vegetative cover that restores the affected area to some acceptable steady-state condition for secondary land use (Norland and Veith, 1995).

The soil-conditioning ability of biochar can enhance plant growth and promote revegetation of contaminated sites. This, in addition to its metal immobilization ability, suggests that biochar may be suitable for enhancing the phytostabilization of contaminated soils. Despite this promising prospect, there has been limited research into the long-term potential of biochar-assisted phytostabilization. Short-term studies using pot experiments have demonstrated the effectiveness of biochar to ameliorate soil nutrients and water retention while also reducing the bioavailability and root-to-shoot translocation of trace metals (Fellet et al., 2014, 2011; Houben et al., 2013a,b; Prapagdee et al., 2014). Plants tested were bioenergy crops such as silvergrass (*Miscanthus*) and rapeseed, as well as metal-tolerant plants including *Anthyllis vulneraria* and *Vigna radiata*. In general, the results seem to favor phytostabilization over phytoextraction mainly because the metal immobilization ability of biochar impedes plant metal uptake, as evidenced by the low bioconcentration factors of metals in the plants. This group of plants that are recommended for assisted phytostabilization are often described as metal excluders (Karer et al., 2018; Wei et al., 2005). In a 2-year experiment, Karer et al. (2018) carried out both greenhouse and field experiments to confirm the metal-excluding ability of miscanthus and its suitability for the production of renewable biomass on biochar-amended metal-contaminated soils. They found that poplar-derived biochar addition reduced cadmium, lead, and zinc concentrations by 75%, 86%, and 92%, respectively. Furthermore, biomass production of miscanthus increased more than 100-fold between the first and second year of the field study. Besides their significant energy production potential, the perennial growth cycle of bioenergy crops offers additional economic and environmental advantages due to their lower pesticide requirement and higher contribution to soil carbon storage compared to annual crops (Fernando et al., 2018). An argument can thus be made for the selection of metal-excluding bioenergy crops, over agricultural food crops, for application in biochar-assisted phytostabilization.

The viability of biochar-assisted phytostabilization remains to be evaluated in long-term and large-scale field studies. An often-mentioned benefit of using biochar amendment is its considerable potential to remain stable in soil over a long period of time due to its biochemically recalcitrant nature (Lehmann et al., 2009). However, researchers agree that the long-term fate of metals in biochar-amended soils is uncertain and in need of further inquiry (Egene et al., 2018; Puga et al., 2015). This makes it difficult to speculate on the long-term interactions that may occur among soil metals, biochar, and phytostabilizing plants.

6.6.2 Biochar to Reduce Uptake of Hazardous Elements to Vegetable Crops

Agricultural fields can become contaminated with trace metals through irrigation with wastewater or the application of sewage sludge, fertilizers, and pesticides (Adriano, 2001). The cultivation of vegetables on or near metal-affected agricultural soils poses a serious risk to human health due to the high metal uptake by these plants (Balsberg Påhlsson, 1989) and the potential dietary exposure. The potential of biochar amendment to ameliorate the metal concentrations and improve the quality of contaminated agricultural soils has been investigated in laboratory scale and greenhouse experiments with encouraging results. A recent meta-analysis of data obtained from 97 articles about metal uptake in edible plants showed that the addition of biochar effectively suppressed the bioaccumulation of trace elements, except for arsenic (Peng et al., 2018). The low suppression of arsenic was attributed to the similarity between phosphates and arsenates, which allows both elements to enter the plants through a common transporter. In contrast, biochar amendment had the greatest effect on decreasing the cadmium and lead bioaccumulation concentrations, suggesting that biochar could significantly contribute to the alleviation of cadmium- and lead-affected soils in practice (Peng et al., 2018).

Rizwan et al. (2016) summarized the mechanisms of biochar-mediated remediation of trace metals in agricultural soils, which include (1) immobilization via sorption on the surfaces and increase in soil pH, (2) alteration of metals redox state in the soil, (3) modification in soil physical and biological properties, and (4) modification of antioxidant enzymes in plants. However, several factors such as biochar type and pyrolysis conditions, target elements, plant species, and other environmental conditions influence the soil-biochar-plant interactions in the contaminated soil. As a result, the effects of using biochar for remediation of contaminated soils may vary widely from one case to another.

Despite the observed significant reductions in the uptake of metals by food crops due to biochar addition, the strict standards for maximum allowable metal concentrations in food and feed are not always satisfied. For example, Nie et al. (2018) found that biochar addition led to a reduction in the shoot concentrations of cadmium and lead in Chinese cabbage from 0.32 and 1.1 mg kg^{-1}, respectively, in the control to ≤ 0.12 and ≤ 0.86 mg kg^{-1}, respectively, in the biochar treatments. These values fell significantly below the maximum allowable concentrations of 0.2 and 1.0 mg kg^{-1} for cadmium and lead, respectively, set by the Chinese Ministry of Health. In contrast, Rizwan et al. (2018) found that addition of 5.0% biochar resulted in significant decreases of 61% and 53% in cadmium concentrations in the shoots and roots of rice, respectively, compared to the control. This corresponded to cadmium concentrations of 0.53 and 1.21 mg kg^{-1}, respectively. These values clearly exceed the upper limit set by the European Commission in Regulation No. 1881/2006 of 0.10 and 0.20 mg kg^{-1} wet weight in stem vegetables and fresh herbs, respectively. Therefore, great caution must be exercised when using biochar for remediation of agricultural soils.

6.7 FIELD EXPERIENCE TO DATE

The Terra Preta soils evidence the long-term beneficial effects of the addition of char-like materials to soils. This loamy, very dark, fertile anthropogenic soil is found in the Amazon Basin. The black color is due to the highly weathered charcoal components in these soils that accumulated there as a result of indigenous soil management, where charred trash was added to the soil. Some of these soils are believed to be 7000 years old (Marris, 2006).

Field experience specifically on the use of biochar to reduce the mobility of metals currently is very limited. A 2018 review focused on field experiences where crops were grown on contaminated land amended with biochar, where 29 studies were located (O'Connor et al., 2018). The most frequently studied elements included arsenic, cadmium, copper, nickel, lead, and zinc, and cadmium was featured in 22 of the 29 studies (O'Connor et al., 2018). Plot sizes were between 1 and 1000 m^2. The equilibration time, i.e., the time between biochar application and planting the crops, ranged from 0 to 6 months, and the entire study period from 1.5 to 60 months. Thus, these trials were rather short in duration and therefore do not yet allow firm conclusions to be drawn. Whereas sufficient data allowed to perform a meta-study on plant concentrations, insufficient data was available to statistically investigate the effects of biochar on soil properties (O'Connor et al., 2018). The field studies came to different conclusions about the potential value of biochar in the remediation of contaminated land, ranging from positive over neutral to negative. This is due to a large variation in influencing factors related to the characteristics of the biochar used and field conditions, i.e., soil properties, climatological and weather conditions, and land use/land use history (O'Connor et al., 2018).

Field data generally suggest that the effectiveness of biochar may initially improve with time as the biochar equilibrates with the soil environment (O'Connor et al., 2018). In a 5-year field experiment, Cui et al. (2011) observed between 10% and 50% reduction in the bioavailable ($CaCl_2$-extractable) fractions of cadmium and lead. On the longer term, the effects of biochar may fade. The alkalinity of the biochar will be gradually neutralized and lost through leaching, eventually resulting in a decrease in pH.

6.8 CONCLUSIONS

Biochar is a promising amendment for application in the management of metal- and metalloid-contaminated land. It provides the immediate effects of liming and organic matter in increasing or buffering the soil pH and contributing binding capacity to the soil for the retention of the inorganic contaminants. Because of its stability, it provides a stable pool of organic carbon for extended periods, contributing to the beneficial effects of organic matter in soils. It thus can reduce contaminant solubility, mobilization, and leaching from the soil solids, and reduce transfer of contaminants to other ecosystem compartments and dispersion to surrounding areas. It may also become an important tool that allows marginally contaminated lands to be kept in use for edible crop production, or to provide more certainty that strict standards for trace element contents in edible crops are not exceeded on plots with normal baseline contents.

The precise extent of biochar effects is highly dependent on various factors, including properties of soil and of the biochar, conditions of the application, soil characteristics, weather and climate conditions, soil management, and land use. Specific field experience is still very scarce. In this context, standardization of methods of biochar characterization, data collection, and assessment of end points will benefit the comparability of field study results. Based on adequate and comparable scientific evidence, prediction of effects upon application in specific situations will become possible, which will lead to more adequate use of biochar optimally designed for the specific situation.

Next to the remaining uncertainties on long-term effects, availability and price is currently hampering more widespread use of biochar for soil amendment. As society increasingly moves to a bio-based economy, pyrolysis may become a pivotal technology to convert biomass into base chemicals for transport fuel and energy and for supplying biochar as a resource that can be used to manage a legacy of metal- and metalloid-contaminated soils.

References

Adriano, D.C., 2001. Trace Elements in Terrestrial Environments: Biogeochemistry, Bioavailability, and Risks of Metals. Springer-Verlag, New York, NY.

Ahmad, M., Rajapaksha, A.U., Lim, J.E., Zhang, M., Bolan, N., Mohan, D., et al., 2014. Biochar as a sorbent for contaminant management in soil and water: a review. Chemosphere 99, 19–33. Available from: https://doi.org/10.1016/j.chemosphere.2013.10.071.

Ahmed, M.B., Zhou, J.L., Ngo, H.H., Guo, W., 2016. Insight into biochar properties and its cost analysis. Biomass Bioenergy 84, 76–86. Available from: https://doi.org/10.1016/j.biombioe.2015.11.002.

Alam, M., Allinson, G., Stagnitti, F., Tanaka, A., Westbrooke, M., 2002. Arsenic contamination in Bangladesh groundwater: a major environmental and social disaster. Int. J. Environ. Health Res. 12, 235–253.

Arienzo, M., Adamo, P., Cozzolino, V., 2004. The potential of Lolium perenne for revegetation of contaminated soil from a metallurgical site. Sci. Total Environ. 319, 13–25. Available from: https://doi.org/10.1016/S0048-9697(03)00435-2.

Augusto, L., Turpault, M.-P., Ranger, J., 2000. Impact of forest tree species on feldspar weathering rates. Geoderma 96, 215–237.

Bain, D.C., Langan, S.J., 1995. Weathering rates in catchments calculated by different methods and their relationship to acidic inputs. Water Air Soil Pollut. 85, 1051–1056.

Baker, A., Brooks, R., 1989. Terrestrial higher plants which hyperaccumulate metallic elements. A review of their distribution, ecology and phytochemistry. Biorecovery 1, 81–126.

Balsberg Påhlsson, A.-M., 1989. Toxicity of heavy metals (Zn, Cu, Cd, Pb) to vascular plants. Water Air Soil Pollut. 47, 287–319.

Beesley, L., Moreno-Jiménez, E., Gomez-Eyles, J.L., 2010. Effects of biochar and greenwaste compost amendments on mobility, bioavailability and toxicity of inorganic and organic contaminants in a multi-element polluted soil. Environ. Pollut. 158, 2282–2287. Available from: https://doi.org/10.1016/j.envpol.2010.02.003.

Beesley, L., Moreno-Jiménez, E., Gomez-Eyles, J.L., Harris, E., Robinson, B., Sizmur, T., 2011. A review of biochars' potential role in the remediation, revegetation and restoration of contaminated soils. Environ. Pollut. 159, 3269–3282. Available from: https://doi.org/10.1016/j.envpol.2011.07.023.

Bernhard, M., Brinckman, F.E., Sadler, P.J., 1986. The Importance of Chemical "Speciation" in Environmental Processes. Springer-Verlag, Berlin.

Bian, R., Joseph, S., Cui, L., Pan, G., Li, L., Liu, X., et al., 2014. A three-year experiment confirms continuous immobilization of cadmium and lead in contaminated paddy field with biochar amendment. J. Hazard. Mater. 272, 121–128. Available from: https://doi.org/10.1016/j.jhazmat.2014.03.017.

Birk, J.J., Steiner, C., Teixiera, W.C., Zech, W., Glaser, B., 2009. Microbial response to charcoal amendments and fertilization of a highly weathered tropical soil. Amazonian Dark Earths: Wim Sombroek's Vision. Springer Dordrecht, The Netherlands, pp. 309–324.

Bolan, N.S., Duraisamy, V.P., 2003. Role of inorganic and organic soil amendments on immobilisation and phytoavailability of heavy metals: a review involving specific case studies. Soil Res. 41, 533–555. Available from: https://doi.org/10.1071/sr02122.

Brady, K.U., Kruckeberg, A.R., Bradshaw Jr., H.D., 2005. Evolutionary ecology of plant adaptation to serpentine soils. Annu. Rev. Ecol. Evol. Syst. 36, 243–266. Available from: https://doi.org/10.1146/annurev.ecolsys.35.021103.105730.

Cao, X., Ma, L.Q., Shiralipour, A., 2003. Effects of compost and phosphate amendments on arsenic mobility in soils and arsenic uptake by the hyperaccumulator, Pteris vittata L. Environ. Pollut. 126, 157–167.

Certini, G., 2005. Effects of fire on properties of forest soils: a review. Oecologia 143, 1–10.

Chen, R., Blagodatskaya, E., Senbayram, M., Blagodatsky, S., Myachina, O., Dittert, K., et al., 2012. Decomposition of biogas residues in soil and their effects on microbial growth kinetics and enzyme activities. Biomass Bioenergy 45, 221–229. Available from: https://doi.org/10.1016/j.biombioe.2012.06.014.

Chen, W., Hrudey, S.E., Rousseaux, C., 1995. Bioavailability in Environmental Risk Assessment. CRC Press, Boca Raton, FL.

Cheng, C.-H., Lehmann, J., Engelhard, M.H., 2008. Natural oxidation of black carbon in soils: Changes in molecular form and surface charge along a climosequence. Geochim. Cosmochim. Acta 72, 1598–1610. Available from: https://doi.org/10.1016/j.gca.2008.01.010.

Cheng, C.-H., Lehmann, J., Thies, J.E., Burton, S.D., Engelhard, M.H., 2006. Oxidation of black carbon by biotic and abiotic processes. Org. Geochem. 37, 1477–1488. Available from: https://doi.org/10.1016/j.orggeochem.2006.06.022.

Chowdhury, A.M.R., 2004. Arsenic crisis in Bangladesh. Sci. Am. 86–91.

Chun, Y., Sheng, G., Chiou, C.T., Xing, B., 2004. Compositions and sorptive properties of crop residue-derived chars. Environ. Sci. Technol. 38, 4649–4655. Available from: https://doi.org/10.1021/es035034w.

Conesa, H.M., Faz, Á., Arnaldos, R., 2006. Heavy metal accumulation and tolerance in plants from mine tailings of the semiarid Cartagena–La Unión mining district (SE Spain). Sci. Total Environ. 366, 1–11. Available from: https://doi.org/10.1016/j.scitotenv.2005.12.008.

Cottenie, A., Verloo, M., 1984. Analytical diagnosis of soil pollution with heavy metals. Fresenius. J. Anal. Chem. 317, 389–393.

Cui, L., Li, L., Zhang, A., Pan, G., Bao, D., Chang, A., 2011. Biochar amendment greatly reduces rice Cd uptake in a contaminated paddy soil: a two-year field experiment. BioResources 6, 2605–2618. Available from: https://doi.org/10.15376/biores.6.3.2605-2618.

Cui, L., Yan, J., Yang, Y., Li, L., Quan, G., Ding, C., et al., 2013. Influence of biochar on microbial activities of heavy metals contaminated paddy fields. BioResources 8, 5536–5548. Available from: https://doi.org/10.15376/biores.8.4.5536-5548.

Dickinson, N.M., Baker, A.J.M., Doronila, A., Laidlaw, S., Reeves, R.D., 2009. Phytoremediation of inorganics: realism and synergies. Int. J. Phytorem. 11, 97–114.

Downie, A., Crosky, A., Munroe, P., 2009. Physical properties of biochar. In: Lehmann, J., Joseph, S. (Eds.), Biochar for Environmental Management: Science and Technology. Earthscan, London; Sterling, VA, pp. 13–32.

Egene, C.E., Van Poucke, R., Ok, Y.S., Meers, E., Tack, F.M.G., 2018. Impact of organic amendments (biochar, compost and peat) on Cd and Zn mobility and solubility in contaminated soil of the Campine region after three years. Sci. Total Environ. 626, 195–202. Available from: https://doi.org/10.1016/j.scitotenv.2018.01.054.

Fellet, G., Marchiol, L., Delle Vedove, G., Peressotti, A., 2011. Application of biochar on mine tailings: effects and perspectives for land reclamation. Chemosphere 83, 1262–1267. Available from: https://doi.org/10.1016/j.chemosphere.2011.03.053.

Fellet, G., Marmiroli, M., Marchiol, L., 2014. Elements uptake by metal accumulator species grown on mine tailings amended with three types of biochar. Sci. Total Environ. 468–469, 598–608. Available from: https://doi.org/10.1016/j.scitotenv.2013.08.072.

Ferguson, C., Darmendrail, D., Freier, K., Jensen, B.K., Jensen, J., Kasamas, H., et al., (Eds.), 1998. Risk Assessment for Contaminated Sites in Europe. Volume 1. Scientific Basis. LQM Press, Nottingham, United Kingdom.

Fernando, A.L., Rettenmaier, N., Soldatos, P., Panoutsou, C., 2018. 8—Sustainability of perennial crops production for bioenergy and bioproducts. In: Alexopoulou, E. (Ed.), Perennial Grasses for Bioenergy and Bioproducts. Academic Press, London, United Kingdom, pp. 245–283. Available from: https://doi.org/10.1016/B978-0-12-812900-5.00008-4.

Field, J.L., Keske, C.M.H., Birch, G.L., DeFoort, M.W., Cotrufo, M.F., 2013. Distributed biochar and bioenergy coproduction: a regionally specific case study of environmental benefits and economic impacts. GCB Bioenergy 5, 177–191. Available from: https://doi.org/10.1111/gcbb.12032.

Fischer, D., Glaser, B., 2012. Synergisms between compost and biochar for sustainable soil amelioration. Management of Organic Waste. InTech, London, United Kingdom.

Fitz, W.J., Wenzel, W.W., 2002. Arsenic transformations in the soil–rhizosphere–plant system: fundamentals and potential application to phytoremediation. J. Biotechnol. 99, 259–278. Available from: https://doi.org/10.1016/S0168-1656(02)00218-3.

Fornes, F., Belda, R.M., Lidón, A., 2015. Analysis of two biochars and one hydrochar from different feedstock: focus set on environmental, nutritional and horticultural considerations. J. Clean. Prod. 86, 40–48. Available from: https://doi.org/10.1016/j.jclepro.2014.08.057.

Granatstein, D., Kruger, C., Collins, H., Garcia-Perez, M., Yoder, J., 2009. Use of biochar from the pyrolysis of waste organic material as a soil amendment. Final project report. Center for Sustaining Agriculture and Natural Resources, Washington State University, Wenatchee, WA.

Gupta, S.K., Herren, T., Wenger, K., Krebs, R., Han, T., 1999. 17 In situ gentle remediation measures for heavy metal-polluted soils. In: Terry, N., Bañuelos, G. (Eds.), Phytoremediation of Contaminated Soil and Water. CRC Press LLC, Boca Raton, Florida, p. 303.

Herath, I., Kumarathilaka, P., Navaratne, A., Rajakaruna, N., Vithanage, M., 2015. Immobilization and phytotoxicity reduction of heavy metals in serpentine soil using biochar. J. Soils Sediments 15, 126–138. Available from: https://doi.org/10.1007/s11368-014-0967-4.

Houben, D., Evrard, L., Sonnet, P., 2013a. Beneficial effects of biochar application to contaminated soils on the bioavailability of Cd, Pb and Zn and the biomass production of rapeseed (Brassica napus L.). Biomass Bioenergy 57, 196–204. Available from: https://doi.org/10.1016/j.biombioe.2013.07.019.

Houben, D., Evrard, L., Sonnet, P., 2013b. Mobility, bioavailability and pH-dependent leaching of cadmium, zinc and lead in a contaminated soil amended with biochar. Chemosphere 92, 1450–1457. Available from: https://doi.org/10.1016/j.chemosphere.2013.03.055.

Inyang, M., Dickenson, E., 2015. The potential role of biochar in the removal of organic and microbial contaminants from potable and reuse water: a review. Chemosphere 134, 232–240. Available from: https://doi.org/10.1016/j.chemosphere.2015.03.072.

Jirka, S., Tomlinson, T., 2014. 2013 State of the Biochar Industry. International Biochar Initiative.

Kabata-Pendias, A., Dudka, S., Chlopecka, A., Gawinowska, T., 1992. Background levels and environmental influences on trace metals in soils of the temperate humid zone of Europe. Biogeochemistry of Trace Metals. Lewis Publishers, Boca Raton, pp. 61–84.

Karer, J., Zehetner, F., Dunst, G., Fessl, J., Wagner, M., Puschenreiter, M., et al., 2018. Immobilisation of metals in a contaminated soil with biochar-compost mixtures and inorganic additives: 2-year greenhouse and field experiments. Environ. Sci. Pollut. Res. 25, 2506–2516. Available from: https://doi.org/10.1007/s11356-017-0670-2.

Kasozi, G.N., Zimmerman, A.R., Nkedi-Kizza, P., Gao, B., 2010. Catechol and humic acid sorption onto a range of laboratory-produced black carbons (biochars). Environ. Sci. Technol. 44, 6189–6195. Available from: https://doi.org/10.1021/es1014423.

Keiluweit, M., Nico, P.S., Johnson, M.G., Kleber, M., 2010. Dynamic molecular structure of plant biomass-derived black carbon (biochar). Environ. Sci. Technol. 44, 1247–1253. Available from: https://doi.org/10.1021/es9031419.

Khan, S.I., Ahmed, A.K.M., Yunus, M., Rahman, M., Hore, S.K., Vahter, M., et al., 2010. Arsenic and cadmium in food-chain in Bangladesh—an exploratory study. J. Health Popul. Nutr. 28, 578–584.

Kookana, R.S., Sarmah, A.K., Van Zwieten, L., Krull, E., Singh, B., 2011. Biochar application to soil. Advances in Agronomy. Elsevier, Amsterdam, The Netherlands, pp. 103–143. Available from: https://doi.org/10.1016/B978-0-12-385538-1.00003-2.

Korentajer, L., Stern, R., Aggasi, M., 1993. Slope effects on cadmium load of eroded sediments and runoff water. J. Environ. Qual. 22, 639–645.

Laird, D.A., Fleming, P., Davis, D.D., Horton, R., Wang, B., Karlen, D.L., 2010. Impact of biochar amendments on the quality of a typical Midwestern agricultural soil. Geoderma 158, 443–449. Available from: https://doi.org/10.1016/j.geoderma.2010.05.013.

Lalor, G.C., 2008. Review of cadmium transfers from soil to humans and its health effects in the Jamaican environment. Sci. Total Environ. 400, 162–172. Available from: https://doi.org/10.1016/j.scitotenv.2008.07.011.

Lalor, G.C., Rattray, R., Simpson, P., Vutchkov, M., 1998. Heavy metals in Jamaica. Part 3: the distribution of cadmium in Jamaican soils. Rev. Int. Contam. Ambient. 14, 7–12.

Lehmann, J., Czimczik, C., Laird, D., Sohi, S., 2009. Stability of biochar in the soil. In: Lehmann, J., Joseph, S. (Eds.), Biochar for Environmental Management: Science and Technology. Earthscan, London; Sterling, VA, pp. 183–206.

Lehmann, J., Gaunt, J., Rondon, M., 2006. Bio-char sequestration in terrestrial ecosystems—a Review. Mitig. Adapt. Strateg. Glob. Change 11, 403–427. Available from: https://doi.org/10.1007/s11027-005-9006-5.

Lehmann, J., Joseph, S. (Eds.), 2009. Biochar for Environmental Management: Science and Technology. Earthscan, London; Sterling, VA.

Li, H., Dong, X., da Silva, E.B., de Oliveira, L.M., Chen, Y., Ma, L.Q., 2017. Mechanisms of metal sorption by biochars: biochar characteristics and modifications. Chemosphere 178, 466–478. Available from: https://doi.org/10.1016/j.chemosphere.2017.03.072.

Li, Y.-T., Rouland, C., Benedetti, M., Li, F., Pando, A., Lavelle, P., et al., 2009. Microbial biomass, enzyme and mineralization activity in relation to soil organic C, N and P turnover influenced by acid metal stress. Soil Biol. Biochem. 41, 969–977. Available from: https://doi.org/10.1016/j.soilbio.2009.01.021.

Liang, B., Lehmann, J., Solomon, D., Kinyangi, J., Grossman, J., O'Neill, B., et al., 2006. Black carbon increases cation exchange capacity in soils. Soil Sci. Soc. Am. J. 70, 1719–1730.

Lu, H., Li, Z., Fu, S., Méndez, A., Gascó, G., Paz-Ferreiro, J., 2015. Combining phytoextraction and biochar addition improves soil biochemical properties in a soil contaminated with Cd. Chemosphere 119, 209–216. Available from: https://doi.org/10.1016/j.chemosphere.2014.06.024.

Lucchini, P., Quilliam, R.S., DeLuca, T.H., Vamerali, T., Jones, D.L., 2014a. Does biochar application alter heavy metal dynamics in agricultural soil? Agric. Ecosyst. Environ. 184, 149–157. Available from: https://doi.org/10.1016/j.agee.2013.11.018.

Lucchini, P., Quilliam, R.S., DeLuca, T.H., Vamerali, T., Jones, D.L., 2014b. Increased bioavailability of metals in two contrasting agricultural soils treated with waste wood-derived biochar and ash. Environ. Sci. Pollut. Res. 21, 3230–3240. Available from: https://doi.org/10.1007/s11356-013-2272-y.

Marris, E., 2006. Putting the carbon back: black is the new green. Nature 442, 620–623.

Matschullat, J., Ottenstein, R., Reimann, C., 2000. Geochemical background—can we calculate it? Environ. Geol. 39, 990–1000.

Mohan, D., Sarswat, A., Ok, Y.S., Pittman, C.U., 2014. Organic and inorganic contaminants removal from water with biochar, a renewable, low cost and sustainable adsorbent—a critical review. Bioresour. Technol. 160, 191–202. Available from: https://doi.org/10.1016/j.biortech.2014.01.120.

Moreno-Jiménez, E., Fernández, J.M., Puschenreiter, M., Williams, P.N., Plaza, C., 2016. Availability and transfer to grain of As, Cd, Cu, Ni, Pb and Zn in a barley agri-system: impact of biochar, organic and mineral fertilizers. Agric. Ecosyst. Environ. 219, 171–178. Available from: https://doi.org/10.1016/j.agee.2015.12.001.

Naidu, R., Semple, K., Megharaj, M., Juhasz, A., Bolan, N., Gupta, S., et al., 2011. Bioavailability: definition, assessment and implications for risk assessment. Chemical Bioavailability in Terrestrial Environments. Elsevier, London, United Kingdom, pp. 39–51.

Naidu, R., Wong, M.H., Nathanail, P., 2015. Bioavailability—the underlying basis for risk-based land management. Environ. Sci. Pollut. Res. 22, 8775–8778. Available from: https://doi.org/10.1007/s11356-015-4295-z.

Namgay, T., Singh, B., Singh, B.P., 2010. Influence of biochar application to soil on the availability of As, Cd, Cu, Pb, and Zn to maize (*Zea mays* L.). Aust. J. Soil Res. 48, 638. Available from: https://doi.org/10.1071/SR10049.

Nie, C., Yang, X., Niazi, N.K., Xu, X., Wen, Y., Rinklebe, J., et al., 2018. Impact of sugarcane bagasse-derived biochar on heavy metal availability and microbial activity: a field study. Chemosphere 200, 274–282. Available from: https://doi.org/10.1016/j.chemosphere.2018.02.134.

Nigussie, A., Kissi, E., Misganaw, M., Ambaw, G., 2012. Effect of biochar application on soil properties and nutrient uptake of lettuces (Lactuca sativa) grown in chromium polluted soils. Am.-Eurasian J. Agric. Environ. Sci. 12, 369–376.

Nogawa, K., 1981. Itai-Itai disease and follow-up studies. Cadmium in the Environment. Part 11: Health Effects. John Wiley and Sons, New York, pp. 1–37.

Norland, M.R., Veith, D.L., 1995. Revegetation of coarse taconite iron ore tailing using municipal solid waste compost. J. Hazard. Mater., Selected papers presented at the Conference on Hazardous Waste Remediation 41, 123–134. https://doi.org/10.1016/0304-3894(94)00115-W

Novak, J.M., Cantrell, K.B., Watts, D.W., Busscher, W.J., Johnson, M.G., 2014. Designing relevant biochars as soil amendments using lignocellulosic-based and manure-based feedstocks. J. Soils Sediments 14, 330–343. Available from: https://doi.org/10.1007/s11368-013-0680-8.

Novak, J.M., Lima, I., Xing, B., Gaskin, J.W., Steiner, C., Das, K.C., et al., 2009. Characterization of designer biochar produced at different temperatures and their effects on a loamy sand. Ann. Env. Sci. 3, 195–206.

Nriagu, J.O., 1979. Global inventory of natural and anthropogenic emissions of trace metals to the atmosphere. Nature 279, 409–411. Available from: https://doi.org/10.1038/279409a0.

Nriagu, J.O., Pacyna, J.M., 1988. Quantitative assessment of worldwide contamination of air, water and soils by trace metals. Nature 333, 134–139. Available from: https://doi.org/10.1038/333134a0.

O'Connor, D., Peng, T., Zhang, J., Tsang, D.C.W., Alessi, D.S., Shen, Z., et al., 2018. Biochar application for the remediation of heavy metal polluted land: a review of in situ field trials. Sci. Total Environ. 619–620, 815–826. Available from: https://doi.org/10.1016/j.scitotenv.2017.11.132.

Oguntunde, P.G., Abiodun, B.J., Ajayi, A.E., van de Giesen, N., 2008. Effects of charcoal production on soil physical properties in Ghana. J. Plant Nutr. Soil Sci. 171, 591–596. Available from: https://doi.org/10.1002/jpln.200625185.

Oliveira, F.R., Patel, A.K., Jaisi, D.P., Adhikari, S., Lu, H., Khanal, S.K., 2017. Environmental application of biochar: current status and perspectives. Bioresour. Technol. 246, 110–122. Available from: https://doi.org/10.1016/j.biortech.2017.08.122.

Onwubuya, K., Cundy, A., Puschenreiter, M., Kumpiene, J., Bone, B., Greaves, J., et al., 2009. Developing decision support tools for the selection of "gentle" remediation approaches. Sci. Total Environ. 407, 6132–6142. Available from: https://doi.org/10.1016/j.scitotenv.2009.08.017.

Park, J.H., Choppala, G.K., Bolan, N.S., Chung, J.W., Chuasavathi, T., 2011. Biochar reduces the bioavailability and phytotoxicity of heavy metals. Plant Soil 348, 439–451. Available from: https://doi.org/10.1007/s11104-011-0948-y.

Peng, X., Deng, Y., Peng, Y., Yue, K., 2018. Effects of biochar addition on toxic element concentrations in plants: a meta-analysis. Sci. Total Environ. 616–617, 970–977. Available from: https://doi.org/10.1016/j.scitotenv.2017.10.222.

Prapagdee, S., Piyatiratitivorakul, S., Petsom, A., Tawinteung, N., 2014. Application of biochar for enhancing cadmium and zinc phytostabilization in *vigna radiata* l. cultivation. Water Air Soil Pollut. 225, 2233. Available from: https://doi.org/10.1007/s11270-014-2233-1.

Puga, A.P., Abreu, C.A., Melo, L.C.A., Beesley, L., 2015. Biochar application to a contaminated soil reduces the availability and plant uptake of zinc, lead and cadmium. J. Environ. Manage. 159, 86–93. Available from: https://doi.org/10.1016/j.jenvman.2015.05.036.

Pulford, I.D., Watson, C., 2003. Phytoremediation of heavy metal-contaminated land by trees—a review. Environ. Int. 29, 529–540.

Reimann, C., Garrett, R.G., 2005. Geochemical background–concept and reality. Sci. Total Environ. 350, 12–27.

Renella, G., Mench, M., Landi, L., Nannipieri, P., 2005. Microbial activity and hydrolase synthesis in long-term Cd-contaminated soils. Soil. Biol. Biochem. 37, 133–139. Available from: https://doi.org/10.1016/j.soilbio.2004.06.015.

Rizwan, M., Ali, S., Abbas, T., Adrees, M., Zia-ur-Rehman, M., Ibrahim, M., et al., 2018. Residual effects of biochar on growth, photosynthesis and cadmium uptake in rice (Oryza sativa L.) under Cd stress with different water conditions. J. Environ. Manage. 206, 676–683. Available from: https://doi.org/10.1016/j.jenvman.2017.10.035.

Rizwan, M., Ali, S., Qayyum, M.F., Ibrahim, M., Zia-ur-Rehman, M., Abbas, T., et al., 2016. Mechanisms of biochar-mediated alleviation of toxicity of trace elements in plants: a critical review. Environ. Sci. Pollut. Res. 23, 2230–2248. Available from: https://doi.org/10.1007/s11356-015-5697-7.

Robinson, B.H., Banuelos, G., Conesa, H.M., Evangelou, M.W.H., Schulin, R., 2009. The phytomanagement of trace elements in soil. Crit. Rev. Plant Sci. 28, 240–266.

Ronsse, F., Jørgensen, H., Schüßler, I., Gebart, R., 2014. Transportfuel. In: Pelkonen, P., Mustonen, M., Asikainen, A., Egnell, G., Kant, P., Leduc, S., Pettenella, D. (Eds.), Forest Bioenergy for Europe. What Science. Can Tell Us. European Forest Institute, Joensuu, Finland, pp. 52–58.

Salt, D., Smith, R., Raskin, I., 1998. Phytoremediation. Annu. Rev. Plant. Physiol. Plant. Mol. Biol. 49, 643–668.

Seneviratne, M., Weerasundara, L., Ok, Y.S., Rinklebe, J., Vithanage, M., 2017. Phytotoxicity attenuation in Vigna radiata under heavy metal stress at the presence of biochar and N fixing bacteria. J. Environ. Manage. 186, 293–300. Available from: https://doi.org/10.1016/j.jenvman.2016.07.024.

Shackley, S., Hammond, J., Gaunt, J., Ibarrola, R., 2011. The feasibility and costs of biochar deployment in the UK. Carbon Manag. 2, 335–356. Available from: https://doi.org/10.4155/cmt.11.22.

Sohi, S., Lopez-Capel, E., Krull, E., Bol, R., 2009. Biochar, climate change and soil: a review to guide future research. CSIRO Land Water Sci. Rep. 5, 17–31.

Sohi, S.P., Krull, E., Lopez-Capel, E., Bol, R., 2010. A review of biochar and its use and function in soil. In: Advances in Agronomy, Volume 105. Elsevier, Burlington, MA, pp. 47–82. Available from: https://doi.org/10.1016/S0065-2113(10)05002-9.

Steiner, C., Geraldes Teixeira, W., Zech, W., 2004. Slash and Char—An Alternative to Slash and Burn Practiced in the Amazon Basin. In: Glaser, B., Woods, W. (Eds.), Amazonian Dark Earths. Springer, Heidelberg, pp. 182–193.

Tack, F.M.G., 2010. Trace elements: general soil chemistry, principles and processes. Trace Elements in Soils. Wiley-Blackwell, Chichester, UK, pp. 9–37.

Tack, F.M.G., Meers, E., 2010. Assisted phytoextraction: helping plants to help us. Elements 6, 383–388.

Tondel, M., Rahman, M., Magnuson, A., Chowdhury, I.A., Faruquee, M.H., Ahmad, S.A., 1999. The relationship of arsenic levels in drinking water and the prevalence rate of skin lesions in Bangladesh. Environ. Health Perspect. 107, 727.

Uchimiya, M., Cantrell, K.B., Hunt, P.G., Novak, J.M., Chang, S., 2012. Retention of heavy metals in a Typic Kandiudult amended with different manure-based biochars. J. Environ. Qual. 41, 1138. Available from: https://doi.org/10.2134/jeq2011.0115.

Uchimiya, M., Lima, I.M., Klasson, K.T., Wartelle, L.H., 2010. Contaminant immobilization and nutrient release by biochar soil amendment: roles of natural organic matter. Chemosphere 80, 935–940. Available from: https://doi.org/10.1016/j.chemosphere.2010.05.020.

Vanholme, B., Desmet, T., Ronsse, F., Rabaey, K., Van Breusegem, F., De Mey, M., et al., 2013. Towards a carbon-negative sustainable bio-based economy. Front. Plant Sci. 4. Available from: https://doi.org/10.3389/fpls.2013.00174.

Verheijen, F., Jeffery, S., Bastos, A.C., European Commission, Joint Research Centre, Institute for Environment andSustainability, 2010. Biochar Application to Soils: A Critical Scientific Review of Effects on Soil Properties, Processes and Functions. Publications Office, Luxembourg.

Wang, M., Zhu, Y., Cheng, L., Andserson, B., Zhao, X., Wang, D., et al., 2018. Review on utilization of biochar for metal-contaminated soil and sediment remediation. J. Environ. Sci. 63, 156–173. Available from: https://doi.org/10.1016/j.jes.2017.08.004.

Wei, S., Zhou, Q., Wang, X., 2005. Identification of weed plants excluding the uptake of heavy metals. Environ. Int. 31, 829–834. Available from: https://doi.org/10.1016/j.envint.2005.05.045.

Xu, G., Lv, Y., Sun, J., Shao, H., Wei, L., 2012. Recent advances in biochar applications in agricultural soils: benefits and environmental implications. CLEAN—Soil Air Water 40, 1093–1098. Available from: https://doi.org/10.1002/clen.201100738.

Yang, X., Liu, J., McGrouther, K., Huang, H., Lu, K., Guo, X., et al., 2016. Effect of biochar on the extractability of heavy metals (Cd, Cu, Pb, and Zn) and enzyme activity in soil. Environ. Sci. Pollut. Res. 23, 974–984. Available from: https://doi.org/10.1007/s11356-015-4233-0.

Zhang, R.-H., Li, Z.-G., Liu, X.-D., Wang, B., Zhou, G.-L., Huang, X.-X., et al., 2017. Immobilization and bioavailability of heavy metals in greenhouse soils amended with rice straw-derived biochar. Ecol. Eng. 98, 183–188. Available from: https://doi.org/10.1016/j.ecoleng.2016.10.057.

Zhang, X., Wang, H., He, L., Lu, K., Sarmah, A., Li, J., et al., 2013. Using biochar for remediation of soils contaminated with heavy metals and organic pollutants. Environ. Sci. Pollut. Res. 20, 8472–8483. Available from: https://doi.org/10.1007/s11356-013-1659-0.

Zhu, X., Chen, B., Zhu, L., Xing, B., 2017. Effects and mechanisms of biochar-microbe interactions in soil improvement and pollution remediation: a review. Environ. Pollut. 227, 98–115. Available from: https://doi.org/10.1016/j.envpol.2017.04.032.

CHAPTER

7

Biochar and Its Composites for Metal(loid) Removal From Aqueous Solutions

Lukáš Trakal[1], *Martina Vítková*[1], *Barbora Hudcová*[1], *Luke Beesley*[2] *and Michael Komárek*[1]

[1]Department of Environmental Geosciences, Faculty of Environmental Sciences, Czech University of Life Sciences Prague, Prague, Czech Republic [2]The James Hutton Institute, Environmental and Biochemical Sciences Group, Aberdeen, United Kingdom

7.1 METAL SORPTION ON VARIOUS BIOCHARS

The conversion of surplus biomass into biosorbent biochars (BC) can be described as a "win–win" solution for the production of a new material with enhanced environmental value (Cao et al., 2009; Zheng et al., 2010) for the treatment of contaminated waters (Tan et al., 2015). Many studies have described effective removal of metals from waters using biochars originating from different organic wastes (Table 7.1). For example, the high performance of pristine biochars to remove Cd and/or Pb from aqueous solutions has been found to be within the range of one order (from 2.87 to 2872 mmol kg^{-1}) (Inyang et al., 2011; Jiang et al., 2012; Han et al., 2013; Kim et al., 2013; Xu et al., 2013a; Trakal et al., 2014a; Zama et al., 2017; Sekulić et al., 2018) Thus, it would appear that modifications to biochar-based sorbents, aimed at stimulating specific geochemical mechanisms, could further enhance their efficiency.

7.1.1 Effect of Biochar Characteristics

The origin and methods of biochar preparation (pyrolysis) are ultimately greatly responsible for its final physical and chemical characteristics (Table 7.1). Thus, biochar properties significantly affect metal-sorption performance.

Biochar from Biomass and Waste
DOI: https://doi.org/10.1016/B978-0-12-811729-3.00007-8

TABLE 7.1 Comparison of Cd(II) and Pb(II) Sorption Onto Different Pristine Biochars Originated From Contrasting Agricultural Waste Materials (From Selected Recent Studies)

Biochar Origin	Temperature of Pyrolysis (°C)	Maximum Metal Sorption		Reference
		Cd (mmol kg^{-1})	Pb (mmol kg^{-1})	
Grape stalks	600	450	2872	Trakal et al. (2014a)
Canna indica	500	1679	/	Cui et al. (2016)
Celery stems	500	/	1467	Zhang et al. (2017a)
Celery stems	350	/	1390	Zhang et al. (2017a)
Wheat straws	600	400	1322	Trakal et al. (2014a)
Canna indica	600	1246	/	Cui et al. (2016)
Sugarcane bagasse	250	/	1240	Ding et al. (2014)
Alternanthera philoxeroides	600	/	1240	Yang et al. (2014)
Chemically pretreated pig manure	600	1041	1110	Kołodyńska et al. (2012)
Mechanically separated dairy cow manure	600	1050	1062	Kołodyńska et al. (2012)
Mechanically separated dairy cow manure	400	1023	1028	Kołodyńska et al. (2012)
Chicken bone	600	967.0	/	Park et al. (2015)
Chemically pretreated pig manure	400	951.9	839.8	Kołodyńska et al. (2012)
Prunus armeniaca	500	941.2	866.3	Sekulić et al. (2018)
Canna indica	400	941.0	/	Cui et al. (2016)
Celery leaves	500	/	907.3	Zhang et al. (2017a)
Grape hulls	600	259.8	859.1	Trakal et al. (2014a)
Sugarcane bagasse	500	/	830.1	Ding et al. (2014)
Poultry manure	450	801.4	/	Idrees et al. (2016)
Dairy manure	350	559.6	791.5	Xu et al. (2013a)
Sesame straw	700	765.1	492.3	Park et al. (2016a)
Farmyard manure	450	681.7	/	Idrees et al. (2016)
Residue of biogas production	600	679.1	/	Bogusz et al. (2017)
Sugarcane bagasse + aerobical digestion	600	/	661.2	Inyang et al. (2011)
Pepper stem	600	593.4	635.6	Park et al. (2016b)
Ipomoea fistulosa	400	635.4	/	Goswami et al. (2016)

(Continued)

TABLE 7.1 (Continued)

Biochar Origin	Temperature of Pyrolysis (°C)	Maximum Metal Sorption		Reference
		Cd (mmol kg^{-1})	Pb (mmol kg^{-1})	
Residue of biogas production	400	610.5	/	Bogusz et al. (2017)
Rosa damascena	450	590.3	255.6	Khare et al. (2017)
Canna indica	300	563.3	/	Cui et al. (2016)
Ipomoea fistulosa	500	556.0	/	Goswami et al. (2016)
Saccharina japonica	700	540.0	/	Poo et al. (2018)
Ipomoea fistulosa	350	493.7	/	Goswami et al. (2016)
Dairy manure	350	490.2	/	Xu et al. (2013b)
Sugarcane bagasse	500	/	419.7	Abdelhafez and Li (2016)
Ipomoea fistulosa	550	370.7	/	Goswami et al. (2016)
Sargassum fusiforme	700	330.9	/	Poo et al. (2018)
Phyllostachys pubescens	700	130.6	325.5	Zhang et al. (2017b)
Sida hermaphrodita	700	317.7	/	Bogusz et al. (2017)
Rice straw	400	299.8	/	Han et al. (2013)
Maize cods	600	293.6	/	Moyo et al. (2016)
Wheat straw	700	289.7	/	Bogusz et al. (2017)
Dairy manure	200	280.2	/	Xu et al. (2013b)
Anaerobic digestion sludge	600	/	260.4	Ho et al. (2017)
Peanut shell	350	/	254.8	Wang et al. (2015b)
Digested animal waste	600	30.34	250.0	Inyang et al. (2012)
Peanut shell	400	/	240.8	Wang et al. (2015b)
Saliburg (UK broadleaf hardwood)	600	/	230.0	Shen et al. (2015)
Peanut shell	300	/	206.1	Wang et al. (2015b)
Anaerobically digested dairy waste (sugar beet)	600	49.55	199.8	Inyang et al. (2012)
Phyllostachys pubescens	450	116.9	196.2	Zhang et al. (2017b)
Peanut shell	600	/	180.5	Wang et al. (2015b)
Sewage sludge + tea waste	300	177.9	/	Fan et al. (2018)
Peanut shell	500	/	166.0	Wang et al. (2015b)
Nut shells	600	40.03	150.1	Trakal et al. (2014a)

(Continued)

TABLE 7.1 (Continued)

Biochar Origin	Temperature of Pyrolysis (°C)	Maximum Metal Sorption		Reference
		Cd (mmol kg^{-1})	Pb (mmol kg^{-1})	
Sewage sludge	550	/	150.1	Lu et al. (2012)
Rice husk	350	70.01	140.0	Xu et al. (2013a)
Sugarcane straw	700	139.7	/	Melo et al. (2013)
Orange peel	500	/	134.5	Abdelhafez and Li (2016)
Miscanthus sacchariflorus	500	120.1	/	Kim et al. (2013)
Miscanthus sacchariflorus	600	120.1	/	Kim et al. (2013)
Miscanthus sacchariflorus	400	110.3	/	Kim et al. (2013)
Plum stones	600	40.03	110.0	Trakal et al. (2014a)
Miscanthus sacchariflorus	300	99.64	/	Kim et al. (2013)
Sugarcane straw	500	79.98	/	Melo et al. (2013)
Garden green waste residues	500	69.39	/	Frišták et al. (2015)
Sugarcane straw	600	50.00	/	Melo et al. (2013)
Pinewood sawdust	700	44.48	/	Poo et al. (2018)
Sugarcane bagasse/peanut hull	600	40.03	19.98	Zhou et al. (2013)
Sugarcane straw	400	40.03	/	Melo et al. (2013)
Spruce cone	600	25.44	38.51	Saletnik et al. (2017)
Larch cone	500	21.26	37.69	Saletnik et al. (2017)
Spruce cone	500	20.02	38.42	Saletnik et al. (2017)
Larch cone	600	7.206	36.58	Saletnik et al. (2017)
Sugarcane bagasse	600	/	30.02	Inyang et al. (2011)
Hickory wood	600	20.02	19.98	Zhou et al. (2013)
Bamboo	600	20.02	9.990	Zhou et al. (2013)
Beech wood chips	500	17.70	/	Frišták et al. (2015)
Peanut shells	350	14.82	6.593	Zama et al. (2017)
Mulberry wood	550	8.086	12.55	Zama et al. (2017)
Mulberry wood	450	6.352	12.06	Zama et al. (2017)
Mulberry wood	350	3.416	12.07	Zama et al. (2017)
Poultry manure	650	2.695	12.01	Zama et al. (2017)
Peanut shells	450	11.12	9.778	Zama et al. (2017)

(Continued)

TABLE 7.1 (Continued)

Biochar Origin	Temperature of Pyrolysis (°C)	Maximum Metal Sorption		Reference
		Cd (mmol kg^{-1})	Pb (mmol kg^{-1})	
Poultry manure	450	11.12	9.653	Zama et al. (2017)
Poultry manure	550	11.12	8.084	Zama et al. (2017)
Peanut shells	550	11.12	6.033	Zama et al. (2017)
Poultry manure	350	6.352	10.14	Zama et al. (2017)
Mulberry wood	650	5.925	9.894	Zama et al. (2017)
Buckwheat husk	550	3.558	9.653	Zama et al. (2017)
Buckwheat husk	350	0.605	9.556	Zama et al. (2017)
Buckwheat husk	450	8.896	6.940	Zama et al. (2017)
Buckwheat husk	650	8.896	6.897	Zama et al. (2017)
Peanut shells	650	7.410	8.523	Zama et al. (2017)
Corn cobs	450	5.231	8.137	Zama et al. (2017)
Corn cobs	650	4.679	7.703	Zama et al. (2017)
Corn cobs	550	3.701	6.892	Zama et al. (2017)
Corn cobs	350	2.865	5.314	Zama et al. (2017)

Firstly, the volume of micropores, which is very strongly related to the Brunauer–Emmett–Teller (BET) surface (Harvey et al., 2011), can affect metal sorption due to the effective reactive surface. This reactive surface [shown here by scanning electron microscopy (SEM) images; Fig. 7.1] shows the various surface morphology of tested biochars. Woody (high lignin) biochars show well-developed structures with high macro- and microporosity (Fig. 7.1A; Chen et al., 2011; Mohan et al., 2014a). On the other hand, poorly structurally developed biochars (Fig. 7.1B) show limited presence of these pores. This is in agreement with the lower BET surface and lower volume of micropores in these cases (Table 7.2). The resultant metal sorption is higher for those biochars with poorly developed structures represented by lower BET surface, so the morphology of biochars is not a crucial factor affecting metal-sorption efficiency (later in Fig. 7.2).

The pH of biochars is usually alkaline in the range of 7.2–10.0 (Verheijen et al., 2010; Trakal et al., 2014a). Generally, biochar production, by pyrolysis, causes pH increases in comparison to the pH of the source material (Lu et al., 2012). The pH also increases after the pyrolysis process because the alkali salts separate from the organic matrix (at pyrolysis temperature ranging from 300 to 600°C; Chen et al., 2011). This physicochemical characteristic of biochar is usually a key factor in bivalent metal sorption from waters, mainly due to the pH-buffering effect of the biochar (Trakal et al., 2014a) affecting not only the surface charge, but also the speciation of metal ions in solution.

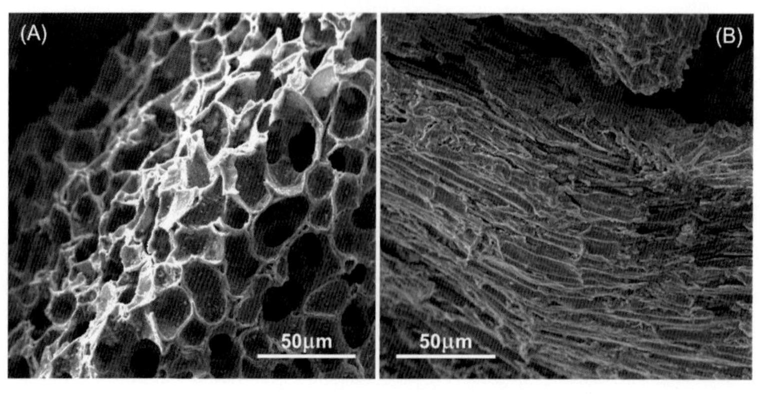

FIGURE 7.1 SEM images of (A) biochar with well-developed structure (nut shields) and (B) biochar with poorly developed structure (grape husks; Trakal et al., 2014a). *SEM*, scanning electron microscopy.

TABLE 7.2 Biochar Characteristics: Yield of Biochar From the Waste Material (Y), Bulk Density (ρ) BET Surface, Volume of Micropores (V_{micro}), pH, and Cation Exchange Capacity (CEC; Trakal et al., 2014a)

	Y	ρ	BET	V_{micro}	pH	CEC
Biochar	**(%)**	**(g cm^{-3})**	**(m^2 g^{-1})**	**(mm^3 g^{-1})**	**(–)**	**(mmol kg^{-1})**
Nut shields	21.8	0.17 ± 0.002	465	180	8.63 ± 0.04	84.4 ± 3.0
Wheat straw	18.9	0.26 ± 0.005	364	130	9.86 ± 0.05	334 ± 2
Grape stalks	30.6	0.16 ± 0.004	72	30	10.0 ± 0.1	402 ± 3
Grape husks	31.6	0.21 ± 0.003	77	32	9.98 ± 0.01	187 ± 4
Plum stones	24.7	0.22 ± 0.002	443	172	7.36 ± 0.12	121 ± 7

The cation-exchange capacity (CEC) of biochar samples is often an underestimated parameter that has considerable influence on metal-sorption efficiency, as reported by Lu et al. (2012), Zhang et al. (2013a), Ahmad et al. (2014), and Trakal et al. (2014a). Specifically, the CEC value of biochars is closely related with: (1) the content of carboxylic groups (Harvey et al., 2011); and (2) the mineralogical composition, mainly the content of K^+, Ca^{2+}, and Mg^{2+}, which is thus important in cation release (Trakal et al., 2014a).

Fig. 7.2 shows the relationship between the metal-removal efficiency (Cd and Pb) and selected biochar characteristics. The results of various published studies were analyzed to obtain complex information about the characteristics that may be responsible for the metal-removal efficiency. Specifically, increasing the CEC significantly enhances the Pb-removal efficiency ($R = 0.90$) of various biochars. This supports the hypothesis that CEC may be one of the most important characteristics in biochar selection for metal sorption because other tested biochar characteristics (such as BET surface or temperature of pyrolysis) showed no significant relationship to the sorption efficiency of metals (represented here by Cd and Pb; Fig. 7.2; Trakal et al., 2014a).

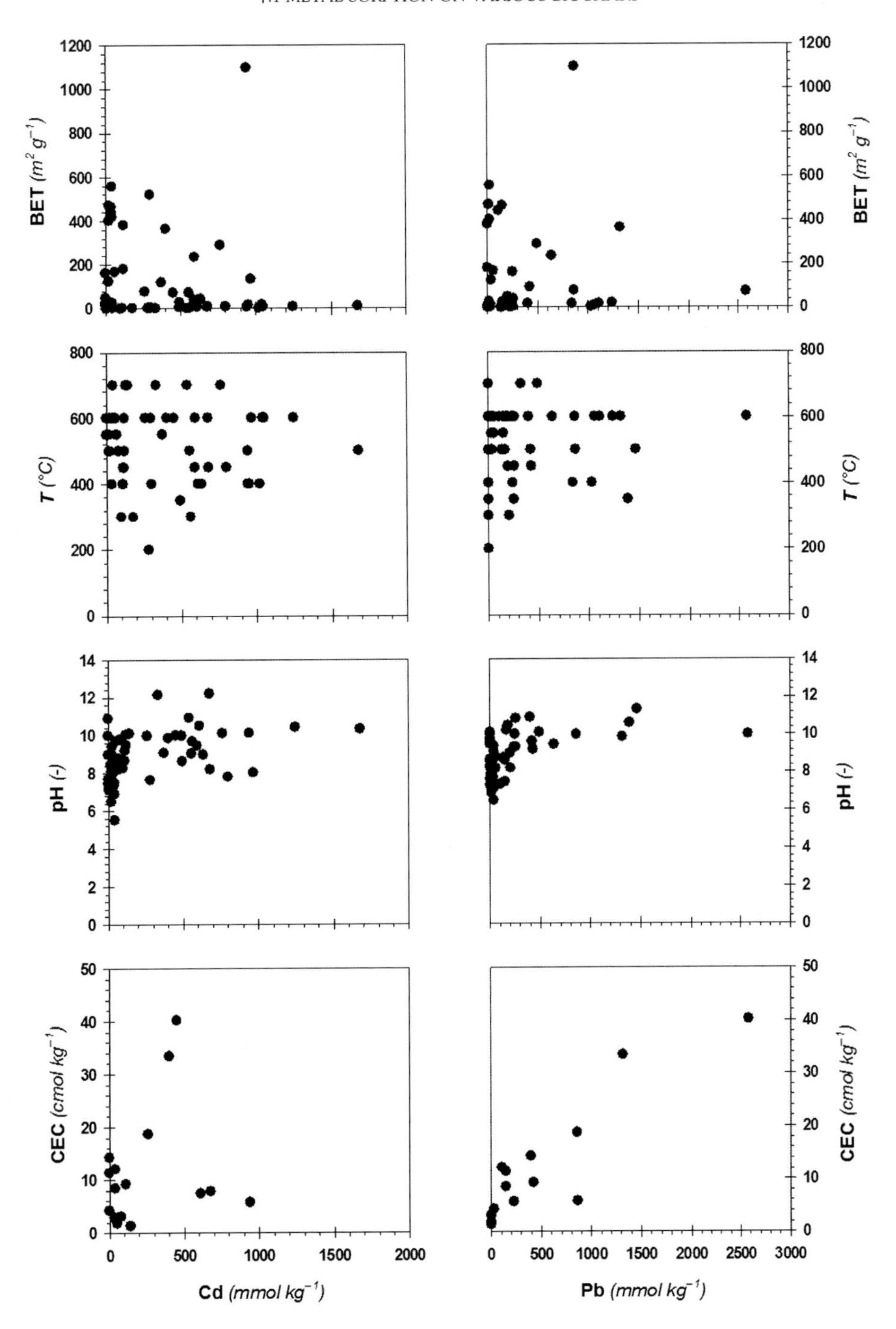

FIGURE 7.2 Relationship between maximum Cd and Pb sorption on different biochars and selected biochars characteristics.

7.1.2 Optimization of Metal Sorption

The optimization of metal sorption can be implemented using batch-sorption procedures designed according to the response surface methodology (RSM). One of the main objectives of RSM is to determine optimum parameters for the control variables that result in a maximum (or a minimum) response over a certain region of interest (Khuri and Mukhopadhyay, 2010). Response surfaces could then be used to determine an optimum and to graphically illustrate the relationship between different experimental variables and their responses (Fig. 7.3). In order to determine an optimum, it is necessary for the polynomial function to contain quadratic terms. The following quadratic model [Eq. (7.1)] could be used to fit experimental data (Montgomery, 2008; Khuri and Mukhopadhyay, 2010):

$$y = \beta_0 + \sum_{i=1}^{k} \beta_i x_i \sum_{i=1}^{k} \beta_{ii} x_i^2 + \sum_{1 \leq i \leq j}^{k} \beta_{ij} x_i x_j + \varepsilon \tag{7.1}$$

where x_1, x_2,... are numbers of associated control (or input) variables, y is a response of interest, β are constant coefficients referred to as parameters, and ε is a random experimental error assumed to have a zero mean. The selected independent variables X_i were coded as x_i according to the following relationship [Eq. (7.2)]:

$$x_i = \left(\frac{X_i - X_0}{\Delta x} \right) \tag{7.2}$$

where X_0 is the uncoded value of X_i at the center point and Δx presents the step change.

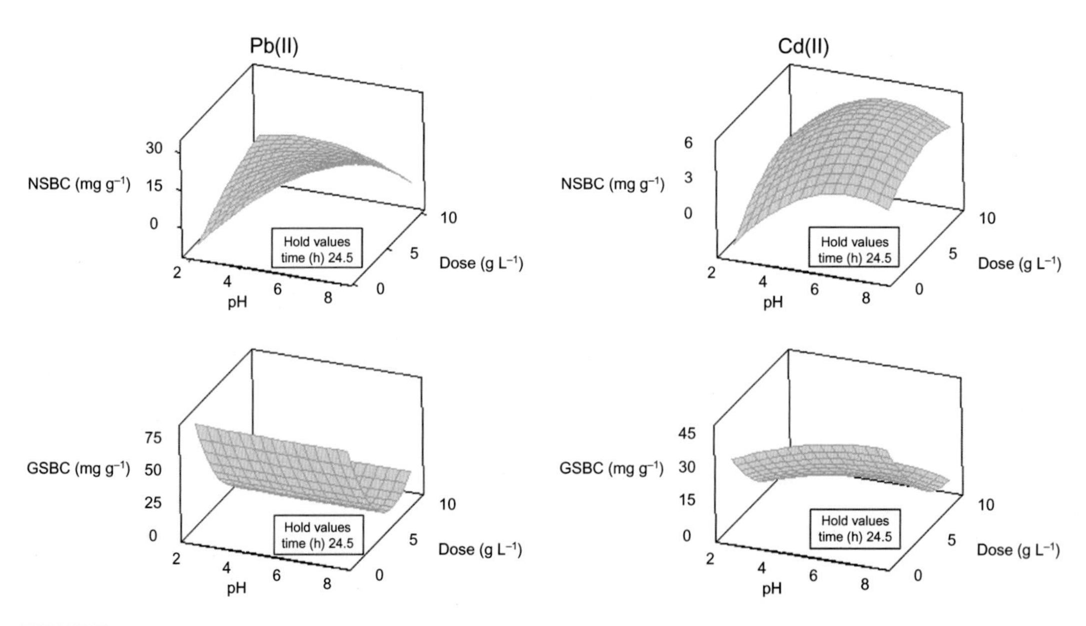

FIGURE 7.3 Response surface plots of Cd(II) and Pb(II) sorption for nut shields and grape stalks biochars, respectively.

In the case of metal-sorption optimization, a conventional batch-sorption experiment is employed before an experiment is designed using central composite design. The influence factors and response(s) are established and the exact number of runs (combing established factors) is then determined by Minitab software. The advantage of this approach is a limited number of runs for in the following example; a design consisting of 20 experiments was used to assess the influence of three factors (pH of solution, dose of applied biochar, and contact time) on two responses (sorbed amount of two contrasting metals) for each biochar.

7.1.3 Metal-Sorption Mechanisms

The metal-sorption process has been described as a result of three different mechanisms (Sohi et al., 2010; Lu et al., 2012): (1) ion exchange (Ca^{2+}, K^{+}, Mg^{2+}, Na^{+}); (2) metal complexation onto free and complexed carbonyl, carboxyl, alcoholic, hydroxyl, or phenolic hydroxyl functional groups; and (3) physical adsorption or surface precipitation caused by sorptive interaction involving delocalization of π electrons of organic carbon (Inyang et al., 2011; Lu et al., 2012; Xu et al., 2013b; Bernardo et al., 2013). As described, ion exchange, complexation, and/or physical adsorption are responsible for metal sorption to biochars. These metal-sorption mechanisms vary according to the type of biochar (group of biochars originating from similar materials) and the metal(loid).

Biochars with well-developed structures (e.g., woody biochars) sorb metals predominantly on the surface. By contrast, biochars with poorly developed structure (represented by low BET surface) exhibit metals not solely sorbed on the surface, but also bonded inside the biochar structure (in higher quantities). Fig. 7.4 shows the exact sorption

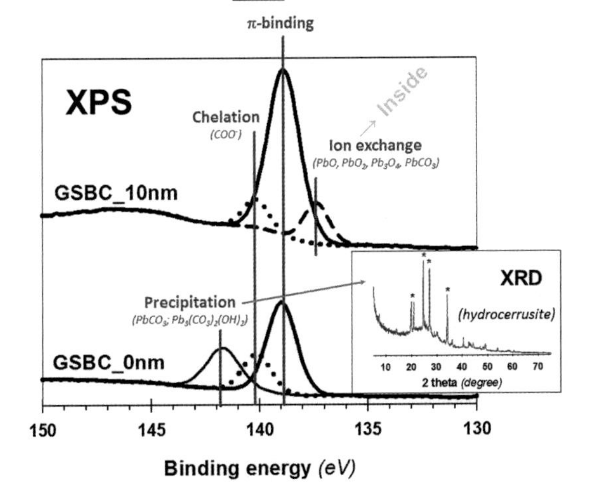

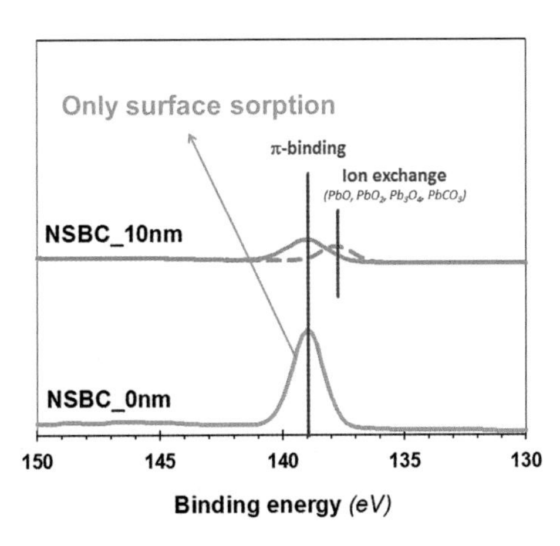

FIGURE 7.4 X-ray photoelectron spectroscopy (XPS) analyses (deconvolution) of the grape stalks and nut shields biochars (on the surface and 10 nm deep) after Pb sorption and X-ray diffraction pattern of the grape stalks biochar confirming precipitation of $Pb_3(CO_3)_2(OH)_2$ (hydrocerrusite; Trakal et al., 2014a).

mechanisms of Pb on/in two contrasting biochars. The weak π-bindings of Pb with poly-organic chains (bonding with electron-rich domains on graphene-like structures; Harvey et al., 2011) has demonstrated physical adsorption (Lu et al., 2012). The weakness of the bond was confirmed by the postdesorption test of fully metal-loaded biochars (Trakal et al., 2014a). The presence of residual CO_3^{2-} (eventually PO_4^{3-}) in biochar together with its alkaline pH are often responsible for the formation of metal-(hydro)carbonates (and/or metal-phosphates), reflecting their surface precipitation [usually confirmed by the X-ray diffraction (XRD) patterns of metal-loaded biochars; see Fig. 7.4; Xu et al., 2013a,b; Trakal et al., 2014a].

Ion exchange, another significant sorption mechanism, was also detected in all tested biochars (Fig. 7.4). Such cation release of metals is limited predominantly to that coming from internal sites of those biochars with poorly developed structures. This could be explained by the low BET surface versus the high quantity of K and Ca (responsible for the ion exchange). The complexation with carboxylic functional groups, as the third sorption mechanism, is demonstrated in Fig. 7.4 (by the binding energy at 141 eV for the Pb). This sorption mechanism (metal-chelate creation) is nevertheless only in those biochars with poorly developed structure where the complexation is a limiting mechanism (Trakal et al., 2014a), which explains why the metal-chelate creation is insignificant in metal sorption.

7.2 BIOCHAR MODIFICATIONS

To further improve their metal-sorption efficiency, biochars can be modified before, during, and/or after the pyrolysis process (Fig. 7.5). As reviewed by Ahmed et al. (2016), there are established surface-modification methodologies such as steam activation, heat treatment, acidic/alkaline modification (chemical activation), and impregnation methods for char-based materials. Steam activation can introduce enhanced porous structures and

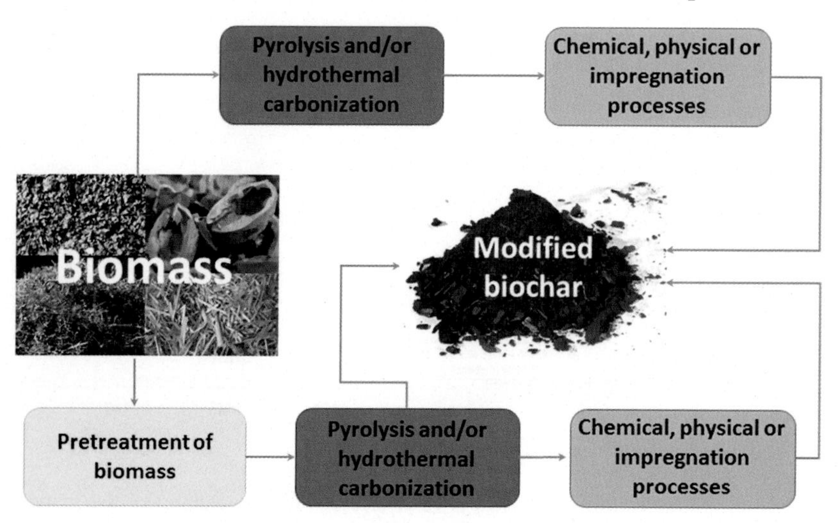

FIGURE 7.5 Schematic of various biochar modifications (inspired by Ahmed et al., 2016).

oxygen-containing functional groups (e.g., carboxylic), whilst heat treatment can provide more basic surface functional groups for hydrocarbon sorption (Shen et al., 2008). Such modifications can effectively be regarded as turbo-boosting the existing sorption characteristics of the biochars. Acidic modification of biochars is applied by different oxidants to increase the acidic property of sorbents by removing or washing mineral elements, thereby increasing the hydrophilic nature of BC (Shen et al., 2008). Conversely, alkaline modification produces negative surface charges that, in turn, assist in adsorbing negatively charged species (Ahmed et al., 2016). Lastly, impregnation methods, by metal salts or oxides mixed with biochars, can facilitate additional physical or chemical attachment of metal ions (Ahmed et al., 2016).

7.2.1 Chemical Activation

Chemical activation of biochars can be performed with an acidic solution or hydroxides as shown by Regmi et al. (2012) and Trakal et al. (2014b), where BC was mixed and stirred with 2 M KOH solution for 1 h at 1:250 ratio (w/v). The resulting solution was then filtered, mixed with ultraclean water, and the pH value of BC then adjusted. Alkaline modification can improve both the physical and chemical properties of BCs. Specifically, 2 M KOH treated brewers draff BC provide significantly increased total pore volume due to the change in pore distribution, but the BET surface area was identical for both biochar samples before and after chemical activation (Table 7.3).

In the context of their use as biosorbents, Cu sorption increased after this alkaline activation from 8.77 to 10.3 mg g^{-1} (Trakal et al., 2014b) as a result of improved physical sorption. Furthermore, the effect of chemical activation on metal-sorption efficiency was even more significant in column-leaching tests (representing a dynamic system). Here, breakthrough curves with an early increase in the c/c_i ratio (Fig. 7.6) were observed for the biochar not previously treated by 2 M KOH. The activated biochar breakthrough curve point (BTCP; the final time step for the maximum sorption efficiency of biochar, when the ratio $c/c_i \approx 0$) occurred at 780 min, as opposed to 50 min for the pristine BC (Fig. 7.6). Such significant differences in BTCP are most likely caused by the presence of free micropores (<6 nm) in the chemically activated biochar (BC_{act}). These pores, created predominantly during high-temperature pyrolysis (Tsai et al., 2012), are formed only in the case of BC_{act} due to the leaching out of the tar particles during alkali solution stirring. In this case, it can be assumed that Cu is trapped inside these free micropores, thus replacing the previously removed tar particles.

TABLE 7.3 Surface Area and Porosity of the Pristine and Chemically Activated BCs Using Alkali Modification by 2 M KOH (Trakal et al., 2014b)

Sample	BET Surface (m^2 g^{-1})	Pores Distribution (%) <6 nm	6–20 nm	20–80 nm	>80 nm	Total Pore Volume (mL g^{-1})
BC	9.80 ± 0.62	/	8.23	59.8	32.0	0.01 ± 0.002
BC_{act}	11.6 ± 0.40	18.7	18.1	38.8	24.5	8.74 ± 0.18

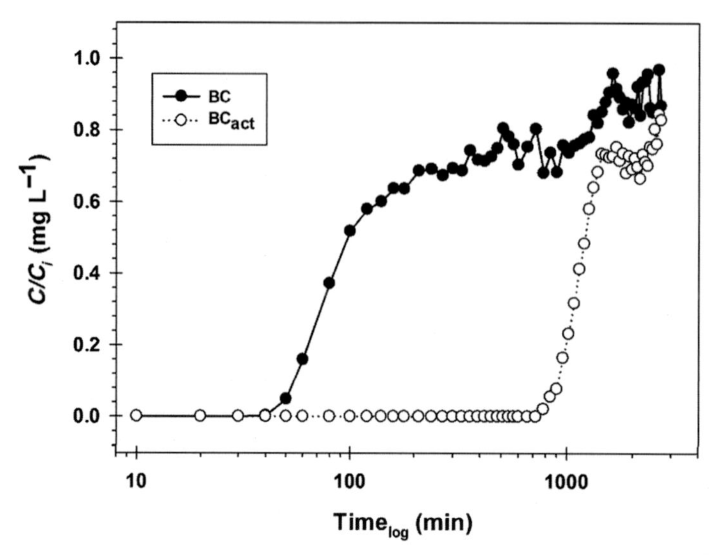

FIGURE 7.6 Breakthrough curves of copper retention, obtained by a column leaching procedure (Trakal et al., 2014b).

7.2.2 Iron Modifications

7.2.2.1 Magnetic Impregnation

Magnetic biochars are produced by mixing pristine biochars with a suspension of various Fe-(hydro)oxides typically when stirred under very alkaline pH (10–12; Mohan et al., 2015). Alternatively, other techniques such as microwave heating and chemical coprecipitation can also be applied (Mubarak et al., 2014; Trakal et al., 2016). For example, $FeSO_4 \cdot 7H_2O$ solution can be alkalinized (at pH $\approx$ 12) in order to precipitate Fe-hydroxides (Trakal et al., 2016). Such created suspension is then put into a microwave, and this microwave-assisted synthesis results in the formation of iron-oxide nanoparticles with diameters ranging between 25 and 100 nm, consisting mainly of nonstoichiometric magnetite. Subsequently, ground biochar is then thoroughly mixed with the iron-oxide nanoparticles (Trakal et al., 2016). The efficiency of magnetically modified biochars as biosorbents is related to their altered surfaces, relative to pristine biochars. Firstly, magnetic modification decreases the active BET surface due to the formation of secondary iron (hydro)oxides on the surface (Mohan et al., 2015; Wang et al., 2015a), which is more apparent in biochars with well-developed structures. Conversely, in biochars with poorly developed structures, the BET surface significantly increases (Trakal et al., 2016) due to the presence of iron oxides, which have: (1) a smaller surface area (36.6–66.0 $m^2\ g^{-1}$; Oliveira et al., 2002; Michálková et al., 2014) than that of biochars with a well-developed structure, but (2) a similar surface area to that of biochars with poorly developed structure. The magnetizing process also modified the pH value as well as the CEC, which could be explained by the presence of Fe oxides, which usually have neutral pH and rather high CEC values (in the range of 61.4–219 mmol kg^{-1}; Trakal et al., 2016).

Secondly, the presence of Fe minerals can enhance the sorption efficiency of various metals (Reddy and Lee, 2014; Wang et al., 2014; Mohan et al., 2015) or metalloids (Zhang et al., 2013b; Baig et al., 2014; Wang et al., 2015a), typically fixed/sorbed on the surface of these precipitated (hydro)oxides. Therefore, magnetic biochars usually have significantly higher sorption, compared with pristine ones, of Cd(II) Pb(II) in those biochars with well-developed structure (Table 7.4), which can be explained by the presence of Fe oxides inside the structure of the biochars (Mohan et al., 2007, 2015; Zhou et al., 2013, 2014; Han et al., 2015). The improvement of metal sorption by magnetic modification is thus related

TABLE 7.4 Cadmium- and Lead-Removal Efficiency of Biochars Before and After Magnetic Modification From Selected Studies (Trakal et al., 2016)

		Biochar Sorption Improvement by Magnetic Modification (%)		
Biochar	**Magnetic Modification**	**Cd**	**Pb**	**References**
Oak bark	Fe^{2+}/Fe^{3+} SO_4 solution	37.0	131	Mohan et al. (2007, 2015)
Oak wood	Fe^{2+}/Fe^{3+} SO_4 solution	676	287	Mohan et al. (2007, 2015)
Bamboo	Fine sized ZVI of <850 μm	/	67.3	Zhou et al. (2013, 2014)
Pine bark	Ferrite ($CoFe_2O_4$)	152	/	Reddy and Lee, (2014)
Nut shield	$FeSO_4$•$7H_2O$	997	461	Trakal et al. (2014a, 2016)
Wheat straw	$FeSO_4$•$7H_2O$	218	14.0	Trakal et al. (2014a, 2016)
Grape stalk	$FeSO_4$•$7H_2O$	121	− 58.4	Trakal et al. (2014a, 2016)
Grape husk	$FeSO_4$•$7H_2O$	316	14.3	Trakal et al. (2014a, 2016)
Plum stone	$FeSO_4$•$7H_2O$	1592	741	Trakal et al. (2014a, 2016)
Palm oil empty fruit bunch	$FeCl_3$	/	− 25.7	Ruthiraan et al. (2015)
Energy cane biochar	Ferric sulfate ($Fe_2(SO_4)_3$•nH_2O)	/	− 11.2	Mohan et al. (2015)
Romchar (Harghita, Romania)	Magnetite/maghemite	/	− 50.5	Han et al. (2015)
Oxford Biochar Ltd. (Dorset, UK)	Magnetite/maghemite	/	139	Han et al. (2015)
Rice hull	$Fe(acac)_3$ + calcination	/	− 50.3	Yan et al. (2015)
Rice hull	$Fe(acac)_3$ + calcination + ZnS	/	665	Yan et al. (2015)

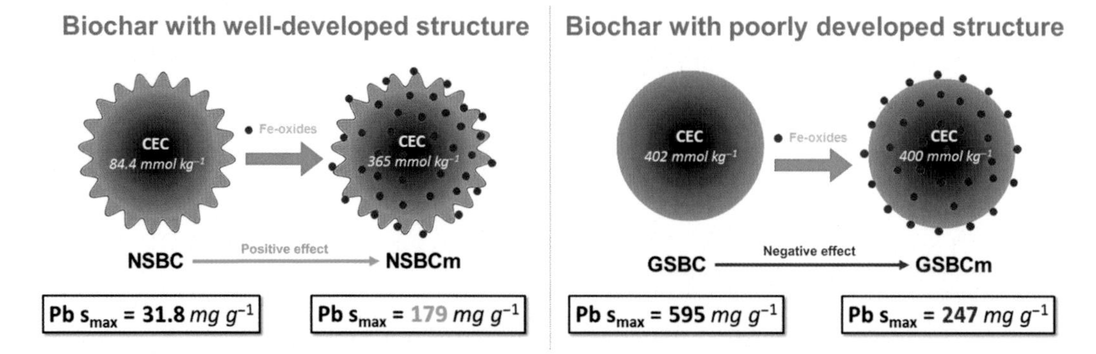

FIGURE 7.7 Examples of magnetic modification on two contrasting biochars (Trakal et al., 2016).

to the origin of the material (especially for Pb(II)), with the well-structured biochars being most suitable for these modifications (Mohan et al., 2007, 2015; Trakal et al., 2014a; Fig. 7.7). However, the sorption efficiency of Pb^{2+} in/on biochars with poorly developed structures changed little or was reduced compared with that of the pristine biochars (Fig. 7.7). Metal sorption is not limited to the biochar surface (Trakal et al., 2014a), but also occurs in the biochar structure. This effect is more significant for Cd(II) than for Pb(II) because of the higher affinity of Cd(II) to sorb on Fe oxides (Adriano, 2001).

The results of XPS analyses (Fig. 7.8) have confirmed similar Cd(II) and Pb(II) loadings for both surface and 10 nm depth against pristine biochars, where the amount of sorbed metals varied significantly between surface and subsurface (Trakal et al., 2014a). This could be explained through the attaching of Fe oxides to the biochar matrix when no clusters of these oxides appear on the surface of the biochar (Han et al., 2015). Fig. 7.8 shows that cation release and/or precipitation, represented by Pb–O binding (Xi et al., 2010), is the predominant sorption mechanism, whereas the confirmed presence of precipitated metal (hydro)carbonates were limited to the surface of pristine biochars only (Lu et al., 2012; Xu et al., 2013a,b; Trakal et al., 2014a). The slightly weak π-binding of metals with polyorganic chains demonstrated physical adsorption as a further sorption mechanism. However, this metal-sorption mechanism was suppressed at the expense of chemisorption (in comparison to the pristine biochar), mainly caused by the presence of Fe oxides in the structure of the magnetic biochars (Mohan et al., 2015). Finally, the observation of metal chelates not only on the biochar surface, but also within the interior, is in agreement with the results of Lu et al. (2012), where the magnetic modification usually enhanced the overall complexation sorption mechanism because of the presence of Fe oxides in the structure of the tested biochars.

7.2.2.2 *Nano Zero-Valent Iron Modification*

Nano zero-valent iron (nZVI) has been shown to be a highly effective sorbent for various inorganic and organic contaminants in aqueous solutions. A typical core–shell structure of nZVI plays a key role in this process with Fe^0 core as a donor of electrons, offering strong reduction capability, and Fe (hydr)oxide shell (formed from the Fe^0 oxidation), allowing electron transfer that is responsible for adsorption of various

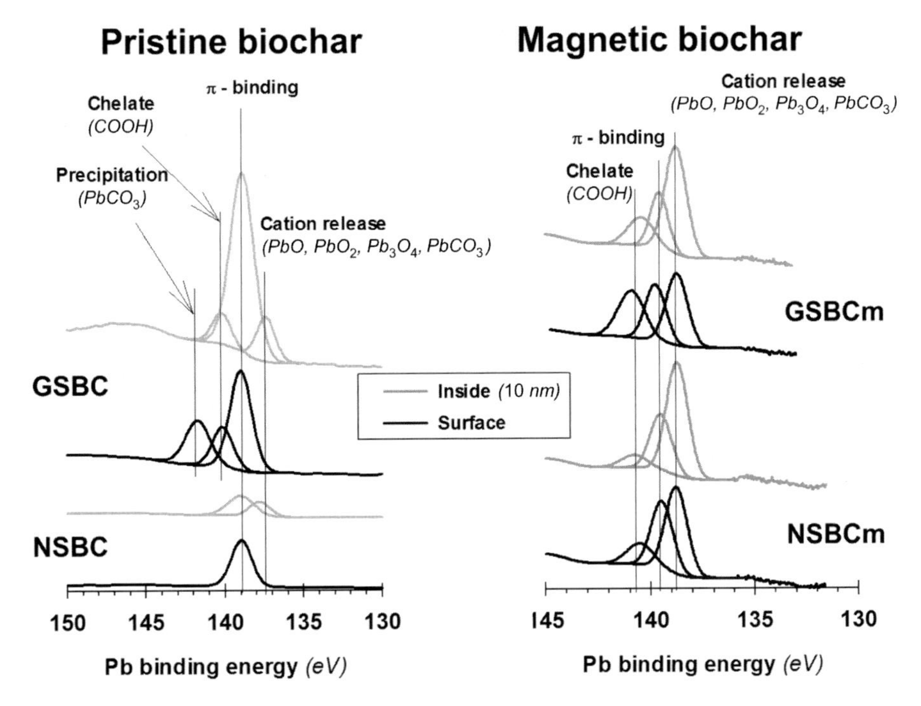

FIGURE 7.8 XPS analyses of metal-loaded biochars (deconvolution of XPS bands for various Pb binding energies); GSBC – grape stalks (poorly developed) biochar; NSBC – nut shields (well-developed) biochar before and after magnetic impregnation (Trakal et al., 2016).

contaminants (Li and Zhang, 2007; O'Carroll et al., 2013; Tosco et al., 2014; Stefaniuk et al., 2016). The nZVI particles are remarkable in their large specific surface area, high reduction capacity, low operational costs, and simple subsequent separation due to magnetic properties. On the other hand, nZVI particles have a tendency to aggregate, which decreases their specific surface area and thus their reactivity, mobility, and reduction capacity (Devi and Saroha, 2014; Tosco et al., 2014; Stefaniuk et al., 2016). When applied to soils, the high reactivity of nZVI particles could have a negative impact on soil properties, and the presence of several metal(loid)s could also limit the immobilization of nZVI (Gil-Díaz et al., 2014, 2017; Su et al., 2016; Vítková et al., 2017). The combination of biochar with Fe nanoparticles may give the resulting biosorbent unique properties beneficial for selective sorption as well as for the stability of the nano-sized amendments in soils and waters (Tan et al., 2016a; Dong et al., 2017).

The nZVI-biochar composite can be produced by different synthesis methods. Generally, there are two approaches: (1) precoating of biomass with nZVI particles before pyrolysis and (2) biomass pyrolysis followed by impregnation of biochar with nZVI particles (Tan et al., 2016a) (Fig. 7.9). In either case, the whole synthesis needs to be conducted in an inert atmosphere to avoid O_2 dissolution during preparation (Yan et al., 2015). The synthesis of the biochar (matrix) with functional nanomaterials can provide a unique combination of the advantages of both materials and should represent an ideal synthesis with

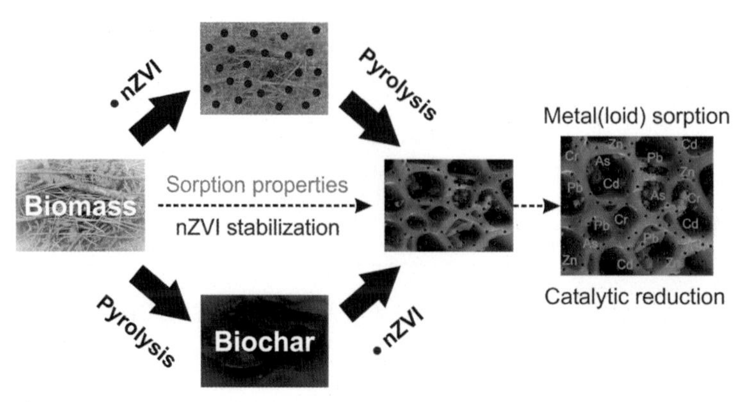

FIGURE 7.9 Schematic of synthesis of the nano zero-valent iron and biochar composite and its improved functions.

nZVI due to: (1) better cost efficiency compared to conventional amendments, nontoxicity, and local availability of biochar; (2) ability to disperse the nanoparticles and thus enhance their stability; and (3) increasing number of oxygen-containing functional groups, an improvement in pore properties, surface active sites, and catalytic degradation ability when combined with nZVI (Fig. 7.9). Therefore, this composite should exhibit an excellent ability to remove a wide range of contaminants (e.g., metals, organochlorine, or nitroaromatic compounds) from aqueous solutions, exerting simultaneous adsorption and catalytic degradation function (Peng et al., 2017; Tan et al., 2016a). Improved magnetic properties of the resulting material should allow the effective separation of the composite together with the adsorbed contaminants from the treated matrix (Trakal et al., 2016; Wang et al., 2017). Although magnetic modification may decrease the active BET surface (see Section 7.2.2.1) due to the formation of secondary iron (hydro)oxides on the surface of biochar, some metals (Cd; Trakal et al., 2016) or metalloids (As; Wang et al., 2015a) can be adsorbed more efficiently on the surface of these precipitated (hydro)oxides.

Impregnation of biochars with nZVI prevents early and excessive reaction of the nanoparticles, but allows them to use their original functionality (reduction and sorption) later in contact with aqueous solutions (Devi and Saroha, 2014; Dong et al., 2017; Peng et al., 2017). A distinct disadvantage of pristine biochars are their low efficiency for anion sorption (e.g., As) and complicated subsequent separation from the treated medium, which limits their applicability for soil and water remediation. The drawbacks of nZVI and biochar thus can be addressed by appropriate modifications. In particular, biochar supported by nZVI results in the increased stability of nZVI and improves sorption efficiency and magnetic properties of the final material. Several recent studies have reported the use of nZVI-BC composite for efficient removal of As, Cr, or organic contaminants and others from (waste)waters or aqueous solutions (Dong et al., 2017; Peng et al., 2017; Wang et al., 2017); in particular, arsenate (Zhou et al., 2014; Wang et al., 2016a, 2017), hexavalent chromium (Zhou et al., 2014; Dong et al., 2017; Qian et al., 2017), lead (Zhou et al., 2014), pentachlorophenol and trichloroethylene (Devi and Saroha, 2014; Yan et al., 2015; Li et al., 2017), and bromate (Wu et al., 2013). However, studies dealing with nZVI-modified biochar in soil environments are scarce (Su et al., 2016).

7.2.3 Layered Double-Hydroxide Modification

Although biochars have shown high sorption efficiency for various metal cations, their usability in waters and soils contaminated by anions, for example, arsenate, phosphate, or chromate, is limited (Wan et al., 2017). Therefore, it is necessary to provide sufficient surface modification of pristine biochars using highly effective sorbents such as layered double hydroxides (LDHs), if simultaneous sorption of cations and anions is to be achieved. Previously, LDHs have shown high sorption efficiencies for various inorganic species, for example, arsenate (Hudcová et al., 2017), arsenite (Jiang et al., 2014), chromate (Zhang et al., 2012), phosphate (Drenkova-Tuhtan et al., 2013), sulfate (Sepehr et al., 2014) as well as metal cations (Hudcová et al., 2018). In general, anion-sorption mechanisms using LDHs are mainly influenced by surface complexation and/or anion exchange (Hudcová et al., 2017), while metals are predominantly immobilized by surface-induced precipitation, surface complexation, and/or cation exchange (Hudcová et al., 2018). Despite the excellent sorption properties of LDHs, they have some drawbacks; aside from their relatively high cost, their stability in acidic conditions is greatly reduced. However, this could be improved by their immobilization on biochar surfaces but data are scarce as the use of LDH/biochar composites is a new topic with fewer than 15 peer-reviewed articles existing, dealing mainly with phosphate (Li et al., 2016; Wan et al., 2017; Zhang et al., 2013c, 2014), arsenate (Wang et al., 2016b,c), organics (Tan et al., 2016b,c), nitrate (Xue et al., 2016), and metal cations (Wang et al., 2018; Wang and Wang, 2018) (ad)sorption.

7.2.3.1 Synthesis of LDH/Biochar Composites

LDH/biochar composites can be synthesized before or after pyrolysis. The prepyrolysis procedure consists of LDH coprecipitation on the biomass surface and subsequent pyrolysis for 2 h at approx. 500–600 °C (Tan et al., 2016b,c; Wang et al., 2016b,c). Alternatively, LDH precoated biomass can also be hydrothermally treated, resulting in the production of LDH/hydrochar composites (Zhang et al., 2014). The postpyrolysis procedure, which is more widely used, consists of pyrolysis of pristine biomass at conditions mentioned above and subsequent coprecipitation of LDHs on the biochar surfaces (Li et al., 2016; Wan et al., 2017, 2016b,c, 2018; Wang and Wang, 2018; Xue et al., 2016; Zhang et al., 2013c). The synthesis prepyrolysis procedure strongly influences the specific surface area of final LDH/biochar composites, namely resulting in a decrease of the specific surface area (nearly twofold), in a higher amount of nonreacted metal species, and in a lower As(V) adsorption capacity compared to postpyrolysis products (Wang et al., 2016b,c). In general, LDH/biochar composites exhibit a lower specific surface area compared to pristine biochar caused by filling/clogging of pores after the modification procedure, which also results in a notably rougher surface of LDH/biochar composites (Wan et al., 2017; Xue et al., 2016).

To demonstrate the structure of LDH/biochar composites, diffractograms of pristine woody biochar, Mg–Fe LDH (molar ratio Mg/Fe = 4) and Mg–Fe LDH/biochar composites are shown in Fig. 7.10A and the SEM images of LDH/biochar composites are shown in Fig. 7.10B. Diffractograms of pristine biochar (with minor phases corresponding to carbonates and/or silicates); pristine LDHs (with characteristic peaks of given intensities); and LDH/biochar composites confirmed the structure of individual materials and the successful synthesis of LDH/biochar composites. The SEM images demonstrated

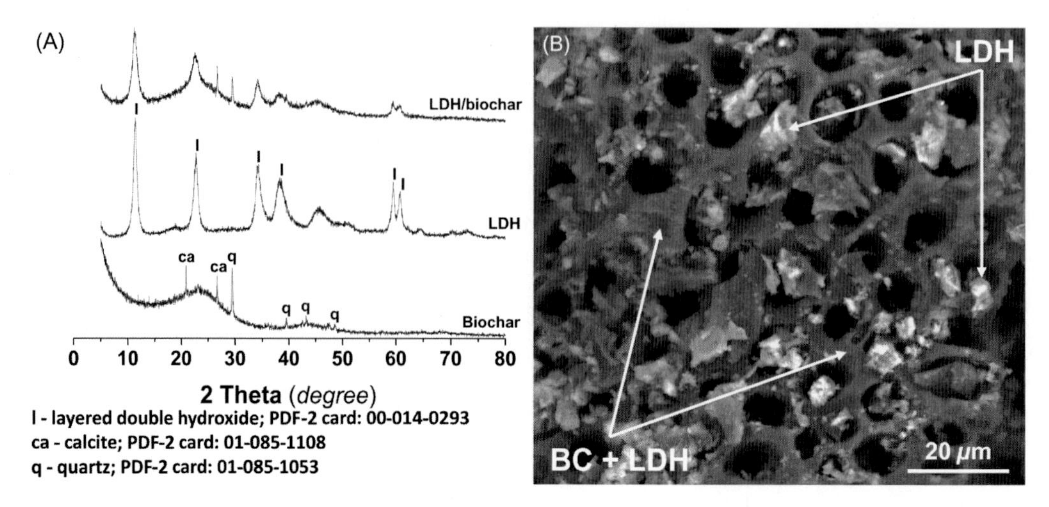

FIGURE 7.10 Diffractograms of pristine woody biochar, Mg–Fe LDH and Mg–Fe LDH/biochar composites (A) and the SEM image of Mg–Fe LDH/biochar composites (B). *LDH*, Layered double hydroxides; *SEM*, scanning electron microscopy.

nonhomogeneous distribution of LDH particles on the biochar surface probably caused by the coprecipitation synthesis procedure of LDHs. The morphology of these LDH particles corresponded to pristine LDHs as described in previous studies (Hudcová et al., 2017, 2018). LDHs were also successfully impregnated on the whole biochar surface, as confirmed by the EDX results, suggesting a high affinity of LDHs with the biochar surface that could positively influence the stability/sorption properties of LDH/biochar composites.

7.2.3.2 *Adsorption Properties of LDH/Biochar Composites*

The adsorption properties of LDH/biochar composites are strongly influenced by the presence of LDHs on the biochar surface. The adsorbed amounts of different inorganic species by various LDH/biochar composites are given in Table 7.5. In general, LDH/biochar composites appeared to be more effective materials for phosphate removal compared to pristine biochar, modified biochar, and many other conventional adsorbents (Li et al., 2016; Wan et al., 2017; Zhang et al., 2013a–c, 2014). The LDH/biochar composites showed even higher adsorbed amounts of phosphate compared to pristine LDHs (adsorption rate per unit weight) suggesting a synergistic effect between biochars and LDHs (Wan et al., 2017), with biochars providing an effective matrix for LDHs that increases accessibility of active adsorption sites (Zhang et al., 2013c). The adsorption mechanism of phosphate using LDH/biochar composites includes surface complexation and anion exchange with interlayered anions in the structure of LDHs. At higher initial phosphate concentration precipitation with released bivalent metal cations from the surface/structure of LDHs was also observed (Li et al., 2016; Wan et al., 2017; Zhang et al., 2013c). In the case of arsenate adsorption, LDHs significantly improved the adsorption properties compared to pristine biochars and some other pristine LDHs. Nevertheless, arsenate adsorption is greatly

TABLE 7.5 The Adsorbed Amounts of Inorganic Species Using LDH/Biochar Composites

Biomass Type	LDH Type	Adsorbate	q_{MAX} (mg g^{-1})[a]	Reference
Bamboo	Mg–Al	Phosphate	172	Wan et al. (2017)
Sugarcane leaves	Mg–Al	Phosphate	82	Li et al. (2016)
Cottonwood	Mg–Al	Phosphate	410	Zhang et al. (2013c)
Cottonwood	Mg–Al	Phosphate	386	Zhang et al. (2014)
Pinus taeda	Ni–Fe	Arsenate	4.4	Wang et al. (2016b)
Pinus taeda	Ni–Fe[b]	Arsenate	1.6	Wang et al. (2016b)
Pinus taeda	Ni–Mn	Arsenate	6.5	Wang et al. (2016c)
Pinus taeda	Ni–Mn[b]	Arsenate	0.6	Wang et al. (2016c)
Wheat straw	Mg–Fe	Nitrate	25	Xue et al. (2016)
Camellia oleifera	Mn–Al	Copper	74	Wang et al. (2018)
Leaves[c]	Mg–Al	Lead	147	Wang and Wang (2018)

[a] *Adsorbed amount (Langmuir model).*
[b] *Prepyrolysis synthesis of LDHs.*
[c] *Nonspecified leaves.*
LDH, layered double hydroxide.

influenced by the pyrolysis procedure, exhibiting significantly higher adsorbed amounts using postpyrolysis LDH/biochar composites. The arsenate-adsorption mechanism follows the same processes as described by phosphate adsorption, except for precipitation. However, the dominant mechanism differs for prepyrolysis (physical adsorption) and postpyrolysis (chemical adsorption/anion exchange) LDH/biochar composites (Wang et al., 2016b,c). Nitrate adsorption was also significantly improved using LDH/biochar composites compared to some biochars and activated carbon. Higher selectivity for nitrate in multianion solutions containing sulfate and phosphate was also observed. The whole adsorption mechanism follows the same processes as noted for phosphate adsorption except for precipitation (Xue et al., 2016). LDH/biochar composites were found to be more effective copper and lead sorbents compared to some biochars, modified biochars, activated carbons, and other conventional sorbents. The dominant sorption mechanism has been shown to be surface precipitation caused by the buffering effect of LDH/biochar composites and the formation of chelation complexes in the case of ethylenediaminetetraacetic acid (EDTA) functionalized LDHs. Even so, the cation exchange of bivalent metal cations in the structure of LDHs by copper was also observed (Wang et al., 2018; Wang and Wang, 2018).

To demonstrate arsenate-adsorption efficiency, the As(V) adsorption kinetics using pristine woody biochar, pristine Mg–Fe LDHs, and Mg–Fe LDH/biochar composites at controlled pH 5.5 (As(V) initial concentration of 1 mM) is given in Fig. 7.11. As previously mentioned, pristine biochar showed a very low As(V) adsorption efficiency (3.1%) compared to highly effective Mg–Fe LDHs (100%) and Mg–Fe LDH/biochar composites

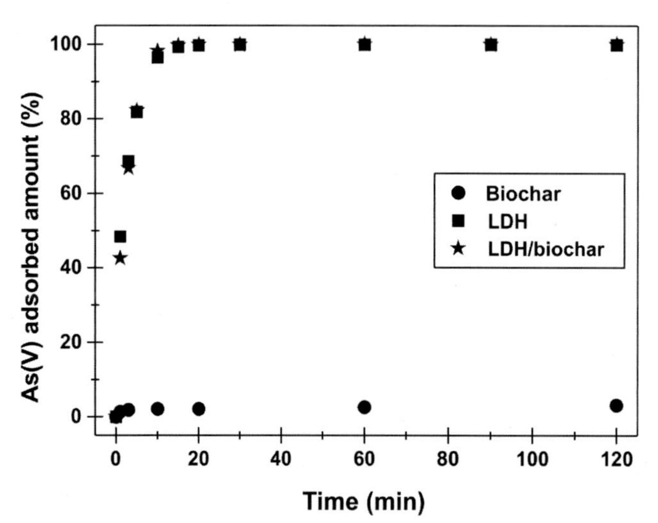

FIGURE 7.11 As(V) adsorption kinetics using pristine woody biochar, Mg–Fe LDHs, and Mg–Fe LDH/biochar composites at controlled pH 5.5. *LDH*, Layered double hydroxide.

(100%). It is worth nothing that Mg–Fe LDH/biochar composites contained only approximately 40% of LDHs on the surface/in the structure compared to pristine LDHs, suggesting a high efficiency of such composites for arsenate removal. In summary, LDH/biochar composites are efficient sorbents, but further studies focused on the stability of LDH/biochar composites at different conditions, detailed mechanistic/modeling approaches, and alternative LDH synthesis procedures to make their production more economical are needed.

7.2.4 Manganese-Oxide Coating

Mn oxides have also been demonstrated to have very high immobilization potential for divalent metals through adsorption processes onto amphoteric surface groups of the Mn oxide, and also metalloid oxyanions due to the initial oxidation/reduction process and consequent sorption, surface complexation, and/or coprecipitation with birnessite and hydrous manganese oxide (Lenoble et al., 2004; Komárek et al., 2013). Among the various types of Mn oxides the modified sol–gel procedure of Ching et al. (1997) was used for the preparation of amorphous manganese oxide (AMO; Della Puppa et al., 2013). The AMO has previously been established as a very suitable sorbent for various metal(loid)s, but associated dissolution of AMO has resulted in excessive Mn leaching, limiting its applicability to remediation scenarios due to, amongst other concerns, Mn phytotoxicity (see Della Puppa et al., 2013; Michálková et al., 2014; Ettler et al. 2015). AMO-modified biochar, combining the advantageous properties of both AMO and pristine biochar with decreased Mn leaching and thus improved sorbent longevity, appears to be an ideal biosorbent for multimetal(loid)-contaminated wastewater treating. Preparation of AMOchar composite involves biochar impregnated by AMO (Della Puppa et al., 2013), with the biochar added directly into the reaction mixture of 0.4 M $KMnO_4$ solution with 1.4 M glucose solution.

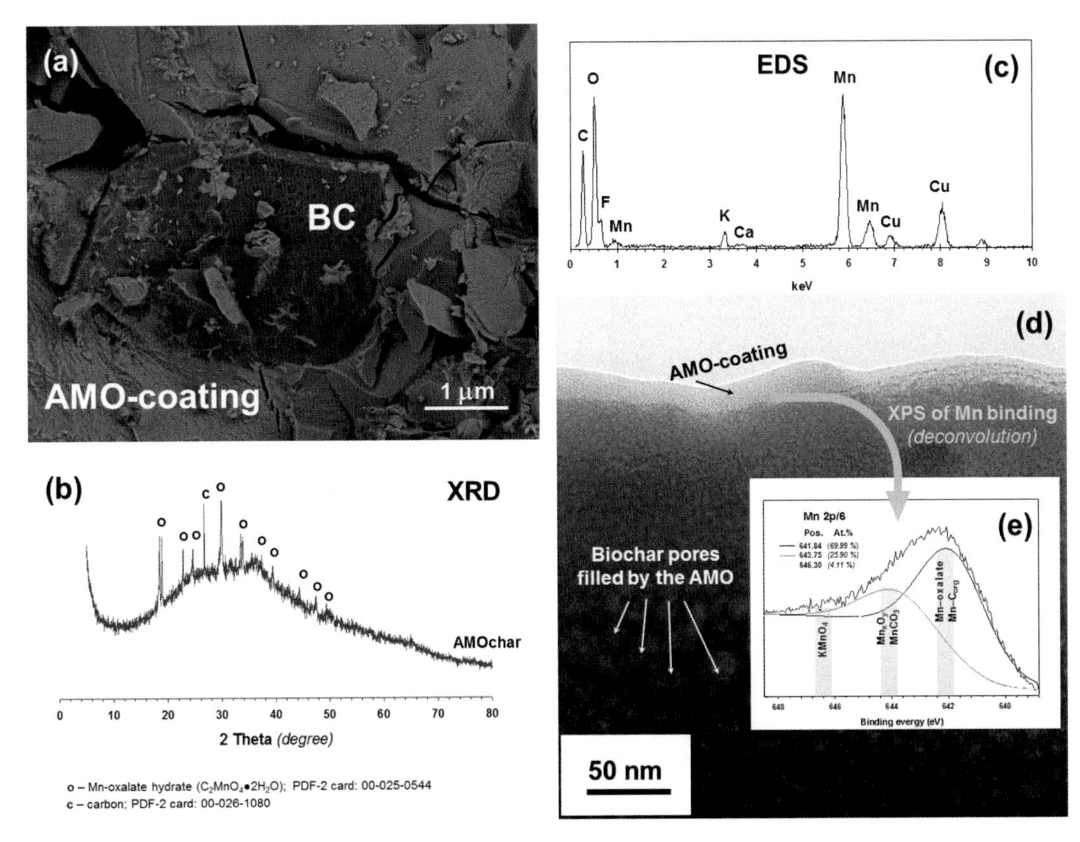

FIGURE 7.12 (A) High-resolution SEM of AMOchar, (B) XRD patterns, (C) energy-dispersive X-ray spectroscopy (EDS) spectra where peaks of Cu originate from the transmission electron microscopy (TEM) metal grid, (D) TEM cross-section of the AMO coating, and (E) XPS of Mn binding and the bond deconvolution of the AMOchar before sorption (Trakal et al., 2018). *SEM*, scanning electron microscopy; *XRD*, X-ray diffraction; *AMO*, amorphous manganese oxide.

XRD analysis (Fig. 7.12B) of the AMOchar confirmed its amorphous character, where all peaks were represented by Mn-oxalate hydrate and by carbon showing the presence of the biochar skeleton. The presence of rhodochrosite was confirmed, which has been explained by Ettler et al. (2014) as a reaction of the removed Mn from the oxalates of the AMO with atmospheric CO_2. Consequently, such leached Mn could react with CO_3^{2-} groups sorbed at the surface of biochar, originating from the meaningful CO_2 sorption (Xu et al., 2016) during synthesis under atmospheric conditions. Fig. 7.12 shows the structure and surface morphology of AMOchar where the AMO with Mn-oxalates is coating the surface of the pristine biochar, even though porous biochar fragments associated with irregular AMO clusters were also observed. The attraction of the AMO on the surface of pristine biochar was confirmed by deconvolution of Mn bond at binding energy 641.84 eV, which is here responsible for the $Mn-C_{org}$ bond. The pH value of the AMOchar composite was alkaline and showed a theoretically positively

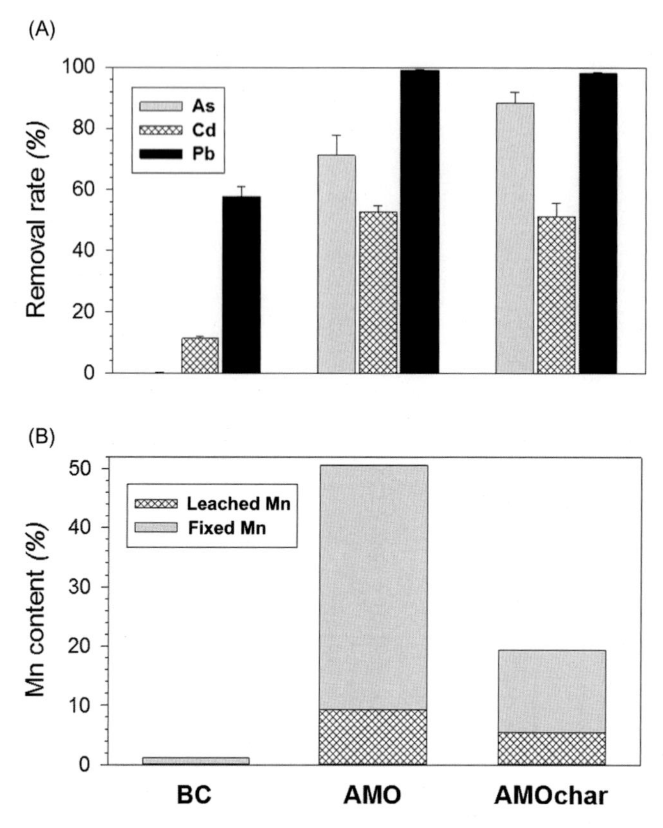

FIGURE 7.13 (A) Maximum removal efficiency of As, Cd, and Pb (obtained from kinetic experiment at equilibrium state; data in triplicates are presented as means ± SD; and Mn leaching of pristine biochar, pure AMO and AMOchar; Trakal et al., 2018). *AMO*, amorphous manganese oxide.

charged surface. The CEC of the composite was similar and/or even higher than that of pure AMO (Trakal et al., 2018).

In terms of metal(loid)s removal from solution (Fig. 7.13), AMOchar was able to remove up to 99%, 51%, and 91% of dissolved Pb(II), Cd(II), and As(V), respectively, in comparison to pristine biochar where the removal rates were limited. The leaching of Mn during metal(loid)s sorption from the AMOchar composite was significantly reduced in comparison to the pure AMO. This demonstrates that total Mn content in AMOchar composite was lower compared to the AMO (without any limitation to metal sorption) and/or stabilization had occurred in the Mn-oxalates on the structure of AMOchar. In addition, the AMOchar composite has also been tested using a dynamic test when the cumulative leachate of As(V) from CCA (chromate-copper-arsenate) ash was significantly reduced (by sixfold compared to untreated variant after 100 min of intensive leaching; see the study of Trakal et al., 2018). Ongoing testing of AMOchar mixed into a variety of metal(loid) contaminated soils confirmed reductions in leaching of metal(loid)s but continued instability of Mn; further refinements to the synthesis methods will be required to achieve a material with sorptive longevity.

7.3 ENGINEERING IMPLICATIONS OF BIOCHAR AND ITS MODIFICATIONS

The reutilization of agrowastes for biochar production appears to be an environmentally friendly technique for metal removal from aqueous solutions, e.g., industrial wastewaters or mine drainage (Lu et al., 2012; Zhang et al., 2013a). For that to be fully realized these materials need to be proven safe, efficient, and cost effective for environmental applications. Though the removal capacity for various metals (e.g., Cd vs Pb) was different for each tested biochar, in general those biochars with well-developed structures (reflected by high BET surface) removed lower quantities of metals in comparison to biochars with poorly developed structures, where sorbed metals are very strongly bound. By contrast, while the biochars with well-developed structure were found to be less effective for the metal sorption they proved more suitable for removal of organic pollutants (e.g., phenol; Han et al., 2013), mainly due to their high BET surface. Biochars with high BET surface are potentially more suitable for consequent modifications (impregnations) mainly due to "sufficient space" for the impregnation, thus resulting in the production of sorbents efficient both for organic and inorganic contaminant removal from waters [and soils].

As has been discussed, chemical activation using various alkali solutions (e.g., 2 M KOH) can improve biochar structure by releasing tar particles, especially from micropores. This improvement could be applicable when the BET surface needs to be enlarged (e.g., for more efficient magnetic impregnation and more efficient biochars with well-developed structure).

Magnetic modification of biochar can improve sorption efficiency even for those biochars with well-developed structures. From a technological point of view, such magnetically modified biochars are generally ferromagnetic and therefore can be recovered more easily from contaminated waters after a filtration process using a magnet (Zhou et al., 2014). This property is highly desirable for the convenient recycling of contaminant-laden magnetic biochar adsorbents when their useful life is expended. Overall, magnetic impregnation significantly affected various biochar characteristics as well as its metal-sorption efficiency. Trakal et al. (2016) has demonstrated that magnetic biochar could (1) sufficiently remove Cd(II) and Pb(II) and (2) be easily separated from the solution with a magnet (Fig. 7.14).

Although biochar modified by Mn oxides does not provide a ferromagnetic material, the AMOchar represents a very efficient and complex biosorbent for treatment of multimetal(loid)s contaminated waters at various pH. Usually, pristine biochars are suitable for selective sorption of particular bivalent metals as well as for multimetal loading, mainly due to their overall basic pH and negatively charged active surface (Mohan et al., 2014b). However, when different bivalent metals are sorbed to the same biochar the sorption efficiency of some metals (e.g., Cd(II) or Zn(II)) can be diminished by the sorption of metals with high affinity to organic matter (e.g., Pb(II) or Cu(II); Trakal et al., 2012, 2014a). Additionally, the efficiency of the pristine biochars to sorb oxyanions (e.g., As(V) or Cr (VI)) is usually limited, especially under alkaline conditions when such oxyanions are very mobile. In order to ensure complexity of this material to sorb various metal(loid)s as well

FIGURE 7.14 Separation of magnetic biochar (from nut shields) previously loaded by Pb(II).

as to ensure its universality (adsorbent that is effective under a wide range of pH and/or Eh conditions), the coating of biochar by Mn oxides represents a step forward for the treatment of waters contaminated with bivalent metals (Cd(II) Pb(II), Zn(II), etc.) and oxyanions (e.g., As(V) or Cr(VI)). Another practical advantage of this composite lies in the sole use of waste materials; for example, biochar can be produced from waste biomass (prunings, off-cuts, etc.), while AMO can be produced from other organic waste materials (e.g., molasses from sugar factories) reacting with potassium permanganate. In comparison to the pure AMO, AMOchar composite represents an equally efficient sorbent with reduced Mn leaching. When improvements are made to the synthesis to further stabilize Mn the resulting biosorbent appears applicable to a wide range of environmental remediation scenarios.

Acknowledgments

The authors are grateful for financial support from Operational Programme Prague – Growth Pole of the Czech Republic (no. CZ.07.1.02/0.0/0.0/16_040/0000368), project TAČR TJ01000015 (Technical Agency of Czech Republic), and project GAČR 18-24782Y (Czech Science Foundation), and the support of the Scottish Government's Rural and Environmental Sciences and Analytical Services (RESAS).

References

Abdelhafez, A.A., Li, J., 2016. Removal of Pb(II) from aqueous solution by using biochars derived from sugar cane bagasse and orange peel. J. Taiwan Inst. Chem. Eng. 61, 367–375.

Adriano, D.C., 2001. Trace Elements in the Terrestrial Environments: Biogeochemistry, Bioavailability, and Risks of Metals, second ed. Springer, New York.

Ahmad, M., Rajapaksha, A.U., Lim, J.E., Zhang, M., Bolan, N., Mohan, D., et al., 2014. Biochar as a sorbent for contaminant management in soil and water: a review. Chemosphere 99, 19–33.

Ahmed, M.B., Zhou, J.L., Ngo, H.H., Guo, W., Chen, M., 2016. Progress in the preparation and application of modified biochar for improved contaminant removal from water and wastewater. Bioresour. Technol. 214, 836–851.

Baig, S.A., Zhu, J., Muhammad, N., Sheng, T., Xu, X., 2014. Effect of synthesis methods on magnetic Kans grass biochar for enhanced As (III, V) adsorption from aqueous solutions. Biomass Bioenerg. 71, 299–310.

Bernardo, M., Mendes, S., Lapa, N., Goncalves, M., Mendes, B., Pinto, F., et al., 2013. Removal of lead (Pb^{2+}) from aqueous medium by using chars from co-pyrolysis. J. Colloid. Interface Sci. 409, 158–165.

Bogusz, A., Nowak, K., Stefaniuk, M., Dobrowolski, R., Oleszczuk, P., 2017. Synthesis of biochar from residues after biogas production with respect to cadmium and nickel removal from wastewater. J. Environ. Manage. 201, 268–276.

Cao, X., Ma, L., Gao, B., Harris, W., 2009. Dairy-manure derived biochar effectively sorbs lead and atrazine. Environ. Sci. Technol. 43, 3285–3291.

Chen, X., Chen, G., Chen, L., Chen, Y., Lehmann, J., McBride, M.B., et al., 2011. Adsorption of copper and zinc by biochars produced from pyrolysis of hardwood and corn straw in aqueous solution. Bioresour. Technol. 102, 8877–8884.

Ching, S., Petrovay, D.J., Jorgensen, M.L., 1997. Sol–gel synthesis of layered birnessite-type manganese oxides. Inorg. Chem. 36, 883–890.

Cui, X., Fang, S., Yao, Y., Li, T., Ni, Q., Yang, X., et al., 2016. Potential mechanisms of cadmium removal from aqueous solution by *Canna indica* derived biochar. Sci. Total Environ. 562, 517–525.

Della Puppa, L., Komárek, M., Bordas, F., Bollinger, J.C., Joussein, E., 2013. Adsorption of copper cadmium, lead and zinc onto a synthetic manganese oxide. J. Colloid. Interface Sci. 399, 99–106.

Devi, P., Saroha, A.K., 2014. Synthesis of the magnetic biochar composites for use as an adsorbent for the removal of pentachlorophenol from the effluent. Bioresour. Technol. 169, 525–531.

Ding, W., Dong, X., Ime, I.M., Gao, B., Ma, L.Q., 2014. Pyrolytic temperatures impact lead sorption mechanisms by bagasse biochars. Chemosphere 105, 68–74.

Dong, H., Deng, J., Xie, Y., Zhang, C., Jiang, Z., Cheng, Y., et al., 2017. Stabilization of nanoscale zero-valent iron (nZVI) with modified biochar for Cr(VI) removal from aqueous solution. J. Hazard. Mater. 332, 79–86.

Drenkova-Tuhtan, A., Mandel, K., Paulus, A., Meyer, C., Hutter, F., Gellermann, C., et al., 2013. Phosphate recovery from wastewater using engineered superparamagnetic particles modified with layered double hydroxide ion exchangers. Water Res. 47, 5670–5677.

Ettler, V., Knytl, V., Komárek, M., Della Puppa, L., Bordas, F., Mihaljevič, M., et al., 2014. Stability of a novel synthetic amorphous manganese oxide in contrasting soils. Geoderma 214–215, 2–9.

Ettler, V., Tomášová, Z., Komárek, M., Mihaljevič, M., Šebek, O., Michálková, Z., 2015. The pH-dependent long-term stability of an amorphous manganese oxide in smelter-polluted soils: implication for chemical stabilization of metals and metalloids. J. Hazard. Mater. 286, 386–394.

Fan, S., Li, H., Wang, Y., Wang, Z., Tang, J., Tang, J., et al., 2018. Cadmium removal from aqueous solution by biochar obtained by co-pyrolysis of sewage sludge with tea waste. Res. Chem. Intermed. 44, 135–154.

Frišták, V., Pipíška, M., Lesný, J., Soja, G., Friesl-Hanl, W., Packová, A., 2015. Utilization of biochar sorbents for Cd^{2+}, Zn^{2+}, and Cu^{2+} ions separation from aqueous solutions: comparative study. Environ. Monit. Assess. 187, 4093.

Gil-Díaz, M., Alonso, J., Rodríguez-Valdés, E., Pinilla, P., Lobo, M.C., 2014. Reducing the mobility of arsenic in brownfield soil using stabilised zero-valent iron nanoparticles. J. Environ. Sci. Health A 49, 1361–1369.

Gil-Díaz, M., Pinilla, P., Alonso, J., Lobo, M.C., 2017. Viability of a nanoremediation process in single or multi-metal(loid)contaminated soils. J. Hazard. Mater. 321, 812–819.

Goswami, R., Shim, J., Deka, S., Kumari, D., Kataki, R., Kumar, M., 2016. Characterization of cadmium removal from aqueous solution by biochar produced from *Ipomoea fistulosa* at different pyrolytic temperatures. Ecol. Eng. 97, 444–451.

Han, Y., Boateng, A.A., Qi, P.X., Lima, I.M., Chang, J., 2013. Heavy metal and phenol adsorptive properties of biochars from pyrolyzed switchgrass and woody biomass in correlation with surface properties. J. Environ. Manage. 118, 196–204.

Han, Z., Sani, B., Mrozik, W., Obst, M., Beckingham, B., Karapanagioti, H.K., et al., 2015. Magnetite impregnation effects on the sorbent properties of activated carbons and biochars. Water Res. 70, 394–403.

Harvey, O.R., Herbert, B.E., Rhue, R.D., Kuo, L.-J., 2011. Metal interactions at the biochar-water interface: energetics and structure–sorption relationships elucidated by flow adsorption microcalorimetry. Environ. Sci. Technol. 45, 5550–5556.

Ho, S.H., Chen, Y.D., Yang, Z.K., Nagarajan, D., Chang, J.S., Ren, N.Q., 2017. High-efficiency removal of lead from wastewater by biochar derived from anaerobic digestion sludge. Bioresour. Technol. 246, 142–149.

Hudcová, B., Veselská, V., Filip, J., Číhalová, S., Komárek, M., 2017. Sorption mechanisms of arsenate on Mg-Fe layered double hydroxides: a combination of adsorption modeling and solid state analysis. Chemosphere 168, 539–548.

Hudcová, B., Veselská, V., Filip, J., Číhalová, S., Komárek, M., 2018. Highly effective Zn(II) and Pb(II) removal from aqueous solutions using Mg–Fe layered double hydroxides: comprehensive adsorption modeling coupled with solid state analyses. J. Clean Prod. 171, 944–953.

Idrees, M., Batool, S., Hussain, Q., Ullah, H., Al-Wabel, M.I., Ahmad, M., et al., 2016. High-efficiency remediation of cadmium (Cd^{2+}) from aqueous solution using poultry manure-and farmyard manure-derived biochars. Sep. Sci. Technol. 51, 2307–2317.

Inyang, M., Gao, B., Ding, W., Pullammanappallil, P., Zimmerman, A.R., Cao, X., 2011. Enhanced lead sorption by biochar derived from anaerobically digested sugarcane bagasse. Sep. Sci. Technol. 46, 1950–1956.

Inyang, M., Gao, B., Yao, Y., Xue, Y., Zimmerman, A.R., Pullammanappallil, P., et al., 2012. Removal of heavy metals from aqueous solution by biochars derived from anaerobically digested biomass. Bioresour. Technol. 110, 50–56.

Jiang, J.Q., Ashekuzzaman, S.M., Hargreaves, J.S.J., McFarlane, A.R., Badruzzaman, A.B.M., Tarek, M.H., 2014. Removal of arsenic(III) from groundwater applying a reusable Mg–Fe–Cl layered double hydroxide. J. Chem. Technol. Biotechnol. 90, 1160–1166.

Jiang, T.-Y., Jiang, J., Xu, R.-K., Li, Z., 2012. Adsorption of Pb(II) on variable charge soils amended with rice-straw derived biochar. Chemosphere 89, 249–256.

Khare, P., Dilshad, U., Rout, P.K., Yadav, V., Jain, S., 2017. Plant refuses driven biochar: application as metal adsorbent from acidic solutions. Arab. J. Chem. 10, S3054–S3063.

Khuri, A.I., Mukhopadhyay, S., 2010. Response surface methodology. In: Wegman, E.J., Said, Y.H., Scott, D.W. (Eds.), Wiley Interdisciplinary Reviews – Computational Statistics, vol. 2, No. 2. Wiley, Hoboken, NJ, pp. 128–149.

Kim, W.-K., Shim, T., Kim, Y.-S., Hyun, S., Ryu, C., Park, Y.-K., et al., 2013. Characterization of cadmium removal from aqueous solution by biochar produced from a giant *Miscanthus* at different pyrolytic temperatures. Bioresour. Technol. 138, 266–270.

Komárek, M., Vaněk, A., Ettler, V., 2013. Chemical stabilization of metals and arsenic in contaminated soils using oxides – a review. Environ. Pollut. 172, 9–22.

Kołodyńska, D., Wnętrzak, R., Leahy, J.J., Hayes, M.H.B., Kwapiński, W., Hubicki, Z., 2012. Kinetic and adsorptive characterization of biochar in metal ions removal. Chem. Eng. J. 197, 295–305.

Lenoble, V., Laclautre, C., Serpaud, B., Deluchat, V., Bollinger, J.-C., 2004. As(V) retention and As(III) simultaneous oxidation and removal on a MnO_2-loaded polystyrene resin. Sci. Total Environ. 326, 197–207.

Li, R., Wang, J.J., Zhou, B., Awasthi, M.K., Ali, A., Zhang, Z., et al., 2016. Enhancing phosphate adsorption by Mg/Al layered double hydroxide functionalized biochar with different Mg/Al ratios. Sci. Tot. Environ. 559, 121–129.

Li, S., Wang, W., Liang, F., Zhang, W.X., 2017. Heavy metal removal using nanoscale zero-valent iron (nZVI): Theory and application. J. Hazard. Mater. 322, 163–171.

Li, X.Q., Zhang, W.X., 2007. Sequestration of metal cations with zerovalent iron nanoparticles – a study with high resolution X-ray photoelectron spectroscopy (HR-XPS). J. Phys. Chem. C 111, 6939–6946.

Lu, H., Zhang, W., Yang, Y., Huang, X., Wang, S., Qiu, R., 2012. Relative distribution of Pb^{2+} sorption mechanisms by sludge-derived biochar. Water Res. 46, 854–862.

Melo, L.C.A., Coscione, A.R., Abreu, C.A., Puga, A.P., Camargo, O.A., 2013. Influence of pyrolysis temperature on cadmium and zinc sorption capacity of sugar cane straw-derived biochar. BioResources 8, 4992–5004.

Michálková, Z., Komárek, M., Šillerová, H., Della Puppa, L., Joussein, E., Bordas, F., et al., 2014. Evaluating the potential of three Fe- and Mn-(nano)oxides for the stabilization of Cd, Cu and Pb in contaminated soils. J. Environ. Manage. 146, 226–234.

Mohan, D., Pittman Jr., C.U., Bricka, M., Smith, F., Yancey, B., Mohammad, J., et al., 2007. Sorption of arsenic, cadmium, and lead by chars produced from fast pyrolysis of wood and bark during bio-oil production. J. Colloid Interface Sci. 310, 57–73.

Mohan, D., Sarswat, A., Ok, Y.S., Pittman Jr., C.U., 2014a. Organic and inorganic contaminants removal from water with biochar, a renewable, low cost and sustainable adsorbent—a critical review. Bioresour. Technol. 160, 191–202.

Mohan, D., Kumar, H., Sarswat, A., Alexandre-Franco, M., Pittman Jr., C.U., 2014b. Cadmium and lead remediation using magnetic oak wood and oak bark fast pyrolysis biochars. Chem. Eng. J. 236, 513–528.

Mohan, D., Kumar, H., Sarswat, A., Alexandre-Franco, M., Pittman Jr., C.U., 2015. Cadmium and lead remediation using magnetic oak wood and oak bark fast pyrolysis bio-chars. Chem. Eng. J. 236, 513–528.

Montgomery, D.C., 2008. Design and analysis of experiments, seventh ed. John Wiley & Sons, New York, 680s.

Moyo, M., Lindiwe, S.T., Sebata, E., Nyamunda, B.C., Guyo, U., 2016. Equilibrium, kinetic, and thermodynamic studies on biosorption of Cd(II) from aqueous solution by biochar. Res. Chem. Intermed. 42, 1349–1362.

Mubarak, N., Kundu, A., Sahu, J., Abdullah, E., Jayakumar, N., 2014. Synthesis of palm oil empty fruit bunch magnetic pyrolytic char impregnating with $FeCl_3$ by microwave heating technique. Biomass Bioenerg. 61, 265–275.

Oliveira, L.C.A., Rios, R.V.R.A., Fabris, J.D., Garg, V., Sapag, K., Lago, R.M., 2002. Activated carbon/iron oxide magnetic composites for the adsorption of contaminants in water. Carbon. N. Y. 40, 2177–2183.

O'Carroll, D., Sleep, B., Krol, M., Boparai, H., Kocur, C., 2013. Nanoscale zero valent iron and bimetallic particles for contaminated site remediation. Adv. Water Resour. 51, 104–122.

Park, J.H., Cho, J.S., Ok, Y.S., Kim, S.H., Kang, S.W., Choi, I.W., et al., 2015. Competitive adsorption and selectivity sequence of heavy metals by chicken bone-derived biochar: Batch and column experiment. J. Environ. Sci. Health A. 50, 1194–1204.

Park, J.H., Ok, Y.S., Kim, S.H., Cho, J.S., Heo, J.S., Delaune, R.D., et al., 2016a. Competitive adsorption of heavy metals onto sesame straw biochar in aqueous solutions. Chemosphere 142, 77–83.

Park, J.H., Cho, J.S., Ok, Y.S., Kim, S.H., Heo, J.S., Delaune, R.D., et al., 2016b. Comparison of single and competitive metal adsorption by pepper stem biochar. Arch. Agron. Soil Sci. 62, 617–632.

Peng, X., Liu, X., Zhou, Y., Peng, B., Tang, L., Luo, L., et al., 2017. New insights into the activity of a biochar supported nanoscale zerovalent iron composite and nanoscale zero valent iron under anaerobic or aerobic conditions. RSC Adv. 7, 8755–8761.

Poo, K.-M., Son, E.-B., Chang, J.-S., Ren, X., Choi, Y.-J., Chae, K.-J., 2018. Biochars derived from wasted marine macro-algae (*Saccharina japonica* and *Sargassum fusiforme*) and their potential for heavy metal removal in aqueous solution. J. Environ. Manage. 206, 364–372.

Qian, L., Zhang, W., Yan, J., Han, L., Chen, Y., Ouyang, D., et al., 2017. Nanoscale zero-valent iron supported by biochars produced at different temperatures: synthesis mechanism and effect on Cr(VI) removal. Environ. Pollut. 223, 153–160.

Reddy, D.H.K., Lee, S.-M., 2014. Magnetic biochar composite: Facile synthesis, characterization, and application for heavy metal removal. Colloid Surf., A—Physicochem. Eng. Asp 454, 96–103.

Regmi, P., Garcia Moscoso, J.L., Kumar, S., Cao, X., Mao, J., Schafran, G., 2012. Removal of copper and cadmium from aqueous solution using switchgrass biochar produced via hydrothermal carbonization process. J. Environ. Manage. 109, 61–69.

Saletnik, B., Zaguła, G., Grabek-Lejko, D., Kasprzyk, I., Bajcar, M., Czernicka, M., et al., 2017. Biosorption of cadmium(II), lead(II) and cobalt(II) from aqueous solution by biochar from cones of larch (*Larix decidua* Mill. *subsp. decidua*) and spruce (*Picea abies* L. H. Karst). Environ. Earth Sci. 76, 574.

Sekulić, M.T., Pap, S., Stojanović, Z., Bošković, N., Radonić, J., Knudsen, T.Š., 2018. Efficient removal of priority, hazardous priority and emerging pollutants with *Prunus armeniaca* functionalized biochar from aqueous wastes: experimental optimization and modelling. Sci. Total Environ. 613–614, 736–750.

Sepehr, M.N., Yetilmezsoy, K., Marofi, S., Zarrabi, M., Ghaffari, H.R., Fingas, M., et al., 2014. Synthesis of nanosheet layered double hydroxides at lower pH: optimization of hardness and sulfate removal from drinking water samples. J. Taiwan Inst. Chem. Eng. 45, 2786–2800.

Shen, W., Li, Z., Liu, Y., 2008. Surface chemical functional groups modification of porous carbon. Recent Patents Chem. Eng. 1, 27–40.

Shen, Z., Jin, F., Wang, F., McMillan, O., Al-Tabbaa, A., 2015. Sorption of lead by salisbury biochar produced from British broadleaf hardwood. Bioresour. Technol. 193, 553–556.

Sohi, S.P., Krull, E., Lopez-Capel, E., Bol, R., 2010. A review of biochar and its use and function in soil. Adv. Agron. 105, 47–82.

Stefaniuk, M., Oleszczuk, P., Ok, Y.S., 2016. Review on nano zerovalent iron (nZVI): from synthesis to environmental applications. Chem. Eng. J. 287, 618–632.

Su, H., Fang, Z., Tsang, E.T., Zheng, L., Cheng, W., Fang, J., et al., 2016. Remediation of hexavalent chromium contaminated soil by biochar-supported zero-valent iron nanoparticles. J. Hazard. Mater. 318, 533–540.

Tan, X., Liu, Y., Zeng, G., Wang, X., Hu, X., Gu, Y., et al., 2015. Application of biochar for the removal of pollutants from aqueous solutions. Chemosphere 125, 70–85.

Tan, X.F., Liu, Y.G., Gu, Y.L., Xu, Y., Zeng, G.M., Hu, X.J., et al., 2016a. Biochar-based nano-composites for the decontamination of wastewater: a review. Bioresour. Technol. 212, 318–333.

Tan, X.F., Liu, Y.G., Gu, Y.L., Liu, S.B., Zeng, G.M., Cai, X., et al., 2016b. Biochar pyrolyzed from MgAl-layered double hydroxides pre-coated ramie biomass (*Boehmeria nivea* (L.) Gaud.): characterization and application for crystal violet removal. J. Environ. Manage. 184, 85–93.

Tan, X.F., Liu, S.B., Liu, Y.G., Gu, Y.L., Zeng, G.M., Cai, X.X., et al., 2016c. One-pot synthesis of carbon supported calcined-Mg/Al layered double hydroxides for antibiotic removal by slow pyrolysis of biomass waste. Sci. Rep. 6, 39691.

Tosco, T., Papini, M.P., Viggi, C.C., Sethi, P., 2014. Nanoscale zerovalent iron particles for groundwater remediation: a review. J. Clean. Prod. 77, 10–21.

Trakal, L., Komárek, M., Száková, J., Tlustoš, P., Tejnecký, V., Drábek, O., 2012. Sorption behaviour of Cd, Cu, Pb and Zn and their interactions in phytoremediated soil. Int. J. Phytoremediat. 14, 806–819.

Trakal, L., Bingöl, D., Pohořelý, M., Hruška, M., Komárek, M., 2014a. Geochemical and spectroscopic investigations of Cd and Pb sorption mechanisms on contrasting biochars: engineering implications. Bioresour. Technol. 171, 442–451.

Trakal, L., Šigut, R., Šillerová, H., Faturíková, D., Komárek, M., 2014b. Copper removal from aqueous solution using biochar: effect of chemical activation. Arab. J. Chem. 7, 43–52.

Trakal, L., Veselská, V., Šafařík, I., Vítková, M., Číhalová, S., Komárek, M., 2016. Lead and cadmium sorption mechanisms on magnetically modified biochars. Bioresour. Technol. 203, 318–324.

Trakal, L., Michálková, Z., Beesley, L., Vítková, M., Ouředníček, P., Barceló, A.P., et al., 2018. AMOchar: amorphous manganese oxide coating of biochar improves its efficiency at removing metal(loid)s from aqueous solutions. Sci. Total Environ. 625, 71–78.

Tsai, W.T., Liu, S.C., Chen, H.R., Chang, Y.M., Tsai, Y.L., 2012. Textural and chemical properties of swine-manure-derived biochar pertinent to its potential use as a soil amendement. Chemosphere 89, 198–203.

Verheijen, F.G.A., Jeffery, S., Bastos, A.C., van der Velde, M., Diafas, I., 2010. Biochar application to soils—a critical scientific review of effects on soil properties. In: Processes and Functions. EUR 24099 EN, Luxembourg.

Vítková, M., Rákosová, S., Michálková, Z., Komárek, M., 2017. Metal(lo-id)s behaviour in soils amended with nano zero-valent iron as a function of pH and time. J. Environ. Manage. 186, 268–276.

Wan, S., Wang, S., Li, Y., Gao, B., 2017. Functionalizing biochar with Mg–Al and Mg–Fe layered double hydroxides for removal of phosphate from aqueous solutions. J. Ind. Eng. Chem. 47, 246–253.

Wang, C., Wang, H., 2018. Pb(II) sorption from aqueous solution by novel biochar loaded with nano-particles. Chemosphere 192, 1–4.

Wang, S., Gao, B., Zimmerman, A.R., Li, Y., Ma, L., Harris, W.G., et al., 2015a. Removal of arsenic by magnetic biochar prepared from pinewood and natural hematite. Bioresour. Technol. 175, 391–395.

Wang, S., Gao, B., Li, Y., Creamer, A.E., He, F., 2016a. Adsorptive removal of arsenate from aqueous solutions by biochar supported zero-valent iron nanocomposite: batch and continuous flow tests. J. Hazard. Mater. 322, 172–181.

Wang, S., Gao, B., Li, Y., Zimmerman, A.R., Cao, X., 2016b. Sorption of arsenic onto Ni/Fe layered double hydroxide (LDH)-biochar composites. RSC Adv. 6, 17792–17799.

Wang, S., Gao, B., Li, Y., 2016c. Enhanced arsenic removal by biochar modified with nickel (Ni) and manganese (Mn) oxyhydroxides. J. Ind. Eng. Chem. 37, 361–365.

Wang, S., Gao, B., Li, Y., Creamer, A.E., He, F., 2017. Adsorptive removal of arsenate from aqueous solutions by biochar supported zero-valent iron nanocomposite: batch and continuous flow tests. J. Hazard. Mater. 322, 172–181.

Wang, S.-y, Tang, Y.-k, Li, K., Mo, Y.-y, Li, H.-f, Gu, Z.-q, 2014. Combined performance of biochar sorption and magnetic separation processes for treatment of chromium-contained electroplating wastewater. Bioresour. Technol. 174, 67–73.

Wang, T., Li, C., Wang, C., Wang, H., 2018. Biochar/MnAl-LDH composites for Cu(II) removal from aqueous solution. Colloids Surf., A.—Physicochem. Eng. Asp. 538, 443–450.

Wang, Z., Liu, G., Zheng, H., Li, F., Ngo, H.H., Guo, W., et al., 2015b. Investigating the mechanisms of biochar's removal of lead from solution. Bioresour. Technol. 177, 308–317.

Wu, X., Yang, Q., Xu, D., Zhong, Y., Luo, K., Li, X., et al., 2013. Simultaneous adsorption/reduction of bromate by nanoscale zerovalent iron supported on modified activated carbon. Ind. Eng. Chem. Res. 52, 12574–12581.

Xi, Y., Mallavarapu, M., Naidu, R., 2010. Reduction and adsorption of Pb2 + in aqueous solution by nano-zero-valent iron—a SEM, TEM and XPS study. Mater. Res. Bull. 45, 1361–1367.

Xu, X., Cao, X., Zhao, L., 2013a. Comparison of rice husk- and dairy manure-derived biochars for simultaneously removing heavy metals from aqueous solutions: role of mineral components in biochars. Chemosphere 92, 955–961.

Xu, X., Cao, X., Zhao, L., Wang, H., Yu, H., Gao, B., 2013b. Removal of Cu, Zn, and Cd from aqueous solutions by the dairy manure-derived biochar. Environ. Sci. Pollut. Res. 20, 358–368.

Xu, X., Kan, Y., Zhao, L., Cao, X., 2016. Chemical transformation of CO_2 during its capture by waste biomass derived biochars. Environ. Pollut. 213, 533–540.

Xue, L., Gao, B., Wan, Y., Fang, J., Wang, S., Li, Y., et al., 2016. High efficiency and selectivity of MgFe-LDH modified wheat-straw biochar in the removal of nitrate from aqueous solutions. J. Taiwan Inst. Chem. Eng. 63, 312–317.

Yan, J., Han, L., Gao, W., Xue, S., Chen, M., 2015. Biochar supported nanoscale zerovalent iron composite used as persulfate activator for removing trichloroethylene. Bioresour. Technol. 175, 269–274.

Yang, Y., Wei, Z., Zhang, X., Chen, X., Yue, D., Yin, Q., et al., 2014. Biochar from *Alternanthera philoxeroides* could remove Pb(II) efficiently. Bioresour. Technol. 171, 227–232.

Zama, E.F., Zhu, Y.-G., Reid, B.J., Sun, G.-X., 2017. The role of biochar properties in influencing the sorption and desorption of Pb(II), Cd(II) and As(III) in aqueous solution. J. Clean. Prod. 148, 127–136.

Zhang, C., Shan, B., Tang, W., Zhu, Y., 2017b. Comparison of cadmium and lead sorption by *Phyllostachys pubescens* biochar produced under a low-oxygen pyrolysis atmosphere. Bioresour. Technol. 238, 352–360.

Zhang, J., Li, Y., Zhou, J., Chen, D., Qian, G., 2012. Chromium(VI) and zinc(II) waste water co-treatment by forming layered double hydroxides: mechanism discussion via two different processes and application in real plating water. J. Hazard. Mater. 205–206, 111–117.

Zhang, M., Gao, B., Varnoosfaderani, S., Hebard, A., Yao, Y., Inyang, M., 2013b. Preparation and characterization of a novel magnetic biochar for arsenic removal. Bioresour. Technol. 130, 457–462.

Zhang, M., Gao, B., Yao, Y., Inyang, M., 2013c. Phosphate removal ability of biochar/MgAl-LDH ultra-fine composites prepared by liquid-phase deposition. Chemosphere 92, 1042–1047.

Zhang, M., Gao, B., Fang, J., Creamer, A.E., Ullman, J.L., 2014. Self-assembly of needle-like layered double hydroxide (LDH) nanocrystals on hydrochar: characterization and phosphate removal ability. RSC Adv. 4, 28171–28175.

Zhang, T., Zhu, X., Shi, L., Li, J., Li, S., Lü, J., et al., 2017a. Efficient removal of lead from solution by celery-derived biochars rich in alkaline minerals. Bioresour. Technol. 235, 185–192.

Zhang, X., Wang, H., He, L., Lu, K., Sarmah, A., Li, J., et al., 2013a. Using biochar for remediation of soils contaminated with heavy metals and organic pollutants. Environ. Sci. Pollut. Res. 20, 8472–8483.

Zheng, W., Guo, M., Chow, T., Bennett, D.N., Rajagopalan, N., 2010. Sorption properties of greenwaste biochar for two triazine pesticides. J. Hazard. Mater. 181, 121–126.

Zhou, Y., Gao, B., Zimmerman, A.R., Fang, J., Sun, Y., Cao, X., 2013. Sorption of heavy metals on chitosan-modified biochars and its biological effects. Chem. Eng. J. 231, 512–518.

Zhou, Y., Gao, B., Zimmerman, A.R., Chen, H., Zhang, M., Cao, X., 2014. Biochar-supported zerovalent iron for removal of various contaminants from aqueous solutions. Bioresour. Technol. 152, 538–542.

Further Reading

Cha, J.S., Park, S.H., Jung, S.C., Ryu, C., Jeon, J.K., Shin, M.C., et al., 2016. Production and utilization of biochar: a review. J. Ind. Eng. Chem. 40, 1–15.

CHAPTER

8

Biochar for Anionic Contaminants Removal From Water

Xiaodian Li, Cheng Zhao and Ming Zhang

Department of Environmental Engineering, P.R. China Jiliang University, Hangzhou, Zhejiang, P.R. China

8.1 ANIONIC CONTAMINANTS IN WATER/WASTEWATER

Water pollution is one of the biggest global issues today due to intensification of human activities, including industry, agriculture, and daily life, in the past century. It was estimated that over 10 million tons of anthropogenic chemicals are released into the aquatic environment every year, and more than 700 contaminants have been detected in water environments (Ali and Gupta, 2006). Anionic contaminants are ionized in water with negative charge, easily dissolved in water, and most are anthropogenic. Some of these contiminants, such as fluoride, come from geological environments and enter water systems. While anions are needed to maintain the growth of organisms they must be limited to a certain concentration range (such as NO_3^-, PO_4^{3-}, and F^-), become harmful and toxic (such as CN^- and AsO_3^{3-}/AsO_4^{3-}), or wastes from domestic use (such as ClO^- and anionic surfactants) (Fang et al., 2017).

Discharged domestic sewage and industrial effluents can contain large amounts of anionic contaminants that could cause harm to organisms and the environment, even at low concentrations. For instance, excess fluoride in drinking water has both short-term and long-term negative effects on human health (Steenbergen et al., 2011). Perchlorate ingestion may result in lower IQ levels or neurological conditions (Ginsberg et al., 2007). Anionic dye Congo Red may cause toxicity to humans due to the generation of carcinogenic amines during degradation and metabolic conversion (Sponza and Işik, 2005). Cr (VI), existing as anionic, is very toxic to organisms because of its oxidizing properties and ability to penetrate biological membranes (Mearns et al., 1976). It can cause serious diseases such as cancer (Zhitkovich, 2011).

Biochar from Biomass and Waste
DOI: https://doi.org/10.1016/B978-0-12-811729-3.00008-X

In order to prevent any possible adverse effects due to anions, strict guidelines have been established by individual countries and organizations. For example, the World Health Organization (WHO) suggested 1.5 mg L^{-1} as threshold of F^- in drinking water (World Health Organization, 2006), above which may cause dental and skeletal fluorosis (Harrison, 2005). The US Environmental Protection Agency (US EPA) set the maximum contaminant levels for cyanide (CN^-, 0.2 mg L^{-1}), fluoride (4.0 mg L^{-1}), nitrate (NO_3^-, measured as nitrogen, 10 mg L^{-1}), and nitrite (NO_2^-, measured as nitrogen, 1 mg L^{-1}) in drinking water (USEPA National Primary Drinking Water Regulations, 1998). A maximum of 0.025 mg L^{-1} PO_4^{3-} was also set by the US EPA for lakes or reservoirs to prevent eutrophication caused by excessive phosphorus (United States Environmental Protection Agency, 1986).

Some anions such as PO_4^{3-} and NO_3^- can be removed by biological processes in conventional wastewater treatment plants, since P and N are essential elements of organisms. However, most aninons are not easily treated and removed from water and thus can enter the food chain and threaten human and environmental safety even at extremely low concentrations.

In order to effectively remove anions from water, physical, chemical, and biological processes have been used by both researchers and engineers (Van der Bruggen and Vandecasteele, 2003; Gabriele and Siglinda, 2003; Sica et al., 2014; Min et al., 2004; Pulkka et al., 2014; Kapoor and Viraraghavan, 1997). Sorption is one preferred approach to removing anionic contaminants from water because of its low cost and environment-friendly nature. Activated carbon (AC) is currently the most widely used adsorbent, but its use is restricted due to high cost (Ma et al., 2012). Bentonite is another widely used natural sorbent because of its high cationic exchange capability, especially for heavy metals and cationic contaminants. However, because of the repulsion between the anions and the negatively charged inner layer, the effect of bentonite on the adsorption of anionic contaminants is weak. As a consequence, bentonites pillared with metal hydroxide have been developed to remove phosphate in water (Yan et al., 2010). Biochar, a stable carbon (C)-rich byproduct synthesized through carbonization of biomass in an oxygen-limited environment, has gained more attention by researchers in recent years and has been recognized as a multifunctional material for environmental applications due to its high sorption capability and low cost (Lehmann and Joseph, 2015). With its negatively charged surface, biochar is also favorable for the sorption of heavy metals (Yan et al., 2015; Wang et al., 2015a) and of cation organics (Bordoloi et al., 2018; Lonappan et al., 2016; Xu et al., 2016; Yang et al., 2016). Sorption of nonionic organics by biochar via partition or adsorption is also widely reported (Ahmad et al., 2014). Due to the possible static repulsion between the negatively charged surface of biochar and anions, sorption of anionic contaminants in water by biochar was once considered unpromising. However, more recent results on the sorption of the anionic contaminants by biochar or engineered/modified biochar have been more promising. In this chapter, the interactions between anionic contaminants and biochar are summarized. The sorption properties of biochar, as well as the potential application of biochar in the treatment of water/wastewater containing complex contaminants, are also covered.

8.2 SORPTION PROPERTIES OF BIOCHAR

In recent years, researchers have reported on biochar's capability to adsorb anionic contaminants including anionic nutrients, anionic heavy metals, and other anions. Some of the sorption properties of biochars are summarized in Table 8.1.

8.2.1 Anionic Nutrients in Water

The increase of anionic nutrients in water, mainly phosphorus (P) and nitrogen (N), is one of the main causes of eutrophication in rivers, lakes, and even seas. Traditionally biological processes have been used to remove P and N from water because they are cost effective and ecofriendly. However, biological processes are slow, particularly for water/wastewater containing high concentrations of anionic nutrients. Physical-chemical methods for N and P removal include ion-exchange, reverse osmosis treatment, adsorption, and so on (Chatterjee and Woo, 2009).

The anionic forms of P and N in water are mainly phosphate (PO_4^{3-}) and nitrate (NO_3^-). Biochar produced from different biomass has been studied in recent years for its ability to adsorb nutrients in water. Some of these studies are discussed in the following.

8.2.1.1 *Phosphate* (PO_4^{3-})

Jung et al. (2015) studied macroalgae root-derived biochar pyrolyzed at different temperatures and found that the biochar could adsorb P in the range of 12.97–18.21 mg g^{-1} at pyrolysis temperature of 400–800°C. However, sorption of P was not observed by raw material of macroalgar root and its biochar produced at 200°C. Orange peel-derived biochar pyrolyzed at 250°C and 400°C had very weak sorption of phosphate, as reported by Chen et al. (2011). However, after modification by Fe, the adsorption capabilities of the orange peel biochar significantly increased. The nanosized magnetite particles in orange peel biochar facilitate the adsorption of P (Chen et al., 2011). Similarly, Zhang and Gao (2013) found that $AlCl_3$-modified biochar nanocomposite (biochar/AlOOH) could successfully adsorb P. The Langmuir model used in the study estimated the sorption capacity (Q_m) to be about 135 mg g^{-1}. A biochar/Mg–Al-assembled nanocomposite using Al as electrode and $MgCl_2$ as electrolyte was created by Jung et al. (2015), and the Q_m was as high as 887 mg g^{-1} at 30°C.

8.2.1.2 *Nitrate* (NO_3^-)

Bamboo (Mizuta et al., 2004), wheat straw (Mishra and Patel, 2009), and mustard straw (Mishra and Patel, 2009) were used as precursors of biochar, and the sorption capabilities were evaluated. Mizuta et al. (2004) found that the Q_m of bamboo biochar was 1.25 mg g^{-1}, which was 15% higher than that of commercial AC (CAC, 1.09 mg g^{-1}). The Q_m of wheat straw biochar, mustard straw biochar, and CAC were also compared by Mishra and Patel (2009), and they were 1.10, 1.30, and 1.22 mg g^{-1}, respectively. Mustard straw biochar showed better sorption of nitrate than CAC (Mishra and Patel, 2009).

TABLE 8.1 Capability of Biochar Sorption of Anionic Contaminants

Biochar Feedstock	Pyrolysis Temperature (°C)	Modification	Target Sorbate	Initial Concentration	Adsorption Performance	Reference
Orange peel	250/400/700	Modified by $FeCl_2$ and $FeCl_3$	PO_4^{3-}	2.4 mg(P) L^{-1}	OB250: 7.50 ± 0.10% MOB250: 67.3 ± 4.0% OB400: 11.0 ± 8.0% MOB400: 9.3 ± 1.6% OB700: 83.3 ± 0.7% MOB700: 99.4 ± 0.4%	Chen et al. (2011)
Brown marine macroalgae	450(N_2)	Electrochemical modification using an aluminum electrode and NaCl electrolyte	PO_4^{3-}	5–200 mg(P) L^{-1}	BM: Q_m:8.23 mg(P) g^{-1} MBM: Q_m:31.28 mg(P) g^{-1}	Jung et al. (2015)
Brown marine macroalgae	600	Electrochemical modification using an aluminum electrode and $MgCl_2$ electrolyte	PO_4^{3-}	1000 mgPO_4^{3-} L^{-1}	245.6 mg(P) g^{-1}	Jung et al. (2015)
U. pinnatifida roots	200/400/600/800 (N_2)	/	PO_4^{3-}	50 mg(P) L^{-1}	UB200:0 mg(P) g^{-1} UB400:18.21 mg(P) g^{-1} UB600:15.69 mg(P) g^{-1} UB800:12.97 mg(P) g^{-1}	Jung et al. (2016)
Oak sawdust	500	Impregnated with $LaCl_3$	PO_4^{3-}	61.3 mg PO_3^- L^{-1}	OB:3.93 mg(P) g^{-1} MOB:7.75 mg(P) g^{-1}	Wang et al. (2015c)
Cottonwood	600(N_2)	Impregnated with $AlCl_3$	PO_4^{3-}	1–1600 mg(P) L^{-1}	Q_m:135 mg(P) g^{-1}	Zhang and Gao (2013)
Oak sawdust	600	Impregnated with $LaCl_3$	NO_3^-	20 mg NO_3^- L^{-1}	OB:0.63 mg N g^{-1} MOB:1.96 mg N g^{-1}	Wang et al. (2015c)
Bamboo powder	900	/	NO_3^-	10 mg N L^{-1}	1.25 mg N g^{-1}	Mizuta et al. (2004)
Wheat straw/ Mustard straw	300	/	NO_3^-	25 mg N L^{-1}	/	Mishra and Patel (2009)
Sugar beet tailings **Peanut shells**	600(N_2)	Impregnated with $MgCl_2$	NO_3^-	20 mg N L^{-1}	MSBT:3.6% MPS:11.7%	Zhang et al. (2012)

Sugar beet bagasse	500/600/700(N2)	Modified by $ZnCl_2$	NO_3^-	100 mg N L^{-1}	41.20%	Demiral and Gunduzoglu (2010)
Corn stover **Ponderosa pine wood chips** **Switchgrass**	650	Mixed with concentrated HCl was heated on hot plate at 200°C for 24 h	NO_3^-	80 mg N L^{-1}	CSB: Q_m:11.35 mg N g^{-1} MCSB: Q_m:22.75 mg N g^{-1} PWCB: Q_m:5.42 mg N g^{-1} MPWCB: Q_m:21.39 mg N g^{-1} SB: Q_m:11.16 mg N g^{-1} MSB: Q_m:26.29 mg N g^{-1}	Chintala et al. (2013)
Coconut granular	500	Modified by $ZnCl_2$	NO_3^-	5–200 mg N L^{-1}	CB: Q_m:1.7 mg N g^{-1} MCB: Q_m:10.2 mg N g^{-1}	Bhatnagar et al. (2008)
Ramie residues	300/450/600(N_2)	/	Cr(VI)	20–800 mg L^{-1}	RB300: Q_m:82.23 mg g^{-1} RB450: Q_m:72.32 mg g^{-1} RB600: Q_m:61.18 mg g^{-1}	Zhou et al. (2016)
Rice straw	Burned in the air	/	Cr(VI)	50 mg L^{-1}	pH = 1.0:100%; pH = 2.0:100%; pH = 3.0:98.4% pH = 4.0:66.6%; pH = 5.0:34.7%	Hsu et al. (2009)
Sugar beet tailing	300(N_2)	/	Cr(VI)	50–800 mg L^{-1}	Q_m:123 mg g^{-1}	Dong et al. (2011)
Coconut fibers **Coconut shells**	200/400/600/800	Impregnated with concentrated sulfuric acid and keeping them in an oven maintained at 150–165°C for a period of 24 h	Cr(VI)	1–100 mg L^{-1}	CFB: Q_m:21.75 mg g^{-1} CSB: Q_m:9.53 mg g^{-1} MCFB: Q_m:9.86 mg g^{-1} MCSB: Q_m:11.51 mg g^{-1}	Mohan et al. (2005)

(Continued)

TABLE 8.1 (Continued)

Biochar Feedstock	Pyrolysis Temperature (°C)	Modification	Target Sorbate	Initial Concentration	Adsorption Performance	Reference
Rubber wood sawdust	400	Impregnated with phosphoric acid	Cr(VI)	200 mg L^{-1}	MRB30: 44.05 mg g^{-1} MRB40: 59.17 mg g^{-1} MRB50: 65.78 mg g^{-1}	Karthikeyan et al. (2005)
Waste tyres **Sawdust**	WB:900(N_2) SB:650(N_2)	After the pyrolysis process, the product was activated at the same temperature for 2 h by using CO_2 as an oxidizing agent	Cr(VI)	60 mg L^{-1}	WB:29.93 mg g^{-1} (pH = 2.0) SB: 24.65 mg g^{-1} (pH = 2.0)	Hamadi et al. (2001)
Pine wood feedstock	600	Modified by $MnCl_2$	As(V)	10 mg L^{-1}	MPB:0.59 g kg^{-1}	Wang et al. (2015b)
Waste products (paper, textile, organic wastes, etc.)	500	Modified by KOH	As(V)	50 mg L^{-1}	WB:24.49 mg g^{-1} MSB:30.98 mg g^{-1}	Jin et al. (2014)
Chestnut shell	450(N_2)	Modified with magnetic gelatin	As(V)	0.2–50 mg L^{-1}	CB: Q_m:17.50 mg g^{-1} (pH = 7.0) MCB: Q_m:45.80 mg g^{-1} (pH = 4.0)	Zhou et al. (2016)
Corn straw	600	Impregnated with $FeCl_3$	As(V)	0.25–100 mg L^{-1}	CB: Q_m:0.017 mg g^{-1} MCB: Q_m:6.80 mg g^{-1}	He et al. (2018)
Paper mill sludge	270–720(CO_2)	/	As(V)	20.9–189.5 mg L^{-1}	Q_m:22.76 mg g^{-1}	Yoon et al. (2017)
Perilla leaf	300/700	/	As(V)	0.05–7.0 mg L^{-1}	PB300: Q_m:3.85 mg g^{-1} PB700: Q_m:7.21 mg g^{-1}	Niazi et al. (2018b)
Japanese oak wood	500(N_2)	/	As(V)	0.05–7.0 mg L^{-1}	Q_m:3.89 mg g^{-1}	Niazi et al. (2018a)
Rice straw	600(N_2)	Modified by red mud suspensions	As(V)	1–50 mg L^{-1}	RB: Q_m:0.552 mg g^{-1} MRB: Q_m:5.924 mg g^{-1}	Wu et al. (2017)
Corn stem	620(N_2)	Modified by $KMnO_4$ and Fe $(NO_3)_3$	As(III)	0.2–50 mg L^{-1}	CB: Q_m:2.89 mg g^{-1} MCB: Q_m:8.25 mg g^{-1}	Liu et al. (2015)
Wheat straw	500(N_2)	Impregnated with Bismuth	As(III)	5–200 mg L^{-1}	MWB: Q_m:16.21 mg g^{-1}	Zhu et al. (2016)

Perilla leaf	300/700	/	As(III)	0.05–7.0 mg L^{-1}	PB300: Q_m:4.71 mg g^{-1} PB700: Q_m:11.01 mg g^{-1}	Niazi et al. (2018b)
Japanese oak wood	500(N_2)	/	As(III)	0.05–7.0 mg L^{-1}	Qm:3.16 mg g^{-1}	Niazi et al. (2018a)
Rice straw	600(N_2)	Modified by Red mud suspensions	As(III)	1–50 mg L^{-1}	RB: Q_m:0.448 mg g^{-1} MRB: Q_m:0.520 mg g^{-1}	Wu et al. (2017)
Mushroom compost	500	Modified by $Al(OH)_3$	Fluoride	5–100 mg L^{-1}	Q_m:36.5 mg g^{-1}	Chen et al. (2016)
Cocos nucifera shell	700	Steam activation of the carbonized char (900°C)	Fluoride	10 mg L^{-1} (artificial solution) 7 mg L^{-1} (real sample)	83.04% 79.3%	Halder et al. (2016)
Water treatment sludge	400/600/700	/	Fluoride	42 mg L^{-1}	WB400:71.9% WB600:57.7% WB700:76.4%	Oh et al. (2012)
Corn stover	500	Modified by $FeCl_3$ and $FeSO_4$	Fluoride	1–100 mg L^{-1}	CB: Q_m:6.42 mg g^{-1} MCB: Q_m:4.11 mg g^{-1}	Mohan et al. (2014)
Mongolian scotch pine tree sawdust	550	Soaked in phosphoric acid soak and radiated in microwave	Fluoride	/	Q_m: 885 mg kg^{-1}	Guan et al. (2015)
Lignin	700	Impregnated with cobalt	BrO_3^-	9.7 mg L^{-1}	89.70%	Cho et al. (2017)
Fir wood chips	500/600/700	/	ClO_4^-	/	Fb500: Q_m:2.5 mg g^{-1} Fb600: Q_m:5.0 mg g^{-1} Fb700: Q_m:5.0 mg g^{-1}	Fang et al. (2014)
Commercial cellulose	980	Modified by Fe $(NO_3)_3$ and $FeSO_4$	Congo Red	/	MCB: Q_m:66.09 mg g^{-1}	Zhu et al. (2011)
Vermicompost	300/500/700	/	Congo Red	5–200 mg L^{-1}	VB300: Q_m:11.63 mg g^{-1} VB500: Q_m:20.00 mg g^{-1} VB700: Q_m:31.28 mg g^{-1}	Yang et al. (2016)

Modification of biochar by metal ions can also significantly increase biochar's sorption of nitrate. For instance, after modification by Lanthanum (La), oak sawdust biochar showed significantly increased nitrate removal (Wang et al., 2015c). Its maximum sorption capacity (estimated by the Langmuir isotherm model) was 100 mg g^{-1}, which is only 8.94 mg g^{-1} for the biochar without modification as reference (Wang et al., 2015c). Modification of biochar with $ZnCl_2$ and $MgCl_2$ was also reported, and their sorption of nitrate was several times higher than the biochar without modification (Demiral and Gunduzoglu, 2010; Bhatnagar et al., 2008; Zhang et al., 2012).

8.2.2 Anionic Heavy Metals in Water

Chromium and arsenic are two typical heavy metals that exist in anionic form in nature. Thus, the process of removing chromium and arsenic in water/wastewater is different than for other heavy metal cations. As discussed in the following, sorption of anionic heavy metals by biochar has also been widely studied in recent years.

8.2.2.1 Hexavalent Chromium

The anionic form of chromium is hexavalent dichromate [$Cr_2O_7^{2-}$, Cr(VI)]. Ramie residue (Zhou et al., 2016), rice straw (Hsu et al., 2009), sugar beet tailings (Dong et al., 2011), coconut fiber and shells (Mohan et al., 2005), rubber wood sawdust (Karthikeyan et al., 2005), and waste tyres (Hamadi et al., 2001) were used to produce biochar for Cr(VI) removal from water and wastewater. The ramie residue biochar produced at 300°C had maximum sorption capacity for Cr(VI) of 82.32 mg/g (Zhou et al., 2016), while it was 123 mg g^{-1} for sugar beet tailing biochar (Dong et al., 2011). Q_m for rubber wood sawdust biochar at 30, 40, and 50°C were 44.05, 59.17, and 65.78 mg g^{-1}, respectively (Hamadi et al., 2001).

Mohan et al. (2005) found that the sorption capacity of biochar derived from coconut fiber and shell of Cr(VI) was weaker than CAC fiber cloth. However, Hamadi et al. (2001) found that the waste tire-derived biochar had comparable performance for the removal of Cr(VI) in water (Q_m = 48.08–58.48 mg g^{-1}) with CAC (Q_m = 44.44–53.19 mg g^{-1}), while the biochar obtained from sawdust was much weaker (Q_m = 1.93–2.29 mg g^{-1}).

8.2.2.2 Arsenic

The anionic forms of arsenic are arsenate [As(V)] and arsenite [As(III)], and their removal by biochars derived from different biomass have been studied. For example, paper mill sludge was used to produce biochar in CO_2 environment, and its maximum sorption capacity of As(V) was 22.76 mg g^{-1} (Yoon et al., 2017). Due to the use of $FeSO_4$ as coagulant during paper mill wastewater treatment, this biochar also had magnetism and can be easily separated after sorption (Yoon et al., 2017). Niazi et al. (2018a,b) studied the sorption capabilities of biochars obtained from Japanese oak wood and perilla leaf of As (V) and As(III), and found that Q_m of Japanese oak wood biochar for As(V) and As(III) was 3.89 and 3.16 mg g^{-1}, respectively, while that for perilla leaf biochar at 700°C was 7.21 and 11.01 mg g^{-1} for As(V) and As(III), respectively.

Different modifications of biochar, such as strong base (Jin et al., 2014) and metal ions (Wang et al., 2015b; He et al., 2018), have been used to improve biochar's sorption capacity of As(V). Although significant enhancement was observed in each study, the modified biochar's overall sorption capacity of As(V) was not significantly different than that of the original biochar. For example, the sorption capacity of As(V) by iron-impregnated biochar was 6.8 mg g^{-1}, but it was only 0.017 mg g^{-1} for the original biochar derived from corn straw. However, sorption of 6.8 mg g^{-1} of As(V) by iron-impregnated biochar was similar to other original biochars without modification.

Wu et al. (2017) studied the effects of adding red mud to rice straw before pyrolysis of biochar to increase the metal-oxide content in biochar and subsequently increase the sorption of As(V) and As(III). The results showed that Q_m of As(V) increased from 0.552 to 5.92 mg g^{-1}, and that of As(III) increased from 0.448 to 0.520 mg g^{-1}.

8.2.3 Other Anionic Contaminants in Water

Sorption of other anionic contaminants including fluoride (F^-) (Oh et al., 2012; Halder et al., 2016; Chen et al., 2016; Mohan et al., 2014; Guan et al., 2015), bromate (BrO_3^-) (Cho et al., 2017), perchlorate (ClO_4^-) (Fang et al., 2014), and Congo Red (Yang et al., 2016; Zhu et al., 2011) by biochars and/or modified biochars has also been evaluated.

Fluoride is an essential element but excessive concrentations of fluoride can be hazardous to human health (Halder et al., 2016). Flouride mainly comes from nature as well as from anthropogenic activities, and is spread widely via water/groundwater use. Sorption is one of the most challenging methods for the removal of fluoride in water. Halder et al. (2016) found that Cocos nucifera shell-derived biochar had a 83.04% removal rate for artificial solution with fluoride concentration of 10 mg L^{-1}, while it was 79.3% for the real sample containing 7 mg L^{-1} of fluoride.

Perchlorate, a newly discovered contaminant, is now being regulated by the US EPA (Kounaves et al., 2010) because it affects the thyroid gland in humans and its ability to regulate hormones and adsorb iodine (Fang et al., 2014). Fang et al. (2014) found that biochar produced from wood at high temperature (700°C) can effectively adsorb perchlorate, reaching a Q_m of 10.55 mg g^{-1}.

The wide use of dyes in the textile industry has caused major environmental and human health problems because many of the dyes are toxic, mutagenic, or carcinogenic (Yang et al., 2016). Some of the dyes contain azo compounds that are hard to degrade using conventional wastewater treatment processes. For example, the sorption capacity of Congo Red by vermicompost-derived biochar at different pyrolysis temperature range from 11.63 to 31.28 mg g^{-1}, showing a significant increase of sorption with production temperature (Yang et al., 2016). Zhu et al. (2011) studied the sorption of Congo Red by Fe_3O_4-modified biochar derived from cellulose, and found that the Q_m for the biochar was 66.09 mg g^{-1}, which was comparable to other reports from literature.

8.3 BIOCHAR SORPTION OF ANIONIC CONTAMINANTS

8.3.1 Pore Filling

The decomposition of organic components of biomass during pyrolysis creates macropore, mesopore, and micropore structures in biochar. Generally, the surface area of biochar increases with increase in pyrolysis temperature, and the sorption capability of biochar also increases with increase of its surface area (Fig. 8.1).

Jung et al. (2015) found that with an increase of pyrolysis temperature from 200 to 400°C, the BET surface area and total pore volume of marine macroalgae root-derived biochar increased significantly, enhancing the sorption of phosphate. With further increase of pyrolytic temperature to 600 and 800°C, the surface area and pore volume decreased significantly due to pore block. As a result, sorption of phosphate on the biochar decreased accordingly. In this study, the adsorption capacity of phosphate onto biochar was linearly correlated to the Brunauer–Emmett–Teller (BET) surface area and pore volume, indicating that the sorption of phosphate significantly depends on the pore structure of biochar.

Modification of biochar with metals can increase pore structure and enhance sorption capacity. For instance, Zhang et al. (2012) studied the adsorption of the magnesium-oxide biochar composite towards anions (PO_4^{3-}, NO_3^-), and found that the nanoporous structure was formed on the modified biochar composite, thus further enhancing the surface area and porosity and greatly increasing its adsorption capacity.

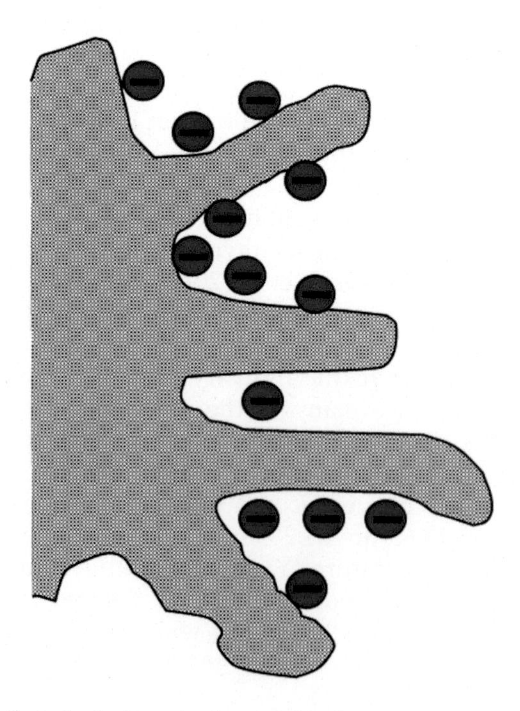

FIGURE 8.1 Pore-filling effect of anionic contaminants into micro- or mesopores of biochar.

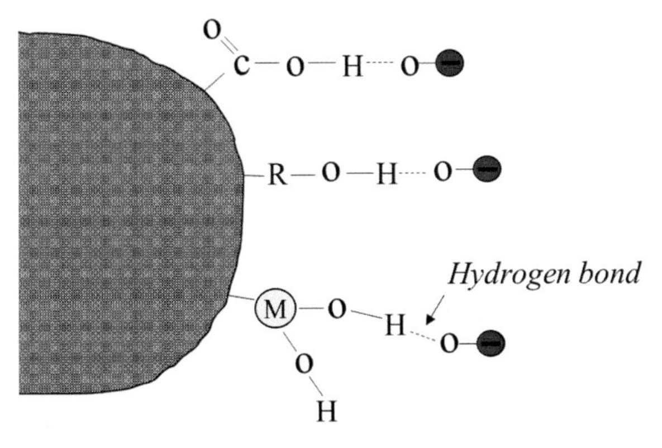

FIGURE 8.2 Hydrogen bonding between hydrogen-containing functional group on biochar surface and oxygen-containing group in anionic contaminants. M refers to the metals on the biochar surface.

8.3.2 Hydrogen Bonding

Hydrogen bonding is considered the main mechanism of the sorption of anionic contaminants such as perchlorate (Fang et al., 2014) (Fig. 8.2). In this study, at certain pH level (around pH_{zpc}), the surface of biochar became neutral, and the hydrogen bonding between the oxygen atoms of perchlorate and the hydrogen of oxygen-containing group was considered the main mechanism. With a further increase of pH in solution, the hydrogen in the oxygen-containing group was ionized, and the anionic surface of the biochar adsorbed almost no perchlorate.

8.3.3 Surface Complexation/Precipitation

Surface complexation or precipitation is also thought to increase the sorption of anionic contaminants by biochar (Fig. 8.3). For example, Halder et al. (2016) proved the formation of fluoride-metal complex on biochar surface by X-ray diffraction (XRD), and found that mineral K_2MgF_4 and potassium iron fluoride ($KFeF_3$) was formed when the biochar adsorbed fluoride. The potassium, iron, and magnesium in biomass were reactive, which facilitated the formation of chemical complex with fluoride.

Zinc chloride-modified biochar from coconut pellet was evaluated for its sorption capacity of NO_3^- by Bhatnagar et al. (2008), and the results showed that although the surface area of modified carbon was lower than unmodified carbon, the formation of ZnO in the macropores and mesopores of the carbon facilitated the sorption of NO_3^-. Consequently, the optimized biochar surface with active adsorption site had better sorption of NO_3^- (Bhatnagar et al., 2008).

As reported by Chen et al. (2011), orange peel-derived biochar pyrolized at 250 and 400°C had weak sorption of phosphate. However, after modification by Fe, the adsorption of phosphate by magnetic orange peel biochar increased significantly. The nanosized magnetite particles in orange peel biochar facilitate the adsorption process of phosphate (Chen et al., 2011). Similarly, Zhang and Gao (2013) found that $AlCl_3$-modified biochar nanocomposite (biochar/AlOOH) showed very high sorption of phosphate. The Langmuir model estimated sorption capacity (Q_m) reached 135 mg g^{-1} after modification.

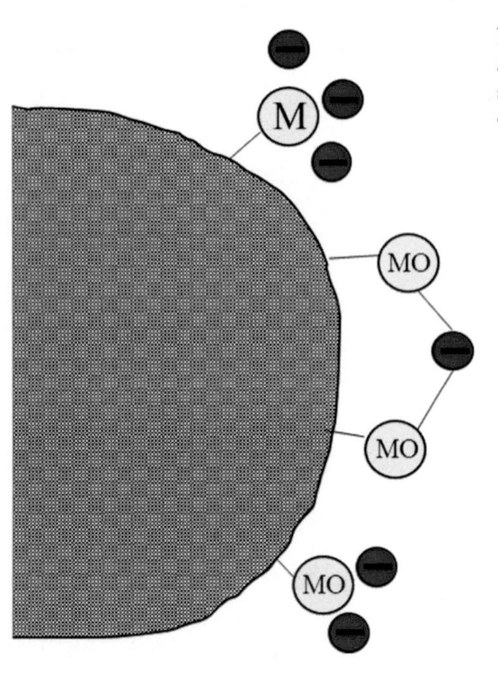

FIGURE 8.3 Surface complexation/precipitation of anionic contaminants with metals or metal oxide on biochar surface. M and MO refer to metals and metal oxides on biochar surface, respectively.

8.3.4 Electrostatic Attraction

The surface of biochar is generally considered negatively charged, especially in basic solution, which is not feasible for the removal of anionic contaminants. However, in low pH environment containing excessive H^+, the oxygen-containing group on biochar surface was protonated, and electrostatic attraction between anions in solution and protonated functional groups on biochar surface may control the sorption of anions by biochar. For instance, Fang et al. (2014) showed that at lower pH range, the functional groups on the surface of wood-derived biochar was protonated, and its sorption of perchlorate was significantly higher than in basic solution.

A similar phenomenon was also observed by Zhu et al. (2011) and Yang et al. (2016) in the sorption of anionic dye Congo Red by biochar. Lower pH had higher sorption capacity, and the protonated surface of biochar was a positively charged surface with electrostatic attraction with anionic dye (Fig. 8.4).

8.3.5 $\pi-\pi$ Interaction

With increase of pyrolysis temperature, the polar functional groups containing H and O decreased along with the increase of aromatic structure. The aromatic graphene-like surface of biochar can act as π-electron donor and the aromatic structure of contaminants can act as π-electron acceptor. This $\pi-\pi$ electron donor–acceptor (EDA) interaction strongly bound contaminants by sorbents, which facilitate the removal of aromatic-containing contaminants (Fig. 8.5). Yang et al. (2016) observed that with increase of pyrolysis temperature, the mole ratio of hydrogen to carbon (H/C ratio) decreased, indicating the increased

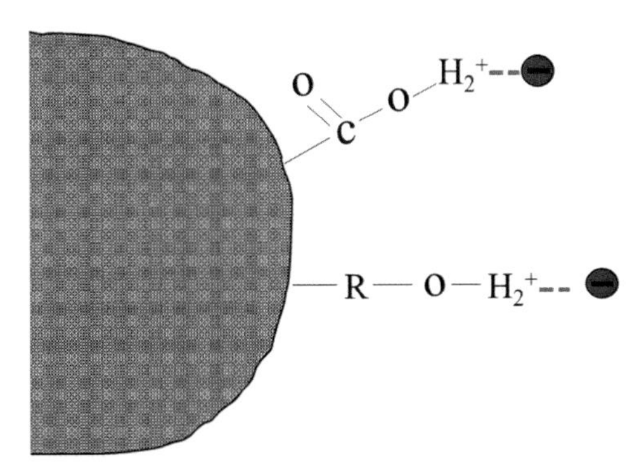

FIGURE 8.4 Electrostatic attraction of protonated biochar surface and anionic contaminants.

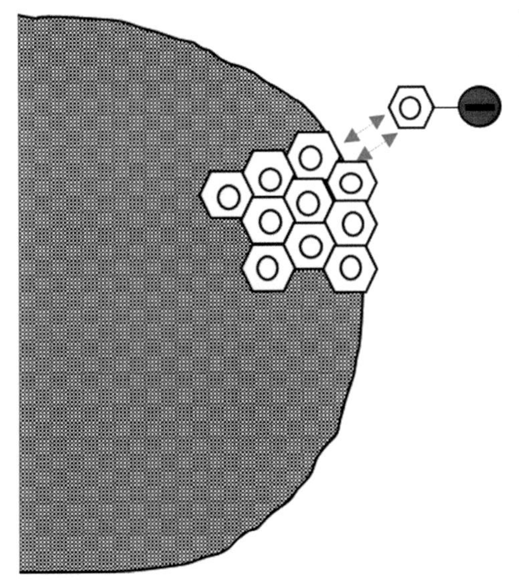

FIGURE 8.5 $\pi-\pi$ EDA interaction between graphene-like structure of biochar surface and aromatic group in anionic contaminants. *EDA*, Electron donor–acceptor.

aromaticity of the biochar. The aromatic structure in biochar surface provides the possibility to form $\pi-\pi$ EDA interaction with Congo Red-containing aromatic groups, thus increasing the sorption capacity. However, in the same study, there was no similar phenomenon observed for the sorption of Methylene Blue (Yang et al., 2016).

8.4 FACTORS INFLUENCING THE SORPTION OF ANIONIC CONTAMINANTS

8.4.1 Pyrolysis Temperature

Most of the organic components in the biomass used to produce biochar consist of mainly cellulose, hemicellulose, and lignin, as well as a small amount of lipid and protein. Decomposition of hemicellulose mainly occurs at relatively lower temperatures in the range of 250–350°C, followed by cellulose at 325–400°C (Stefanidis et al., 2014; Williams and Besler, 1996). Lignin is the most stable component and decomposes at a relatively higher temperature range of 300–550°C (Jung et al., 2016). Different biomass has different organic component content, which leads to different characteristics of the produced biochar. However, the general trends of biochar characteristics are almost the same: decrease of polar functional groups (H-, O-, N-containing groups) and increase of aromaticity with an increase in pyrolysis temperature. Furthermore, decomposition of organic components leads to the development of pore structure and surface area, which may significantly affect the sorption of contaminants.

Demiral and Gunduzoglu (2010) studied the sorption of NO_3^- by sugar beet bagasse biochar modified by $ZnCl_2$, and found that biochar produced at higher temperature (700°C) had the greatest sorption capacity, due to its higher pore volume and surface area than biochar produced at lower temperatures (Demiral and Gunduzoglu, 2010). Sorption of Congo Red by Vermicompost-derived biochar showed a significant increase in sorption capacity with an increase in pyrolytic temperature from 300 to 700°C (Yang et al., 2016). The increased aromaticity due to the increase of pyrolytic temperature was believed to be the main mechanism.

However, the opposite phenomenon has also been observed. Jung et al., (2016) found that the increase of pyrolysis temperature decreased total pore volume and surface area of waste-marine macroalgae biochar caused by the blockage of pores, and thus significantly decreased the sorption of PO_4^{3-}. Zhou et al. (2016) also found that lower pyrolysis temperature produced biochar with better sorption of Cr(V), which was attributed to the abundant carboxyl and hydroxyl groups at lower pyrolysis temperature to facilitate the sorption of Cr(V).

8.4.2 pH of the Solution

As noted in Section 8.3.2, the increase of pH in solution decreased the sorption of perchlorate since the oxygen-containing functional groups on the biochar surface was protonated at lower pH environment, providing electrostatic attraction between perchlorate and biochar. The surface of biochar became neutral with increase of pH, and the hydrogen bonding in this case played a dominant role in the sorption of perchlorate (Fang et al., 2014). With further increase of pH, there was almost no sorption due to electrostatic repulsion between the ionized and negatively charged surface of biochar (Fang et al., 2014). Sorption of NO_3^- by corn stover-derived biochar was also found to be affected by pH, and the lower pH improved sorption capacity due to the electrostatic attraction between the protonated biochar surface and nitrate.

The effect of pH on the sorption of F^- had a similar effect as that of perchlorate. However, at lower pH values in acidic solution, weakly ionized hydrofluoric acid was formed, which decreased the sorption of F^- onto biochar. The maximum sorption amount was observed at pH value around 6.25. Further increase of pH led to the competition of sorption site on biochar surface between fluoride and hydroxyl ions (Halder et al., 2016).

8.4.3 Coexisting Ions

Due to the complexity of natural water or wastewater, contaminants are not exist individually. Coexisting ions, especially these with similar properties, should be considered when evaluating the sorption capacity of biochars.

Pakade et al. (2016) found that the removal efficiency of Cr(VI) by quaternized AC decreased in the presence of coexisting ions, and the comptition of positively charged carbon surface by other negatively charged ions, such as SO_4^{2-}, PO_4^{3-}, NO_3^-, and $Cr_2O_7^{2-}$ was the main reason for the competing adsorption. Furthermore, the size of the anions played an important role in the suppression of the adsorption of Cr(VI) (Pakade et al., 2016). Sorption of NO_3^- by corn stover-derived biochar was also found to be significantly influenced by the presence of highly negative charged ions such as PO_4^{3-} and SO_4^{2-} in aqueous solution.

8.4.4 Temperature

Sorption of a sorbate by sorbent was energy-related processes, including endothermic and exothermic. Thus, change of environmental temperature may significantly affect sorption capability. To investigate the effect of temperature on the sorption of anions by biochar, Jung et al. (2015) studied the adsorption of phosphate by biochar/Mg–Al-assembled nanocomposites at 10–30°C, and found that the maximum phosphate adsorption capacity of biochar was 335, 480, and 727 mg g^{-1}, respectively. This suggests that adsorption of phosphate on this biochar is an endothermic process (Jung et al., 2015). Adsorption of Cr (VI) by used tire biochar, sawdust biochar, and CAC were also found to be endothermic in nature by the thermodynamic parameters obtained by Hamadi et al. (2001)

8.5 CONCLUSIONS AND PERSPECTIVES

Biochar is as an effective sorbent for the removal of contaminants in water/wastewater. In addition to heavy metals and nonionic organic contaminants, biochar also has the potential to remove anionic contaminants. The pore filling effect, hydrogen bonding, surface complexation/precipitation, electrostatic attraction, and $\pi-\pi$ EDA interactions are the main mechanisms in the adsorption of anionic contaminants on biochar surface. Environmental conditions, including pH values, coexisting ions, temperature, as well as pyrolysis temperature of biochar, affect the adsorption process significantly.

However, biochar's adsorption of most anionic contaminants is still weak compared with heavy metals and organic contaminants. In most cases, the adsorption capacity of biochars for anionic contaminants is less than 10 mg g^{-1}. As a result, new approaches are needed to improve adsorption capacity of biochars of anionic contaminants. However, biochar also contains anions such as PO_4^{3-} and NO_3^- since P and N are major elements in biomass, and PO_4^{3-} and NO_3^- may form during the pyrolysis process. Thus, dissolution of these anions into water should be quantitatively evaluated before possible application of biochar into water/wastewater treatment.

References

Ahmad, M., Rajapaksha, A.U., Lim, J.E., Zhang, M., Bolan, N., Mohan, D., et al., 2014. Biochar as a sorbent for contaminant management in soil and water: a review. Chemosphere 99, 19–33.

Ali, I., Gupta, V.K., 2006. Advances in water treatment by adsorption technology. Nat. Protoc. 1 (6), 2661–2667.

Bhatnagar, A., Ji, M., Choi, Y.H., Jung, W., Lee, S.H., Kim, S.J., et al., 2008. Removal of nitrate from water by adsorption onto zinc chloride treated activated carbon. Sep. Sci. Technol. 43 (4), 886–907.

Bordoloi, N., Dey, M.D., Mukhopadhyay, R., Kataki, R., 2018. Adsorption of Methylene blue and Rhodamine B by using biochar derived from Pongamia glabra seed cover. Water Sci. Technol. 77 (3), 638–646.

Chatterjee, S., Woo, S.H., 2009. The removal of nitrate from aqueous solutions by chitosan hydrogel beads. J. Hazard. Mater. 164 (2-3), 1012–1018.

Chen, B.L., Chen, Z.M., Lv, S.F., 2011. A novel magnetic biochar efficiently sorbs organic pollutants and phosphate. Bioresour. Technol. 102 (2), 716–723.

Chen, G.J., Peng, C.Y., Fang, J.Y., Dong, Y.Y., Zhu, X.H., Cai, H.M., 2016. Biosorption of fluoride from drinking water using spent mushroom compost biochar coated with aluminum hydroxide. Desalin Water Treat. 57 (26), 12385–12395.

Chintala, R., Mollinedo, J., Schumacher, T.E., Papiernik, S.K., Malo, D.D., Clay, D.E., et al., 2013. Nitrate sorption and desorption in biochars from fast pyrolysis. Microporous Mesoporous Mater. 179, 250–257.

Cho, D.W., Kwon, G., Ok, Y.S., Kwon, E.E., Song, H., 2017. Reduction of bromate by cobalt-impregnated biochar fabricated via pyrolysis of lignin using CO_2 as a reaction medium. ACS Appl. Mater. Interfaces 9 (15), 13142–13150.

Demiral, H., Gunduzoglu, G., 2010. Removal of nitrate from aqueous solutions by activated carbon prepared from sugar beet bagasse. Bioresour. Technol. 101 (6), 1675–1680.

Dong, X.L., Ma, L.N.Q., Li, Y.C., 2011. Characteristics and mechanisms of hexavalent chromium removal by biochar from sugar beet tailing. J. Hazard. Mater. 190 (1-3), 909–915.

Fang, C., Dharmarajan, R., Megharaj, M., Naidu, R., 2017. Gold nanoparticle-based optical sensors for selected anionic contaminants. TrAC, Trends Anal. Chem. 86, 143–154.

Fang, Q.L., Chen, B.L., Lin, Y.J., Guan, Y.T., 2014. Aromatic and hydrophobic surfaces of wood-derived biochar enhance perchlorate adsorption via hydrogen bonding to oxygen-containing organic groups. Environ. Sci. Technol. 48 (1), 279–288.

Gabriele, C., Siglinda, P., 2003. Remediation of water contamination using catalytic technologies. Appl. Catal., B: Environ. 41 (1-2), 15–29.

Ginsberg, G.L., Hattis, D.B., Zoeller, R.T., Rice, D.C., 2007. Evaluation of the U.S. EPA/OSWER preliminary remediation goal for perchlorate in groundwater: focus on exposure to nursing infants. Environ. Health Perspect. 115 (3), 361–369.

Guan, X.J., Zhou, J., Ma, N., Chen, X.Y., Gao, J.Q., Zhang, R.Q., 2015. Studies on modified conditions of biochar and the mechanism for fluoride removal. Desalin Water Treat. 55 (2), 440–447.

World Health Organization, 2006. Guidelines for Drinking Water Quality, third ed., World Health Organization.

Halder, G., Khan, A.A., Dhawane, S., 2016. Fluoride sorption onto a steam-activated biochar derived from *Cocos nucifera* shell. Clean: Soil, Air, Water 44 (2).

Hamadi, N.K., Chen, X.D., Farid, M., Lu, M.G.Q., 2001. Adsorption kinetics for the removal of chromium(VI) from aqueous solution by adsorbents derived from used tyres and sawdust. Chem. Eng. J. 84, 95–105.

Harrison, P.T.C., 2005. Fluoride in water: a UK perspective. J. Fluor. Chem. 126, 1448–1456.

He, R.Z., Peng, Z.Y., Lyu, H.H., Huang, H., Nan, Q., Tang, J.C., 2018. Synthesis and characterization of an iron-impregnated biochar for aqueous arsenic removal. Sci. Total Environ. 612, 1177–1186.

Hsu, N.H., Wang, S.L., Liao, Y.H., Huang, S.T., Tzou, Y.M., Huang, Y.M., 2009. Removal of hexavalent chromium from acidic aqueous solutions using rice straw-derived carbon. J. Hazard. Mater. 171 (1-3), 1066–1070.

Jin, H.M., Capareda, S., Chang, Z.Z., Gao, J., Xu, Y.D., Zhang, J.Y., 2014. Biochar pyrolytically produced from municipal solid wastes for aqueous As(V) removal: adsorption property and its improvement with KOH activation. Bioresour. Technol. 169, 622–629.

Jung, K.W., Jeong, T.U., Hwang, M.J., Kim, K., Ahn, K.H., 2015. Phosphate adsorption ability of biochar/Mg–Al assembled nanocomposites prepared by aluminum-electrode based electro-assisted modification method with MgCl(2) as electrolyte. Bioresour. Technol. 198, 603–610.

Jung, K.W., Hwang, M.J., Jeong, T.U., Ahn, K.H., 2015. A novel approach for preparation of modified-biochar derived from marine macroalgae: dual purpose electro-modification for improvement of surface area and metal impregnation. Bioresour. Technol. 191, 342–345.

Jung, K.W., Kim, K., Jeong, T.U., Ahn, K.H., 2016. Influence of pyrolysis temperature on characteristics and phosphate adsorption capability of biochar derived from waste-marine macroalgae (*Undaria pinnatifida* roots). Bioresour. Technol. 200, 1024–1028.

Kapoor, A., Viraraghavan, T., 1997. Nitrate removal from drinking water—review. J. Environ. Eng. 123 (4), 371–380.

Karthikeyan, T., Rajgopal, S., Miranda, L.R., 2005. Chromium(VI) adsorption from aqueous solution by Hevea Brasilinesis sawdust activated carbon. J. Hazard. Mater. 124 (1-3), 192–199.

Kounaves, S.P., Stroble, S.T., Anderson, R.M., Moore, Q., Catling, D.C., Douglas, S., et al., 2010. Discovery of natural perchlorate in the antarctic dry valleys and its global implications. Environ. Sci. Technol. 44 (7), 2360–2364.

Lehmann, J., Joseph, S., 2015. Biochar for Environmental Management: Science and Technology, first ed. Earthscan, London.

Liu, C.H., Chuang, Y.H., Chen, T.Y., Tian, Y., Li, H., Wang, M.K., et al., 2015. Mechanism of arsenic adsorption on magnetite nanoparticles from water: Thermodynamic and spectroscopic studies. Environ. Sci. Technol. 49 (13), 7726–7734.

Lonappan, L., Rouissi, T., Das, R.K., Brar, S.K., Ramirez, A.A., Verma, M., et al., 2016. Adsorption of methylene blue on biochar microparticles derived from different waste materials. Waste Manage 49, 537–544.

Ma, J.F., Qi, J., Yao, C., Cui, B.Y., Zhang, T.L., Li, D.L., 2012. A novel bentonite-based adsorbent for anionic pollutant removal from water. Chem. Eng. J. 200, 97–103.

Mearns, A.J., Oshida, P.S., Sherwood, M.J., Young, R., Reish, D.J., 1976. Chromium effects on coastal organisms. J. Water Pollut. Control Fed. 48 (8), 1929–1939.

Min, B., Evans, P.J., Chu, A.K., Logan, B.E., 2004. Perchlorate removal in sand and plastic media bioreactors. Water Res. 38 (1), 47–60.

Mishra, P.C., Patel, R.K., 2009. Use of agricultural waste for the removal of nitrate-nitrogen from aqueous medium. J. Environ. Manage. 90 (1), 519–522.

Mizuta, K., Matsumoto, T., Hatate, Y., Nishihara, K., Nakanishi, T., 2004. Removal of nitrate-nitrogen from drinking water using bamboo powder charcoal. Bioresour. Technol. 95 (3), 255–257.

Mohan, D., Kumar, S., Srivastava, A., 2014. Fluoride removal from ground water using magnetic and nonmagnetic corn stover biochars. Ecol. Eng. 73, 798–808.

Mohan, D.S., Singh, K.P., Singh, V.K., 2005. Removal of hexavalent chromium from aqueous solution using low cost activated carbons derived from agricultural waste materials and activated carbon fabric cloth. Ind. Eng. Chem. Res. 44 (4), 1027–1042.

Niazi, N.K., Bibi, I., Shahid, M., Ok, Y.S., Shaheen, S.M., Rinklebe, J., et al., 2018a. Arsenic removal by Japanese oak wood biochar in aqueous solutions and well water: investigating arsenic fate using integrated spectroscopic and microscopic techniques. Sci. Total Environ. 621, 1642–1651.

Niazi, N.K., Bibi, I., Shahid, M., Ok, Y.S., Burton, E.D., Wang, H., et al., 2018b. Arsenic removal by perilla leaf biochar in aqueous solutions and groundwater: an integrated spectroscopic and microscopic examination. Environ. Pollut. 232, 31–41.

Oh, T.K., Choi, B., Shinogi, Y., Chikushi, J., 2012. Effect of pH conditions on actual and apparent fluoride adsorption by biochar in aqueous phase. Water, Air, Soil Pollut. 223 (7), 3729–3738.

Pakade, V.E., Maremeni, L.C., Ntuli, T.D., Tavengwa, N.T., 2016. Application of quaternized activated carbon derived from macadamia nutshells for the removal of hexavalent chromium from aqueous solutions. S. Afr. J. Chem. 69, 180–188.

Pulkka, S., Martikainen, M., Bhatnagar, A., Sillanpaa, M., 2014. Electrochemical methods for the removal of anionic contaminants from water—a review. Sep. Purif. Technol. 132, 252–271.

Sica, M., Duta, A., Teodosiu, C., Draghici, C., 2014. Thermodynamic and kinetic study on ammonium removal from a synthetic water solution using ion exchange resin. Clean Technol. Environ. 16 (2), 351–359.

Sponza, D.T., Işik, M., 2005. Toxicity and intermediates of C.I. Direct Red 28 dye through sequential anaerobic/aerobic treatment. Process Biochem. 20, 2735–2744.

Steenbergen, F., Haimanot, R.T., Sideil, A., 2011. High fluoride, modest fluorosis: Investigation in drinking water supply in Halaba (SNNPR, Ethiopia). J. Water Resour. Protect 3, 120–126.

Stefanidis, S.D., Kalogiannis, K.G., Iliopoulou, E.F., Michailof, C.M., Pilavachi, P.A., Lappas, A.A., 2014. A study of lignocellulosic biomass pyrolysis via the pyrolysis of cellulose, hemicellulose and lignin. J. Anal. Appl. Pyrol. 105, 143–150.

USEPA National Primary Drinking Water Regulations, 1998. ⟨https://www.epa.gov/ground-water-and-drinking-water/national-primary-drinking-water-regulations⟩ (April 11th).

United States Environmental Protection Agency, 1986. Quality criteria for water, EPA 440/5-86-001.

Van der Bruggen, B., Vandecasteele, C., 2003. Removal of pollutants from surface water and groundwater by nanofiltration: overview of possible applications in the drinking water industry. Environ. Pollut. 122 (3), 435–445.

Wang, H., Gao, B., Wang, S., Fang, J., Xue, Y., Yang, K., 2015a. Removal of Pb(II), Cu(II), and Cd(II) from aqueous solutions by biochar derived from KMnO4 treated hickory wood. Bioresour. Technol. 197, 356–362.

Wang, S.S., Gao, B., Li, Y.C., Mosa, A., Zimmerman, A.R., Ma, L.Q., et al., 2015b. Manganese oxide-modified biochars: preparation, characterization, and sorption of arsenate and lead. Bioresour. Technol. 181, 13–17.

Wang, Z.H., Guo, H.Y., Shen, F., Yang, G., Zhang, Y.Z., Zeng, Y.M., et al., 2015c. Biochar produced from oak sawdust by Lanthanum (La)-involved pyrolysis for adsorption of ammonium (NH_4^+), nitrate (NO_3^-), and phosphate (PO_4^{3-}). Chemosphere 119, 646–653.

Williams, P.T., Besler, S., 1996. The influence of temperature and heating rate on the slow pyrolysis of biomass. Renew. Energy 7 (3), 233–250.

Wu, C., Huang, L., Xue, S.G., Huang, Y.Y., Hartley, W., Cui, M.Q., et al., 2017. Arsenic sorption by red mud-modified biochar produced from rice straw. Environ. Sci. Pollut. Res. 24 (22), 18168–18178.

Xu, Y., Liu, Y.G., Liu, S.B., Tan, X.F., Zeng, G.M., Zeng, W., et al., 2016. Enhanced adsorption of methylene blue by citric acid modification of biochar derived from water hyacinth (*Eichornia crassipes*). Environ. Sci. Pollut. Res. 23 (23), 23606–23618.

Yan, L.G., Xu, Y.Y., Yu, H.Q., Xin, X.D., Wei, Q., Du, B., 2010. Adsorption of phosphate from aqueous solution by hydroxy-aluminum, hydroxy-iron and hydroxy-iron-aluminum pillared bentonites. J. Hazard. Mater. 179 (1-3), 244–250.

Yan, L.L., Kong, L., Qu, Z., Lo, L., Shen, G.Q., 2015. Magnetic biochar decorated with ZnS nanocrytals for Pb (II) removal. ACS Sustainable Chem. Eng. 3 (1), 125–132.

Yang, G., Wu, L., Xian, Q.M., Shen, F., Wu, J., Zhang, Y.Z., 2016. Removal of Congo red and methylene blue from aqueous solutions by vermicompost-derived biochars. PLoS ONE 11 (5), e0154562. Available from: https://doi.org/10.1371/journal.pone.0154562.

Yoon, K., Cho, D.W., Tsang, D.C.W., Bolan, N., Rinklebe, J., Song, H., 2017. Fabrication of engineered biochar from paper mill sludge and its application into removal of arsenic and cadmium in acidic water. Bioresour. Technol. 246, 69–75.

Zhang, M., Gao, B., 2013. Removal of arsenic, methylene blue, and phosphate by biochar/AlOOH nanocomposite. Chem. Eng. J. 226, 286–292.

Zhang, M., Gao, B., Yao, Y., Xue, Y.W., Inyang, M., 2012. Synthesis of porous MgO-biochar nanocomposites for removal of phosphate and nitrate from aqueous solutions. Chem. Eng. J. 210, 26–32.

Zhitkovich, A., 2011. Chromium in drinking water: sources, metabolism, and cancer risks. Chem. Res. Toxicol. 24 (10), 1617–1629.

Zhou, L., Liu, Y.G., Liu, S.B., Yin, Y.C., Zeng, G.M., Tan, X.F., et al., 2016. Investigation of the adsorption-reduction mechanisms of hexavalent chromium by ramie biochars of different pyrolytic temperatures. Bioresour. Technol. 218, 351–359.

Zhu, H.Y., Fu, Y.Q., Jiang, R., Jiang, J.H., Xiao, L., Zeng, G.M., et al., 2011. Adsorption removal of Congo Red onto magnetic cellulose/Fe_3O_4/activated carbon composite: equilibrium, kinetic and thermodynamic studies. Chem. Eng. J. 173 (2), 494–502.

Zhu, N., Yan, T., Qiao, J., Cao, H., 2016. Adsorption of arsenic, phosphorus and chromium by bismuth impregnated biochar: adsorption mechanism and depleted adsorbent utilization. Chemosphere 164, 32–40.

CHAPTER

9

Biochar for Soil Water Conservation and Salinization Control in Arid Desert Regions

Xiaodong Yang[1] *and Arshad Ali*[2]

[1]Institute of Resources and Environment Science, Xinjiang University, Urumqi, Xinjiang, China [2]Spatial Ecology Lab, School of Life Science, South China Normal University, Guangzhou, Guangdong, China

9.1 ARID DESERT ECOSYSTEM

An arid desert (including semiarid) ecosystem is one of the most widely distributed and vulnerable ecosystems in the world, accounting for a quarter to one-fifth of global land (Whitford and Wade, 2002). It is mainly distributed in west and central Asia, Mongolia, northwest India, Africa, midwestern Australia, and the southwestern United States, but is found in more than 100 countries and regions (Fig. 9.1). Extreme water scarcity, uncertainty of precipitation, soil desertification, and salinization are the most prominent features of arid deserts (Whitford and Wade, 2002). Despite desert ecosystems suffering from serious environmental stress, they still support 15% of the world's population. However, due to climate change and human disturbances, water resources shortage and soil salinization have created a bottleneck and restrict the sustainable development and ecosystem maintenance of arid deserts (Akkad, 1989; Houerou, 1996). At present, about two thirds of arid desert's soils are degraded by salt and water stresses in arid desert regions (Maliva and Missimer, 2012; Thomas and Middleton, 1993). Thus, in the context of climate change and human disturbances, water resources protection and conservation must be urgently addressed in arid desert regions (Akkad, 1989; Houerou, 1996). Agricultural production as one of the biggest human activities for water consumption in arid desert regions, the search of a much more effective methods of water conservation

Biochar from Biomass and Waste
DOI: https://doi.org/10.1016/B978-0-12-811729-3.00009-1

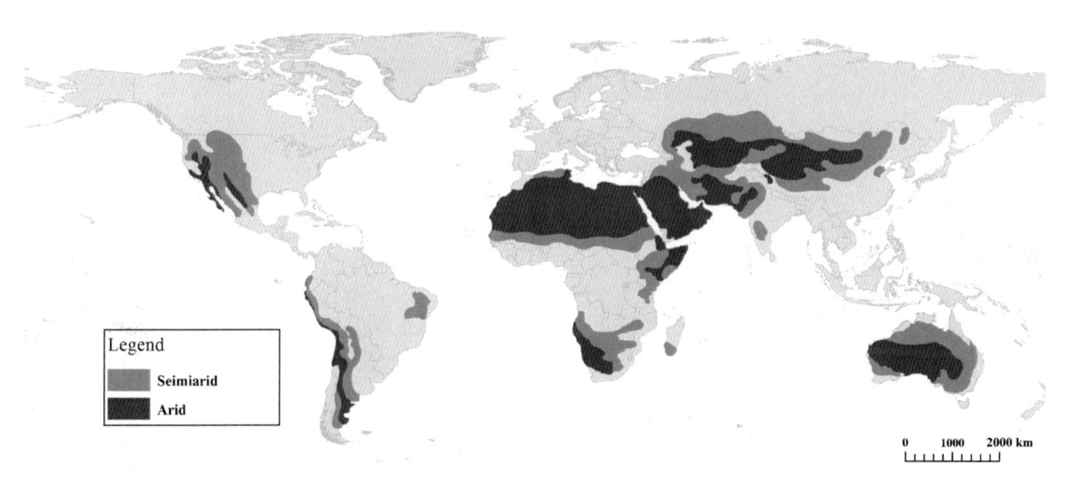

FIGURE 9.1 Global desert distribution.

and soil salinization control in this activities have drawn great concern worldwide (Akkad, 1989; Houerou, 1996; Li, 2010).

With the development of theories and technologies, biochar is widely used in agricultural and environmental activities in arid desert areas (Abel et al., 2013; Akhtar et al., 2015a; Gao et al., 2011; Qian et al., 2015). For example, the addition of biochar into soils are used to reduce the leaching of soil nitrogen and phosphorus, soil heavy metal pollution, soil release amount of CO_2, as well as to improve the photosynthetic physiological characteristics, microbial structure and diversity, crop yield, and soil enzyme activities in arid desert regions (Abel et al., 2013; Gao et al., 2011; Lehmann and Joseph, 2015; Qian et al., 2015). Interestingly these effects of biochar to soil ecosystem are all verified in the farmland, lawn, and forest ecosystem in arid desert regions (Abel et al., 2013; Dugan et al., 2010; Lehmann and Joseph, 2015). In addition, agricultural workers also explore a series of technologies, such as the reduction in greenhouse gas emissions, the improvement in crop yield, saline–alkaline soil amelioration, and soil pollution control, in order to apply biochar theories in arid and desert regions (Abel et al., 2013; Akhtar et al., 2015b; Lehmann and Joseph, 2015; Pan et al., 2017; Wang et al., 2014). Biochar has already become one of the important ways of agricultural and environmental activities in arid desert regions.

9.2 METHODS FOR WATER CONSERVATION AND SALINIZATION CONTROL IN ARID DESERT REGIONS

Using water-holding agents to improve soil water conservation is one of the most common and widely used methods to improve agricultural production in arid desert regions (Han et al., 2004). Water-holding agents are divided into three categories based on the manufacturing material they are produced with: starch, fiber, or polymer. After applying the water-holding agent in soils, it usually absorbs a large quantity of soil capillary water and gravitational water, sometimes even the air moisture, and then gradually provides nearby plant roots via its water absorption and conservation features. Due to its super absorbent and maintaining capacity of water, water-holding agent slowly releases water into soils in order to promote plant growth and maintain transpiration in drought seasons after it obtained soil and air moisture (Han et al., 2004; Li and Huang, 1996). However, water-holding agents are not completely formed from natural soil materials, and thus their use for soil water conservation in agricultural land of arid desert regions has the following disadvantages: (1) the application of a water-holding agent may cause secondary chemical soil pollution, and may reduce the duration of soil planting (Dang et al., 2006; Gong and Yin, 2009); and (2) in the case of extreme soil water scarcity, the water-holding agent may cause water competition between itself and plants and thus reduce subsequent crop yields (Dang et al., 2006; Gong and Yin, 2009). Hence, the development of a new water conservation method becomes a hot spot of agricultural technology problems in arid desert regions at present.

Soil salinity determines the salt content in soil, whereas soil salinization is the process through which salt content is increasing in soil (Li, 2010). Soil salinization can be caused by natural and artificial processes, such as mineral weathering and irrigation (Li, 2010; Salama et al., 1999). In arid desert regions, soil salinization usually occurs when the water table is shallow from the surface of the soil. Salts from the groundwater are usually raised through the process of capillary action to the surface of the soil (Salama et al., 1999). Soil salinization is thus mainly determined by the irrigation in arid desert regions (Li, 2010). For example, irrigation may cause salts to accumulate over time. During the process of water uptake by plants or evaporation, the salts are left behind in the soils (Rhoades and Loveday, 1990; Salama et al., 1999). Since soil salinity makes it more difficult for plants to absorb soil moistures, these salts must be leached out of the plant roots by applying additional irrigation water (Rhoades and Loveday, 1990; Salama et al., 1999). The consequences of soil salinization are therefore detrimental to ecosystem stability, water quality, plant growth, and production in arid desert regions (Li, 2010; Rhoades and Loveday, 1990; Salama et al., 1999).

Yet, a wide variety of methods has been developed and applied in order to control soil salinization in the agricultural lands of arid desert regions (Li, 2010; Rhoades and Loveday, 1990). These methods include the application of chemical amendments, planting salt-tolerant plants, leaching irrigation, optimizing drainage systems, and soil tillage (Li, 2010; Rhoades and Loveday, 1990). The major disadvantage of chemical amendments is that they are not natural soil materials, and hence can easily cause secondary chemical soil pollution (Li, 2010). As such, planting salt-tolerant plant in order to control salinization in practice take an enormous amount of time, and is limited to some large-scale implementation are not possible (Li, 2010). Leaching irrigation, drainage system optimization, and soil tillage methods may only shift salinity from shallow to deep soil with no reduction in the salinity content (Li, 2010; Rhoades and Loveday, 1990). Due to the disadvantages associated with these methods, development of a simple and nearly natural method is needed to help improve agricultural production in arid desert regions.

9.3 APPLICATION OF BIOCHAR TO SOILS

Biochar is a carbon-rich solid obtained by heating biomass, such as wood or manure with a little or no oxygen, that can be applied to soil for both agricultural production and carbon sequestration (Lehmann and Joseph, 2015). Biochar can reduce carbon release to the atmosphere from burning or degrading by carbon stabilization into a form resembling charcoal. Biochar has significant effects on soil physicochemical properties and can enhance soil structure, increase pH, and augmented soil aeration and moisture content (Lehmann and Joseph, 2015; Ok et al., 2015).

Web of Science indexed journal titles that include the word "biochar" have steadily increased from 3 to 735 over the past 10 years (Fig. 9.2), indicating a growing interest in biochar research. This increasing interest has resulted in multidisciplinary areas for scientific research. In recent years, a large number of studies have highlighted the benefits of the addition of biochar into soil to mitigate global warming, to improve crop production, and to increase soil carbon storage (Lehmann and Joseph, 2015; Ok et al., 2015; Qian et al., 2015; Tan et al., 2015). As has been shown by the research, because of its absorptivity and permeability, adding biochar to agricultural soil can change soil water percolating capacity, increase retention time of water transportation in soils, and increase water flow. Thus, applying biochar to agricultural soil is advantageous for water conservation (Abel et al., 2013; Chen et al., 2013; Pan et al., 2017; Wang et al., 2014). To illustrate, the number of publications on the Web of Science indexed journals containing the keywords "biochar+ water" also increased significantly in the last ten years, and now account for approximately one-twentieth of all "biochar" publications (Fig. 9.1). Additionally, biochar addition to agricultural soil can also decrease the salinity content of soil due to its use for adsorption and soil amendment (Akhtar et al., 2015a; Kanwal et al., 2018). Considering the wide availability of feedstock and its low cost, biochar has great potential for use in water

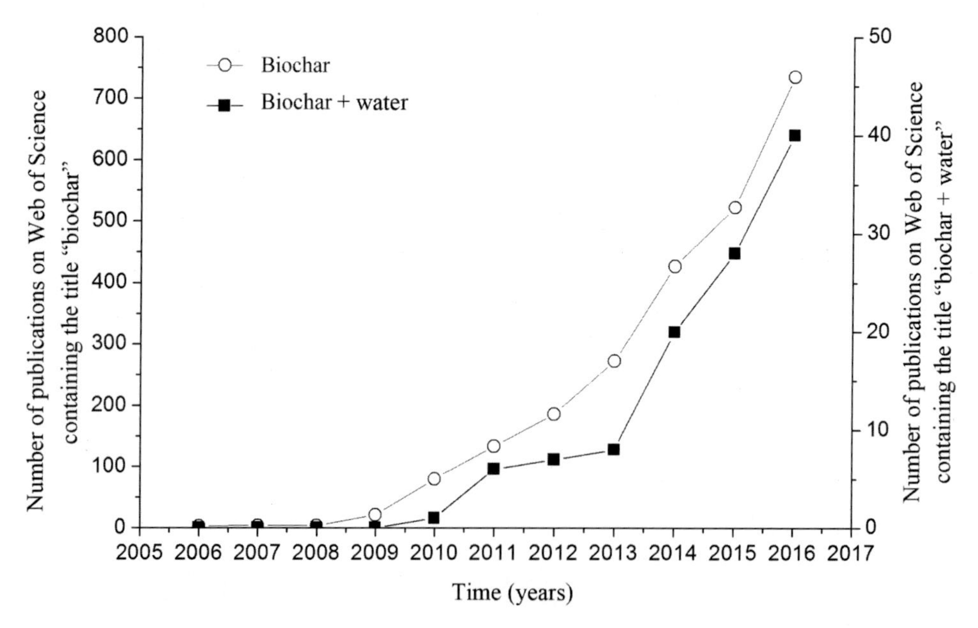

FIGURE 9.2 Number of journal titles on Web of Science containing "biochar" and "biochar + water" from 2006 to 2016.

conservation and for salinization control (Abel et al., 2013; Akhtar et al., 2015a; Kanwal et al., 2018; Wang et al., 2014).

9.3.1 Application of Biochar for Water Conservation in Arid Desert Regions

Water is the principal limited factors in arid desert regions, and largely determines the development and stability of local ecosystems (Akkad, 1989; Whitford and Wade, 2002). In soil environments, water content mainly depends on the soil's water-holding capacity and precipitation (Schlesinger and Pilmanis, 1998). Due to the scarcity of precipitation, water-holding capacity is considered an important key to the success of the abundance of soil water content and agricultural production in arid desert regions (Rotini, 1970; Schlesinger and Pilmanis, 1998). Here, water-holding capacity includes the amounts of soil capillary water and soil gravitational water (Rotini, 1970). Soil water-holding capacity has benefits for water content and water-saving efficiency in agricultural lands of arid desert regions (Rotini, 1970).

As has been reported, many factors affect water-holding capacity, such as total soil porosity, soil density, soil texture, soil organic matter content, and soil capillary porosity (Rotini, 1970). Since the density of biochar is lower than that of soil (Ok et al., 2015), biochar addition to agricultural soil can reduce soil bulk density but improve soil aggregates, soil porosity, and soil texture (Abel et al., 2013; Chen et al., 2013; Pan et al., 2017; Qian et al., 2015; Wang et al., 2014). Thus, biochar addition to agricultural soil can contribute to increase water-holding capacity and retention time of water transportation in soils (Lehmann and Joseph, 2015; Ok et al., 2015; Qian et al., 2015; Wang et al., 2014). In addition, previous studies have indicated that the spatial distance and connectivity among soil pores are positively to soil porosity, soil organic matter content, soil structure characteristics, and the surface area of soil particles (Rotini, 1970). After application of biochar into soils, the improvements in spatial distance and connectivity among soil pores could increase the above underlying soil physicochemical properties, and could in turn increase water-holding capacity in agricultural land (Abel et al., 2013; Pan et al., 2017; Wang et al., 2014). For example, G1aser et al. (2002) and Karhu et al. (2011) both showed that the water-holding capacity of biochar-treated soils is at least 11% higher than that of untreated soils in field experiments. Ogutunde et al. (2008) showed that biochar addition to agricultural soil can increase soil-saturated water conductivity, soil porosity and soil permeability, while decreasing soil density.

In addition, owning to the oxidation and carboxyl groups in biochar, the addition of biochar into agricultural soil can improve the capabilities of water absorption and preservation, and thereby increasing soil water-holding capacity (Abel et al., 2013; Chen et al., 2013; Karhu et al., 2011). Specifically, biochar reduces the rates of water evaporation and permeation in soil because the water molecules tightly adsorbed on the oxidation and carboxyl groups of biochar surface. Then, the adsorbed water on the oxidation and carboxyl groups is also gradually released during the soil desiccation process (Abel et al., 2013; Karhu et al., 2011; Wang et al., 2014). Thus, biochar addition to agricultural soil can reduce water losses, thus increasing the water-holding capacity of soil (Abel et al., 2013; Karhu et al., 2011; Wang et al., 2014).

It's worth noting that soil water-holding capacity does not grow with the increased addition of biochar into soils because biochar can produce hydrophobic effects and thus reduce the water-holding capacity when the addition of biochar exceeding a certain limit (Dugan et al., 2010). Thus, when biochar is added to agricultural soil, the negative effects of hydrophobic groups on water-holding capacity can be higher than the positive effects of water conservation (Peng et al., 2011). In this case, water will stay in the soil surface in the form of liquid beads, and cannot quickly infiltrate from shallow soils to deep soils. Hence, water evaporates from the soil surface and subsequently decreases soil water content (Dugan et al., 2010; Peng et al., 2011). Using the horizontal column infiltration method in an arid desert region, Gao et al. (2011) reported that the water-holding capacity of sandy and loamy soils was highest when the amount of biochar was 80 t h m^{-2} (Gao et al., 2011). Overall, biochar improves water conservation of agricultural lands in arid desert regions, but its effectiveness depends on the amount of biochar added (Gao et al., 2011; Peng et al., 2011; Singh et al., 2010).

9.3.2 Application of Biochar for Soil Salinization Control in Arid Desert Regions

Drought and evaporation are the main causes of soil salinization in the arid desert regions (Thomas and Middleton, 1993). In recent decades, soil salinization has been recognized as one of the main factors affecting agricultural productivity due to climate change, anthropogenic disturbance, and reduction in water resources in arid desert regions (Li, 2010). As noted above, biochar addition to soils can effectively reduce the amount of soil evaporation, thus decreasing the amount of salinity transferred from the deep soils or groundwater to shallow soils via water transportation (Abel et al., 2013; Gao et al., 2011; Wang et al., 2014). Furthermore, due to its strong adsorbability and superlarge surface area, biochar has good potential for adsorbing the active soil salt ions, that is, Na^{+}, K^{+}, Ca^{2+}, and then in reducing the amount of salt in both soil and solution (Akhtar et al., 2015b; Glaser et al., 2002; Kanwal et al., 2018). This can help reduce the soil salinization of agricultural lands in arid desert regions.

Previous studies have suggested that soil salinization control is a complicated process (Li, 2010; Rhoades and Loveday, 1990), but there are several remediation methods available (Li, 2010). In arid desert regions, leaching irrigation is the most commonly used method to control soil salinization (Rhoades and Loveday, 1990). Namely, this method is a washing process that uses the irrigation system to rinse soil salinity from shallow soils into deep soils or into rivers and other water bodies (Rhoades and Loveday, 1990). The efficiency of salt-leaching desalination is influenced by many factors such as soil texture, saturated hydraulic conductivity, organic matter content, and soil porosity (Rhoades and Loveday, 1990; Thomas and Middleton, 1993). As discussed, biochar can improve soil-saturated hydraulic conductivity, soil texture, organic matter content, and soil porosity in agricultural lands (Karhu et al., 2011; Ok et al., 2015; Pan et al., 2017). Combining biochar and leaching irrigation can help control soil salinization in arid desert regions.

As with water conservation, soil salinization does not decrease when more biochar is added (Akhtar et al., 2015a; Kanwal et al., 2018), because biochar contains base cations that are released into the soil (Ok et al., 2015). With the increasing of the addition of biochar into soil, an increasing number of base cations are released from biochar surface. This can gradually increase soil pH and salinity, and then counteracting the positive effects of biochar on salinization control (Glaser et al., 2002; Kanwal et al., 2018; Ok et al., 2015). Thus, the effect on soil salinization control of biochar is dependent by its applied amount.

9.4 OTHER ADVANTAGES OF BIOCHAR APPLICATION IN ARID DESERT REGIONS

Besides aiding in water conservation and salinization control, biochar can also increase the availability of nutrients to plants and subsequently improve crop yields in arid desert regions (Lehmann and Joseph, 2015; Ok et al., 2015; Qian et al., 2015). This may be due to (1) the charges of chemical functional groups on biochar surface can catch and accumulate fertilizers nutrients by electrostatic adsorption, and thereby increasing the soil nutrient-holding capacity (Ok et al., 2015); (2) biochar acting as a sustained-release carrier decreasing nutrition loss and maintaining nutrient balance (Gao et al., 2011); (3) biochar contains elements such as nitrogen, phosphorus, and potassium that are beneficial to soil (Gao et al., 2011; Lehmann and Joseph, 2015; Ok et al., 2015); (4) biochar can promote the activities of soil microorganisms to enhance the process of nitration, as well as the mineralization of organic phosphorus (Ok et al., 2015; Qian et al., 2015).

In addition, due to the loose macropore, superlarge surface area, and abundant carboxyl, phenolic hydroxyl, and anhydride groups, biochar has a good ability to reduce organic pollutants (Ok et al., 2015; Tan et al., 2015). Biochar can also help reduce the health risks of the soil environment and crops (Lehmann and Joseph, 2015; Ok et al., 2015).

9.5 CONCLUSIONS

In arid desert regions, applying biochar to soils can improve soil adsorption capacity, organic matter content, soil porosity, and soil aeration, but it can also decrease soil bulk density. This subsequently changes the patterns of water permeation and salinity transportation in the soil profile as well as between groundwater and soils, hence enhancing water conservation and salinization in soil. In addition to improving soil physical properties, biochar can also help plant roots uptake nutrients, thereby improving the crop production of agricultural lands in arid desert regions.

References

Abel, S., Peters, A., Trinks, S., Schonsky, H., Facklam, M., Wessolek, G., 2013. Impact of biochar and hydrochar addition on water retention and water repellency of sandy soil. Geoderma 202-203, 183–191.

Akhtar, S.S., Andersen, M.N., Liu, F., 2015a. Biochar mitigates salinity stress in potato. J. Agron. Crop. Sci. 201, 368–378.

Akhtar, S.S., Andersen, M.N., Naveed, M., Zahir, Z.A., Liu, F., 2015b. Interactive effect of biochar and plant growth-promoting bacterial endophytes on ameliorating salinity stress in maize. Funct. Plant Biol. 42, 770–781.

Akkad, A.A., 1989. Water conservation in arid and semi-arid regions. Desalination 72, 185–205.
Chen, J., Li, L.Q., Zheng, J.W., Yu, X.Y., Pan, G.X., Lin, Z.H., 2013. Research on water retention capacity of water-retaining agent of PAM-biochar. Bull. Soil, Water Conserv. 33, 302–307 (in Chinese with English abstract).
Dang, X., Zhang, Y., Huang, Y., 2006. Research status and prospect on application of water holding agents in agriculture. Chin J. Soil Sci. 37, 146–149 (in Chinese with English abstract).
Dugan, E., Verhoef, A., Robinson, S., Sohi, S., Gilkes, R.J., Prakpongep, N., 2010. Bio-char from sawdust, maize stover and charcoal: impact on water holding capacities (WHC) of three soils from Ghana. In: World Congress of Soil Science: Soil Solutions for A Changing World.
Gao, H., He, X., Geng, Z., She, D., Yin, J., 2011. Effects of biochar and biochar-based nitrogen fertilizer on soil water-holding capacity. Agric. Sci. Bull. China 27, 207–213.
Glaser, B., Lehmann, J., Zech, W., 2002. Ameliorating physical and chemical properties of highly weathered soils in the tropics with charcoal—a review. Biol. Fert. Soil 35, 219–230.
Gong, L., Yin, Z., 2009. Absorbent polymers in agricultural production on the applied research. Agric. Sci. Bull. China 25, 174–177.
Han, E.X., Han, G., Ying, S.B., Zhang, X.P., Zhang, W.B., Yin, L.U., 2004. Utilization of water-retaining agent in plantation in semi-arid region. J. Northwest For. Univ. 19, 50–52.
Houerou, H.N.L., 1996. Climate change, drought and desertification. J. Arid Environ. 34, 133–185.
Kanwal, S., Iiyas, N., Shabir, S., Saeed, M., Gul, R., Zahoor, M., 2018. Application of biochar in mitigation of negative effects of salinity stress in wheat (*Triticum aestivum* L.). J. Plant Nutr. 41, 528–538.
Karhu, K., Mattila, T., Bergström, I., Regina, K., 2011. Biochar addition to agricultural soil increased CH 4 uptake and water holding capacity—results from a short-term pilot field study. Agric. Ecosyst. Environ 140, 309–313.
Lehmann, J., Joseph, S., 2015. Biochar for Environmental Management. Science, Technology and Implementation, 2nd Edition. Routledge, London.
Li, B., 2010. Soil Salinization. Springer, Berlin, Heidelberg.
Li, J., Huang, Y., 1996. Present status of soil water-holding agent study. J. Desert Res. 16, 86–91 (in Chinese with English abstract).
Maliva, R., Missimer, T., 2012. Arid Lands Water Evaluation and Management. Springer, Berlin, Heidelberg.
Oguntunde, P., Abiodun, B., Ajayi, A., van, dG.N., 2008. Effects of charcoal production on soil physical properties in Ghana. J. Plant Nutr. Soil Sci. 171, 591–596.
Ok, Y., Uchimiya, S., Chang, S., Bolan, N., 2015. Biochar: Production, Characterization, and Applications. CRC Press, Boca Raton, Florida, US.
Pan, Q., Chen, K., Song, T., Xu, X.N., Peng, Z.X., 2017. Influences of biochar and biochar-based compound fertilizer on soil water retention in brown soil. Res. Soil, Water, Conserv. 24, 115–121.
Peng, X., Ye, L.L., Wang, C.H., Zhou, H., Sun, B., 2011. Temperature- and duration-dependent rice straw-derived biochar: characteristics and its effects on soil properties of an Ultisol in southern China. Soil Till. Res. 112, 159–166.
Qian, K.Z., Kumar, A., Zhang, H.L., Bellmer, D., Huhnke, R., 2015. Recent advances in utilization of biochar. Renewable Sustainable Energy Rev. 42, 1055–1064.
Rhoades, J.D., Loveday, J., 1990. Salinity in irrigated agriculture. Agronomy 1089–1142.
Rotini, O.T., 1970. Structure, Water-Holding Capacity, and Stability of Soil. Ital. Agric. 107, 691–705.
Salama, R.B., Otto, C.J., Fitzpatrick, R.W., 1999. Contributions of groundwater conditions to soil and water salinization. Hydro. J. 1, 46–64.
Schlesinger, W.H., Pilmanis, A.M., 1998. Plant–Soil Interactions in Deserts. Springer, Netherlands.
Singh, B.P., Hatton, B.J., Balwant, S., Cowie, A.L., Kathuria, A., 2010. Influence of biochars on nitrous oxide emission and nitrogen leaching from two contrasting soils. J. Environ. Qual 39, 1224–1235.
Tan, X., Liu, Y., Zeng, G., Wang, X., Hu, X., Gu, Y., 2015. Application of biochar for the removal of pollutants from aqueous solutions. Chemosphere 125, 70–85.
Thomas, D.S.G., Middleton, N.J., 1993. Salinization: new perspectives on a major desertification issue. J. Arid Environ. 24, 95–105.
Wang, D.Y., Yan, D.H., Song, X.S., Wang, H., 2014. Impact of biochar on water holding capacity of two Chinese agricultural soil. Adv. Mater. Res. 941-944, 952–955.
Whitford, W., Wade, E.L., 2002. Ecology of desert systems. Academic Press, New York and London.

CHAPTER

10

Biochars and Biochar Composites: Low-Cost Adsorbents for Environmental Remediation

Rizwan Tareq[1], *Nahida Akter*[2] *and Md. Shafiul Azam*[2]

[1]Department of Materials and Metallurgical Engineering, Bangladesh University of Engineering and Technology (BUET), Dhaka, Bangladesh [2]Department of Chemistry, Bangladesh University of Engineering and Technology (BUET), Dhaka, Bangladesh

10.1 INTRODUCTION

Adsorption is a surface phenomenon that causes accumulation of atoms, molecules, or ions at the interface of two phases like solid and liquid or solid and gas. The substance being adsorbed is the adsorbate and the solid on which adsorption occurs is the adsorbent. An unbalanced force of attraction on the surface of the solid is generally responsible for the adsorption to occur. The adsorption process where the adhesion between adsorbent and adsorbate molecules is physical in nature like weak van der Waals forces of attraction is called physisorption (or physical adsorption). On the other hand, adsorption resulting from chemical bonding between adsorbent and adsorbate molecules is known as chemisorption (or chemical adsorption). Since the surface area of the adsorbent has significant influence on the adsorption, high porosity and small particle size are two important requirements for good industrial adsorbents. An efficient adsorbent also must have improved mechanical properties leading to abrasion resistance and good kinetic properties so that it can adsorb the adsorbate molecules readily using its adsorption sites (Thomas and Crittenden, 1998).

Adsorption involves mass transfer accompanied by physical or chemical change, a well understanding of which is necesary for predicting adsorption behavior. In addition to physisorption, adsorption is often controlled by various chemisorption such as ionic exchange,

DOI: https://doi.org/10.1016/B978-0-12-811729-3.00010-8

surface complexation, and precipitation. Ionic exchange of pollutants takes place by a selective replacement of positively charged ions on biochar surfaces with target species. During ionic exchange based sorption, efficiency depends largely on size of the pollutant and surface chemistry of biochar. Electrostatic interaction between charged surface of biochars and oppositely charged pollutant ions depends primarily on solution pH and point of zero charge (P_{ZC}) of adsorbent (Dong et al., 2011; Mukherjee et al., 2011). Complexation is formation of multiatom structures or complexes involving specific metal-ligand interactions. This binding mechanism is crucial for transition metals with partially filled d-orbitals with a high affinity for ligands (Crabtree, 2005a,b). Formation of solids either in a solution or on a surface during the adsorption process is known as precipitation. The contributions of these interactions to a particular adsorption play the most important role in determining the adsorption capacity of the adsorbent and thus the remediation process.

Adsorption processes pave the way for a wide range of applications in pollution control, separation and purification of liquid and gas mixtures, drying of organic solvents, refining of mineral oil, removal of odors from gases, etc. The release of textile dye, paints, pesticides, pharmaceuticals, personal care products, surfactants, and preservatives into water bodies has long been a serious concern, since it affects drinking water and its safety. Although various techniques are available for water purification and wastewater treatment, the adsorption process has gained popularity in terms of simplicity of operation and cost effectiveness. Chromatography, a well-known separation technique that utilizes adsorbents for the separation of components from a mixture and the softening of hard water by ion exchangers, is also based on the principle of adsorption. In gas masks the function of adsorbents is to supply pure air for breathing by adsorbing poisonous gases. Some adsorbents also act as catalysts, catalyst supports, indicators, desiccants, etc., in numerous applications (Rouquerol et al., 1999).

10.2 COMMON ADSORBENT MATERIALS

There are various types of adsorbents used in different applications. Most of the adsorbents are specially manufactured materials such as activated carbon, and some are naturally available materials like zeolites, clay, etc. Nevertheless, the natural adsorbents often require extensive treatment to reach their most effective form as adsorbent. Examples of some commonly used adsorbent materials include silica, activated alumina, activated carbon, molecular sieve carbon, molecular sieve zeolites, and clay and polymeric adsorbents.

10.2.1 Silica

Silica has an amorphous structure with the general formula $SiO_2 \cdot nH_2O$. High surface area and the presence of SiOH groups at the surface lead to their high adsorption capacity (Yang, 2003a). Silica has basic types of surface silanol (SiOH) groups. Hydrogen bonding between water molecules and these surface SiOH groups promotes water adsorption making silica a powerful desiccant or drying agent. In chromatographic techniques, silica gel is the most common stationary phase used in thin-layer chromatography and column chromatography. Silica adsorbs textile industrial effluents such as organic and inorganic dyes from wastewater (Ahmed and Ram, 1992). Natural sand also shows strong affinity for dyes and can be efficiently used for the removal of textile effluents through adsorption. Moreover, modified forms of silica such as chemically coated or functionalized silica particles have been used as adsorbent materials in the last few decades to remove various types of pollutants. For instance, humic fraction-immobilized silica gel helps to adsorb various agricultural organophosphate pesticides (Lai and Chen, 2013). Such compounds possess intense neurotoxic activity inhibiting enzymes important for nerve functioning. Ammonium functionalized mesoporous silica particles are effective in the removal of important water pollutants such as phosphate and nitrate anions (Saad et al., 2008). Amino or thiol functionalized silica can be applied for acidic gas adsorption, heavy metal removal, etc. (Tahergorabi et al., 2016).

10.2.2 Zeolites

Zeolites are aluminosilicates with variable Si/Al ratios (Yang, 2003b). Both natural and synthetic zeolites are available that have become increasingly important due to their versatile physical and chemical properties (Tomlinson, 1988). Zeolites have extensive applications as adsorbents, molecular sieves, membranes, ion exchangers, and catalysts in pollution control as well as in soil remediation. The high cation-exchange ability and molecular sieve properties of zeolites make their use possible for the removal of toxic metal ions like lead, nickel, zinc, cadmium, copper, chromium, and cobalt from wastewater (Rashad et al., 2012). Zeolites are also able to encapsulate organic dyes within their channels via in situ diazotization and coupling (Huddersman et al., 1998; Richards and Pope, 1996).

10.2.3 Activated Alumina

Activated alumina is a chemical-grade alumina with high porosity and surface activity making it a superior adsorbent (Yang, 2003a). The pore structure and surface chemistry of activated alumina can be tailored for various applications by the acid or alkali treatment and controlled heat treatment. Alumina-based adsorbents are mainly used in wastewater treatment to remove inorganic species like cadmium, lead, arsenic, and fluorides from water (Rathore and Mondal, 2017; Naiya et al., 2009; Kim et al., 2004). The pH of the aqueous system significantly influences the adsorption behavior of alumina. The most effective adsorption of Se(VI) on alumina takes place at pH 5–6 whereas at pH 7 alumina can remove more than 90% of As(V) from surface and groundwater. The chemical binding of As to alumina prevents its further leaching to the environment leading to cost-effective and nonhazardous sludge removal. The effective removal of fluoride from water by alumina is due to a high exchange capacity for this ion, which is not affected strongly by the SO_4^{2-} or Cl^- present in water. Moreover, alumina is efficient in removing phosphorous, which is the major cause of eutrophication in ponds and lakes (Kasprzyk-Hordern, 2004).

10.2.4 Activated Carbon

Activated carbon is a processed form of carbon with small, microscale pores resulting in greater surface area exposed to the adsorbate molecules. Carbonaceous materials like wood, peat, coal, coke, petroleum, bones, coconut shells, etc., are the most common raw materials for activated carbon. Activated carbon is produced by dehydration and carbonization followed by chemical or physical activation (Jaroniec, 2003). The raw materials are first heated at 400–500°C in the absence of air to remove all the volatile matter. Then partial gasification at 800–1000°C with high-temperature steam, air, or CO_2 makes the carbon porous or activated. In the case of chemical activation, the materials are impregnated with strong acid, base, or salts prior to carbonization. The number of pores with the desired pore size is tailored during the activation process. The pore volume usually ranges from 0.20 to 0.60 $cm^3\,g^{-1}$ whereas surface area ranges from 800 to 2000 $m^2\,g^{-1}$ (Jaroniec, 2003). Activated carbon is the most popular adsorbent since it can remove a wide range of pollutants such as heavy metals, dyes, detergents, pesticides, and organic and inorganic contaminants even when present in trace amounts (Leimkuehler and Suppes, 2010; Helbig, 1946).

10.2.5 Polymeric Resins

Synthetic nonionic polymeric resins offer great advances to adsorption technology. These macroporous polymers, also known as macroreticular polymers, can be produced with varied surface area and pore sizes (Jaroniec, 2003). Polymeric resins find their main applications in wastewater treatment, specifically in the removal of dyes, pesticides, and other toxic compounds from effluents. Highly aromatic polymeric surface is often used as a superior sorbent for phenol removal from aqueous solution due to its greater hydrophobicity (Wagner and Schulz, 2001). Other important applications include withdrawal of volatile organic compounds (VOCs) from air, adsorption of moisture, bioseperations, etc. The selective interaction of a resin with a particular solute or pollutant can be achieved by anchoring various functional groups onto the polymer assembly. Such tailored polymeric resins have been used in pharmaceutical and food industries for purifying antibiotics, vitamins, and food products for many years.

10.3 BIOCHAR AS ADSORBENT

In spite of the availability of various commercial adsorbents, their widespread use is often constrained due to high cost, lack of versatility, and limited accessibility. Therefore, researchers have been attempting to develop alternative low-cost yet efficient adsorbents utilizing agricultural and industrial wastes. Biochar, a heterogeneous carbonaceous material obtained from the thermochemical decomposition of biomass, plays a big role in addressing the current demands of adsorbents for various applications. Although availability and application of biochar has reached its peak in this scientific era, its roots go back to the early stages of human history. For example, Native Indians used to pile up wood stocks in pits and produce char by slowly burning them at the Amazon Basin of South America. As a result, a fertile and dark soil called *terra preta* was found in living locations scattering over 20 ha (Thines et al., 2017). However, today is obtained using various thermochemical technologies such as slow and fast pyrolysis, torrefaction, hydrothermal carbonization, flash carbonization, and gasification. Biochar has attracted attention worldwide as a useful, low-cost, environmentally friendly adsorbent for various pollutant remediation due to its large surface area, high adsorption capacity, microporosity, and ion-exchange capacity.

Carbonaceous materials are well-known adsorbents for remediation of both organic and inorganic contaminants from water and soil. Activated carbon is an excellent example of a widely used carbonaceous adsorbent, which is basically a form of oxygen-treated char. In terms of production via pyrolysis, biochar is similar to activated carbon with medium-to-high surface area but biochar is not oxygen treated and contains noncarbonized fraction (Cao and Harris, 2010). As estimated by Ahmed et al. (2015), the cost of different biochar productions is calculated in a range of $0.2–0.5 kg^{-1}, whereas adsorbents such as ion-exchange resins may cost up to $150 kg^{-1}. Hence, biochar is an impressive cost-effective pollutant adsorbent that can be developed with a low-carbon footprint with a beneficial role in water treatment, greenhouse gas mitigation, and soil amelioration. In this chapter, adsorption and treatment of major types of contaminants by biochar are highlighted along with specific properties governing the remediation process.

Variability in physiochemical properties allows biochar to maximize its efficacy to targeted applications. Physicochemical properties that influence the adsorption behavior of biochar include surface area, porosity, pH, surface charge density, functional groups, and mineral contents. A brief discussion on these properties is given to evaluate the biochar–pollutant interaction during the adsorption process.

10.3.1 Surface Area and Porosity

Surface area and porosity are often the most important physical properties that influence the adsorption capacity of biochars. Cellulose, hemicellulose, lignin, fats, starch, etc., are thermally broken during pyrolysis of biomass and micropores are formed in biochar due to the loss of water in the dehydration process. Pores formed via this process are highly variable in size and may range from nano- to micrometers in diameter. Pore size is important for pollutant adsorption because regardless of polarity or charge biochar with small pore size cannot capture large adsorbates. Research shows that elevated temperature pyrolysis generates larger pore size and eventually greater surface area. Chen et al. (2014) observed that for a gradual temperature increase from 500 to 900°C the porosity of biochar increased from 0.056 to 0.099 $cm^3\,g^{-1}$ accompanied by an increase in surface area from 25.4 to 67.6 $m^2\,g^{-1}$ (Chen et al., 2014). Along with pyrolysis, composition of feedstock also governs porosity of biochar. For example, pyrolysis of lignin-rich biomass, like coconut shell and bamboo, produces macroporous biochar. On the contrary, cellulose-rich biomass gets converted into microporous biochar upon pyrolysis. (Joseph et al., 2007).

10.3.2 pH and Surface Charge

Biochar pH also differs from one another depending on pyrolysis temperature and feedstock type. Most biochars are alkaline with some exceptions due to feedstock composition. Biochar pH increases as pyrolysis temperature is increased (Jin et al., 2016; Chen et al., 2014; Al-Wabel et al., 2013; Hossain et al., 2011; Nguyen et al., 2010). Elevated pyrolysis temperature gives more ash content in biochar and decomposes acidic functional groups like −COOH, which in return contributes to higher pH.

Surface charge is the property most important for adsorption of inorganics. As biochar is applied in an aqueous media for remediation of inorganic ions, its surface charge is strongly influenced by the pH of the solution. Point of zero charge (pH_{PZC}) is defined as the pH of the solution when the net surface charge becomes zero. When pH of a solution is greater than pH_{PZC}, biochar surface becomes negatively charged and metal ions are adsorbed while at pH lower than pH_{PZC} biochar is positively charged and binds oxyanions. Increased pyrolysis temperature creates biochar with less negative charges as various negative charge containing functional groups like $-COO^-$, −OH, etc., are vanished.

10.3.3 Functional Groups, Aromaticity, and Polarity

Pyrolysis of biomass at variable temperatures reduces the amount of C, H, and O elements at different proportions introducing varying ratios of O/C, H/C, and N/C into the biochar. These ratios correspond directly to the extent of aromaticity and polarity of a certain biochar. These properties are very important for the removal of organic pollutants as these promote $\pi-\pi$ electron donor interaction, electrophilic interaction, hydrogen bonding, and hydrophobic interaction. For instance, biochars synthesized at higher pyrolysis temperature offers a lower ratio of H/C and O/C than those produced at lower pyrolysis temperature showing increased aromaticity accompanied by decreased polarity.

10.3.4 Mineral Components

Mineral contents like Ca, Mg, K, and P in biochar are very important for metal-ion remediation. The presence of these minerals promotes cationic exchange and precipitation of targeted pollutants (Uchimiya et al., 2010; Cao et al., 2009). Like all other physicochemical properties, the mineral content of biochar is also controlled by feedstock type and pyrolysis temperature. For example, K content in poultry litter and swine manure biochar is high while P content in oak wood biochar is low (Subedi et al., 2016; Chen et al., 2014; Hossain et al., 2011). During pyrolysis of biomass, water-soluble mineral contents increase in biochar with the increase of pyrolysis temperature up to 200°C and then decreases as temperature rises further due to formation of different compound phases (Cao et al., 2009).

In summary, pyrolysis temperature and biomass type govern all the physicochemical properties of biochar. High pyrolysis temperatures result in biochar with relatively high surface area, microporosity, and hydrophobicity, and are more effective in the sorption of organic contaminants (Rajapaksha et al., 2014; Kong et al., 2011b). In contrast, biochar produced at low pyrolysis temperature consists of highly dissolved organic carbon content, relatively low porosity, low C/N ratio, and high O-bearing functional groups (Rajapaksha et al., 2014; Kong et al., 2011b). On the other hand, the pyrolysis temperature is known to affect the surface area of the biochar significantly. For example, biochar obtained from woody biomass and crop residues offer higher surface area than those obtained from municipal waste solids or dairy manures, both produced at the same pyrolysis temperature. Hence, by tuning these two parameters, biomass type and pyrolysis temperature, good control over the properties of biochar that make it more amenable for various applications involving the adsorption process can be gained.

10.4 BIOCHAR FOR ADSORPTION OF ORGANIC MOLECULES

Rapid industrialization and agrochemical-based cultivation produce considerable amounts of organic pollutants and thus affect the environment greatly (Table 10.1). The removal of these pollutants from soil, drinking water sources, and wastewater still remains a challenge. Organic contaminants, which are of special concern, include pesticides, herbicides, fungicides, polycyclic aromatic hydrocarbons (PAHs), polychlorinated biphenyls (PCBS), dyes, antibiotics, etc. (Teixido et al., 2011; Beesley et al., 2010b; Qiu et al., 2009; Xu et al., 2012). They can persist in the environment for long periods of time and badly affect aquatic environments as well as human health via the food chain. Conventional techniques such as chemical precipitation, ion-exchange, adsorption, and membrane separation processes are often costly and yield considerable amounts of chemical waste. In recent years, applications of biochar in organic remediation have attracted significant research interest as a sustainable, environment friendly, and low-cost alternative (Thompson et al., 2016).

In order to exploit the uniqueness of biochar for the adsorption of contaminants, understanding of the surface interaction mechanisms is necessary. Basic mechanisms involved for the adsorption of organics are $\pi-\pi$ interaction, hydrophobic sorption, hydrogen bonding, electrostatic interaction, pore diffusion, surface complexation, and partitioning through functional groups as shown in Fig. 10.1. Organic contaminants are generally sorbed or immobilized on biochar surface during the adsorption process and eventually mineralized via biodegradation. Pyrolysis at temperatures higher than 500°C offers higher surface area and aromatic nature but lower polarity and acidity as oxygen- and hydrogen-containing functional groups are lost and thus promote hydrophobic adsorption. Biochars produced at less than 500°C have more functional groups and are prone to adsorb polar organic compounds by hydrogen bonding.

10.4.1 Adsorption of Antibiotics

The presence of pharmaceuticals, especially antibiotics, in aquatic environments and soil has raised serious health and social concerns. Overuse and limited biodegradability of antibiotics promote the growth of antibiotic-resistant bacteria. (Phillips et al., 2012; Passerat et al., 2011). Biochar can act as a safeguard against the leaching of the antibiotics to the environment by preventing their mobility and bioavailability. Sulfamethazine (SMT) is a veterinary antibiotic widely used to prevent diseases such as bacterial pneumonia,

TABLE 10.1 Summary of Adsorptive Mechanisms for Contaminant Remediation by Biochar

Contaminant	Biochar Feedstock	Contaminated Media	Mechanism During Adsorption	References
Sulfamethazine	Softwood	Soil	π–π electron donor–acceptor interaction and electrostatic cation exchange	Vithanage et al. (2014)
Brilliant blue and Rhodamine B	Rice and wheat straw	Water	Electrostatic attraction and intermolecular hydrogen bonds	Qiu et al. (2009)
Chloropyrifos and carbofuran	Woodchips	Soil	Physiosorption due to high surface area and nanoporosity	Yu et al. (2009b)
Phenanthrene	Pinewood	Soil	Entrapment in micro or meso-pores	Zhang et al. (2010)
Trichloroethylene	Peanut	Water	Hydrophobic partitioning and hydrogen bonding	Ahmad et al. (2012)
Pentachlorophenol	Bamboo	Soil	Reduced leaching due to diffusion and partition	Xu et al. (2012)
Tetracycline	Rice husk	Water	Formation of π–π interactions between ring structure of tetracycline molecule and graphite-like sheets of biochars	Liu et al. (2012)
Cu(II)	Peanut, soybean, and canola straw	Water	Formation of surface complexes with carboxylic and phenolic hydroxyl groups of biochar	Tong et al. (2011b)
Cu(II)	Swine manure	Water	Binding with phosphate and silicate minerals of biochar	Meng et al. (2014)
Pb(II)	Wastewater treatment sludge	Water	Cation exchange with K^+ and Na^+, surface complexation with active functional groups, precipitation as lead-phosphate, and coprecipitation with mineral phase	Lu et al. (2012)
Pb(II)	Dairy manure	Soil	Immobilization due to formation of hydroxypyromorphite	Cao et al. (2011)
Cd(II)	Honey mesquite, cordgrass and loblolly pine	Water	Precipitation and cation exchange	Harvey et al. (2011)
Phenanthrene and Hg(II)	Soyabean stalk	Water	Hg(II) was adsorbed through precipitation, complexation and reduction while phenanthrene underwent hydrophobic partitioning and physisorption	Harvey et al. (2011)

(*Continued*)

TABLE 10.1 (Continued)

Contaminant	Biochar Feedstock	Contaminated Media	Mechanism During Adsorption	References
As(III) and As(V)	Sludge	Water	Electrostatic attraction due to lower solution pH and surface complexation with functional groups of biochar	Wang et al. (2015c)
Cr(VI)	Sugarbeet tailings	Water	Electrostatic attraction between biochar and Cr(VI) and complexation with carboxyl and phenolic groups and reduction to Cr(III)	Kong et al. (2011b)
Cr(III)	Crop wastes and municipal sludge	Water	Surface complexation with active functional groups and cation exchange with Ca^{2+} and Mg^{2+}	Chen et al. (2015a), Pan et al. (2013)
Cr(VI)	Australian common weed	Soil	Reduction to Cr(III)	Choppala et al. (2012)
Cu(II), Cd(II), Ni (II), and Zn(II)	Corn straw, broiler litter, Alfalfa stems, switch grass, corn cob, corn stover, guayule bagasse, and guayule shrubs	Water	Chemisorption onto inorganic fraction of biochar	Lima et al. (2010)
Ni(II), Cu(II), Pb (II), and Cd(II)	Cottonseed hulls	Soil	Metal sequestration due to surface functional groups of biochar	Uchimiya et al. (2011a)
Nitrate	Corn stover, pinewood chips and switchgrass	Water	Electrostatic attraction, cation exchange, and outer sphere complexation	Chintala et al. (2013)
Phosphate	Anaerobically digested sugar beet tailing	Water	Nano MgO particles produced on biochar act as adsorption site	Yao et al. (2011a)
Ammonium	Pig manure, straw, corncob, pomelo peel, and banana stalk	Water	Electrostatic attraction and cation exchange	Cai et al. (2016), Yu et al. (2016)

necrotic pododermatitis, calf diphtheria, acute mastitis, acute metritis, etc. This antibiotic is metabolized very slowly and a large fraction is dumped onto soil from pasturage and manure applications, thus eventually reaching the aquatic environment via surface runoff (Kemper, 2008).

Teixido et al. (2011) reported on the successful application of biochar in SMT remediation and its pH-dependent interaction with biochar. SMT can exist as the cation (SMT^{+}), anion (SMT^{-}), uncharged molecule (SMT^{0}), and zwitterion ($SMT^{\pm}$) in water resulting from ion-exchange at the aromatic amine and sulfonamide group (Lin et al., 1996). At low pH,

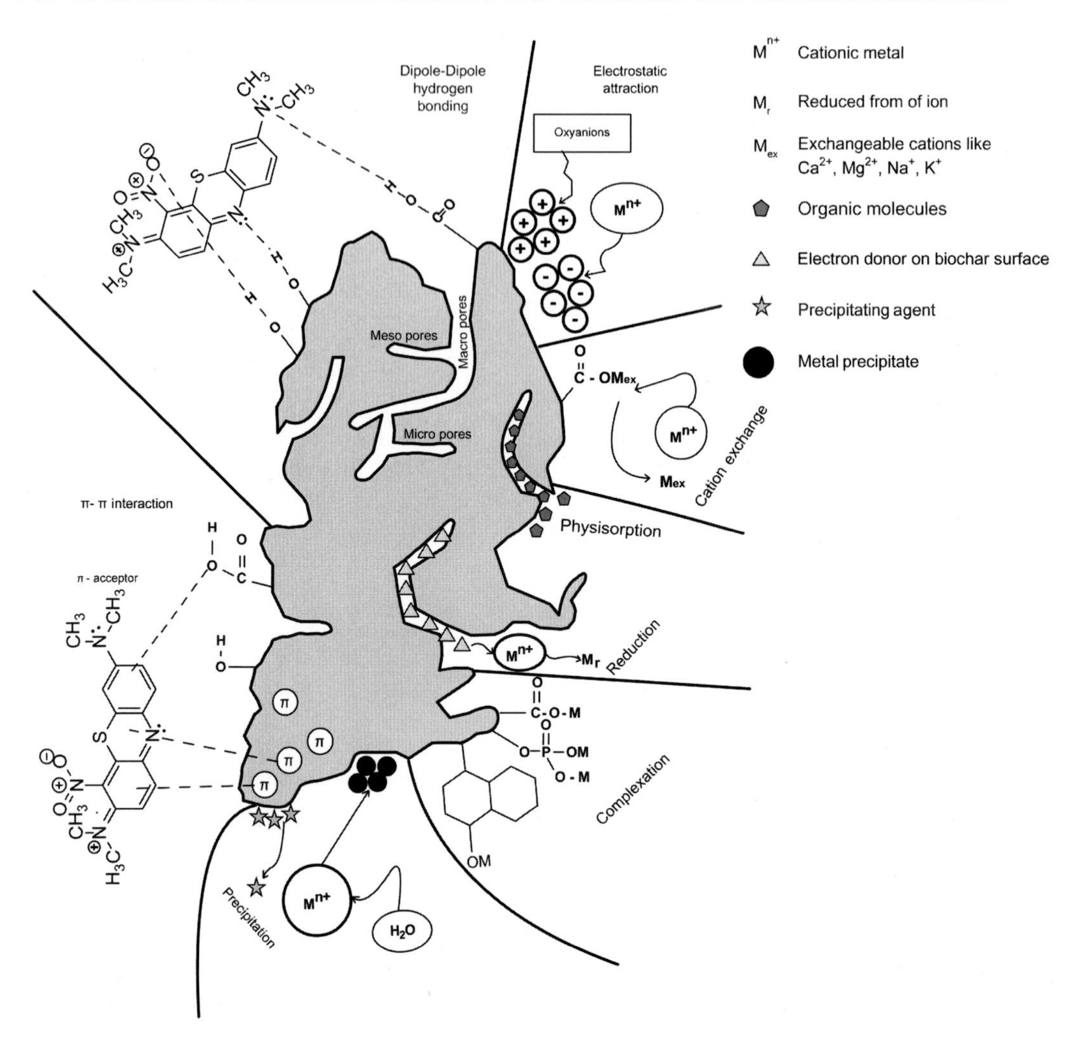

FIGURE 10.1 Illustration of the major mechanisms governing the environmental remediation of organic and inorganic pollutants.

$\pi^{+}-\pi$ electron donor–acceptor (EDA) interaction facilitates the adsorption of SMT^{+} onto biochar. The charged p-amino-sulfonamide ring acts as the acceptor and the graphene π-system as the donor. Since the positive charge of the acceptor lies within an arene unit and there is also charge-quadruple (cation-π) interaction, the overall interaction is called $\pi^{+}-\pi$ EDA. At higher pH SMT^{-} (surface is negative) predominates and undergoes proton-exchange reaction with water resulting in strong H-bonding between SMT^{0} and surface carboxylate or phenolate, considered as a negative charge-assisted H-bond as follows:

$$SMT^{-} + H_2O \rightarrow SMT^{0} + OH^{-}$$
$$SMT^{0} + BC \rightarrow SMT^{0}\text{---}BC$$

Sulfamethoxazole (SMX), a type of sulfonamide antibiotic, is widely used for the treatment of both human and animal diseases. Due to its ubiquity in reclaimed water it can easily reach the soil system via irrigation. If released into water bodies from wastewater treatment plants it can produce drug-resistant pathogens and toxic effects on aquatic organisms. Applications of biochar can reduce SMX-mediated contamination of groundwater and agricultural lands. Biochar soil amelioration also reduces the mobility and bioavailability of SMX. Biochars produced from bamboo, Brazilian pepper wood, sugarcane bagasse, and hickory wood at two different pyrolysis temperatures (i.e., 450 and 600°C) were applied to examine their sorption capacity for SMX. Lower temperature biochars showed better adsorption ability for SMX since they have a greater number of surface functional groups and thus result in enhanced interactions with SMX (Yao et al., 2012).

Biochar produced by fast pyrolyzing pine chips also shows better adsorption capacity for acetaminophen and naproxen (94.1% and 97.7%, respectively) than commercially available activated carbon (81.6% and 94.1%, respectively) (Jung et al., 2015). The polar–π interaction along with the π stacking between the antibiotics and the benzene bundles on the biochar are responsible for the greater adsorption onto biochar. The slightly higher adsorption affinity of naproxen is attributed to its hydrophobicity as well as its polarizability. The more polar sites of naproxen induce greater interaction with the quadrupole moment of benzene bundles on the biochar compared to acetaminophen. Moreover, greater adsorption is eminent for naproxen since it has a naphthalene aromatic ring whereas acetaminophen has a single benzene ring only.

10.4.2 Adsorption of Pesticides, Herbicides, and Fumigants

Farmers use pesticides and herbicides during cultivation to kill unwanted insects and weeds that reduce crop yield. But the surface run-off and leaching of these toxic chemicals from agricultural fields causes pollution of surface and groundwater. Biochars produced from woodchips at 450 and 850°C were applied to soil to investigate their effectiveness in reducing the bioavailability of pesticides such as chloropyrifos and carbofuran (Yu et al., 2009a). A marked decrease in the bioavailability and plant uptake of pesticides was observed in biochar-amended soil indicating reduced degradation and increased sequestration of pesticides. Due to higher surface area, nanoporosity, and affinity for organic compounds, biochar produced at 850°C showed even greater effectiveness. Sorption of such nonpolar organic pollutants from aqueous phase depends largely on surface polarity and aromaticity. Deisopropylatrazine is a metabolite of the widely used herbicide atrazine that stays in groundwater and surface water for a long time. Uchimiya et al. (2010) applied broiler litter-derived biochar produced at 350 and 700°C for the sorption of deisopropylatrazine from water. Biochars obtained at higher temperature showed increased aromaticity and decreased polarity leading to greater adsorption of deisopropylatrazine.

The adsorption behavior of two functionally different herbicides such as norflurazon (NORO) and fluridone (FLUN) on biochar has been studied (Sun et al., 2011). Biochars produced from grass and wood at a wide range of temperatures from 200 to 600°C were applied to investigate their capacity in reducing the bioavailability of FLUN and NORO. Application of biochar limits the leaching of herbicide leading to reduced contamination of water bodies. Low-temperature biochars with amorphous nature showed better sorption capacities than high-temperature biochars. The effect of pH on the sorption behavior of NORO and FLUN was also investigated. Adsorption of FLUN by grass and wood biochars increases with decreasing pH. At low pH, protonation of the negatively charged surface functional groups of biochar induces hydrophobic sites leading to enhanced interactions with protonated FLUN via H-bonding and/or nonspecific London forces. Changes in pH do not show significant effects on NORO sorption since it is polar but nonionic. It interacts with the neutral sorption sites of biochar via London forces and with aromatic structures via π-EDA interactions. Biochar contains π-electron rich aromatic rings whereas the presence of electronegative CF_3 substituent makes the aromatic ring of NORO electron deficient.

Biochars have also been applied to adsorb various fumigants, which are important in controlling soil-borne pests including nematodes, fungi, bacteria, insects, and weeds. Fumigants are volatile and are readily emitted from the soil to air and spread causing adverse health effects such as coughing, nausea, vomiting, sweating, eye irritation, itchy skin, etc. 1,3-Dichloropropene (DCP) helps to control root-knot nematodes acting as a nematocide (Wang et al., 2016a). Its emissions need to be strictly controlled since it can cause chest pain, irritation of the eyes, and respiratory tract, liver, and kidney damage, cardiac arrhythmias, etc. Application of biochar can effectively reduce DCP emissions from fumigated soil, which in turn depends on the properties of biochar, i.e., specific surface area, water content, carbon content, feedstock type, etc. In a laboratory column study, it was found that, using >0.5% (w/w) biochar into soil can reduce DCP loss by >92% (Wang et al., 2016b, 2014b). Moreover, the porous structure of biochar provides shelter for soil microorganisms, which are affected adversely due to the broad biocidal activity of fumigants, thus promoting their growth and developing healthy soil.

10.4.3 Adsorption of Color/Dyes

Dyes have extensive use in the textile, printing, dyeing, food, and paper industries. The discharge of dye-containing wastewater into streams and rivers has always been a severe problem as dyes are toxic to aquatic life and badly hamper the aesthetic nature of the environment. Although different techniques such as sedimentation, chemical analysis, biological methods, oxidation methods, etc., are used to treat wastewater these techniques are often time-consuming, costly, and hence less effective. In addition, textile dyes are more troublesome to remediate due to their stability against light or oxidizing agents.

Electrostatic interaction between pollutant species and biochar surface is the prime mechanism responsible for the removal of aromatic cationic dyes. Biochars obtained from crop residue and rice and wheat straw can effectively remove dyes like methyl violet, methylene blue, and rhodanine utilizing electrostatic interaction (Xu et al., 2011; Qiu et al., 2009). Biochars produced at or less than 400°C contain aromatic π-systems and functional groups capable of withdrawing electrons. Since these are electron deficient they can act as π-electron acceptor and interact with the electron-donating functional groups present in the dyes. On the other hand, high temperature-derived biochars contain both electron withdrawing and donating functional groups and thus can interact with both electron donors and acceptors (Sun et al., 2012). The type and extent of interaction between biochar and organic dyes are also influenced by the pH of the aqueous media. At higher pH, the phenolic OH groups of biochar surface undergo dissociation leading to net negative charge and enhanced electrostatic interaction with methyl violet. With decrease in pH, $\pi-\pi$ EDA interaction increases that promotes H-bonding with methylene blue (Li et al., 2016a; Xu et al., 2011).

10.4.4 Adsorption of Polycyclic Aromatic Hydrocarbons

PAHs are ubiquitous environment pollutants generated during the incomplete combustion of coal, oil, petrol, wood, tobacco, garbage, or other organic materials. From the combustion sources, PAHs are first emitted to the atmosphere and then are deposited on the earth, i.e., on soil and water. PAHs are highly soluble in lipid and thus have a tendency to find places in body fat. They also show moderate toxicity to birds and aquatic life, affecting their immunity, development, and reproduction (Abdel-Shafy and Mansour, 2016). Several investigations have been conducted to understand the adsorption of *N*-nitrosomodimethylamin, naphthalene, naphthenic acids, nonylphenol, and phenanthrene by biochar produced from bamboo, wood chips, softwood bark, aspen wood, and rice straw (Frankel et al., 2016; Lou et al., 2015; Reddy et al., 2014). In most cases, removal mechanisms were attributed to hydrogen bonding, hydrophobic interaction, and partitioning of PAHs. Hardwood and sewage sludge-derived biochars, when applied to the PAH-contaminated soil, reduced the bioavailable concentration as well as the total concentration of PAHs attributing the sorption capacity of biochar and its influence on PAH partitioning. Furthermore, it was reported that the addition of biochar increases the microbial activity in soil stimulating biodegradation of PAHs (Khan et al., 2013; Beesley et al., 2010b).

10.4.5 Adsorption of Polychlorinated Biphenyls

PCBs are manmade hydrophobic organic compounds consisting of carbon, hydrogen, and chlorine. Due to their chemical stability, nonflammability, high boiling point, and electrical insulating properties these were once widely used as insulators in capacitors and transformers, hydraulic fluids, caulks, plasticizers, additives, fire retardants, etc. (Erickson and Kaley, 2011). Production of PCBs was banned in the late 1980s due to their toxicity and bioaccumulation capability, but they are still present in the environment due to the disposal of PCBs-containing products (Melnyk et al., 2015). Biochar sorption of PCBs has turned out to be a promising technology for the remediation of PCBs in recent years.

Three different PCBs—PCB3 (planar, nonortho-substituted), PCB4 (nonplanar, diortho-substituted), PCB5 (nonplanar, monoortho-substituted)—were selected to study the effects of hydrophobicity and planarity on PCB sorption onto corn straw-derived biochars produced at 200 and 700°C (Wang et al., 2016a). Since biochar produced at low temperature (<300°C) contains a large number of amorphous carbon (Chen et al., 2008), the sorption of PCBs on biochar produced at 200°C is dominated by hydrophobic partitioning. On the other hand, highly aromatic and microporous biochar obtained at 700°C facilitates the adsorption of PCBs via specific interactions and pore-filling mechanism in addition to partitioning. The electron-withdrawing chlorine atoms in the benzene rings of PCBs make them act as π-acceptor giving rise to $\pi-\pi$ interactions with the electron-rich aromatic moieties of the biochar (Chun et al., 2004). The planarity of the sorbate molecules is also believed to play an important role in determining the extent of interaction as indicated by the weaker interaction between nonplanar PCBs and biochar produced at 700°C.

10.4.5.1 Adsorption of Volatile Organic Compounds

An important example of VOCs is trichloroethylene (TCE), which is commonly used as industrial solvent, metal degreasing agent, refrigerant, etc. It acts as a major source of groundwater pollution because a minute amount of TCE can pollute large volumes of groundwater (Aggarwal et al., 2006). TCE contamination usually results from its discharge as industrial waste as well as evaporative loss during use. High solubility of TCE in water (1100 mg L^{-1} at 25°C), strong binding ability to soil, and resistance to biodegradation make it a potential threat to the environment (Wei and Seo, 2010). Different adsorbents have been applied to remove TCE from groundwater (Wei and Seo, 2010; Erto et al., 2010; Karanfil and Dastgheib, 2004), but biochar has gained much attention as a natural adsorbent. Biochars derived from soybean stover and peanut shell at two different carbonization temperatures of 300 and 700 °C were applied by Ahmad et al. (2012) for TCE remediation. It was observed that at 700°C, the greater extent of carbonization and removal of volatile materials resulted in microporous biochar with higher surface area and hence greater diffusion of TCE into these micropores. The removal of O- and N-containing functional groups at higher temperature raised the hydrophobicity of biochar surface facilitating the adsorption of hydrophobic organic contaminants like TCE.

10.5 BIOCHAR FOR ADSORPTION OF INORGANIC SPECIES

Inorganic contaminants are nonbiodegradable and can be passed down the ecological food chain via bioaccumulation generating chronic toxicity in higher tropic levels. Wide mining, fertilizers, power plants, leaded gasoline, wastewater, sewage sludge, battery manufacture, animal manure, pesticides, smelting, sewage sludge, metal finishing, etc., are the root causes of expedition of inorganic pollutants as these anthropogenic activities disrupt biogeochemical cycling of elements within nature. Major inorganic species that cause ecotoxicological effects to in natural environments are heavy metal ions, phosphates, nitrates, fluorides, hydrogen sulfide gas, and other greenhouse gases. Table 10.1 summarizes the adsorption of heavy metals and other inorganic species by various biochars.

10.5.1 Adsorption of Heavy Metal Ions

Among the inorganic species, heavy metals pose the biggest threat to the environment as well as to living beings because of their greater involvement in industrial areas and agricultural practices. Microarray analysis of human DNA indicates that the biological actions of six heavy metals (i.e., arsenic, cadmium, nickel, antimony, mercury, and chromium) are distinctly related to the generation of reactive oxygen species that can trigger

DNA damage (Kawata et al., 2007). Numerous studies have been made on biochar absorption of Cr, Cu, Pb, Cd, Hg, Zn, and As ions. Heavy metals interact with biochar depending on the pyroslysis temperature, feedstock type of the biochar, and the pH of the media. Major heavy toxic metals like As, Cd, Cr, Hg, Cu, Pb, etc., are generally discharged from road runoff, manufacture of batteries, electronics and alloys, pigments, dyes, etc. Most of these toxic elements exist as divalent cations in the environment. Cd and Zn have lower ionic potential than Cu, Hg, and Pb due to their larger ionic radius. As a result, Cu, Hg, and Pb easily bind with the functional groups present on biochar surface and hence show less mobility than Cd and Zn (Chen et al., 2011b). The five mechanisms believed to govern metal adsorption by biochar are complexation, cation exchange, precipitation, electrostatic interactions, and chemical reduction as illustrated in Fig. 10.1. However, the contribution of a certain mechanism to a particular metal adsorption by biochar depends primarily on the nature and valence states of the target metals. The next section of this chapter discusses the main mechanisms involving the adsorption of As, Cr, Pb, Cd, Cu, and Hg individually by biochars.

10.5.1.1 Adsorption of Heavy Metal Ions From Water

The adsorption capacity of biochar ranges between 2.4 and 147 $mg\ g^{-1}$, and 0.3 and 39.1 $mg\ g^{-1}$ respectively for Pb, and Cd (Inyang et al., 2016). On the other hand, activated carbon shows a metal adsorption capacity of 255 and 91.4 $mg\ g^{-1}$ for Pb and Cd, respectively (Wilson et al., 2006). In terms of the production process and availability of raw materials, biochar is a low-cost adsorbent for heavy metal remediation due to the abundance of biowastes and economic synthetics.

Biochar with specific porous structure, higher organic carbon content at noncarbonized fraction, and a number of functional groups interact with heavy metal contaminants in several ways. Among divalents, copper showed greater affinity toward biochar due to the formation of surface complexes with carboxylic and phenolic hydroxyl groups on hardwood- and crop straw-derived biochars (Tong et al., 2011a). Furthermore, XRD analysis has shown that binding with minerals like phosphate and silicate particles present on biochar surface derived from swine manure is the adsorption mechanism for copper ion removal from water (Meng et al., 2014).

Adsorption mechanisms governing remediation of lead from water are surface complexation, cation exchange, and precipitation. Lu et al. (2012) performed analysis of the relative distribution of mechanisms for Pb^{2+} adsorption on sludge biochar in a pilot plant scale study for wastewater treatment. In this case study, exchange with K^+ and Na^+ contributed up to 4.8%–8.5% adsorption whereas surface complexation contributed 42% of the adsorption. The adsorption was also governed by precipitation as lead phosphate, surface complexation with active carboxyl and hydroxyl functional groups, and coprecipitation of lead with organic matter and mineral phases of biochar. Chemical speciation, infrared spectroscopy and X-ray diffraction studies by Cao et al. (2009) showed that 84%–87% removal of lead took place through precipitation in the form of $\beta\text{-}Pb_9(PO_4)_6$ and $Pb_3(CO_3)_2(OH)_2$ on biochars prepared from dairy manure at 200 and 350°C, respectively. The biochar treated at 350°C was six times more effective at lead sorption than commercial-activated carbon (<150 μm) (Cao et al., 2009). Xu et al. (2014) separated biochar into organic and inorganic fraction and showed that lead adsorption was higher onto inorganic fraction. This observation indicates that cation exchange and precipitation are the dominant mechanisms while complexation with functional groups has a small contribution (Baig et al., 2014).

Batch adsorption of cobalt onto swinebone char produced by calcination of small fragments of swinebone was studied by Pan et al. (2009). The study showed that cobalt uptake was rapid during the first 5 min and was governed by ion exchange through calcium release, as was confirmed by XRD analysis and calcium concentration in the solution. Cobalt adsorption had a competitive disadvantage due to the presence of co-ions like zinc and cooper. These co-ions showed higher affinity toward swinebone char (Pan et al., 2009).

Among divalent heavy metals, the most studied element after lead for remediation purposes is cadmium. Like other divalent cations it also shows a strong tendency to get hydrated in aqeous solution. Hence, the removal of cadmium by adsorption is dominated by cation exchange and precipitation. Harvey et al. (2011) prepared and classified a number of plant biochars into two groups, namely low exchange capacity and high exchange capacity biochars. Cation exchange was found to be predominant at eliminating cadmium for high cation-exchange biochars (Harvey et al., 2011). When mineral content of biochar is higher, the main mechanism for the adsorption of cadmium is precipitation. For example, dairy manure biochar prepared by Xu et al. (2013b) at 200 and 350°C had a highly soluble concentration of both phosphate and carbonate. They used Visual MINTEQ modeling coupled with FTIR experiments to reveal the adsorption mechanism of cadmium onto dairy manure biochar. The findings showed that 88% of cadmium adsorbed on biochar produced at 350°C was the result of precipitation of metal phosphates and metal carbonates. Biochar produced at 200°C showed 100% precipitation-based removal of cadmium.

Although mercury retains oxidation states 0, +1, and +2 in aqueous media, most species of mercury is Hg^{2+} at pH 3–7. Even at lower environmental concentration, mercury can pass the blood barrier and damage the nervous system. Cation exchange, surface complexation with functional groups or π electrons, and precipitation have been cited as the remediation mechanisms controlling adsorption of mercury from water. Surface complexation with carboxylic, phenolic hydroxyl, and thiol groups is the most dominating mechanism as confirmed by XPS spectra and X-ray absorption fine-structure analyses. For biochars derived at higher temperature or where functional groups are less or absent, binding with the π electron system becomes the major contributor (Xu et al., 2016; Dong et al., 2013). The presence of sulphur and chlorine species on alkaline biochar obtained from soyabean stalk and dairy manure effectively removed above 90% of toxic Hg^{2+} in the form of $HgCl_2$ or $Hg(OH)_2$ through precipitation (Liu et al., 2016; Kong et al., 2011a). However, Lloyd-Jones et al. (2004) noted that Hg precipitates with Cl^- species during adsorption in the form of Hg_2Cl_2. Therefore, more research concerning the precipitation mechanism for the remediation of mercury from water has been recommended (Li et al., 2017).

Arsenic and chromium are redox-sensitive species and show different toxicity and mobility in different redox environments. For instance, discharged effluents of tanneries contain Cr(VI) species usually in the form of chromate (CrO_4^{2-}) or dichromate ($Cr_2O_7^{2-}$), which are strong oxidizing agents and exhibit carcinogenic behavior in biological systems. Although Cr(III) is less mobile than hexavalent chromium species and acts as a micronutrient in humans by enhancing insulin production, its excessive disposal in the environment may lead to hexavalent state under oxidizing condition leading to serious health risks (Fendorf et al., 2000). The three major mechanisms in Cr(III) removal are complexation with oxygenated functional groups, cation exchange, and electrostatic interaction between negatively charged biochar and positively charged Cr(III) ions. Pan et al. (2013) prepared biochar from peanut, soybean, canola, and rice husk and showed that adsorption capacity increases as functional groups on biochar increases, suggesting complexation with functional groups as the controlling remediation mechanism for trivalent chromium. Again, Chen et al. (2015a) observed release of Ca^{2+} and Mg^{2+} cations into solution during adsorption. These released cations correlated well with adsorbed Cr(III) indicating cation exchange as the main removal mechanism (Chen et al., 2015a). At lower pH of solution, cation exchange between Cr(III) and minerals of biochar was hindered due to higher concentration of H^+. When solution pH is 2–5, surface biochar obtained from crop straw becomes negatively charged and trivalent chromium species remains positively charged leading to electrostatic interaction and the removal of the chromium. Electrostatic attraction between positively charged biochar and negatively charged Cr(VI) species, reduction of Cr(VI) to Cr(III) by oxygenated functional groups, and subsequent Cr(III) complexation with functional groups of biochar are the dominating mechanisms for remediation of hexavalent chromium species. Biochar produced from pyrolysis of oak wood and oak bark was employed for hexavalent chromium remediation from water by Mohan et al. (2011). Due to reactive surface functional groups such as unsaturated anhydrosugars, diols, catechol, and substituted catechol structures originating from pyrolysis of lignin, cellulose, and hemicellilose, Cr(VI) is easily reduced to Cr(III) at lower pH($<$2.5) by accepting the π-electrons of disordered PAHs sheets. Although the surface area of these biochars was very small compared to those of commercial-activated carbons, adsorption of hexavalent chromium per unit surface area for the biochars was higher than that of activated carbons due to the swelling effect. The biochars showed interesting swelling properties that led to additional adsorption sites by increasing internal char/water contact with maximum chromium removal at pH 2.

Unlike trivalent chromium species, trivalent arsenite (AsO_3^{3-}) is toxic to humans and occurs in anaerobic conditions. In the Indian subcontinent, arsenic poisoning has become an alarming issue due to the use of natural arsenic contaminated drinking water (Sizmur et al., 2017). Arsenic also retains a pentavalent state in the form of arsenate (AsO_4^{3-}) under aerobic condition and is less mobile and toxic than arsenite. Complexation and electrostatic interaction are the major mechanisms governing remediation of arsenic from aqueous solution. Samsuri et al. (2013) prepared biochar from empty fruit bunch and rice husk and used it for remediation of both As(III) and As(V) from water. Although the surface area of empty fruit bunch biochar was found to be considerably lower than that of rice husk biochar adsorption capacity of both biochars were quite similar. The authors attributed such adsorption capacity of empty fruit bunch biochar to the presence of oxygenated functional groups that counterbalanced the impact of lower surface area. Biochars have also been prepared from different feedstock at lower and higher pyrolysis temperature to compare adsorption capacity. The findings indicated that biochar produced at lower pyrolysis temperature had higher adsorption capacity than that produced at higher temperature (Zhang et al., 2015b; Wang et al., 2015b). For biochars produced at elevated pyrolysis temperature, electrostatic interaction becomes a vital removal mechanism for the adsorption of As(V). For example, pinewood biochar obtained at 600°C with pH_{PZC} greater than 7 was employed by Wang et al. (2015b) to study remediation of As(V). The study showed that when the pH of solution is near 7, As(V) mainly existed in the form of $HAsO_4^{2-}$. Since the solution pH was less than the pH_{PZC} of the biochar due to protonation of some functional groups, the biochar surface became positively charged and electrostatic interaction between As(V) oxyanion and the biochar took place.

10.5.1.2 Adsorption of Heavy Metals From Soil

Biochar has the potential to undergo in situ remediation by affecting metal behavior in soils through alteration of spatial distribution, availability, solubility, and transport. Understandably so, biochar can improve crop productivity by increasing soil organic carbon pools. Performance of biochar in terms of carbon sequestration and soil conditioning depends on the source material of the biochar. For example, plant-derived biochars are considered to be soil conditioners rather than fertilizers, whereas manure-derived biochars can release nutrients and be used as both soil fertilizer and conditioner (Uchimiya et al., 2011b). The soil surface has various types of hydroxyl groups with different levels of reactivity. Generally, binding of metals within the soil surface takes place through formation of inner surface complexes, outer surface complexes, and multidentate complexes. Various factors such as soil pH, temperature, soil type, dominant cations, ionic strength, etc., affect the formation of organometallic complexes. Among these factors, soil pH is the most significant factor influencing metal–soil chemistry and the mobility of metals within soil surface. Biochars are mostly alkaline and thus induce a limiting effect in soil causing immobilization of metals and mobilization of oxyanions (Almaroai et al., 2013).

However, Beesley et al. (2010a) suggested a different explanation. For example, mobility of copper in soil is highly influenced by organic carbon content of biochar. Biochars produced at less than 500°C carry highly dissolved organic carbon responsible for the formation of soluble Cu complexes (Beesley et al., 2010a). On the other hand, Hartley et al. (2009) found increased mobility of arsenic in biochar-amended soil due to increase in soil pH and competition of arsenic with soluble phosphorous in the biochar. Electrostatic repulsion between the cationic Sb^{3+} and anionic surfaces of broiler litter-derived biochar is the main dominating cause for increased mobility of Sb^{3+} ion within soil surface.

Nelissen et al. (2014) studied the effect of feedstock type and pyrolysis temperature on biochar characteristics and response of soil and crop using willow- and pine-derived biochar. The study showed that removal rates of Na^+, Ca^{2+}, Cu^{2+}, Zn^{2+}, B^+, K^+, and Sr^{2+} from soil were higher for willow biochar than for pine biochar (Nelissen et al., 2014). Biochars obtained from wood chips, coconut shell, sewage sludge, cottonseed hull, chicken manure, and green waste at a temperature of $\leq 550°C$ were used to examine the effective removal of Pb^{2+}, Cu^{2+}, As^{3+}, Zn^{2+}, Ni^{2+}, Cr^{2+}, Sb^{3+}, Hg^{2+}, Cd^{2+}, etc., from soil (Zhang et al., 2013c; Lu et al., 2012; Uchimiya et al., 2012, 2011a; Park et al., 2011; Beesley et al., 2010b; Hartley et al., 2009). Metal complexation on hard wood-derived biochar was found to be the most effective mechanism at removal of Cu^{2+}, Cd^{2+}, and As^{3+}. This effect is attributed to the high soil pH caused by the blending of soil with biochar containing high organic carbon content (Uchimiya et al., 2012).

The role of oxygenated functional groups on biochar surfaces in metal binding under soil surface was predicted by Uchimiya et al. (2011b). The authors reported that cottonseed hull-derived biochar produced at 350°C contains high functional groups resulting in reduced bioaccumulation of Cu, Ni, Cd, and Pb in plants. Biochars obtained at 500°C from six species of wetland plants were used to study Cd^{2+}/NH^{4+} adsorption. Surface area and microporosity of biochar did not influence removal of these inorganic pollutants. Remediation mechanisms were controlled by complexation with oxygenated functional groups at basic pH, precipitation as metal-phosphate complex and cation exchange.

Surface coprecipitation and inner-complexation taking place on organic structure and mineral oxides of dairy-manure and sewage sludge-derived biochar are the main mechanisms of Pb^{2+} removal from contaminated soil (Lu et al., 2012; Lima et al., 2010; Cao et al., 2009). Biochar derived from wood chip and coconut shell removed 64%–92% Pb^{2+} from contaminated soil mostly via precipitation of the pollutant in the forms of phosphates, carbonates, oxides, and hydroxides (Reddy et al., 2014; Paranavithana et al., 2016).

Biochar contains several functional groups capable of providing protons to reduce Cr^{6+} to Cr^{3+} during adsorption. According to Wang et al., numerous PAHs also provide π-electrons for reduction of Cr^{6+}. The resulting Cr^{3+} either adsorbs or participates in surface complexation with organic amendments (Wang et al., 2010). However, alkaline biochar prevents dissociation of hydroxyl and carboxyl groups limiting availability of the protons required for reduction of Cr^{6+} to Cr^{3+} in soil. Therefore, acidic biochar with a high number of functional groups is feasible for chromium remediation in soil (Choppala et al., 2012).

10.5.2 Adsorption of Anions and Other Inorganic Pollutants

Unplanned application of fertilizer and effluents from sewage increases concentration of NH_4^+, PO_4^{3-}, and NO_3^- not only in surface water but also in groundwater (Mueller and Helsel, 1996). Since elevated nutrients get discharged into oceans, coastal zones suffer from ecological damage due to reduction of dissolved oxygen and formation of algal blooms (Conley et al., 2009). NH_4^+ has a low charge-to-radius ratio similar to alkali metal ions and hence does not form a stable bond with biochar surface like divalent metal ions do. The most influential factor for the adsorption of NH_4^+ ions is the amount of acid functional groups present on the biochar surface and not the pore structure and micromorphology. Biochars prepared from different agricultural wastes at varying temperatures were found to be capable of retaining over 90% of NH_4^+ (Cai et al., 2016). In this case, adsorption of ammonium was governed by electrostatic interaction and hydrogen bonding with biochar surface containing carboxylic acid and ketone groups. Ion exchange was reported as another mechanism involved in the remediation of ammonium ion by biochar derived from pig manure and straw (Yu et al., 2016).

On the other hand, oxyanions such as PO_4^{3-} and NO_3^- show distinct mobility resulting in different binding behavior to biochar. NO_3^- is more mobile than PO_4^{3-} and binds readily with colloids in water. Generally, adsorption of oxyanions from aqueous solution takes place on positively charged biochar surface through electrostatic attraction. Hence, solution pH is an important factor in NO_3^- adsorption as biochar surface demonstrates an amphoteric nature in response to solution pH. For example, Chintala et al., used corn stover-, pinewood-, chips-, and switchgrass-derived biochar to remove NO_3^- from water.

The removal was primarily due to outer-sphere complexation arising from the electrostatic interaction but cationic exchange also contributed to some extent (Chintala et al., 2013). However, the majority of investigations concerning nitrate remediation from water have focused on the application of acid-activated biochar and metal oxide-biochar composite. The reason is that after such modification it is easy to create positive surface charge on biochar surfaces. This type of adsorption on modified biochar will be discussed later in this chapter.

Retention of nitrate has been more often observed for biochars obtained at elevated temperature (>650°C) than for those produced at lower pyrolysis temperature because the anion-exchange capacity of biochar is 2–3 orders of magnitude lower than the cation-exchange capacity (Novak et al., 2009; Clough et al., 2013; Silber et al., 2010). Kammann et al. (2015) cocomposted biochar and observed improved plant growth due to retention of nitrates. The authors suggested that this retention was due to the interaction of nitrate with the functional groups and mineral oxides present on biochar surface (Kammann et al., 2015). They also suggested the retention was due to the formation of unconventional hydrogen bonding originating from donation of electrons from the π cloud of the polyaromatic systems in biochar to the electron-deficient hydrogens of water in soil.

Anaerobically digested sugar-beet tailings were pyrolyzed at 600°C by Yao et al. (2011b) to study adsorption of aqueous phosphate. Postadsorption characterizations using FTIR, SEM–EDS, and XRD showed the presence of colloidal and nano-sized MgO particles on biochar surface (Yao et al., 2011b). These MgO particles acted as adsorption sites for phosphate, confirmed by postadsorption characterization of the biochar. MgO surface becomes hydroxylated in water forming positively charged surface when the solution pH is lower than the point of zero charge of MgO, eventually attracting negatively charged phosphate species and creating mono and polynuclear complexes.

Another major hazardous inorganic pollutant is fluoride. According to the World Health Organization (WHO) the concentration of fluoride in drinking water should not exceed 1.5 mg L^{-1} because ingestion above this level can lead to dental and skeletal fluorosis (Fawell et al., 2006; Harrison, 2005; Groves, 2001). Oh et al. (2012) aimed evaluated the effect of pH level on fluoride adsorption by biochar prepared from orange peels and water treatment sludge at 400, 600, and 700°C. The results showed that acidic condition is more favorable for fluoride uptake due to higher positive charges on biochar surface (Oh et al., 2012). Oxidized aluminum and iron species on the surface of sludge-based biochar were positively charged at low pH and thus induced greater electrostatic attraction with fluoride ions. Greater adsorption of F^- ions at acidic pH has also been reported by Mohan et al. (2012), who employed pine char for the defluoridation of drinking water. However, the mechanism in this case was different. The chars swelled in water due to the high oxygen content (8%–11%) exposing internal pore structure, where the F^- ions were diffused (Mohan et al., 2012). In addition to various hazardous heavy metal ions and anions, biochar has also been employed effectively in the removal of toxic gases. Discharge of anthropogenic toxic gases into the atmosphere includes compounds of combustion, industrial flue gases along with deliberate emission of chemical warfare agents. For example, H_2S gas is an odiferous gas found in wastewater treatment plants, coal gasification plants, manmade fiber paper plants, petrochemical plants, landfill sites, and so on. Biochars prepared from rice hull, bamboo, hardwood chips, camphor, sludge, and pig manure were found to efficiently adsorb H_2S with removal efficiency of over 95% and adsorption capacity ranging from 100 to 380 mg g^{-1} (Kanjanarong et al., 2017; Xu et al., 2014; Shang et al., 2013). Available surface area, biochar pH (>7), moisture content (>80%, v/w), and chemical bonding with surface groups like carboxylic and hydroxyl regulate the adsorption of H_2S. The ionic interaction between these biochar functional groups and H_2S in the presence of oxygen and moisture produce $(K, Na)_2SO_4$, which remain in soil without causing any secondary pollution (Xu et al., 2014).

10.6 MODIFIED BIOCHAR AS ADSORBENT

Highly heterogeneous and complex physicochemical composition of biochar is an exciting possibility for the removal of pollutants through adsorption. The heterogeneity and complexity on biochar surface originate from widely available precursor materials in the form of biowaste. Low-cost pyrolysis of these biowastes provides the physicochemical surface properties feasible for this adsorption phenomenon. However, raw biochar often has limited ability to adsorb pollutants from aqueous solution especially at elevated concentrations of pollutants for three major reasons: (1) low functionalities derived from precursor after pyrolysis (Yao et al., 2013a); (2) low surface area and pore volume for biochars obtained at low temperature; and (3) difficult separation of biochar due to the small size of the powdered biochar. Extensive research on modification of biochar for introducing new surface properties is therefore needed to enhance adsorption efficacy of biochar for a wide variety of pollutants in terms of origin, toxicity, and chemical and biological activity.

Modification techniques involve both physical activation and chemical routes. Chemical approaches include loading with organic functional groups, minerals, reductants, and functional nanoparticles. In this section, surface functionalized biochar and biochar-based composites are discussed focusing on the following aspects:

1. Improved surface and morphological characteristics of modified biochar with respect to pristine form;
2. Functional amelioration in remediation of pollutants; and
3. Mechanisms regulating improved adsorption of pollutants from a defined media.

10.6.1 Surface Functionalized Biochar as Adsorbent

From various mechanisms governing adsorption phenomena it is clear that functional groups like carboxyl, hydroxyl, amide, ether, alkyl, alkyne, alkene, amine, carbonyl, etc., are responsible for the accumulation of pollutants on biochar surface. Functional groups can be chemically modified or new functionalities introduced on the surface to obtain biochars tailored to specific applications such as treatment of wastewater or soil amelioration. Atomic ratios like H/C, O/C, and N/C are related to the specific properties of biochar like aromaticity and polarity. Steam activation, chemical treatment either in the form of acidic or alkaline modification, and heat treatment are the methods used to enrich biochar in terms of surface chemistry. Modification can be either pre- or postpyrolysis depending

TABLE 10.2 Summary on Adsorptive Removal by Surface Functionalized Biochar

Biochar Feedstock	Treatment	Contaminant	Advantage	Adsorption Mechanism	Reference
Mung bean husk	Steam activation	Ibuprofen	Greater aromatic nature than pristine form	Hydrophobic partitioning	Mondal et al. (2016)
Bamboo	Heat treatment	Furfural	Greater aromatic nature than pristine form	π electron donor–acceptor interaction and hydrogen bonding	Li et al. (2014)
Wheat straw	Acidic treatment by HCl	Methylene blue	Incorporation of acidic functional groups and development of negative surface charge on biochar	Electrostatic attraction, π electron donor–acceptor interaction and hydrogen bonding	Li et al. (2016b)
Peanut hull	Acidic treatment by H_2O_2	Pb(II), Cd(II), Ni (II), and Cu(II)	Increased oxygenated functional groups acting as cation exchange site than pristine form	Cation exchange and surface complexation	Xue et al. (2012)
Bamboo	Alkali treatment by NaOH	Chloramphenicol	Higher aromaticity and increased surface area and pore volume than raw biochar	π–π interaction, hydrogen bond and hydrophobic interaction	Fan et al. (2010)
Bamboo, sugarcane bagasse, hickory wood, and peanut hull	Chitosan modification	Pb(II), Cu(II), Cd (II)	Incorporation of amino groups on biochar	Formation of chemical bond between amino group and metal ions	Zhou et al. (2013)
Rice husk	Esterification	Tetracycline	Increased oxygenated functional groups	Electron donor–acceptor interaction	Jing et al. (2014)

on the properties required for the biochar. The adsorption of different pollutants by biochars are summarized in Table 10.2.

10.6.1.1 Steam-Activated Biochar

In the steam activation process, biochar obtained after pyrolysis of biomass with limited or oxygen-free state undergoes partial gasification with steam at 800–900°C (Rajapaksha et al., 2015; Ippolito et al., 2012). As a result, partial devolatization takes place and formation of crystalline carbon appears on biochar surface. The activated biochar consists of oxygenated functional groups like carbonyl, carboxylic, phenolic, ether, etc., leading to increased hydrophilicity. Trapped products generated from incomplete pyrolysis are removed during steam activation and carbon surface is oxidized due to the generation of syngas H_2 in most cases (Rajapaksha et al., 2016b). As a result, new porosities are formed and the diameter of smaller pores enlarges resulting in increased surface area.

Henceforth, increased surface area, higher aromaticity (decreased value of H/C), and reduced polarity are marked properties of steam-activated biochar. For example, Shim et al. (2015) found that the Cu^{2+} sorption capacity of biochar produced by slow pyrolysis of local invasive grass at 500°C was not significantly changed by activation with steam at 800°C. They found that although steam activation doubled the surface area, abundance of functional groups decreased along with an increase in aromaticity (Shim et al., 2015). However, enhanced adsorption of organic molecules has been cited in most reports (Mondal et al., 2016; Rajapaksha et al., 2015, 2014). For example, the increased aromatic nature of biochar can be considered responsible for the greater hydrophobic partitioning and thus greater adsorption of ibuprofen antibiotic (Mondal et al., 2016).

10.6.1.2 Heat-Treated Biochar

Changes in surface functionality of porous carbons upon various treatments were reviewed by Shen et al. (2010) and mostly included studies on activated carbon. They summarized that hydrophobic porous carbon are generated upon thermal treatment of activated carbon at 800–900°C in the presence of hydrogen gas as acidic functional groups and heteroatoms on carbon surface are removed through the formation of C–H bonds and gasifying the unstable reactive carbon located on the edges of crystallites (Shen et al., 2010). Basic porous carbon is effective for the adsorption of nonpolar organic compounds as suggested by Radovic et al. (1997) through experimental and theoretical study of adsorption of nitro-benzene from aqueous phase. Both theoretical modeling and experimental findings showed that dispersive interaction between nitro-benzene molecule and π electron of heat-treated basic carbon surface was responsible for the adsorption. Similar results were also demonstrated by Li et al. (2014) in which bamboo biochar was acid treated, base treated, and heat treated to remediate furfural, a skin irritant, from water. The study revealed that heat-treated bamboo char showed maximum uptake of furfural with a removal efficiency of 100% for a concentration of $10\ g\ L^{-1}$ (Li et al., 2014). Porosity and surface area did not appear to be contributing functions whereas chemisorption was dominated by dispersive π–π interaction, EDA mechanism, and hydrogen bonding. Formation of hydrogen bond with water and hydrophilic acidic functional groups of biochar help the physisorption of furfural by blocking the motion of furfural molecules within the micropores of biochar.

10.6.1.3 Acid-Treated Biochar

Treatment with strong acids provides acidic functional groups on the carbonized surface that augment metal-sorption potential through cation exchange and surface complexation with the additional active sites obtained. In general, treatment is done by soaking the biochar in acid solution at a ratio of 1:10 (biochar:acid) for variable soaking (Zhang et al., 2015a; Zhou et al., 2013; Jin et al., 2014a). Moreover, intentional oxidation using potassium permanganate ($KMnO_4$), ammonium persulfate [$(NH_4)_2S_2O_8$], hydrogen per oxide (H_2O_2), and ozone (O_3) in addition to washing with strong acids like hydrochloric, sulfuric, nitric, and phosphoric acid have been reported to modify surface functional groups of biochar (Lin et al., 2012; Uchimiya et al., 2011a; Cho et al., 2010). Among the various chemical agents, phosphoric acid is the most commonly used to modify biochar properties due to its ecofriendly nature. Phosphoric acid forms phosphate and polyphosphate cross-bridges during decomposition of lignocellulosic, aliphatic, and aromatic materials present in the precursor or biochar depending on pre- or posttreatment (Yang et al., 2011; Fierro et al., 2010; Klijanienko et al., 2008). As a result, shrinkage is avoided during the development of porosity. Prepyrolysis treatment of pine-tree sawdust with diluted H_3PO_4 acid increased total pore volume leading to higher surface area and a greater number of micropores in biochar (Zhao et al., 2017). Enhanced lead adsorption capacity by 20% compared to a nontreated sample was reported by the authors due to introduction of P–O–P bonds into carbon structure governed by phosphate precipitation and surface adsorption. Li et al. (2016b) reported that HCl-treated wheat straw biochar showed effective removal of cationic dye, methylene blue, through electrostatic attraction between the positively charged dye and the negative surface of biochar developed from acidic functional groups. Nevertheless, treatment with H_2SO_4 and HNO_3 corroded the micropore walls of the biochar reducing the surface area (Stavropoulos et al., 2008). Moreover, mineral components of biochars like phosphorous, potassium, sodium, magnesium, calcium, etc., may be removed from the biochar matrix through dissolution during modification by acids. These minerals are required for metal ion removal through precipitation (Xu et al., 2013a). Therefore, adsorption by precipitation may decrease due to acid treatment of biochar (Sizmur et al., 2017).

Considering large-scale production of acid-modified biochar and environmental issues regarding discharge of acidification media, modification by H_2O_2 can offer cleaner products for lower cost. Huff and Lee (2016) reported doubled cation-exchange capacity of pinewood biochar treated with 30% H_2O_2 compared to untreated biochar due to the presence of oxygenated functional groups. These biochars with increased oxygenated functional groups provide heavy metals such as mercury and lead with additional sites for cation exchange and surface complexation (Xue et al., 2012).

10.6.1.4 Alkali-Treated Biochar

Alkali-treated biochar provides more positive charges on biochar surface, which obviously favors adsorption of negatively charged contaminants. Numerous treatment methods have also been reported where the concentration of base, treatment temperature, and duration were varied (Li et al., 2014; Ma et al., 2014; Jung et al., 2013; Fan et al., 2010). These treatments also produce biochar with higher surface area with higher aromaticity and increased N/C ratio suitable for organic contaminant removal. Although modification by both potassium hydroxide (KOH) and sodium hydroxide (NaOH) increases surface basicity of biochar, treatment with NaOH is considered more economic and less corrosive (Cazetta et al., 2011). However, NaOH modification at lower temperature (up to 100°C) offers low surface area with few micropores (Fan et al., 2010; Li et al., 2014), whereas KOH-treated biochar develop potassium species like K_2O, K_2CO_3, etc., within the biochar through intercalation of K^+ within crystallites of condensed carbon framework. These species increase the porosity of biochar by diffusing into the internal structure (Mao et al., 2015).

Jin et al. (2014b) found improved adsorption of As(V) from aqueous phase by KOH-treated municipal solid waste biochar where increase in surface area and pore volume and alternation of functional groups on biochar and $\pi-\pi$ EDA interaction were found to be responsible. Within the experimental pH range, As(V) mostly remained in the form of AsO_4^- and developed surface complexes interacting with the positive charge of biochar surface. On the other hand, Fan et al. (2010) reported EDA $\pi-\pi$ interaction as well as formation of hydrogen bond as major contributors to the adsorptive removal of an antibiotic, chloramphenicol, from wastewater by bamboo biochar treated with NaOH.

10.6.1.5 Biochar Modified With Nitrogen-Based Functional Groups

Biochar with N-containing functionalities like amide, imide, lactame, pyrrole, and pyridine show higher adsorption tendency particularly for base metals like Cu, Zn, and Cd due to strong complexation affinity of the biochar (Rajapaksha et al., 2016a). N-containing functional groups are introduced on biochar surface through nitration in the presence of concentrated H_2SO_4 followed by reduction (Yang and Jiang, 2014). Zhou et al. (2013) used chitosan to incorporate N-containing functional groups on biochar to remediate Pb^{2+}, Cu^{2+}, and Cd^{2+} from aqueous solution. The metal ion uptake was attributed to the formation of chemical bond between amino groups of modified biochar and metal ions as confirmed by pre- and postadsorption FTIR analysis (Zhou et al., 2013). Amino-functionalized hydrochar prepared by Chen et al. (2015b) showed high Pb^{2+} removal capacity and high selectivity over other metal ions due to the formation of rod-like crystals of $Pb_5(PO_4)_3(OH)$ on biochar surface. During formation of these crystals, amino groups act as bridges to interact crystal nucleus of Pb and hydrolyze to provide basic functional groups. Therefore, removal of Pb^{2+} is the result of synergistic precipitation of phosphates and amino functionalities (Chen et al., 2015b).

10.6.2 Biochar-Based Composite as Adsorbent

Introduction of foreign materials into biochar matrix is a relatively new practice for combining multifunctional properties into one hybrid composite material. Composites can be markedly different from surface functionalized biochar as they endure new functional physicochemical properties that are absent in feedstock or pyrolyzed biomass. Hence, fabrication of biochar-based composites not only opens the door to improved physicochemical properties but also combines advantages of both the incorporated biochar matrix and foreign material. In general, fabrication of biochar meets four unified goals: waste management, carbon sequestration, contaminant remediation, and energy production. However, toxicity of incorporated foreign materials in terms of biological and environmental response must be taken into consideration. According to the available literature, the improved properties that can be obtained upon introduction of foreign materials into biochar are uniform pore size distribution, higher pore volume, increased surface area, presence of more active adsorption sites, catalytic degradability, easy separation, etc. Table 10.3 summarizes the possible mechanisms for the adsorption of various contaminants by biochar composite.

The foreign materials impregnated to alter the properties of biochar to suit specific applications include metal oxides, clay minerals, carbonaceous materials, organic compounds, and microorganisms. In most cases, biochar functions as a scaffold support for the foreign materials, retaining the original functional properties. In terms of

TABLE 10.3 Summary on Adsorptive Removal by Biochar-Based Composites

Contaminant	Biochar Feedstock	Composite	Advantage	Effect	Reference
Phosphate	Bio-accumulated Mg-enriched plant	Mg-biochar composite	Presence of nanoscale $Mg(OH)_2$ and MgO particles on biochar surface	Biochars with high Mg level removed greater extent of phosphate	Yao et al. (2013c)
Arsenic, methylene blue and phosphate	Not specified	Biochar/AlOOH composite	AlOOH nanoparticles and flakes	Nanosized polycrystalline AlOOH flakes grown on biochar surfaces dramatically increased the reactive area	Zhang and Gao (2013)
As(V) and Pb(II)	Pinewood	Mn-biochar composites	Presence of manganosite	About two times higher for As(V) and Pb(II) than pristine biochar	Wang et al. (2014a)
As(V)	Rice husk	Iron oxide amended biochar	Increased surface area	Nearly two times higher adsorption capacity than those reported for iron oxide amended sand	Cope et al. (2014)
As(V)	Not specified	Biochar/δ-Fe_2O_3 composite	Presence of colloidal or nanosized δ-Fe_2O_3 particles within biochar	The composite showed excellent ferromagnetic property and higher adsorption ability	Zhang et al. (2013a)
As(V) and Cr(VI)	Rice husk and the organic fraction of municipal solid wastes	Fe–Ca/biochar composite	Increased Fe and Ca content in its mineral phase	Much better As(V) removal capacity compared to the non-impregnated biochars	Agrafioti et al. (2014)
Methylene blue	Cotton wood	Graphene-biochar composite	Improved thermal stability of biochar	More than 20 times higher than pristine form	Zhang et al. (2012)
Methylene blue, Pb(II) and sulfapyridine	Hickory chips and sugarcane bagasse	CNT coated biochar	Better thermal stabilities, higher surface areas, and larger pore volumes	Much higher than pristine biochars	Inyang et al. (2014)
Phosphate	Cotton wood	Mg/Al-LDH biochar composite	Increased Mg/Al-LDH particles on biochar	Higher than that any other LDH adsorbents	Zhang et al. (2013b)
Phenol	Chicken droppings, wood, and old car tire	Hydrogel/biochar composite	Neutral hydrogel loaded on biochars	Comparable to absorption characteristics of activated carbons	Karakoyun et al. (2011)
Pentachlorophenol	Paper mill sludge	Zero-valent iron Biochar composite	Impregnation of ZVI with improved pore structures	Much higher compared to biochar and ZVI alone	Devi and Saroha (2015)
Pb(II)	Rice hull	Magnetic biochar/ZnS composite	Biochar impregnated with ZnS nanocrystals	Ten times higher than that of reported magnetic biochar	Yan et al. (2015)

fabrication, Tan et al. (2016) classified biochar-based nanocomposites into the following three categories:

1. Nanometal oxide/hydroxide-biochar composites
2. Magnetic biochar composites
3. Functional nanoparticles-impregnated biochar composites

10.6.2.1 Nanometal Oxide/Hydroxide-Biochar Composites

High surface area, negative surface charge, and elevated pH of prepared biochar make it an outstanding adsorbent for metal ions due to the directional adsorption on oxygenated functionalities, precipitation on mineral contents of biochar, and electrostatic attraction to aromatic portion of biochar. But these properties cannot contribute to the adsorption of oxyanionic pollutants like AsO_4^{3-}, NO_3^-, PO_4^{3-}, etc. Metal oxide-based biochar composites exploit the high surface area of the biochar to embed metal-oxide particles, creating positive surface charge below certain pH so oxyanions can be easily adsorbed. But metal-oxide biochar composite may have decreased surface area due to filling of micropores by metal-oxide particles (Michalekova-Richveisova et al., 2017; Zhou et al., 2017; Frišták et al., 2016; Rajapaksha et al., 2016a). However, an investigation by Frišták et al. (2016) revealed that in spite of reduced surface area due to clogging of pores by Fe particles, adsorption of phosphate from water was improved because of positive surface functionalities at the experimental pH.

In one process of preparing metal oxide-biochar composite, targeted chemical component is bioaccumulated in a plant body so that subsequent thermal treatment provides the engineered biochar. For example, Yao et al. (2013b) irrigated tomato plants with magnesium solution and subsequent pyrolysis of it yielded biochar with $Mg(OH)_2$ and MgO particles in nanosize on the surface employed for phosphate removal from aqueous phase. Improved adsorption took place due to electrostatic attraction between those fine nanoparticles and phosphate, as well as subsequent surface complexation (Yao et al., 2013b). Nevertheless, this method of composite formation may be expensive because it is time consuming.

Pretreating biomass using a chemical reagent before pyrolysis or copyrolysis of biomass and doping agent is another approach to fabrication of biochar metal-oxide nanocomposite. Biomass is treated with a suitable metal salt in such a way that metal ions get into the inland of it. Pyrolysis of as-treated biomass generates required biochar metal oxide/hydroxide nano-composite. In order to improve the functional properties required for higher adsorption of contaminants, biochar has been treated with various precursors of metal oxide to load biochar with ZnO, SiO_2, MnO, CaO, MgO, and AlOOH nanoparticles (Wang et al., 2016c; 2015a; Fang et al., 2015; Gan et al., 2015; Wang et al., 2014a; Zhang et al., 2013b; Zhang and Gao, 2013). All these nanoparticle-incorporated biochars showed improved removal efficiencies over those of the original ones. For example, compared to the original biochar, the SiO_2-biochar nanocomposite showed enhanced sorption of phosphate due to the presence of SiO_2 particles on the surface.

The evaporative method, conventional wet impregnation, heat treatment, rapid reduction, direct hydrolysis, etc., have been used to incorporate metal-oxide particles after pyrolysis of rice husk, corn straw, pinewood, swine manure compost, and dried hickory chips to form biochar metal-oxide composites used to remediate arsenate and Cu^{2+}, Cd^{2+}, and Pb^{2+} contaminants from water (Liang et al., 2017; Wang et al., 2015a; Hu et al., 2015; Song et al., 2014; Cope et al., 2014). All these studies showed improved adsorption capacity that can be attributed to a special feature of biochar composite delivered by iron-oxide, iron-hydroxide, hyrous manganese-oxide, amorphous MnO_2, and micro-MnO_x particles. For example, improved adsorption of arsenate was reported by Cope et al. (2014) with 2.5 times higher surface area of composite biochar, whereas Hu et al. (2015) found a reduction of arsenate to trivalent As as the remediation mechanism. Song et al. (2014) showed inner surface complexation and cation exchange to be responsible for adsorptive removal of Cu^{2+}. Tang et al. (2017) showed reduced bioavailability of chlorpyrifos, a pesticide, within soil using iron-oxide embedded biochar as soil amender.

10.6.2.2 Magnetic Biochar Composites as Adsorbent

Adsorption using magnetic adsorbents has emerged as an exigent water remediation technology particularly for wastewater treatment while eliminating filtration shortcomings of nonmagnetic adsorbents. Magnetic separation not only simplifies isolation but also opens the ground for easy washing followed by redispersion. Moreover, mechanisms controlling the adsorption process are also enhanced. Pyrolysis, coprecipitation, and calcination are the methods frequently used for preparation of good-quality and high yield of magnetic biochar (Thines et al., 2017).

Conventional heating and microwave-assisted heating have been used in laboratory scale to generate magnetic biochar adsorbents. Conventional pyrolysis has been successfully integrated in industrial production of magnetic biochars using modified furnace. Cottonwood, pinewood, date pits, pine needles, hydrochar waste, orange peels, and pine bark underwent conventional pyrolysis after being treated with magnetic precursors like $FeCl_3 \cdot 6H_2O$, $Co(NO_3)_2 \cdot 6H_2O$, natural hematite, $Fe(NO_3)_3 \cdot 9H_2O$, etc., to create magnetic biochars (Yang et al., 2016; Zhu et al., 2014; Zahoor and Ali Khan, 2014; Harikishore Kumar Reddy and Lee, 2014; Wang et al., 2015c; Zhang et al., 2013a,d; Theydan and Ahmed, 2012; Chen et al., 2011a; Liu et al., 2010). All these magnetic biochars used for adsorption of phosphate, arsenate, methylene blue, aflatoxin B1, triclosan, Cd^{2+}, Pb^{2+}, and metallic Hg showed improved performance in magnetic response and adsorptive removal from aqueous phase due to incorporation of the more active sites required for adsorption and enhanced physical properties. This can be attributed to uniform and dispersive reinforcement of γ-Fe_2O_3, Fe_3O_4, and $CoFe_2O_4$ forming strong mechanical bonds with biochar matrix. The oxide particles embedded showed particle size within 20 nm to 1 μm with variable shapes such as cubic or octahedral. However, reduction in surface area (Wang et al., 2015c; Zahoor and Ali Khan, 2014; Chen et al., 2011a) and lowered adsorption capacity upon reinforcement of magnetic oxide (Khan et al., 2015) did not appear significant indicating minimum hindrance in adsorptive removal of pollutants by these composites.

Microwave-assisted pyrolysis has also found its way in the production of magnetic biochars from bamboo and empty fruit branch used for the remediation of Cr(VI), Cd^{2+}, methylene blue, and Pb^{2+} from aqueous phase (Ruthiraan et al., 2015; Mubarak et al., 2014; Zhang et al., 2013d; Wang et al., 2013, 2012, 2011). These magnetic biochars containing hydrous Fe_2O_3, cobalt oxide, binary Co–Fe oxide, and metallic Ni crystals adsorbed these contaminants through electrostatic attraction, ion exchange, inner sphere surface complexation, and physisorption. Superparamagnetic cotton fabric biochars were obtained following both conventional pyrolysis and microwave-assisted pyrolysis by ZHu et al. (2014) in order to compare their properties. The authors found that microwave-heated biochar showed no apparent agglomeration and was characterized by more controlled size and dispersion of oxide particles. Modification of magnetic biochar to further improve its functionality has also been reported. For example, chitosan modification of magnetic biochar obtained from invasive species *Eichhornia crassipes* provided more oxygenated functional groups for greater electrostatic interaction and therefore enhanced Cr(VI) remediation (Zhang et al., 2015a).

Coprecipitation is another process by which magnetic biochar can be fabricated. Yu et al. (2013) employed sugarcane bagasse as the raw material for the production of magnetic-modified sugarcane bagasse through the chemical precipitation of Fe^{2+} and Fe^{3+} over the sugarcane bagasse particles in an ammonia solution under ultrasound irradiation at 60°C. A large amount of carboxyl groups found on the surface of biochar, which made the surface more negatively charged. That's why better adsorption was found for the removal of Pb^{2+} and Cd^{+} due to the ion-exchange mechanism (Yu et al., 2013). A comparison of two synthesis methods including chemical coprecipitation of iron oxides onto biochar after pyrolysis and chemical coprecipitation of iron oxides onto biomass before pyrolysis for preparing magnetic biochars was studied by Baig et al. (2014). The results suggested that the chemical coprecipitation of iron oxides before pyrolysis led to greater Fe_3O_4 loading, higher saturation magnetization, improved thermal stability, and superior As(III, V) adsorption efficiency of the biochars (Baig et al., 2014).

Magnetization in biochar can also be introduced via calcination in which biochar is subjected to heat treatment to remove water and drive off CO_2, SO_2, and other volatile constituents. The simplicity of this process was the main reason behind the wide application of this process in the production of magnetic biochar composites. For instance, calcination of rice hull and ferric acetylacetonate in tube furnace generates magnetic biochar consisting of good dispersion of Fe_3O_4 particles on the surface. The biochar showed improved lead removal performance through hydroxide precipitation followed by suitable magnetic separation.

10.6.2.3 Functional Nanoparticles-Coated Biochar

Coating of functional nanoparticles like chitosan, graphene, graphene oxide, carbon nanotubes (CNTs), ZnS nanocrystals, layered double hydroxides, nanoscale zero-valent iron, and graphitic C_3N_4 on biochar surface can create an affordable composite material capable of removing various contaminants by combining the advantages of biochar matrix and nanoparticles. These functional nanoparticles can improve the surface functional groups, surface area, porosity, and thermal stability of biochar, which contribute to better performance of contaminant removal. In particular, biochar-based nanocomposites with different catalytic and oxidative/reductive nanoparticles such as nanoscale zero-valent iron and graphitic C_3N_4 dispersed on biochar surface can exert simultaneous adsorption and degradation ability for removing organic contaminants.

Nanoparticles coated biochar can also be fabricated by precoating biomass with functional nanoparticles followed by pyrolysis. Several functional nanoparticles were applied to pretreat biomass before pyrolysis. Generally, biomass can be converted to biochar-based functional composites following a dip-coating procedure. Specifically, functional nanoparticle suspensions were prepared by adding functional nanoparticles into deionized water. The resulted suspensions were homogenized by sonication using ultrasonicator. Then the biomass was dipped into the suspension and ovendried for pyrolysis afterward.

A new engineered graphene-coated biochar was created by annealing graphene/pyrene-derivative treated biomass feedstock. Experimental results indicated that the graphene coating could improve the thermal stability and adsorption ability of the biochar (Zhang et al., 2012). Similarly, a graphene/biochar composite was synthesized via slow pyrolysis of graphene pretreated wheat straw, and the results suggested that graphene coating on the surface of biochar offered larger surface area, more functional groups, greater thermal stability, and higher removal efficiency of phenanthrene and mercury compared to pristine biochar (Tang et al., 2015). Multiwalled CNT-wrapped biochars were synthesized by dip-coating biomass in varying concentrations of carboxyl-functionalized CNT solutions prior to slow pyrolysis (Inyang et al., 2015, 2014). The surface area, porosity, and thermal stability of the biochars were enhanced due to the incorporation of CNTs.

10.6.2.4 Impregnation of Functional Nanoparticles After Pyrolysis

Impregnation of functional nanoparticles onto raw biochar after pyrolysis to create biochar-based composites can also combine the advantages of biochar with the properties of functional nanoparticles. Hydrogel (Karakoyun et al., 2011), Mg/Al LDH (Zhang et al., 2013b), chitosan (Zhou et al., 2013; Zhang et al., 2015a), zerovalent iron (Yan et al., 2015; Devi and Saroha, 2015, 2014; Zhou et al., 2014), and ZnS nanocrystals (Yan et al., 2014) are commonly used in this fabrication process.

The particles of functional materials were deposited on the carbon surface within the biochar matrix, which could serve as the active sites to adsorb contaminants from aqueous solutions. The resulted composites have superior functions and properties inherited from both functional nanoparticles and biochar. However, functional nanoparticles coated on the biomass feedstock may also cause the partial blockage of pores of biochar. Fortunately, the superior properties of functional nanoparticles may overcome this potential shortcoming. For instance, chitosan-modified biochars were synthesized by coating chitosan onto biochar surfaces (Zhou et al., 2013). It was found that this modification combined the advantages of the porous network as well as relatively large surface area of biochar with the high chemical affinity of chitosan. However, the surface area of the biochar decreased significantly due to the blockage of partial pores by the chitosan. But this effect was counterbalanced by incorporation of amino functionalities, which played an important role in remediation of multiple metal ions.

10.7 CONCLUDING REMARKS AND FUTURE PERSPECTIVES

Biochar and biochar-based adsorbents have become the focus of many environmental remediation studies today. Their wide availability, low cost, and environmental viability along with relatively high adsorption capacity make them suitable for the adsorption of a wide range of pullutants from heavy metals to pesticides, synthetic dyes to antibiotics. However, their adsorption capacity appreciably depending on the type of pollutant and the synthesis route for the biochar. Hence, selecting the right biochars for the right pollutants is crucial. Furthermore, one single type of biochar may not be appropriate for the removal of all types of contaminants. Physicochemical properties including surface area, porosity, pH, surface charge, functional groups, and mineral contents influence the ability of biochars to adsorb contaminants. Thus, the properties of biochars need to be well understood and planned.

The complex nature of soil systems compared to aquatic systems has constrained the applications of biochar to soil. The relationship between biochar properties and its potential to enhance agricultural soils is still unclear. For this reason, enhanced collaboration among researchers working in different fields such as agriculture, materials science, and soil science is needed to develop specific biochars and measure their environmental and agronomical benefits to agricultural soils. At present, there are few studies of using biochar to remove heavy metals or other pollutants from contaminated wastewater for field application. Contaminated water is more complicated than the simulated water used in current studies. To ensure the suitability of biochar to treat wastewater, simulating the physicochemical conditions of contaminated water or using actual contaminated water for studies is needed.

The adsorption capacity of biochars also needs to be further studied due to a lack of consistency in the literature. Sorption capacities have been reported for different pHs, temperatures, adsorbate concentration ranges, biochar doses, particle sizes, and surface areas. Biochars have been used to treat groundwater, drinking water, synthetic industrial wastewater, and actual wastewater. Moreover, the types and concentrations of interfering ions reported are different and often not reported. Some adsorption capacities have been reported in batch experiments and others in column modes making the comparison of adsorption capacities difficult. Another reason biochar adsorbents are hard to compare is that they are often prepared under different conditions (temperature, time, atmosphere, etc.). Nevertheless, Mohan et al. (2014), keeping all these limitations and inadequacies in mind, selected some of the best biochars with the highest adsorption capacities for selected contaminants and compared them using a bar diagram. This could be a good starting point for the biochar researchers and the International Biochar Initiative (IBI) to propose a common platform for the comparisons of biochars.

Despite biochar being known as a low-cost adsorbent, cost estimation for the production of biochars from different precursors is mostly unknown. Individual biochar costs depend on local precursor availability, processing requirements, pyrolysis conditions, reactor availability, recycling, and product lifetime. Although the costs for biochars vary in different countries these data are not even available in current literature. Costs depend on whether pyrolysis is part of an existing biorefinery and if value-added coproducts are included. Sometimes the biochar is the product and sometimes it is a byproduct as in the case of biorefineries. Therefore, serious economic studies on biochar production in various-sized biorefineries connected with agriculture and forestry is required.

There are also serious concerns for modified biochars or biochar composites. Various modification and impregnation methods have been introduced to increase the adsorption capacity and mechanical strength of biochars and thus economical assessment of modifying or impregnating biochars should be considered. For example, while physical activation of biochar would increase energy consumption costs, chemical activation and production of biochar-based composites would also include the price of reagents (Banerjee et al., 2016). In these cases, a particularly challenging aspect of future optimization is to decrease the quantity of the chemicals required to activate or modify the biochar produced to minimize costs of production while maintaining maximum sorption capacity by optimizing the biochar:modifying agent ratio. The stability of biochar-based composite modifications over time should be monitored in future experiments as some of the materials imbibed in the biochar matrix (e.g., metal compounds, C nanotubes, organic compounds) can leach away from the biochar if they are not tethered strongly. To address this issue, leaching tests are required. One scenario worthy of investigation is the stability of metal-biochar composites at low pH (pH 4–5), to determine whether metals are released from the biochar matrix. Another concern is the ecotoxicology of these new biochar formulations. Biochar itself has been shown to be good for environmental applications, but sometimes it can contain toxic compounds (Soudek et al., 2017). Modifying or activating the biochar leads to chemical and physical changes that could potentially increase the toxicity (reactivity, presence of nanoparticles, metals, etc.). For example, Shim et al. (2015) noted that activated biochars can induce toxicity to *Daphnia magna*. To ensure environmental protection, the toxicity of modified biochars needs to be evaluated to avoid undesirable impacts to aquatic organisms.

References

Abdel-Shafy, H.I., Mansour, M.S.M., 2016. A review on polycyclic aromatic hydrocarbons: source, environmental impact, effect on human health and remediation. Egypt. J. Petrol. 25, 107–123.

Aggarwal, V., Li, H., Boyd, S.A., Teppen, B.J., 2006. Enhanced sorption of trichloroethene by smectite clay exchanged with Cs + . Environ. Sci. Technol. 40, 894–899.

Agrafioti, E., Kalderis, D., Diamadopoulos, E., 2014. Ca and Fe modified biochars as adsorbents of arsenic and chromium in aqueous solutions. J. Environ. Manage. 146, 444–450.

Ahmad, M., Lee, S.S., Dou, X., Mohan, D., Sung, J.K., Yang, J.E., et al., 2012. Effects of pyrolysis temperature on soybean stover- and peanut shell-derived biochar properties and TCE adsorption in water. Bioresour. Technol. 118, 536–544.

Ahmed, M.N., Ram, R.N., 1992. Removal of basic dye from waste-water using silica as adsorbent. Environ. Pollut. 77, 79–86.

Ahmed, M.B., Zhou, J.L., Ngo, H.H., Guo, W., 2015. Adsorptive removal of antibiotics from water and wastewater: progress and challenges. Sci. Total Environ. 532, 112–126.

Al-Wabel, M.I., Al-Omran, A., El-Naggar, A.H., Nadeem, M., Usman, A.R.A., 2013. Pyrolysis temperature induced changes in characteristics and chemical composition of biochar produced from conocarpus wastes. Bioresour. Technol. 131, 374–379.

Almaroai, Y.A., Usman, A.R.A., Ahmad, M., Moon, D.H., Cho, J.-S., Joo, Y.K., et al., 2013. Effects of biochar, cow bone, and eggshell on Pb availability to maize in contaminated soil irrigated with saline water. Environ. Earth Sci. 71, 1289–1296.

Baig, S.A., Zhu, J., Muhammad, N., Sheng, T., Xu, X., 2014. Effect of synthesis methods on magnetic Kans grass biochar for enhanced As(III, V) adsorption from aqueous solutions. Biomass Bioenergy 71, 299–310.

Banerjee, S., Mukherjee, S., Laminka-Ot, A., Joshi, S.R., Mandal, T., Halder, G., 2016. Biosorptive uptake of Fe^{2+}, Cu^{2+} and As^{5+} by activated biochar derived from *Colocasia esculenta*: isotherm, kinetics, thermodynamics, and cost estimation. J. Adv. Res. 7, 597–610.

Beesley, L., Moreno-Jimenez, E., Gomez-Eyles, J.L., 2010a. Effects of biochar and greenwaste compost amendments on mobility, bioavailability and toxicity of inorganic and organic contaminants in a multi-element polluted soil. Environ. Pollut. 158, 2282–2287.

Beesley, L., Moreno-Jiménez, E., Gomez-Eyles, J.L., 2010b. Effects of biochar and greenwaste compost amendments on mobility, bioavailability and toxicity of inorganic and organic contaminants in a multi-element polluted soil. Environ. Pollut. 158, 2282–2287.

Cai, Y., Qi, H., Liu, Y., He, X., 2016. Sorption/desorption behavior and mechanism of NH^{4+} by biochar as a nitrogen fertilizer sustained-release material. J. Agric. Food Chem. 64, 4958–4964.

Cao, X., Harris, W., 2010. Properties of dairy-manure-derived biochar pertinent to its potential use in remediation. Bioresour. Technol. 101, 5222–5228.

Cao, X., Ma, L., Gao, B., Harris, W., 2009. Dairy-manure derived biochar effectively sorbs lead and atrazine. Environ. Sci. Technol. 43, 3285–3291.

Cao, X., Ma, L., Liang, Y., Gao, B., Harris, W., 2011. Simultaneous immobilization of lead and atrazine in contaminated soils using dairy-manure biochar. Environ. Sci. Technol. 45, 4884–4889.

Cazetta, A.L., Vargas, A.M.M., Nogami, E.M., Kunita, M.H., Guilherme, M.R., Martins, A.C., et al., 2011. NaOH-activated carbon of high surface area produced from coconut shell: kinetics and equilibrium studies from the methylene blue adsorption. Chem. Eng. J. 174, 117–125.

Chen, B., Chen, Z., Lv, S., 2011a. A novel magnetic biochar efficiently sorbs organic pollutants and phosphate. Bioresour. Technol. 102, 716–723.

Chen, B., Zhou, D., Zhu, L., 2008. Transitional adsorption and partition of nonpolar and polar aromatic contaminants by biochars of pine needles with different pyrolytic temperatures. Environ. Sci. Technol. 42, 5137–5143.

Chen, T., Zhang, Y., Wang, H., Lu, W., Zhou, Z., Zhang, Y., et al., 2014. Influence of pyrolysis temperature on characteristics and heavy metal adsorptive performance of biochar derived from municipal sewage sludge. Bioresour. Technol. 164, 47–54.

Chen, T., Zhou, Z., Xu, S., Wang, H., Lu, W., 2015a. Adsorption behavior comparison of trivalent and hexavalent chromium on biochar derived from municipal sludge. Bioresour. Technol. 190, 388–394.

Chen, X., Chen, G., Chen, L., Chen, Y., Lehmann, J., Mcbride, M.B., et al., 2011b. Adsorption of copper and zinc by biochars produced from pyrolysis of hardwood and corn straw in aqueous solution. Bioresour. Technol. 102, 8877–8884.

Chen, Y., Chen, J., Chen, S., Tian, K., Jiang, H., 2015b. Ultra-high capacity and selective immobilization of Pb through crystal growth of hydroxypyromorphite on amino-functionalized hydrochar. J. Mater. Chem. A 3, 9843–9850.

Chintala, R., Mollinedo, J., Schumacher, T.E., Papiernik, S.K., Malo, D.D., Clay, D.E., et al., 2013. Nitrate sorption and desorption in biochars from fast pyrolysis. Microporous Mesoporous Mater. 179, 250–257.

Cho, H.H., Wepasnick, K., Smith, B.A., Bangash, F.K., Fairbrother, D.H., Ball, W.P., 2010. Sorption of aqueous Zn [II] and Cd[II] by multiwall carbon nanotubes: the relative roles of oxygen-containing functional groups and graphenic carbon. Langmuir 26, 967–981.

Choppala, G.K., Bolan, N.S., Megharaj, M., Chen, Z., Naidu, R., 2012. The influence of biochar and black carbon on reduction and bioavailability of chromate in soils. J. Environ. Qual. 41, 1175–1184.

Chun, Y., Sheng, G., Chiou, C.T., Xing, B., 2004. Compositions and sorptive properties of crop residue-derived chars. Environ. Sci. Technol. 38, 4649–4655.

Clough, T., Condron, L., Kammann, C., Müller, C., 2013. A review of biochar and soil nitrogen dynamics. Agronomy 3, 275–293.

Conley, D.J., Paerl, H.W., Howarth, R.W., Boesch, D.F., Seitzinger, S.P., Havens, K.E., et al., 2009. Controlling eutrophication: nitrogen and phosphorus. Science 323, 1014–1015.

Cope, C.O., Webster, D.S., Sabatini, D.A., 2014. Arsenate adsorption onto iron oxide amended rice husk char. Sci. Total Environ. 488–489, 554–561.

Crabtree, R.H., 2005a. Complexes of π-bound ligands. The Organometallic Chemistry of the Transition Metals. John Wiley & Sons, Inc.

Crabtree, R.H., 2005b. Metal–ligand multiple bonds. The Organometallic Chemistry of the Transition Metals. John Wiley & Sons, Inc.

Devi, P., Saroha, A.K., 2014. Synthesis of the magnetic biochar composites for use as an adsorbent for the removal of pentachlorophenol from the effluent. Bioresour. Technol. 169, 525–531.

Devi, P., Saroha, A.K., 2015. Simultaneous adsorption and dechlorination of pentachlorophenol from effluent by Ni–ZVI magnetic biochar composites synthesized from paper mill sludge. Chem. Eng. J. 271, 195–203.

Dong, X., Ma, L.Q., Li, Y., 2011. Characteristics and mechanisms of hexavalent chromium removal by biochar from sugar beet tailing. J. Hazard. Mater. 190, 909–915.

Dong, X., Ma, L.Q., Zhu, Y., Li, Y., Gu, B., 2013. Mechanistic investigation of mercury sorption by Brazilian pepper biochars of different pyrolytic temperatures based on X-ray photoelectron spectroscopy and flow calorimetry. Environ. Sci. Technol. 47, 12156–12164.

Erickson, M.D., Kaley II, R.G., 2011. Applications of polychlorinated biphenyls. Environ. Sci. Pollut. Res. Int. 18, 135–151.

Erto, A., Andreozzi, R., Lancia, A., Musmarra, D., 2010. Factors affecting the adsorption of trichloroethylene onto activated carbons. Appl. Surf. Sci. 256, 5237–5242.

Fan, Y., Wang, B., Yuan, S., Wu, X., Chen, J., Wang, L., 2010. Adsorptive removal of chloramphenicol from wastewater by NaOH modified bamboo charcoal. Bioresour. Technol. 101, 7661–7664.

Fang, C., Zhang, T., Li, P., Jiang, R., Wu, S., Nie, H., et al., 2015. Phosphorus recovery from biogas fermentation liquid by Ca–Mg loaded biochar. J. Environ. Sci. (China) 29, 106–114.

Fawell, J., Bailey, K., Chilton, J., Dahi, E., Fewtrell, L., Magara, Y., 2006. Fluoride in Drinking-Water. IWA Publishing.

Fendorf, S., Wielinga, B.W., Hansel, C.M., 2000. Chromium transformations in natural environments: the role of biological and abiological processes in chromium(VI) reduction. Int. Geol. Rev. 42, 691–701.

Fierro, V., Muniz, G., Basta, A.H., El-Saied, H., Celzard, A., 2010. Rice straw as precursor of activated carbons: activation with ortho-phosphoric acid. J. Hazard. Mater. 181, 27–34.

Frankel, M.L., Bhuiyan, T.I., Veksha, A., Demeter, M.A., Layzell, D.B., Helleur, R.J., et al., 2016. Removal and biodegradation of naphthenic acids by biochar and attached environmental biofilms in the presence of co-contaminating metals. Bioresour. Technol. 216, 352–361.

Frišták, V., Micháleková-Richveisová, B., Víglašová, E., ĎUriška, L., Galamboš, M., Moreno-Jimenéz, E., et al., 2016. Sorption separation of Eu and As from single-component systems by Fe-modified biochar: kinetic and equilibrium study. J. Iran. Chem. Soc. 14, 521–530.

Gan, C., Liu, Y., Tan, X., Wang, S., Zeng, G., Zheng, B., et al., 2015. Effect of porous zinc–biochar nanocomposites on Cr(vi) adsorption from aqueous solution. RSC Adv. 5, 35107–35115.

Groves, B., 2001. Fluoride: Drinking Ourselves to Death? Gill & Macmillan (Dublin) Ltd.

Harikishore Kumar Reddy, D., Lee, S.-M., 2014. Magnetic biochar composite: facile synthesis, characterization, and application for heavy metal removal. Colloids Surf. A: Physicochem. Eng. Aspects 454, 96–103.

Harrison, P.T.C., 2005. Fluoride in water: a UK perspective. J. Fluor. Chem. 126, 1448–1456.

Hartley, W., Dickinson, N.M., Riby, P., Lepp, N.W., 2009. Arsenic mobility in brownfield soils amended with green waste compost or biochar and planted with Miscanthus. Environ. Pollut. 157, 2654–2662.

Harvey, O.R., Herbert, B.E., Rhue, R.D., Kuo, L.J., 2011. Metal interactions at the biochar-water interface: energetics and structure-sorption relationships elucidated by flow adsorption microcalorimetry. Environ. Sci. Technol. 45, 5550–5556.

Helbig, W.A., 1946. Activated carbon. J. Chem. Educ. 23, 98.

Hossain, M.K., Strezov, V., Chan, K.Y., Ziolkowski, A., Nelson, P.F., 2011. Influence of pyrolysis temperature on production and nutrient properties of wastewater sludge biochar. J. Environ. Manage. 92, 223–228.

Hu, X., Ding, Z., Zimmerman, A.R., Wang, S., Gao, B., 2015. Batch and column sorption of arsenic onto iron-impregnated biochar synthesized through hydrolysis. Water Res. 68, 206–216.

Huddersman, K., Patruno, V., Blake, G.J., Dahm, R.H., 1998. Azo dyes encapsulated within aluminosilicate microporous materials. J. Soc. Dyers Colour. 114, 155–159.

Huff, M.D., Lee, J.W., 2016. Biochar-surface oxygenation with hydrogen peroxide. J. Environ. Manage. 165, 17–21.

Inyang, M., Gao, B., Zimmerman, A., Zhang, M., Chen, H., 2014. Synthesis, characterization, and dye sorption ability of carbon nanotube–biochar nanocomposites. Chem. Eng. J. 236, 39–46.

Inyang, M., Gao, B., Zimmerman, A., Zhou, Y., Cao, X., 2015. Sorption and cosorption of lead and sulfapyridine on carbon nanotube-modified biochars. Environ. Sci. Pollut. Res. Int. 22, 1868–1876.

Inyang, M.I., Gao, B., Yao, Y., Xue, Y., Zimmerman, A., Mosa, A., et al., 2016. A review of biochar as a low-cost adsorbent for aqueous heavy metal removal. Crit. Rev. Environ. Sci. Technol. 46, 406–433.

Ippolito, J.A., Strawn, D.G., Scheckel, K.G., Novak, J.M., Ahmedna, M., Niandou, M.A., 2012. Macroscopic and molecular investigations of copper sorption by a steam-activated biochar. J. Environ. Qual. 41, 1150–1156.

Jaroniec, M., 2003. Adsorbents: Fundamentals and Applications By Ralph T. Yang (University of Michigan). Wiley-Interscience, Hoboken, NJ, xii + 410, pp. $89.95. ISBN 0-471-29741-0. *J. Am. Chem. Soc.* 125, 12059–12059.

Jin, H., Capareda, S., Chang, Z., Gao, J., Xu, Y., Zhang, J., 2014a. Biochar pyrolytically produced from municipal solid wastes for aqueous As(V) removal: adsorption property and its improvement with KOH activation. Bioresour. Technol. 169, 622–629.

Jin, H., Capareda, S., Chang, Z., Gao, J., Xu, Y., Zhang, J., 2014b. Biochar pyrolytically produced from municipal solid wastes for aqueous As(V) removal: adsorption property and its improvement with KOH activation. Bioresour. Technol. 169, 622–629.

Jin, J., Li, Y., Zhang, J., Wu, S., Cao, Y., Liang, P., et al., 2016. Influence of pyrolysis temperature on properties and environmental safety of heavy metals in biochars derived from municipal sewage sludge. J. Hazard. Mater. 320, 417–426.

Jing, X.-R., Wang, Y.-Y., Liu, W.-J., Wang, Y.-K., Jiang, H., 2014. Enhanced adsorption performance of tetracycline in aqueous solutions by methanol-modified biochar. Chem. Eng. J. 248, 168–174.

Joseph, S.D., Downie, A., Crosky, A., Lehmann, J., Munroe, P., 2007. Biochar for carbon sequestration, reduction of greenhouse gas emissions and enhancement of soil fertility; a review of the materials science. Rend. Circ. Mat. Palermo Suppl. 48, 101–106.

Jung, C., Oh, J., Yoon, Y., 2015. Removal of acetaminophen and naproxen by combined coagulation and adsorption using biochar: influence of combined sewer overflow components. Environmen. Sci. Pollut. Res. 22, 10058–10069.

Jung, C., Park, J., Lim, K.H., Park, S., Heo, J., Her, N., et al., 2013. Adsorption of selected endocrine disrupting compounds and pharmaceuticals on activated biochars. J. Hazard. Mater. 263 (Pt 2), 702–710.

Kammann, C.I., Schmidt, H.P., Messerschmidt, N., Linsel, S., Steffens, D., Muller, C., et al., 2015. Plant growth improvement mediated by nitrate capture in co-composted biochar. Sci. Rep. 5, 11080.

Kanjanarong, J., Giri, B.S., Jaisi, D.P., Oliveira, F.R., Boonsawang, P., Chaiprapat, S., et al., 2017. Removal of hydrogen sulfide generated during anaerobic treatment of sulfate-laden wastewater using biochar: evaluation of efficiency and mechanisms. Bioresour. Technol. 234, 115–121.

Karakoyun, N., Kubilay, S., Aktas, N., Turhan, O., Kasimoglu, M., Yilmaz, S., et al., 2011. Hydrogel–biochar composites for effective organic contaminant removal from aqueous media. Desalination 280, 319–325.

Karanfil, T., Dastgheib, S.A., 2004. Trichloroethylene adsorption by fibrous and granular activated carbons: aqueous phase, gas phase, and water vapor adsorption studies. Environ. Sci. Technol. 38, 5834–5841.

Kasprzyk-Hordern, B., 2004. Chemistry of alumina, reactions in aqueous solution and its application in water treatment. Adv. Colloid. Interface Sci. 110, 19–48.

Kawata, K., Yokoo, H., Shimazaki, R., Okabe, S., 2007. Classification of heavy-metal toxicity by human DNA microarray analysis. Environ. Sci. Technol. 41, 3769–3774.

Kemper, N., 2008. Veterinary antibiotics in the aquatic and terrestrial environment. Ecol. Indic. 8, 1–13.

Khan, A., Rashid, A., Younas, R., 2015. Adsorption of reactive black-5 by pine needles biochar produced via catalytic and non-catalytic pyrolysis. Arab. J. Sci. Eng. 40, 1269–1278.

Khan, S., Wang, N., Reid, B.J., Freddo, A., Cai, C., 2013. Reduced bioaccumulation of PAHs by *Lactuca satuva* L. grown in contaminated soil amended with sewage sludge and sewage sludge derived biochar. Environ. Pollut. 175, 64–68.

Kim, Y., Kim, C., Choi, I., Rengaraj, S., Yi, J., 2004. Arsenic removal using mesoporous alumina prepared via a templating method. Environ. Sci. Technol. 38, 924–931.

Klijanienko, A., Lorenc-Grabowska, E., Gryglewicz, G., 2008. Development of mesoporosity during phosphoric acid activation of wood in steam atmosphere. Bioresour. Technol. 99, 7208–7214.

Kong, H., He, J., Gao, Y., Wu, H., Zhu, X., 2011a. Cosorption of phenanthrene and mercury(II) from aqueous solution by soybean stalk-based biochar. J. Agric. Food Chem. 59, 12116–12123.

Kong, H., He, J., Gao, Y., Wu, H., Zhu, X., 2011b. Cosorption of phenanthrene and mercury(II) from aqueous solution by soybean stalk-based biochar. J. Agric. Food Chem. 59, 12116–12123.

Lai, Y.-S., Chen, S., 2013. Adsorption of organophosphate pesticides with humic fraction-immobilized silica gel in hexane. J. Chem. Eng. Data 58, 2290–2301.

Leimkuehler, E.P., Suppes, G.J., 2010. Production, Characterization, and Applications of Activated Carbon. University of Missouri-Columbia.

Li, G., Zhu, W., Zhang, C., Zhang, S., Liu, L., Zhu, L., et al., 2016a. Effect of a magnetic field on the adsorptive removal of methylene blue onto wheat straw biochar. Bioresour. Technol. 206, 16–22.

Li, G., Zhu, W., Zhang, C., Zhang, S., Liu, L., Zhu, L., et al., 2016b. Effect of a magnetic field on the adsorptive removal of methylene blue onto wheat straw biochar. Bioresour. Technol. 206, 16–22.

Li, H., Dong, X., Da Silva, E.B., De Oliveira, L.M., Chen, Y., Ma, L.Q., 2017. Mechanisms of metal sorption by biochars: biochar characteristics and modifications. Chemosphere 178, 466–478.

Li, Y., Shao, J., Wang, X., Deng, Y., Yang, H., Chen, H., 2014. Characterization of modified biochars derived from bamboo pyrolysis and their utilization for target component (furfural) adsorption. Energy Fuels 28, 5119–5127.

Liang, J., Li, X., Yu, Z., Zeng, G., Luo, Y., Jiang, L., et al., 2017. Amorphous MnO_2 modified biochar derived from aerobically composted swine manure for adsorption of Pb(II) and Cd(II). ACS Sustain. Chem. Eng. 5, 5049–5058.

Lima, I.M., Boateng, A.A., Klasson, K.T., 2010. Physicochemical and adsorptive properties of fast-pyrolysis biochars and their steam activated counterparts. J. Chem. Technol. Biotechnol. 85, 1515–1521.

Lin, C.E., Lin, W.C., Chiou, W.C., Lin, E.C., Chang, C.C., 1996. Migration behavior and separation of sulfonamides in capillary zone electrophoresis. I. Influence of buffer pH and electrolyte modifier. J. Chromatogr. A 755, 261–269.

Lin, Y., Munroe, P., Joseph, S., Henderson, R., Ziolkowski, A., 2012. Water extractable organic carbon in untreated and chemical treated biochars. Chemosphere 87, 151–157.

Liu, P., Liu, W.J., Jiang, H., Chen, J.J., Li, W.W., Yu, H.Q., 2012. Modification of bio-char derived from fast pyrolysis of biomass and its application in removal of tetracycline from aqueous solution. Bioresour. Technol. 121, 235–240.

Liu, P., Ptacek, C.J., Blowes, D.W., Landis, R.C., 2016. Mechanisms of mercury removal by biochars produced from different feedstocks determined using X-ray absorption spectroscopy. J. Hazard. Mater. 308, 233–242.

Liu, Z., Zhang, F.-S., Sasai, R., 2010. Arsenate removal from water using Fe_3O_4-loaded activated carbon prepared from waste biomass. Chem. Eng. J. 160, 57–62.

Lloyd-Jones, P.J., Rangel-Mendez, J.R., Streat, M., 2004. Mercury sorption from aqueous solution by chelating ion exchange resins, activated carbon and a biosorbent. Process Saf. Environ. Protect. 82, 301–311.

Lou, L., Yao, L., Cheng, G., Wang, L., He, Y., Hu, B.-L., 2015. Application of rice-straw biochar and microorganisms in nonylphenol remediation: adsorption-biodegradation coupling relationship and mechanism. PLoS ONE 10, e0137467. Available from: https://doi.org/10.1371/journal.pone.0137467.

Lu, H., Zhang, W., Yang, Y., Huang, X., Wang, S., Qiu, R., 2012. Relative distribution of Pb^{2+} sorption mechanisms by sludge-derived biochar. Water Res. 46, 854–862.

Ma, Y., Liu, W.J., Zhang, N., Li, Y.S., Jiang, H., Sheng, G.P., 2014. Polyethylenimine modified biochar adsorbent for hexavalent chromium removal from the aqueous solution. Bioresour. Technol. 169, 403–408.

Mao, H., Zhou, D., Hashisho, Z., Wang, S., Chen, H., Wang, H., 2015. Preparation of pinewood-and wheat straw-based activated carbon via a microwave-assisted potassium hydroxide treatment and an analysis of the effects of the microwave activation conditions. Bioresour. 10, 809–821.

Melnyk, A., Dettlaff, A., Kuklinska, K., Namiesnik, J., Wolska, L., 2015. Concentration and sources of polycyclic aromatic hydrocarbons (PAHs) and polychlorinated biphenyls (PCBs) in surface soil near a municipal solid waste (MSW) landfill. Sci. Total Environ. 530–531, 18–27.

Meng, J., Feng, X., Dai, Z., Liu, X., Wu, J., Xu, J., 2014. Adsorption characteristics of Cu(II) from aqueous solution onto biochar derived from swine manure. Environ. Sci. Pollut. Res. Int. 21, 7035–7046.

Michalekova-Richveisova, B., Fristak, V., Pipiska, M., Duriska, L., Moreno-Jimenez, E., Soja, G., 2017. Iron-impregnated biochars as effective phosphate sorption materials. Environ. Sci. Pollut. Res. Int. 24, 463–475.

Mohan, D., Rajput, S., Singh, V.K., Steele, P.H., Pittman Jr, C.U., 2011. Modeling and evaluation of chromium remediation from water using low cost bio-char, a green adsorbent. J. Hazard. Mater. 188, 319–333.

Mohan, D., Sharma, R., Singh, V.K., Steele, P., Pittman, C.U., 2012. Fluoride removal from water using bio-char, a green waste, low-cost adsorbent: equilibrium uptake and sorption dynamics modeling. Ind. Eng. Chem. Res. 51, 900–914.

Mohan, D., Sarswat, A., Ok, Y.S., Pittman Jr, C.U., 2014. Organic and inorganic contaminants removal from water with biochar, a renewable, low cost and sustainable adsorbent - a critical review, Bioresour Technol. 160, 191–202.

Mondal, S., Bobde, K., Aikat, K., Halder, G., 2016. Biosorptive uptake of ibuprofen by steam activated biochar derived from mung bean husk: equilibrium, kinetics, thermodynamics, modeling and eco-toxicological studies. J. Environ. Manage. 182, 581–594.

Mubarak, N.M., Kundu, A., Sahu, J.N., Abdullah, E.C., Jayakumar, N.S., 2014. Synthesis of palm oil empty fruit bunch magnetic pyrolytic char impregnating with $FeCl_3$ by microwave heating technique. Biomass Bioenergy 61, 265–275.

Mueller, D.K., Helsel, D.R., 1996. Nutrients in the nation's waters—too much of a good thing? US Geological Survey Circular 1136. <http://water.usgs.gov/nawqa/CIRC-1136.html> (accessed 20.02.03.). https://doi.org/10.3133/cir1136.

Mukherjee, A., Zimmerman, A.R., Harris, W., 2011. Surface chemistry variations among a series of laboratory-produced biochars. Geoderma 163, 247–255.

Naiya, T.K., Bhattacharya, A.K., Das, S.K., 2009. Adsorption of Cd(II) and Pb(II) from aqueous solutions on activated alumina. J. Colloid Interface Sci. 333, 14–26.

Nelissen, V., Ruysschaert, G., Müller-Stöver, D., Bodé, S., Cook, J., Ronsse, F., et al., 2014. Short-term effect of feedstock and pyrolysis temperature on biochar characteristics, soil and crop response in temperate soils. Agronomy 4, 52–73.

Nguyen, B.T., Lehmann, J., Hockaday, W.C., Joseph, S., Masiello, C.A., 2010. Temperature sensitivity of black carbon decomposition and oxidation. Environ. Sci. Technol. 44, 3324–3331.

Novak, J., Busscher, W.J., Laird, D.L., Niandou, M.A.S., 2009. Impact of biochar amendment on fertility of a southeastern coastal plain soil. Soil Science 174, 105–112.

Oh, T.-K., Choi, B., Shinogi, Y., Chikushi, J., 2012. Effect of pH conditions on actual and apparent fluoride adsorption by biochar in aqueous phase. Water Air Soil Pollut. 223, 3729–3738.

Pan, J., Jiang, J., Xu, R., 2013. Adsorption of Cr(III) from acidic solutions by crop straw derived biochars. J. Environ. Sci. 25, 1957–1965.

Pan, X., Wang, J., Zhang, D., 2009. Sorption of cobalt to bone char: kinetics, competitive sorption and mechanism. Desalination 249, 609–614.

Paranavithana, G., Kawamoto, K., Inoue, Y., Saito, T., Vithanage, M., Kalpage, S., et al., 2016. Adsorption of Cd^{2+} and Pb^{2+} onto coconut shell biochar and biochar-mixed soil. Environ. Earth Sci. 75 (2016), 484.

Park, J.H., Choppala, G.K., Bolan, N.S., Chung, J.W., Chuasavathi, T., 2011. Biochar reduces the bioavailability and phytotoxicity of heavy metals. Plant Soil 348, 439–451.

Passerat, J., Ouattara, N.K., Mouchel, J.M., Rocher, V., Servais, P., 2011. Impact of an intense combined sewer overflow event on the microbiological water quality of the Seine River. Water Res. 45, 893–903.

Phillips, P.J., Chalmers, A.T., Gray, J.L., Kolpin, D.W., Foreman, W.T., Wall, G.R., 2012. Combined sewer overflows: an environmental source of hormones and wastewater micropollutants. Environ. Sci. Technol. 46, 5336–5343.

Qiu, Y., Zheng, Z., Zhou, Z., Sheng, G.D., 2009. Effectiveness and mechanisms of dye adsorption on a straw-based biochar. Bioresour. Technol. 100, 5348–5351.

Radovic, L.R., Silva, I.F., Ume, J.I., Menéndez, J.A., Leon, C.A.L.Y., Scaroni, A.W., 1997. An experimental and theoretical study of the adsorption of aromatics possessing electron-withdrawing and electron-donating functional groups by chemically modified activated carbons. Carbon N. Y. 35, 1339–1348.

Rajapaksha, A.U., Chen, S.S., Tsang, D.C., Zhang, M., Vithanage, M., Mandal, S., et al., 2016a. Engineered/designer biochar for contaminant removal/immobilization from soil and water: potential and implication of biochar modification. Chemosphere 148, 276–291.

Rajapaksha, A.U., Chen, S.S., Tsang, D.C.W., Zhang, M., Vithanage, M., Mandal, S., et al., 2016b. Engineered/designer biochar for contaminant removal/immobilization from soil and water: potential and implication of biochar modification. Chemosphere 148, 276–291.

Rajapaksha, A.U., Vithanage, M., Ahmad, M., Seo, D.-C., Cho, J.-S., Lee, S.-E., et al., 2015. Enhanced sulfamethazine removal by steam-activated invasive plant-derived biochar. J. Hazard. Mater. 290, 43–50.

Rajapaksha, A.U., Vithanage, M., Zhang, M., Ahmad, M., Mohan, D., Chang, S.X., et al., 2014. Pyrolysis condition affected sulfamethazine sorption by tea waste biochars. Bioresour. Technol. 166, 303–308.

Rashad, M., Selim, E.M., Assaad, F.F., 2012. Removal of some environmental pollutants from aqueous solutions by Linde zeolite: adsorption and kinetic study. Adv. Environ. Biol. 6, 1716–1724.

Rathore, V.K., Mondal, P., 2017. Competitive adsorption of arsenic and fluoride onto economically prepared aluminum oxide/hydroxide nanoparticles: multicomponent isotherms and spent adsorbent management. Ind. Eng. Chem. Res. 56, 8081–8094.

Reddy, K.R., Xie, T., Dastgheibi, S., 2014. Evaluation of biochar as a potential filter media for the removal of mixed contaminants from urban storm water runoff. J. Environ. Eng. 140, 04014043.

Richards, M.D., Pope, C.G., 1996. Adsorption of methylene blue from aqueous solutions by amorphous aluminosilicate gels and zeolite X. J. Chem. Soc. Faraday Trans. 92, 317–323.

Rouquerol, F., Rouquerol, J., Sing, K., 1999. Chapter 1—introduction. Adsorption by Powders and Porous Solids. Academic Press, London.

Ruthiraan, M., Mubarak, N.M., Thines, R.K., Abdullah, E.C., Sahu, J.N., Jayakumar, N.S., et al., 2015. Comparative kinetic study of functionalized carbon nanotubes and magnetic biochar for removal of Cd^{2+} ions from wastewater. Korean J. Chem. Eng. 32, 446–457.

Saad, R., Hamoudi, S., Belkacemi, K., 2008. Adsorption of phosphate and nitrate anions on ammonium-functionalized mesoporous silicas. J. Porous Mater. 15, 315–323.

Samsuri, A.W., Sadegh-Zadeh, F., Seh-Bardan, B.J., 2013. Adsorption of As(III) and As(V) by Fe coated biochars and biochars produced from empty fruit bunch and rice husk. J. Environ. Chem. Eng. 1, 981–988.

Shang, G., Shen, G., Liu, L., Chen, Q., Xu, Z., 2013. Kinetics and mechanisms of hydrogen sulfide adsorption by biochars. Bioresour. Technol. 133, 495–499.

Shen, W., Li, Z., Liu, Y., 2010. Surface Chemical Functional Groups Modification of Porous Carbon. Recent Patents on Chemical Engineering (Discontinued) 1, 27. Available from: https://doi.org/10.2174/2211334710801010027.

Shim, T., Yoo, J., Ryu, C., Park, Y.-K., Jung, J., 2015. Effect of steam activation of biochar produced from a giant Miscanthus on copper sorption and toxicity. Bioresour. Technol. 197, 85–90.

Silber, A., Levkovitch, I., Graber, E.R., 2010. pH-dependent mineral release and surface properties of cornstraw biochar: agronomic implications. Environ. Sci. Technol. 44, 9318–9323.

Sizmur, T., Fresno, T., Akgül, G., Frost, H., Moreno-Jiménez, E., 2017. Biochar modification to enhance sorption of inorganics from water. Bioresour. Technol. 246, 34–47.

Song, Z., Lian, F., Yu, Z., Zhu, L., Xing, B., Qiu, W., 2014. Synthesis and characterization of a novel MnO_x-loaded biochar and its adsorption properties for Cu^{2+} in aqueous solution. Chem. Eng. J. 242, 36–42.

Soudek, P., Rodriguez Valseca, I.M., Petrová, Š., Song, J., Vaněk, T., 2017. Characteristics of different types of biochar and effects on the toxicity of heavy metals to germinating sorghum seeds. J. Geochem. Explor. 182, 157–165.

Stavropoulos, G.G., Samaras, P., Sakellaropoulos, G.P., 2008. Effect of activated carbons modification on porosity, surface structure and phenol adsorption. J. Hazard. Mater. 151, 414–421.

Subedi, R., Taupe, N., Pelissetti, S., Petruzzelli, L., Bertora, C., Leahy, J.J., et al., 2016. Greenhouse gas emissions and soil properties following amendment with manure-derived biochars: Influence of pyrolysis temperature and feedstock type. J. Environ. Manage. 166, 73–83.

Sun, K., Jin, J., Keiluweit, M., Kleber, M., Wang, Z., Pan, Z., et al., 2012. Polar and aliphatic domains regulate sorption of phthalic acid esters (PAEs) to biochars. Bioresour. Technol. 118, 120–127.

Sun, K., Keiluweit, M., Kleber, M., Pan, Z., Xing, B., 2011. Sorption of fluorinated herbicides to plant biomass-derived biochars as a function of molecular structure. Bioresour. Technol. 102, 9897–9903.

Tahergorabi, M., Esrafili, A., Kermani, M., Shirzad-Siboni, M., 2016. Application of thiol-functionalized mesoporous silica-coated magnetite nanoparticles for the adsorption of heavy metals. Desal. Water Treat. 57, 19834–19845.

Tan, X.F., Liu, Y.G., Gu, Y.L., Xu, Y., Zeng, G.M., Hu, X.J., et al., 2016. Biochar-based nano-composites for the decontamination of wastewater: a review. Bioresour. Technol. 212, 318–333.

Tang, J., Lv, H., Gong, Y., Huang, Y., 2015. Preparation and characterization of a novel graphene/biochar composite for aqueous phenanthrene and mercury removal. Bioresour. Technol. 196, 355–363.

Tang, X.Y., Huang, W.D., Guo, J.J., Yang, Y., Tao, R., Feng, X., 2017. Use of Fe-impregnated biochar to efficiently sorb chlorpyrifos, reduce uptake by *Allium fistulosum* L., and enhance microbial community diversity. J. Agric. Food Chem. 65, 5238–5243.

Teixido, M., Pignatello, J.J., Beltran, J.L., Granados, M., Peccia, J., 2011. Speciation of the ionizable antibiotic sulfamethazine on black carbon (biochar). Environ. Sci. Technol. 45, 10020–10027.

Theydan, S.K., Ahmed, M.J., 2012. Adsorption of methylene blue onto biomass-based activated carbon by $FeCl_3$ activation: equilibrium, kinetics, and thermodynamic studies. J. Anal. Appl. Pyrol. 97, 116–122.

Thines, K.R., Abdullah, E.C., Mubarak, N.M., Ruthiraan, M., 2017. Synthesis of magnetic biochar from agricultural waste biomass to enhancing route for waste water and polymer application: a review. Renew. Sustain. Energy Rev. 67, 257–276.

Thomas, W.J., Crittenden, B., 1998. 2 – Adsorbents. Adsorption Technology & Design. Butterworth-Heinemann, Oxford.

Thompson, K.A., Shimabuku, K.K., Kearns, J.P., Knappe, D.R.U., Summers, R.S., Cook, S.M., 2016. Environmental comparison of biochar and activated carbon for tertiary wastewater treatment. Environ. Sci. Technol. 50, 11253–11262.

Tomlinson, A.A.G., 1988. Modern Zeolites: Structure and Function in Detergents and Petrochemicals. Trans Tech Publications.

Tong, X.-J., Li, J.-Y., Yuan, J.-H., Xu, R.-K., 2011a. Adsorption of Cu(II) by biochars generated from three crop straws. Chem. Eng. J. 172, 828–834.

Tong, X.-J., Li, J.-Y., Yuan, J.-H., Xu, R.-K., 2011b. Adsorption of Cu(II) by biochars generated from three crop straws. Chem. Eng. J. 172, 828–834.

Uchimiya, M., Bannon, D.I., Wartelle, L.H., Lima, I.M., Klasson, K.T., 2012. Lead retention by broiler litter biochars in small arms range soil: impact of pyrolysis temperature. J. Agric. Food Chem. 60, 5035–5044.

Uchimiya, M., Chang, S., Klasson, K.T., 2011a. Screening biochars for heavy metal retention in soil: role of oxygen functional groups. J. Hazard. Mater. 190, 432–441.

Uchimiya, M., Klasson, K.T., Wartelle, L.H., Lima, I.M., 2011b. Influence of soil properties on heavy metal sequestration by biochar amendment: 1. Copper sorption isotherms and the release of cations. Chemosphere 82, 1431–1437.

Uchimiya, M., Lima, I.M., Thomas Klasson, K., Chang, S., Wartelle, L.H., Rodgers, J.E., 2010. Immobilization of heavy metal ions (CuII, CdII, NiII, and PbII) by broiler litter-derived biochars in water and soil. J. Agric. Food Chem. 58, 5538–5544.

Vithanage, M., Rajapaksha, A.U., Tang, X., Thiele-Bruhn, S., Kim, K.H., Lee, S.-E., et al., 2014. Sorption and transport of sulfamethazine in agricultural soils amended with invasive-plant-derived biochar. J. Environ. Manage. 141, 95–103.

Wagner, K., Schulz, S., 2001. Adsorption of phenol, chlorophenols, and dihydroxybenzenes onto unfunctionalized polymeric resins at temperatures from 294.15 K to 318.15 K. J. Chem. Eng. Data 46, 322–330.

Wang, F., Ren, X., Sun, H., Ma, L., Zhu, H., Xu, J., 2016a. Sorption of polychlorinated biphenyls onto biochars derived from corn straw and the effect of propranolol. Bioresour. Technol. 219, 458–465.

Wang, H., Gao, B., Wang, S., Fang, J., Xue, Y., Yang, K., 2015a. Removal of Pb(II), Cu(II), and Cd(II) from aqueous solutions by biochar derived from $KMnO_4$ treated hickory wood. Bioresour. Technol. 197, 356–362.

Wang, M.C., Sheng, G.D., Qiu, Y.P., 2014a. A novel manganese-oxide/biochar composite for efficient removal of lead(II) from aqueous solutions. Int. J. Environ. Sci. Technol. 12, 1719–1726.

Wang, Q., Gao, S., Wang, D., Spokas, K., Cao, A., Yan, D., 2016b. Mechanisms for 1,3-dichloropropene dissipation in biochar-amended soils. J. Agric. Food Chem. 64, 2531–2540.

Wang, Q., Mao, L., Wang, D., Yan, D., Ma, T., Liu, P., et al., 2014b. Emission reduction of 1,3-dichloropropene by soil amendment with biochar. J. Environ. Qual. 43, 1656–1662.

Wang, S., Gao, B., Zimmerman, A.R., Li, Y., Ma, L., Harris, W.G., et al., 2015b. Physicochemical and sorptive properties of biochars derived from woody and herbaceous biomass. Chemosphere 134, 257–262.

Wang, S., Gao, B., Zimmerman, A.R., Li, Y., Ma, L., Harris, W.G., et al., 2015c. Removal of arsenic by magnetic biochar prepared from pinewood and natural hematite. Bioresour. Technol. 175, 391–395.

Wang, W., Wang, X., Wang, X., Yang, L., Wu, Z., Xia, S., et al., 2013. Cr(VI) removal from aqueous solution with bamboo charcoal chemically modified by iron and cobalt with the assistance of microwave. J. Environ. Sci. 25, 1726–1735.

Wang, X.J., Wang, Y., Wang, X., Liu, M., Xia, S.Q., Yin, D.Q., et al., 2011. Microwave-assisted preparation of bamboo charcoal-based iron-containing adsorbents for Cr(VI) removal. Chem. Eng. J. 174, 326–332.

Wang, X.S., Chen, L.F., Li, F.Y., Chen, K.L., Wan, W.Y., Tang, Y.J., 2010. Removal of Cr(VI) with wheat-residue derived black carbon: reaction mechanism and adsorption performance. J. Hazard. Mater. 175, 816–822.

Wang, Y.-Y., Lu, H.-H., Liu, Y.-X., Yang, S.-M., 2016c. Removal of phosphate from aqueous solution by SiO_2–biochar nanocomposites prepared by pyrolysis of vermiculite treated algal biomass. RSC Adv. 6, 83534–83546.

Wang, Y., Wang, X.J., Liu, M., Wang, X., Wu, Z., Yang, L.Z., et al., 2012. Cr(VI) removal from water using cobalt-coated bamboo charcoal prepared with microwave heating. Ind. Crops Prod. 39, 81–88.

Wei, Z., Seo, Y., 2010. Trichloroethylene (TCE) adsorption using sustainable organic mulch. J. Hazard. Mater. 181, 147–153.

Wilson, K., Yang, H., Seo, C.W., Marshall, W.E., 2006. Select metal adsorption by activated carbon made from peanut shells. Bioresour. Technol. 97, 2266–2270.

Xu, R.K., Xiao, S.C., Yuan, J.H., Zhao, A.Z., 2011. Adsorption of methyl violet from aqueous solutions by the biochars derived from crop residues. Bioresour. Technol. 102, 10293–10298.

Xu, T., Lou, L., Luo, L., Cao, R., Duan, D., Chen, Y., 2012. Effect of bamboo biochar on pentachlorophenol leachability and bioavailability in agricultural soil. Sci. Total Environ. 414, 727–731.

Xu, X., Cao, X., Zhao, L., 2013a. Comparison of rice husk- and dairy manure-derived biochars for simultaneously removing heavy metals from aqueous solutions: role of mineral components in biochars. Chemosphere 92, 955–961.

Xu, X., Cao, X., Zhao, L., Sun, T., 2014. Comparison of sewage sludge- and pig manure-derived biochars for hydrogen sulfide removal. Chemosphere 111, 296–303.

Xu, X., Cao, X., Zhao, L., Wang, H., Yu, H., Gao, B., 2013b. Removal of Cu, Zn, and Cd from aqueous solutions by the dairy manure-derived biochar. Environ. Sci. Pollut. Res. Int. 20, 358–368.

Xu, X., Cao, X., Zhao, L., Zhoua, H., Luo, Q., 2014. Interaction of organic and inorganic fractions of biochar with Pb(II) ion: further elucidation of mechanisms for Pb(II) removal by biochar. RSC Adv. 4, 44930–44937.

Xu, X., Schierz, A., Xu, N., Cao, X., 2016. Comparison of the characteristics and mechanisms of Hg(II) sorption by biochars and activated carbon. J. Colloid Interface Sci. 463, 55–60.

Xue, Y., Gao, B., Yao, Y., Inyang, M., Zhang, M., Zimmerman, A.R., et al., 2012. Hydrogen peroxide modification enhances the ability of biochar (hydrochar) produced from hydrothermal carbonization of peanut hull to remove aqueous heavy metals: batch and column tests. Chem. Eng. J. 200–202, 673–680.

Yan, J., Han, L., Gao, W., Xue, S., Chen, M., 2015. Biochar supported nanoscale zerovalent iron composite used as persulfate activator for removing trichloroethylene. Bioresour. Technol. 175, 269–274.

Yan, L., Kong, L., Qu, Z., Li, L., Shen, G., 2014. Magnetic biochar decorated with ZnS nanocrytals for Pb(II) removal. ACS Sustain. Chem. Eng. 3, 125–132.

Yang, G.X., Jiang, H., 2014. Amino modification of biochar for enhanced adsorption of copper ions from synthetic wastewater. Water Res. 48, 396–405.

Yang, J., Zhao, Y., Ma, S., Zhu, B., Zhang, J., Zheng, C., 2016. Mercury removal by magnetic biochar derived from simultaneous activation and magnetization of sawdust. Environ. Sci. Technol. 50, 12040–12047.

Yang, R., Liu, G., Xu, X., Li, M., Zhang, J., Hao, X., 2011. Surface texture, chemistry and adsorption properties of acid blue 9 of hemp (*Cannabis sativa* L.) bast-based activated carbon fibers prepared by phosphoric acid activation. Biomass Bioenergy 35, 437–445.

Yang, R.T., 2003a. Silica gel, MCM, and activated alumina. Adsorbents: Fundamentals and Applications. John Wiley & Sons, Inc.

Yang, R.T., 2003b. Zeolites and molecular sieves. Adsorbents: Fundamentals and Applications. John Wiley & Sons, Inc.

Yao, Y., Gao, B., Chen, H., Jiang, L., Inyang, M., Zimmerman, A.R., et al., 2012. Adsorption of sulfamethoxazole on biochar and its impact on reclaimed water irrigation. J. Hazard. Mater. 209-210, 408–413.

Yao, Y., Gao, B., Chen, J., Yang, L., 2013a. Engineered biochar reclaiming phosphate from aqueous solutions: mechanisms and potential application as a slow-release fertilizer. Environ. Sci. Technol. 47, 8700–8708.

Yao, Y., Gao, B., Chen, J., Zhang, M., Inyang, M., Li, Y., et al., 2013b. Engineered carbon (biochar) prepared by direct pyrolysis of Mg-accumulated tomato tissues: characterization and phosphate removal potential. Bioresour. Technol. 138, 8–13.

Yao, Y., Gao, B., Chen, J., Zhang, M., Inyang, M., Li, Y., et al., 2013c. Engineered carbon (biochar) prepared by direct pyrolysis of Mg-accumulated tomato tissues: characterization and phosphate removal potential. Bioresour. Technol. 138, 8–13.

Yao, Y., Gao, B., Inyang, M., Zimmerman, A.R., Cao, X., Pullammanappallil, P., et al., 2011a. Biochar derived from anaerobically digested sugar beet tailings: characterization and phosphate removal potential. Bioresour. Technol. 102, 6273–6278.

Yao, Y., Gao, B., Inyang, M., Zimmerman, A.R., Cao, X., Pullammanappallil, P., et al., 2011b. Removal of phosphate from aqueous solution by biochar derived from anaerobically digested sugar beet tailings. J. Hazard. Mater. 190, 501–507.

Yu, J.-X., Wang, L.-Y., Chi, R.-A., Zhang, Y.-F., Xu, Z.-G., Guo, J., 2013. Competitive adsorption of Pb^{2+} and Cd^{2+} on magnetic modified sugarcane bagasse prepared by two simple steps. Appl. Surf. Sci. 268, 163–170.

Yu, Q., Xia, D., Li, H., Ke, L., Wang, Y., Wang, H., et al., 2016. Effectiveness and mechanisms of ammonium adsorption on biochars derived from biogas residues. RSC Adv. 6, 88373–88381.

Yu, X.-Y., Ying, G.-G., Kookana, R.S., 2009a. Reduced plant uptake of pesticides with biochar additions to soil. Chemosphere 76, 665–671.

Yu, X.Y., Ying, G.G., Kookana, R.S., 2009b. Reduced plant uptake of pesticides with biochar additions to soil. Chemosphere 76, 665–671.

Zahoor, M., Ali Khan, F., 2014. Adsorption of aflatoxin B1 on magnetic carbon nanocomposites prepared from bagasse. Arab. J. Chem. 11, 729–738.

Zhang, H., Lin, K., Wang, H., Gan, J., 2010. Effect of Pinus radiata derived biochars on soil sorption and desorption of phenanthrene. Environ. Pollut. 158, 2821–2825.

Zhang, M.-M., Liu, Y.-G., Li, T.-T., Xu, W.-H., Zheng, B.-H., Tan, X.-F., et al., 2015a. Chitosan modification of magnetic biochar produced from *Eichhornia crassipes* for enhanced sorption of Cr(vi) from aqueous solution. RSC Ad. 5, 46955–46964.

Zhang, M., Gao, B., 2013. Removal of arsenic, methylene blue, and phosphate by biochar/AlOOH nanocomposite. Chem. Eng. J. 226, 286–292.

Zhang, M., Gao, B., Varnoosfaderani, S., Hebard, A., Yao, Y., Inyang, M., 2013a. Preparation and characterization of a novel magnetic biochar for arsenic removal. Bioresour. Technol. 130, 457–462.

Zhang, M., Gao, B., Yao, Y., Inyang, M., 2013b. Phosphate removal ability of biochar/MgAl-LDH ultra-fine composites prepared by liquid-phase deposition. Chemosphere 92, 1042–1047.

Zhang, M., Gao, B., Yao, Y., Xue, Y., Inyang, M., 2012. Synthesis, characterization, and environmental implications of graphene-coated biochar. Sci. Total Environ. 435–436, 567–572.

Zhang, W., Zheng, J., Zheng, P., Tsang, D.C., Qiu, R., 2015b. Sludge-derived biochar for arsenic(III) immobilization: effects of solution chemistry on sorption behavior. J. Environ. Qual. 44, 1119–1126.

Zhang, X., Wang, H., He, L., Lu, K., Sarmah, A., Li, J., et al., 2013c. Using biochar for remediation of soils contaminated with heavy metals and organic pollutants. Environ. Sci. Pollut. Res. Int. 20, 8472–8483.

Zhang, Z., Wang, X., Wang, Y., Xia, S., Chen, L., Zhang, Y., et al., 2013d. Pb(II) removal from water using Fe-coated bamboo charcoal with the assistance of microwaves. J. Environ. Sci. 25, 1044–1053.

Zhao, L., Zheng, W., Masek, O., Chen, X., Gu, B., Sharma, B.K., et al., 2017. Roles of phosphoric acid in biochar formation: synchronously improving carbon retention and sorption capacity. J. Environ. Qual. 46, 393–401.

Zhou, L., Huang, Y., Qiu, W., Sun, Z., Liu, Z., Song, Z., 2017. Adsorption properties of nano-MnO(2)-biochar composites for copper in aqueous solution. Molecules 22. Available from: https://doi.org/10.3390/molecules22010173.

Zhou, Y., Gao, B., Zimmerman, A.R., Cao, X., 2014. Biochar-supported zerovalent iron reclaims silver from aqueous solution to form antimicrobial nanocomposite. Chemosphere 117, 801–805.

Zhou, Y., Gao, B., Zimmerman, A.R., Fang, J., Sun, Y., Cao, X., 2013. Sorption of heavy metals on chitosan-modified biochars and its biological effects. Chem. Eng. J. 231, 512–518.

ZHu, X., Liu, Y., Zhou, C., Zhang, S., Chen, J., 2014. Novel and high-performance magnetic carbon composite prepared from waste hydrochar for dye removal. ACS Sustain. Chem. Eng. 2, 969–977.

CHAPTER

11

Biochar for Sustainable Agriculture: Nutrient Dynamics, Soil Enzymes, and Crop Growth

Chathuri Peiris[1], *Sameera R. Gunatilake*[1], *Jayani J. Wewalwela*[2] *and Meththika Vithanage*[3]

[1]College of Chemical Sciences, Institute of Chemistry Ceylon, Rajagiriya, Sri Lanka [2]Department of Agricultural Technology, Faculty of Technology, University of Colombo, Colombo, Sri Lanka [3]Ecosphere Resilience Research Center, Faculty of Applied Sciences, University of Sri Jayewardenepura, Nugegoda, Sri Lanka

11.1 INTRODUCTION

With the increasing demands of the ever-changing global population, a pragmatic approach to viable agricultural practices has become necessary. Sustainable agriculture (SA) is an area of growing interest as it focuses on plausible means to produce crops in an environmentally friendly, socially fair, and economically beneficial manner that can be sustained long term (Hester and Harrison, 2005).

Implementation of biochar amendment in agriculture serves to enrich the sustainability of soils in numerous ways. Biochar has the ability to act as a reservoir of macro- and micronutrients. It can also act as a short-term source of highly available nutrients instigating an acceleration of nutrient cycling processes long term (DeLuca et al., 2015). Nutrient dynamics are influenced by altering physiochemical properties and microbial community composition of the soil. Biochar has become a more cost-effective alternative for commercially available, slow-release nutrient sources such as coated- and nanofertilizers. In addition, the long-term stability of biochar in soil avoids the need for multiple periodic applications (Laird et al., 2010b; Novak et al., 2009).

Biochar from Biomass and Waste
DOI: https://doi.org/10.1016/B978-0-12-811729-3.00011-X

From the perspective of community, creating and applying an organic-based amendment such as biochar can be easily taught to farmers. Any toxic effects that can result as a consequence are minimized by enabling the delivery of high-quality foods to the public (Atkinson et al., 2010).

Environmentally friendly effects result from biochar use in place of artificial amendments as it prevents nutrient leaching from soil. If leached into water ways, it can eventually lead to eutrophication (Laird et al., 2010a). The biochar production itself is a carbon-negative process, sequestering the carbon present in waste biomass that would otherwise have being released back into the atmosphere (Lehmann and Joseph, 2015).

This chapter focuses on how biochar benefits two key aspects of a soil ecosystem—nutrient dynamics and soil enzyme dynamics—that show how biochar can play a key role in sustainable agricultural practices (Lehmann et al., 2011; Rillig and Thies, 2012).

11.2 EVOLUTION OF SUSTAINABLE AGRICULTURE

11.2.1 Malthusian Catastrophe and Green Revolution

The British scholar Thomas Malthus in 1798 developed a concept to express the salient effects growing food demands have on agricultural production. Malthus predicted a population decline to an optimal state once the increasing food demand which is resulted by the population growth would surpass the agricultural production, later called the Malthusian catastrophe. (Maltus, 2006). Despite the threat predicted by the Malthusian catastrophe, the lucrative initiatives taken by the agricultural technology toward a green revolution, which expanded its scope during the 1930s and 1960s increased agricultural production globally, saving the world for over a century from catastrophe (Cullather, 2004). Adoption of new technologies including the utilization of chemical fertilizers and pesticides, the engineering of irrigation systems, the introduction of high-yielding and pest-resistant varieties along with the large-scale agricultural operations supported by mechanization can also be considered as significant attributes of the green revolution (Cullather, 2004).

However, one major drawback of commercial farming is the continuous removal of plant nutrients from farm lands requiring regular application of synthetic fertilizers to prevent the depletion of soil fertility and to improve soil health (Laird et al., 2010b). Leaching of applied nutrients from soils can have adverse effects on the quality of ground and surface water causing deleterious impacts on surrounding aquatic ecosystems. The excessive application of fertilizer can also result in increased cost of crop production (Laird et al., 2010a).

The drawbacks of the green revolution set the stage for a new concept called SA, creating a marriage between traditional agricultural systems and modern technology schemes. The SA movement began in the latter part of the 20th century and while there are many different definitions of it and its systems, it is an economically viable, environmentally safe, and socially fair form of agriculture production (Lichtfouse et al., 2009; Abubakar and Attanda, 2013).

11.2.2 Role of Biochar in Sustainable Agriculture

The idea of SA does not imply a strict deviation from technological developments, but rather includes the practices of (1) integrating natural processes in soil and plants that includes soil regeneration, nitrogen fixation, nutrient cycling, competition, predation, and parasitism into agricultural operations; (2) reducing nonrenewable inputs that are harmful to the environment; and (3) substituting expensive external inputs by human capital by enhancing the knowledge and skills of farmers (Pretty, 2008; Hester and Harrison, 2005).

Biochar is a renewable, environment friendly, low-cost material that can be used in agricultural soil amendments (Ahmad et al., 2014). Incorporation of biochar in agriculture has proven to be an excellent method of reaching sustainability as a result of its potential to (1) increase the ability of soils to retain and recycle nutrients (Biederman and Harpole, 2013); (2) increase carbon sequestration (Lehmann and Joseph, 2015); (3) enhance cation-exchange capacity (CEC) in soil (Liang et al., 2006); (4) act as a reservoir of macro- and micronutrients (Gaskin et al., 2010); (5) stimulate microbial activity and potentially contribute to enzyme dynamics in soil (Lehmann et al., 2011, Warnock et al., 2007); and (6) reduce nitrous-oxide and methane emissions (Cabeza et al., 2018). However, the productivity of soil amendments depends on various characteristics of the biochar such as the surface charge at operating pH, surface functionality and morphology, CEC, ash content, etc. (Herath et al., 2015). A rich science lies behind biochar application since these properties are heavily dependent on feedstock type, pyrolysis conditions, and the age of the biochar (Peiris et al., 2017). Thus biochar-based agronomy research has become a popular research topic in the past decade.

11.3 INFLUENCE OF BIOCHAR ON SOIL NUTRIENT DYNAMICS

SA is influenced by the effectiveness with which nutrients are cycled in the environment, and is critical for plant productivity. Both biotic factors such as community composition of plants, microbes, and soil fauna, and abiotic factors such as climate, soil type, and organic matter (OM) affect nutrient dynamics. Biochar can influence nutrient dynamics by increasing bioavailable nutrients, altering physiochemical properties of soil, and affecting soil ecosystems (Eviner and Firestone, 2007).

Continuous application of nutrients to soil in the form of synthetic fertilizer, manure, or other fertility amendment techniques enable the replacement of nutrients lost due to the harvesting of crop residues (Laird et al., 2010b). However, inorganic fertilizers have reduced retention in the soil as a consequence of the low nutrient-holding capacity of infertile soils (Glaser et al., 2001). This issue has been addressed by methods such as using slow-release forms of nutrients, multiple fertilizer applications, and by covering crops that maintain integrated root systems during the offseason despite the high cost associated with these methods (Laird et al., 2010a).

Currently, biochar is gaining acceptance as a relatively inexpensive and efficient alternative due to its high nutrient value that is made available to plants directly or indirectly. The direct contribution is by providing its labile nutrients to the plant whereas the indirect contribution is by improving soil quality that in turn increases the efficiency of fertilizer use (Xu and Chan, 2012; Xu et al., 2013).

11.3.1 Direct Nutrient Values of Biochar

Biochar is capable of enriching the soil nutrient pool by acting as a source of both macro- and micronutrients. Nutrient-rich biochar can function as a slow-release fertilizer (Ding et al., 2016). The production temperature of biochar is a critical factor for determining its nutrient content as a certain fraction of elements in the feedstock can be excessively lost by volatilization. The lowest volatilization temperatures (VTs) are reported in N (VT ~ 200°C) and S (VT ~ 375°C) while P and K have moderate VTs (~700–800°C). Nutrients such as Ca, Mg, and Mn are said to be thermally stable at typical biochar production temperatures (VT > 1000°C) (DeLuca et al., 2015; Laird et al., 2010b). At higher pyrolysis temperatures, the carbon content lost is increased, leaving a higher percentage of thermally stable nutrients behind. For instance, sludge based BC contains P mainly in the form of thermally stable inorganic salts creating a direct relationship between the P percentage and the pyrolysis temperature up to 800 °C. The K content in sludge-based biochar also increases with pyrolysis temperature due to inorganic associations (Hossain et al., 2011).

It is also important to note that only a small fraction of the total nutrient content of biochar is available to plants since a considerable fraction usually exists in recalcitrant forms (Gaskin et al., 2010; Laird et al., 2010b; DeLuca et al., 2015). As an example, the total S content of biochar exists partly as labile inorganic sulfates and partly as organic S, which is not bioavailable (Knudsen et al., 2004; Freney et al., 1975). The pyrolysis temperature can also affect labile and recalcitrant fractions of nutrients in a biochar. For instance, high-temperature biochar contains high ash content where nutrients exist mainly as soluble salts that can be readily liberated in soil (Ding et al., 2016; Zheng et al., 2013b; Irfan et al., 2017). It has been reported that the bioavailable amine-N fraction such as amino acids and amino sugars in the biomass can be lowered during high-temperature pyrolysis (>700°C) due to formation of N-heterocyclic aromatic compounds (Gaskin et al., 2010; Novak et al., 2009). A study by Zheng et al. reported significant enhancement of P content when the pyrolysis temperature was increased from 300 to 600°C. However, biochar produced at 300°C contained a low fraction of crystallized P-associated minerals with higher bioavailability than the highly crystallized P found in high-temperature biochar (Zheng et al., 2013b).

The direct supply of nutrients is reported to be higher in fresh biochar generated from nutrient-rich feedstock (DeLuca et al., 2015). Multiple studies have reported high bioavailable nutrient content of newly prepared biochars and their capabilities to release increased amounts of N and P (Mukherjee and Zimmerman, 2013; Zheng et al., 2013b). Biochar produced from animal waste such as sewage sludge, manure, and broiler litter are reported to contain higher amounts of P and N than plant-based biochar (Xu and Chan, 2012; Irfan et al., 2017). Furthermore, the N content of swine manure biochar was found to be significantly higher than the biochar produced from giant cane at identical pyrolysis temperatures (400°C) (Ding et al., 2016). According to studies reported by Chan et al., increased levels of N, P, S, Na, Ca, and Mg have been detected in radish plants grown in poultry litter biochar-amended soil whereas only P, K, and Ca have been increased in plant-based biochar-amended soil (Chan et al., 2008b, 2008a).

The same type of feedstock can produce biochar with varying nutrient contents despite their identical pyrolysis conditions. As an example, the bioavailability of P in sludge-based biochar depends on the amount and the type of stabilizers applied during sludge treatment (Hossain et al., 2011). Significant variations of total N contents were reported in two biochars made from different poultry litters under the same conditions (Lima and Marshall, 2005).

However, multiple studies have reported a decline in the nutrient values of biochar after 1 year of application, making the unavailability of nutrients for long-term crop growth a major drawback of its direct application (Gaskin et al., 2010; Wu et al., 2011). It has currently become a challenging task to determine the pattern of nutrient bioavailability in the long term. As reported by Ding et al. knowledge of the long-term nutrient availability of biochar is insufficient as the majority of the studies reported are based on short-term column leaching experiments (Ding et al., 2016).

11.3.2 Indirect Nutrient Values of Biochar

Biochar is capable of indirectly influencing the different physiochemical properties of soil such as total organic carbon, pH, CEC, and soil bulk density leading to an elevation of its inherent quality and health (DeLuca et al., 2015).

Cation-exchange capacity: The CEC of soil, which is a measure of the total cations that can be retained by soil-exchangeable sites, is a key contributor of soil quality. This parameter is mainly governed by the mineral content and the soil organic carbon. In soils where aforementioned factors are low, reduced CEC is exhibited-leading to the deleterious consequence of nutrient leaching (Masulili et al., 2010). There are several reports of noteworthy augmentation in the CEC of soils subjected to biochar amendment (Laird et al., 2010b; Jien and Wang, 2013). Enhanced cationic nutrient retention of metal ions such as K, Ca, Na, and Mg in soil upon biochar amendment has also been reported occasions (Wang et al., 2014; Gaskin et al., 2010).

The CEC is mainly due to the negative surface charge on the biochar that arises from both OM and oxygenated surface functional groups (O-SFGs) (Atkinson et al., 2010; Novak et al., 2009). The fulvic and humic substances present in biochar constitute the OM that acts as exchangeable sites for cations (Atkinson et al., 2010). The functional groups present on the biochar surface vary depending on the pyrolysis conditions incorporated. Low-temperature-produced biochar consists of numerous lactonic and carboxylic groups on its surface that get deprotonated in soil to produce negatively charged anions capable of binding to cations via electrostatic attractions. In comparison, high-temperature-produced biochar is low in such O-SFGs, yielding low CECs (Ippolito et al., 2015).

Smaller and highly charged cations in soil show higher affinity toward exchangeable sites in biochar. For instance, Novek et al. reported increased retention of multivalent cations such as Ca, Mg, Zn, and Mn compared to monovalent cations such as Na and K in biochar-amended soil (Novak et al., 2009). Bioavailable nitrogen can exist in ammonium and nitrate forms. Soil CEC is mainly responsible for the retention of $NH_4^+ - N$ whereas anion-exchange capacity (AEC) and porefilling mechanisms govern the retention of

$NO_3^- - N$ (Trindade et al., 1997). In comparison with $NO_3^- - N$, higher retention of $NH_4^+ - N$ is often observed by biochar application due to higher CEC of biochar compared to its AEC (Zheng et al., 2013a). Phosphate retention is analogous to nitrate retention. However, ligand exchange reactions can also contribute to the sorption capacity of anions (Novak et al., 2009).

Total organic carbon: Total organic carbon in soil is a validated parameter to judge the condition of a soil. A soil's productivity is dependent on its capacity to retain water and nutrients as well as on its CEC and soil structure. However, lack of OM in the soil can lead to the decline of afore mentioned facets regardless of the application of synthetic fertilizer as soil amendment (Laird et al., 2010b).

The harvesting period is a time during which much of the OM is removed from the soil. To compensate for this, use of manure and biochar has proven to be effective. However, one of the drawbacks of manure application is its rapid decomposition (Laird et al., 2010b; Jeffery et al., 2011). In Venezuelan rain forests, for example, the average lifespan for the OM applied is less than 4 years. Decomposition of soil OM is accelerated as a result of high temperatures, which leads to increased bacterial degradation of the organic amendments applied. The humidity that results as a consequence of increased rainfall can also be a contributing factor. Manure has minimum stabilizing agents in its structure to confer resistance to degradation and thereby contributes to the lifespan of these organic particulates (Glaser et al., 2001).

Biochar, in contrast, is reportedly more stable than manure due to its carbonaceous structure (Laird et al., 2010b; Downie et al., 2009). For instance, Laird et al. (2016b) reported significant enrichments of total organic C in soil upon biochar amendment when compared to manure application (Laird et al., 2010b). It is important to note that biochars produced at extreme temperatures ($>800°C$) with more graphitized structures are less stable in soil than low-temperature biochars that are generally disordered and recalcitrant in nature (Downie et al., 2009; Rajapaksha et al., 2014).

The C:N ratio: The C:N ratio is another important parameter when considering soil nutrient availability as it serves as a measure of the amount of nitrogen immobilized (Laird et al., 2010b). As reported in the literature, a ratio of greater than 32:1 in organic residues is justification for significant nitrogen immobilization in the soil (Alexander, 1977). In biochars, this parameter can take on values ranging from 7 to 400, with a numerical average of 67. However, the low decomposition rate of biochars despite their high C:N ratios make nitrogen immobilization insignificant, which is an additional advantage over the application of organic amendments such as manure (Lehmann, 2007).

Soil pH: Biochar plays a role in enriching the labile nutrient pool by altering the soil pH, leading to increased bioavailability of nutrients, facilitating microbial activity and also root access to water and nutrients. Upon its application to soil, biochar can elevate the bioavailability of nutrients by the liming effect and by trapping trivalent species (Nigussie et al., 2012).

Weathered soils abundant in iron and aluminum tend to be acidic due to the liberation of hydronium ions upon hydroxide formation (Sato et al., 2009; Novak et al., 2009). Insoluble iron and aluminum phosphates reduce the phosphorous bioavailability in such soils (Novak et al., 2009). The liming effect arises as a result of the calcium oxides present in the biochar that react with the soil phosphorous in order to form calcium phosphates.

The solubility of these complexes make the inorganic P bioavailable to the plant while reducing the soil acidity (Novak et al., 2009). The exchangeable sites of biochar show high affinity for trivalent aluminum and iron reducing their availability to form insoluble complexes with P (Nigussie et al., 2012). The high cost associated with liming makes biochar a more economically viable technique (Masulili et al., 2010).

Furthermore, the presence of elements such as Al, Cu, and Mn in the soil that can cause toxic effects to the plant at acidic pH are also negated by biochar application (Atkinson et al., 2010).

Water-holding capacity: WHC is an integral aspect of agriculture that is dependent on the texture of soil and precipitation rate. Soil OM is a key factor determining its WHC, which influences nutrient movement and leaching, prominently in the rooting zone (Atkinson et al., 2010). Reduced WHC results in poor crop productivity and soil degradation (Amezketa, 1999). Soil WHC can be influenced by biochar application due to humic substances, porous nature, and interactions with roots (Ding et al., 2016).

The pore structure of biochars have different effects on the soil quality (Yuan et al., 2015). Micropores and mesopores are important for the retention of available water content whereas macropores are involved in hydraulic conductivity (Herath et al., 2013). Biochar is reported to influence WHC more significantly in sandy soils than in soils with a high clay fraction (Atkinson et al., 2010).

In addition to water retention, the macroporous structure of biochar assists in improving soil aeration and water infiltration (Yuan et al., 2015). Soil aggregation is a term used to describe the process of soil particles adhering to each other, creating pore spaces for holding water and air. Biochar enhances this aggregation by interacting with soil OM, minerals, and microorganisms (MOs) (Kelly et al., 2017).

Soil bulk density: Bulk density, defined as the weight of soil in a given volume, serves as an indicator of soil compaction and soil health. Bulk density has a noteworthy effect on key soil processes as it affects infiltration, rooting depth, available water capacity, soil porosity, and MO activity. For example, a soil with a bulk density greater than 1.6 g cm^{-3} would restrict root growth. Soil OM, texture, and the packing arrangement are factors governing bulk density (Jury and Stolzy, 2018).

The organic amendments used dictate the value that the soil bulk density will assume. Biochar application, however, has shown to result in a significant reduction in bulk density as compared to manure application. This is attributed to the porous structure of the biochar. Biochar particles are highly porous and thereby have low densities that when applied to soil can lead to a decrease in the overall soil bulk density (Laird et al., 2010b; Herath et al., 2013).

11.4 INFLUENCE OF BIOCHAR ON SOIL ENZYMES

Changes in biological properties must also be considered for prudent evaluation of soil fertility (Sherene, 2017). The enzymatic activity that takes place in the rhizosphere poses a significant impact on the nutrient bioavailability to the plant, which in turn affects plant health and productivity (Abubakar and Attanda, 2013). Soil enzymes are an effective means of appraising soil quality due to their high sensitivity and the rapid responses elicited to changes in the soil environment (Sherene, 2017).

MOs, fauna, and plant roots in the soil are sources from which these enzymes originate and are usually stabilized in the soil matrix by forming complexes with OM, humic colloids, or clay particles (Bandick and Dick, 1999; Laird et al., 2010b). Soil enzymes perform their inherent role of catalysis, taking part in metabolic processes such as degradation of OM, mineralization, and nutrient transformation whose efficiencies are dependent on temperature and pH (Burns et al., 2013).

Soils subjected to organic amendments, crop rotations, and cover crops have shown enhanced enzymatic activity that elevates nutrient cycling processes in the soil (Du et al., 2014). Application of biochar serves to alter physiochemical properties of the soil and the microbial community composition, which affects soil fertility as a consequence (Warnock et al., 2007). The expanding scope of biochar-based soil amendments as a plausible means of managing soil biota provides evidence for the growing interest in the field (Ding et al., 2016).

11.4.1 Influence of Biochar on Microorganism-Derived Soil Enzymes

There have been various reports of variations in MO populations upon biochar application (Anderson et al., 2011). The reported rise in numbers of anaerobic and cellulose hydrolyzing bacteria, for example, has been shown to result from biochar amendments (Lehmann et al., 2011). Three mechanisms can be used to portray this alteration, which results from the influence that biochar has on (Warnock et al., 2007) (1) nutrient availability and soil physiochemical parameters, (2) the activity of MO and the ability to serve as a refuge for colonizing MO, and (3) its effect on the signaling dynamics between plants and MO, discussed as follows.

Influence on nutrient availability and alterations in soil physiochemical properties: MO and their enzymes contribute substantially to the regulation of processes such as nutrient cycling. Some examples of enzymes involved in nutrient cycling are β-glycosidase, β-D-cellobiosidase, and β-xylosidase involved in the carbon cycle; *N*-acetyl-β-glucosaminidase, urease, and leucine aminopeptidase involved in the nitrogen cycle; and phosphomonoesterase involved in the phosphorous cycle (Sherene, 2017; Song et al., 2018). Biochar can have an effect on these enzymatic activities. The genus *Pseudomonas* is an MO involved in the cycling of phosphorous, releasing phosphatase enzymes that catalyze the hydrolysis of ester phosphate bonds. This leads to the inorganic phosphorous becoming solubilized and made bioavailable (Beheshti et al., 2017; Khan et al., 2007).

Application of biochar leads to a change in physiochemical properties of soil that causes MO composition and their functions to be altered significantly by the additional complexity to the extracellular enzyme activities. Biochar is comprised of macropores, mesopores, and micropores of varying sizes that can become a habitat for MO. Saprophytic fungi can form colonies inside these pores leading to the decomposition of the biochar. This makes the nutrients available for the plant, increasing the crop growth. Fungi such as *Trichoderma* and *Penicillium* spp., for example, produce enzymes such as manganese peroxidase and phenol oxidase to depolymerize the biochar (Rillig and Thies, 2012). The intricate porous structure of biochar leads to an enhancement in oxygen diffusion, increasing the respiratory activities of certain aerobic organisms that carry out ammonia oxidation and methane oxidation.

The adverse influence of biochar amendments on microbial enzyme activities have also been reported in the literature. The ability of acidophilic *Thiobacillus* to oxidize sulfur is an example of a mechanism that is diminished due to this practice. As sulfur is not bioavailable in its usual organic form, it has to be oxidized to sulfate by enzymes. However, the ability of the bacterium to carry out this process is hindered by the addition of biochar since it creates an unfavorable environment for sulfur oxidation (DeLuca et al., 2015). Another example would be the reported decrease in maximum velocity of enzymes such as cellobiosidase and glucosidase with higher rates of biochar application (12 t of biochar per hectare or greater) (Akça and Namli, 2015). Furthermore, an inhibition of the enzyme function can occur due to the sorption of organic and inorganic substrates onto biochar (Lehmann et al., 2011).

Influence on the activity of MO and its role in acting as a refuge for MO colonization: Colonization of MO can have an indirect effect on soil enzyme dynamics that is favored by the carbonaceous and nutrient-rich nature of biochar. Examples of these colonizing MO include mycorrhiza helper bacteria (MHB) and phosphorous solubilizing bacteria (PSB) (Riedlinger et al., 2006). Specific conditions induce bacteria to secrete metabolites such as flavonoids and furans that assist the growth of fungal hyphae and subsequent colonization of Ectomycorrhiza (ECM) and Arbuscularmycorrhiza (AMF) in plant roots (Warnock et al., 2007). Raffinose produced by strains of *Paenibacillus* (Hildebrandt et al., 2006) and flavonoids produced by *Rhizobium* and *Bradyrhizobium* (Cohn et al., 1998) species can also contribute to extraradical mycelium growth and to an increase in root colonization of AM fungi.

Influence on the signaling dynamics between plants and MO: Biochar application can modify signaling pathways between microbes and plant roots. These variations can result due to changes in pH and the temperature at which the biochar was produced. Flavonoid signaling compounds are dependent on pH and can elicit excitatory or inhibitory responses in soil biota (Warnock et al., 2007). If excitatory in nature, these responses would yield high fungal populations in the soil. Biochar produced at high temperatures can capture signaling molecules that are not immediately detected by AMF hyphae or spores, thereby promoting signal transduction. Changes in signal dynamics due to these factors can indirectly contribute to soil enzyme dynamics by causing changes in microbial populations.

11.4.2 Faunal Population Response to Biochar in Soil

The involvement of soil fauna in the events of enzymatic action has not been extensively studied. Soil fauna is an essential part of fungal and bacterial energy channels (Cragg and Bardgett, 2001). However, its involvement is significant as fauna is positioned at the top of the food chain and thereby understanding of the effects of biochar application on its characteristics and biology can widen scientific understanding of the various microbial responses produced with biochar amendment to soil. For instance, the N-cycling MO in the guts of earthworms have more pronounced action with biochar amendment (Lehmann et al., 2011).

11.4.3 Plant Root Response to Biochar in Soil

Plant roots secrete many exudates that pose beneficial and harmful effects to rhizosphere microbial populations. Such compounds include inorganic ions and substances, amino acids, volatile aromatic compounds, proteins, and enzymes (Dundek et al., 2014). These exudates can vary according to the different plant species and conditions. Pruned tea bushes, for example, secrete more root exudates that influence microbiological and biochemical properties in the rhizosphere than unpruned tea plants (Pramanik et al., 2017). MO abundance of species such as *Bacillus*, *Pseudomonas*, and *Trichoderma* can alter in response to the induced systemic resistance produced by exudates of plants of tomato, pepper, and bean, respectively (Kolton et al., 2011). Therefore, these diverse plant root exudes including enzymes enable the communication with rhizosphere MO to cope with plant pathogens. However, information on this in regard to biochar amendment is minimal (Akhter et al., 2015).

11.5 EFFECT OF BIOCHAR ON CROP GROWTH

Increased crop yields, seed germination, and crop growth is evident after biochar application to soils. The interface between the plant root and the soil (the rhizosphere) and the root system of the plant are vital components in crop growth since they are involved in water and nutrient uptake, storage, and regulation. The rhizosphere tends to be larger in soils containing biochar (Zheng et al., 2013a).

Since it is a proven fact that the plant roots are attracted to biochar, we know it is involved in the direct uptake of plant nutrients. Rhizosheath size can help determine the efficiency of phosphorous uptake to the plant under phosphorous-deficient conditions (Prendergast-Miller et al., 2014). Biochar addition leads to decreased accumulation of rhizosheath, indicating increased supply of phosphorous to the plant (Brown et al., 2012). Considerable enhancements of root volume, length, and surface area have been reported after biochar amendment (Zheng et al., 2013a).

The amount of sunlight reflected by the earth surface is known as the albedo. The black surfaces of biochar increase the albedo of farmlands leading to enhanced crop growth due to improved rates of photosynthesis (Usowicz et al., 2016).

The contributions of enhanced nutrient cycling and enzyme activity on crop growth have been comprehensively discussed in the previous subsections. Multiple studies have reported high crop yields after biochar application (Irfan et al., 2017; Zheng et al., 2013a,b; Lehmann et al., 2003; Graber et al., 2010; Herath et al., 2015). However, a decrease in plant yield and MO community has been observed by some as a result of high biochar application rates (Herath et al., 2015). This could be as a consequence of the toxic elements and the high percentage of volatile content in the soil, leading to an abatement in nutrient uptake by the plant (Asai et al., 2009).

11.6 CONCLUSIONS

Biochar contributes to soil fertility by either acting as a direct nutrient source or by altering the physiochemical properties in the soil. The nutrient content that constitutes biochar is dependent on the type of feedstock used. Optimum pyrolysis conditions should be employed to ensure minimal volatilization of essential nutrients. A high percentage of the nutrients present in biochar should be in their bioavailable forms. This makes biochar an effective slow-release nutrient source when applied to soil. It, however, is not a long-term contributor to soil fertility.

Significant enhancement in crop productivity can be seen with biochar application to acidic soils as it leads to an increase in the soil pH. Biochar can act as a stable carbon source in the soil and can elevate the soil CEC resulting in the retention of many micro- and macronutrients. The WHC of soil, water infiltration, and soil aeration are governed by macropores of biochar particles. The overall soil enzyme activity that originates from MO, plants, and animals is enhanced by biochar application, which heightens the decomposition of OM and nutrient cycling.

The simplicity, cost effectiveness, and physical and chemical characteristics associated with biochar has attracted research interest worldwide. It has the potentially to expand the success SA practices.

References

Abubakar, M.S., Attanda, M., 2013. The Concept of Sustainable Agriculture: Challenges and Prospects. IOP Conference Series: Materials Science and Engineering. IOP Publishing, , 012001.

Ahmad, M., Rajapaksha, A.U., Lim, J.E., Zhang, M., Bolan, N., Mohan, D., et al., 2014. Biochar as a sorbent for contaminant management in soil and water: a review. Chemosphere 99, 19–33.

Akça, M.O., Namli, A., 2015. Effects of poultry litter biochar on soil enzyme activities and tomato, pepper and lettuce plants growth. Eur. J. Soil Sci. 4, 161.

Akhter, A., Hage-Ahmed, K., Soja, G., Steinkellner, S., 2015. Compost and biochar alter mycorrhization, tomato root exudation, and development of Fusarium oxysporum f. sp. lycopersici. Front. Plant Sci. 6, 529.

Alexander, M., 1977. Introduction to Soil Microbiology, Second ed John Wiley & Sons Inc.

Amezketa, E., 1999. Soil aggregate stability: a review. J. Sustain. Agric. 14, 83–151.

Anderson, C.R., Condron, L.M., Clough, T.J., Fiers, M., Stewart, A., Hill, R.A., et al., 2011. Biochar induced soil microbial community change: implications for biogeochemical cycling of carbon, nitrogen and phosphorus. Pedobiologia 54, 309–320.

Asai, H., Samson, B.K., Stephan, H.M., Songyikhangsuthor, K., Homma, K., Kiyono, Y., et al., 2009. Biochar amendment techniques for upland rice production in Northern Laos: 1. Soil physical properties, leaf SPAD and grain yield. Field Crops Res. 111, 81–84.

Atkinson, C.J., Fitzgerald, J.D., Hipps, N.A., 2010. Potential mechanisms for achieving agricultural benefits from biochar application to temperate soils: a review. Plant Soil 337, 1–18.

Bandick, A.K., Dick, R.P., 1999. Field management effects on soil enzyme activities. Soil Biol. Biochem. 31, 1471–1479.

Beheshti, M., Etesami, H., Alikhani, H.A., 2017. Interaction study of biochar with phosphate-solubilizing bacterium on phosphorus availability in calcareous soil. Arch. Agron. Soil Sci. 63, 1572–1581.

Biederman, L.A., Harpole, W.S., 2013. Biochar and its effects on plant productivity and nutrient cycling: a meta-analysis. GCB Bioenergy 5, 202–214.

Brown, L.K., George, T.S., Thompson, J.A., Wright, G., Lyon, J., Dupuy, L., et al., 2012. What are the implications of variation in root hair length on tolerance to phosphorus deficiency in combination with water stress in barley (Hordeum vulgare)? Ann. Bot. (Lond.) 110, 319–328.

Burns, R.G., Deforest, J.L., Marxsen, J., Sinsabaugh, R.L., Stromberger, M.E., Wallenstein, M.D., et al., 2013. Soil enzymes in a changing environment: current knowledge and future directions. Soil Biol. Biochem. 58, 216–234.

Cabeza, I., Waterhouse, T., Sohi, S., Rooke, J., 2018. Effect of biochar produced from different biomass sources and at different process temperatures on methane production and ammonia concentrations in vitro. Anim. Feed Sci. Technol. 237, 1–7.

Chan, K., Van Zwieten, L., Meszaros, I., Downie, A., Joseph, S., 2008a. Using poultry litter biochars as soil amendments. Soil Res. 46, 437–444.

Chan, K.Y., Van Zwieten, L., Meszaros, I., Downie, A., Joseph, S., 2008b. Agronomic values of greenwaste biochar as a soil amendment. Soil Res. 45, 629–634.

Cohn, J., Day, R.B., Stacey, G., 1998. Legume nodule organogenesis. Trends Plant Sci. 3, 105–110.

Cragg, R.G., Bardgett, R.D., 2001. How changes in soil faunal diversity and composition within a trophic group influence decomposition processes. Soil Biol. Biochem. 33, 2073–2081.

Cullather, N., 2004. Miracles of modernization: the green revolution and the apotheosis of technology. Diplomat. Hist. 28, 227–254.

Deluca, T.H., Gundale, M.J., Mackenzie, M.D., Jones, D.L., 2015. Biochar effects on soil nutrient transformations. In: Biochar for Environmental Management: Science, Technology and Implementation, vol. 2. Taylor & Francis Group, Routledge, New York, pp. 421–454.

Ding, Y., Liu, Y., Liu, S., Li, Z., Tan, X., Huang, X., et al., 2016. Biochar to improve soil fertility. A review. Agron. Sustain. Dev. 36, 36.

Downie, A., Crosky, A., Munroe, P., 2009. Physical properties of biochar. In: Lehmann, J., Joseph, S. (Eds.), Biochar for environmental management. Science and technology. Earthscan, London, UK, pp. 13–32.

Du, Z., Wang, Y., Huang, J., Lu, N., Liu, X., Lou, Y., et al., 2014. Consecutive biochar application alters soil enzyme activities in the winter wheat–growing season. Soil Sci. 179, 75–83.

Dundek, P., Holík, L., Rohlík, T., Hromádko, L., Vranová, V., Rejšek, K., et al., 2014. Methods of plant root exudates analysis: a review. Acta Univ. Agric. Silvic. Mendelianae Brun. 59, 241–246.

Eviner, V.T., Firestone, M.K., 2007. Mechanisms determining patterns of nutrient dynamics. In: Stromberg M. (Ed.), California Grasslands: Ecology and Management. Oxford University Press, pp. 94–106.

Freney, J.R., Melville, G.E., Williams, C.H., 1975. Soil organic matter fractions as sources of plant-available sulphur. Soil Biol. Biochem. 7, 217–221.

Gaskin, J.W., Speir, R.A., Harris, K., Das, K., Lee, R.D., Morris, L.A., et al., 2010. Effect of peanut hull and pine chip biochar on soil nutrients, corn nutrient status, and yield. Agron. J. 102, 623–633.

Glaser, B., Haumaier, L., Guggenberger, G., Zech, W., 2001. The 'Terra Preta' phenomenon: a model for sustainable agriculture in the humid tropics. Naturwissenschaften 88, 37–41.

Graber, E.R., Harel, Y.M., Kolton, M., Cytryn, E., Silber, A., David, D.R., et al., 2010. Biochar impact on development and productivity of pepper and tomato grown in fertigated soilless media. Plant Soil 337, 481–496.

Herath, H., Camps-Arbestain, M., Hedley, M., 2013. Effect of biochar on soil physical properties in two contrasting soils: an Alfisol and an Andisol. Geoderma 209, 188–197.

Herath, I., Kumarathilaka, P., Navaratne, A., Rajakaruna, N., Vithanage, M., 2015. Immobilization and phytotoxicity reduction of heavy metals in serpentine soil using biochar. J. Soil Sediment 15, 126–138.

Hester, R.E., Harrison, R.M., 2005. Sustainability in Agriculture. Royal Society of Chemistry.

Hildebrandt, U., Ouziad, F., Marner, F.-J., Bothe, H., 2006. The bacterium Paenibacillus validus stimulates growth of the arbuscular mycorrhizal fungus Glomus intraradices up to the formation of fertile spores. FEMS Microbiol. Lett. 254, 258–267.

Hossain, M.K., Strezov, V., Chan, K.Y., Ziolkowski, A., Nelson, P.F., 2011. Influence of pyrolysis temperature on production and nutrient properties of wastewater sludge biochar. J. Environ. Manage. 92, 223–228.

Ippolito, J.A., Spokas, K.A., Novak, J.M., Lentz, R.D., Cantrell, K.B., 2015. Biochar elemental composition and factors influencing nutrient retention. In: Lehmann, J., Joseph, S. (Eds.), Biochar for environmental management: science, technology and implementation, second ed. Routledge, Abingdon, UK, pp. 139–163.

Irfan, M., Kaleri, F.N., Rizwan, M., Mehmood, I., 2017. Potential value of biochar as a soil amendment: a review. Pure Appl. Biol. 6, 1494–1502.

Jeffery, S., Verheijen, F.G., Van Der Velde, M., Bastos, A.C., 2011. A quantitative review of the effects of biochar application to soils on crop productivity using meta-analysis. Agric. Ecosyst. Environ. 144, 175–187.

Jien, S.-H., Wang, C.-S., 2013. Effects of biochar on soil properties and erosion potential in a highly weathered soil. Catena 110, 225–233.

Jury, W.A., Stolzy, L.H., 2018. Soil physics. Handbook of Soils and Climate in Agriculture. CRC Press. Available from: <https://www.taylorfrancis.com/books/e/9781351081528/chapters/10.1201%2F9781351073073-3>.

Kelly, C.N., Benjamin, J., Calderón, F.C., Mikha, M.M., Rutherford, D.W., Rostad, C.E., 2017. Incorporation of biochar carbon into stable soil aggregates: the role of clay mineralogy and other soil characteristics. Pedosphere 27, 694–704.

Khan, M.S., Zaidi, A., Wani, P.A., 2007. Role of phosphate-solubilizing microorganisms in sustainable agriculture—a review. Agron. Sustain. Dev. 27, 29–43.
Knudsen, J.N., Jensen, P.A., Lin, W., Frandsen, F.J., Dam-Johansen, K., 2004. Sulfur transformations during thermal conversion of herbaceous biomass. Energy Fuels 18, 810–819.
Kolton, M., Harel, Y.M., Pasternak, Z., Graber, E.R., Elad, Y., Cytryn, E., 2011. Impact of biochar application to soil on the root-associated bacterial community structure of fully developed greenhouse pepper plants. Appl. Environ. Microbiol. 77, 4924–4930.
Laird, D., Fleming, P., Wang, B., Horton, R., Karlen, D., 2010a. Biochar impact on nutrient leaching from a Midwestern agricultural soil. Geoderma 158, 436–442.
Laird, D.A., Fleming, P., Davis, D.D., Horton, R., Wang, B., Karlen, D.L., 2010b. Impact of biochar amendments on the quality of a typical Midwestern agricultural soil. Geoderma 158, 443–449.
Lehmann, J., 2007. Bio-energy in the black. Front. Ecol. Environ. 5, 381–387.
Lehmann, J., Da Silva, J.P., Steiner, C., Nehls, T., Zech, W., Glaser, B., 2003. Nutrient availability and leaching in an archaeological Anthrosol and a Ferralsol of the Central Amazon basin: fertilizer, manure and charcoal amendments. Plant Soil 249, 343–357.
Lehmann, J., Joseph, S., 2015. Biochar for Environmental Management: Science, Technology and Implementation. Taylor & Francis.
Lehmann, J., Rillig, M.C., Thies, J., Masiello, C.A., Hockaday, W.C., Crowley, D., 2011. Biochar effects on soil biota – a review. Soil Biol. Biochem. 43, 1812–1836.
Liang, B., Lehmann, J., Solomon, D., Kinyangi, J., Grossman, J., O'neill, B., et al., 2006. Black carbon increases cation exchange capacity in soils. Soil Sci. Soc. Am. J. 70, 1719–1730.
Lichtfouse, E., Navarrete, M., Debaeke, P., Souchère, V., Alberola, C., Ménassieu, J., 2009. Agronomy for sustainable agriculture: a review, Agron. Sustain. Dev., 29. Springer, pp. 1–6.
Lima, I.M., Marshall, W.E., 2005. Granular activated carbons from broiler manure: physical, chemical and adsorptive properties. Bioresour. Technol. 96, 699–706.
Maltus, T.R., 2006. An Essay on the Principle of Population. Cosimo, Inc.
Masulili, A., Utomo, W.H., Syechfani, M., 2010. Rice husk biochar for rice based cropping system in acid soil. 1. The characteristics of rice husk biochar and its influence on the properties of acid sulfate soils and rice growth in West Kalimantan, Indonesia. J. Agric. Sci. 2, 39.
Mukherjee, A., Zimmerman, A.R., 2013. Organic carbon and nutrient release from a range of laboratory-produced biochars and biochar–soil mixtures. Geoderma 193, 122–130.
Nigussie, A., Kissi, E., Misganaw, M., Ambaw, G., 2012. Effect of biochar application on soil properties and nutrient uptake of lettuces (Lactuca sativa) grown in chromium polluted soils. Am. Eur. J. Agric. Environ. Sci. 12, 369–376.
Novak, J.M., Busscher, W.J., Laird, D.L., Ahmedna, M., Watts, D.W., Niandou, M.A., 2009. Impact of biochar amendment on fertility of a southeastern coastal plain soil. Soil Sci. 174, 105–112.
Peiris, C., Gunatilake, S.R., Mlsna, T.E., Mohan, D., Vithanage, M., 2017. Biochar based removal of antibiotic sulfonamides and tetracyclines in aquatic environments: a critical review. Bioresour. Technol. 246, 150–159.
Pramanik, P., Phukan, M., Ghosh, S., Goswami, A., 2017. Pruned tea bushes secrete more root exudates to influence microbiological properties in soil. Arch. Agron. Soil Sci. 64, 1172–1180.
Prendergast-Miller, M., Duvall, M., Sohi, S., 2014. Biochar–root interactions are mediated by biochar nutrient content and impacts on soil nutrient availability. Eur. J. Soil Sci. 65, 173–185.
Pretty, J., 2008. Agricultural sustainability: concepts, principles and evidence. Philos. Trans. R. Soc. Lond. B Biol. Sci. 363, 447–465.
Rajapaksha, A.U., Vithanage, M., Lim, J.E., Ahmed, M.B.M., Zhang, M., Lee, S.S., et al., 2014. Invasive plant-derived biochar inhibits sulfamethazine uptake by lettuce in soil. Chemosphere 111, 500–504.
Riedlinger, J., Schrey, S.D., Tarkka, M.T., Hampp, R., Kapur, M., Fiedler, H.-P., 2006. Auxofuran, a novel metabolite that stimulates the growth of fly agaric, is produced by the mycorrhiza helper bacterium Streptomyces strain AcH 505. Appl. Environ. Microbiol. 72, 3550–3557.
Rillig, M.C., Thies, J.E., 2012. Characteristics of biochar: biological properties. Biochar for Environmental Management. Routledge.
Sato, S., Neves, E.G., Solomon, D., Liang, B., Lehmann, J., 2009. Biogenic calcium phosphate transformation in soils over millennial time scales. J. Soil Sediment 9, 194–205.

Sherene, T., 2017. Role of soil enzymes in nutrient transformation: a review. Bio Bull. 3 (1), 109–131.

Song, D., Tang, J., Xi, X., Zhang, S., Liang, G., Zhou, W., et al., 2018. Responses of soil nutrients and microbial activities to additions of maize straw biochar and chemical fertilization in a calcareous soil. Eur. J. Soil Biol. 84, 1–10.

Trindade, H., Coutinho, J., Van Beusichem, M., Scholefield, D., Moreira, N., 1997. Nitrate leaching from sandy loam soils under a double-cropping forage system estimated from suction-probe measurements. Plant Soil 195, 247–256.

Usowicz, B., Lipiec, J., ŁUkowski, M., Marczewski, W., Usowicz, J., 2016. The effect of biochar application on thermal properties and albedo of loess soil under grassland and fallow. Soil Tillage Res. 164, 45–51.

Wang, Y., Yin, R., Liu, R., 2014. Characterization of biochar from fast pyrolysis and its effect on chemical properties of the tea garden soil. J. Anal. Appl. Pyrol. 110, 375–381.

Warnock, D.D., Lehmann, J., Kuyper, T.W., Rillig, M.C., 2007. Mycorrhizal responses to biochar in soil – concepts and mechanisms. Plant Soil 300, 9–20.

Wu, H., Yip, K., Kong, Z., Li, C.-Z., Liu, D., Yu, Y., et al., 2011. Removal and recycling of inherent inorganic nutrient species in mallee biomass and derived biochars by water leaching. Ind. Eng. Chem. Res. 50, 12143–12151.

Xu, G., Wei, L., Sun, J., Shao, H., Chang, S., 2013. What is more important for enhancing nutrient bioavailability with biochar application into a sandy soil: direct or indirect mechanism? Ecol. Eng. 52, 119–124.

Xu, Z., Chan, K.Y., 2012. Biochar: nutrient properties and their enhancement. Biochar for Environmental Management. Routledge.

Yuan, H., Lu, T., Huang, H., Zhao, D., Kobayashi, N., Chen, Y., 2015. Influence of pyrolysis temperature on physical and chemical properties of biochar made from sewage sludge. J. Anal. Appl. Pyrol. 112, 284–289.

Zheng, H., Wang, Z., Deng, X., Herbert, S., Xing, B., 2013a. Impacts of adding biochar on nitrogen retention and bioavailability in agricultural soil. Geoderma 206, 32–39.

Zheng, H., Wang, Z., Deng, X., Zhao, J., Luo, Y., Novak, J., et al., 2013b. Characteristics and nutrient values of biochars produced from giant reed at different temperatures. Bioresour. Technol. 130, 463–471.

CHAPTER

12

Biochar Is a Potential Source of Silicon Fertilizer: An Overview

Muhammad Rizwan[1], *Muhammad Zia ur Rehman*[2], *Shafaqat Ali*[1], *Tahir Abbas*[1], *Arosha Maqbool*[1] *and Arooj Bashir*[1]

[1]Department of Environmental Sciences and Engineering, Government College University, Faisalabad, Pakistan [2]Institute of Soil and Environmental Sciences, University of Agriculture, Faisalabad, Pakistan

12.1 INTRODUCTION

Silicon (Si), however, has not been considered as an essential nutrient for plants, but has been considered as a beneficial element for many plants (Epstein, 1994). The use of Si to reduce the toxic effects of heavy metals has increased in various parts of the world and there is an increasing interest in the application of Si for many crops. Numerous beneficial impacts of Si have been shown in crops (Epstein, 1999). Silicon significantly reduced the toxic effect of Cu in wheat seedlings by reducing Cu absorption from solution (Nowakowski and Nowakowska, 1997). The use of Si diminished the toxic impacts of Cd in some plants species (Feng et al., 2010; Rizwan et al., 2012). Man ignored the exogenous use of Si, considering that the soil itself can maintain Si supply; but, most recently, Guntzer et al. (2012) showed that continuous cropping and removal of wheat straw reduced the bioavailable Si from the soil. On the other hand, Si that occurs in soil is mostly in unavailable form for plants and it has to be depolymerized for plant availability.

The question of Si bioavailability in soil is still not well understood. The main form of Si absorbed by plants is silicic acid (H_4SiO_4) (Ding et al., 2005), but most sources of silica are not available for plants (Savant et al., 1997). Exogenous Si sources are considered as silicates of calcium and magnesium, silicate slag, dolomite, as well as rock phosphate, but these contain only a small portion of available silicon (Rizwan et al., 2012). An important consideration is that these silicate sources that are derived from industrial by-products

Biochar from Biomass and Waste
DOI: https://doi.org/10.1016/B978-0-12-811729-3.00012-1

often contain high level of heavy metals (Berthelsen et al., 2001) that are highly toxic to plants. Based upon the above consideration, we need a pure source of silicon with high solubility. Both under a tropical and a temperate forest, phytoliths have been considered as the first Si pool for plants (Alexandré et al., 1997; Ding et al., 2008). Indeed, Fraysse et al. (2006, 2009) demonstrated that phytoliths dissolve faster than other silicates at pH greater than 4. Therefore, in natural ecosystems, soil phytoliths can be a major source of bioavailable Si for plants (Farmer et al., 2005). However, there is no source of easily accessible phytoliths that could be used as fertilizer and little is known regarding the ability of different minerals to release plant available Si.

Biochar is used to enhance soil quality, carbon sequestration, to reduce the toxicity of heavy metal in plants, as well as a source of plant nutrients (Bian et al., 2014; Al-Wabel et al., 2015; Puga et al., 2015; Rizwan et al., 2016; Liu et al., 2017). Recently, studies also reported that biochar also increases plant-available Si in the soil (Abbas et al., 2017). Biochar prepared from agricultural residues, such as rice and wheat straws, sugarcane residues, and their husks may contain a large amount of available Si as these crops are Si-accumulators (Abbas et al., 2017; Wang et al., 2018a). However, the release of Si from the biochars depends upon several factors, such as preparation temperature, feedstock, and soil types (Abbas et al., 2017; Wang et al., 2018b). In the present chapter, the possible sources of Si, its effects on crops, and possibility of biochar as a potential plant available Si have been discussed.

12.2 SILICON

Silicon is considered the 2nd most abundant element on the Earth's crust as well as in soils. Si comprises approximately 28% of the Earth's crust (Ma and Yamaji, 2006; Sommer et al., 2006). Although the contents of Si in plants are comparable to the levels of many macronutrients, Si is not considered an essential nutrient for the growth of plant (Epstein, 1994, 1999). However, undeniable beneficial impacts of this element have been observed in crop plants with respect to plant growth and better resistance to both biotic and abiotic stresses (Ma and Yamaji, 2006). In agricultural soils, Si generally comes due to application of silicate fertilizers, irrigation water, and decomposition of crop residues (Fig. 12.1). Generally, $210-224 \times 10^6$ tons of plant available Si are removed annually (Matichenkov and Bocharnikova, 2001). This removal of Si takes place due to plant exportation, as shown in Fig. 12.1.

Silicon concentration in crop plants ranges from 0.1% to 10% on dry biomass bases (Epstein, 1994, 1999; Hodson et al., 2005; Ma and Yamaji, 2008). Plants with higher than 1% Si in the biomass of dry leaf are considered as accumulators of Si (Epstein, 1994). Generally, graminaceous plants take up higher amount of Si as compared to other plant species and few dicotyledons, including legumes, are considered Si excluders (Ma et al., 2001). It has been widely accepted that wheat and rice are Si accumulating plants (Mayland et al., 1991; Rafi and Epstein, 1999). However, Si contents in plants varies considerably, depending upon the variations among species and genotypes (Epstein, 1994; Ma and Yamaji, 2008). Similarly, a species or cultivar grown under different conditions will

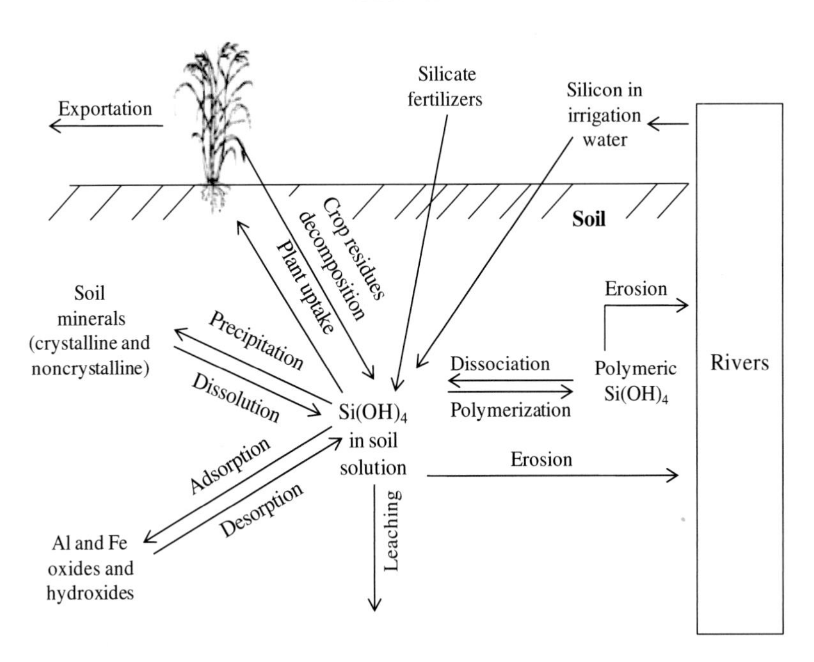

FIGURE 12.1 Biogeochemical Si cycle and processes influencing the Si concentration and solubility in the soil. Where $Si(OH)_4$ indicates a plant available silicon. Source: *Modified from Savant, N.K., Datnoff, L.E., Snyder, G.H., 1997. Depletion of plant available silicon in soils: a possible cause of declining rice yields. Commun. Soil. Sci. Plant. Anal. 28, 1245–1252; Meunier, J.D., Alexandre, A., Colin, F., Braun, J.J., 2001. Intérêt de l'étude du cycle biogéochimique du silicium pour interpréter la dynamique des sols tropicaux. Bull. De La Soc. Geol. De France, 172, 533–538.*

absorb different amounts of Si and different parts of plants generally have varying Si concentrations (Ma and Takahashi, 2002).

12.2.1 Forms of Silicon in Soil

Silicon forms in soil are generally present as silicon dioxide (SiO_2) and in many alumino-silicate forms. The SiO_2 consists about 50%–70% of the soil mass (Ma and Yamaji, 2006). The forms of Si in the soil exist in both liquid and in solid phases (Fig. 12.2). Amorphous and crystalline forms mainly exist in solid phase of Si. Uncharged orthosilicic acid, H_4SiO_4 is mainly present in natural soil solutions (Epstein, 1994; Sommer et al., 2006). The soil solution Si concentration varied from 0.1 to 0.6 mM, which may be the only form available for plants (Epstein, 1994, 1999; Ding et al., 2005). The maximum solubility of $Si(OH)_4$ is 1.7 mM at 25°C at $pH < 9$ (Knight and Kinrade, 2001). The chemical similarity between silicate and phosphate anion may result in a competitive reaction among the various phosphates and monosilicic acid in the soil.

12.2.2 Bioavailable Si in Soil

The question of Si bioavailability in soil is still not well understood. Most sources of silica are not bioavailable (Savant et al., 1997). Solubility of Si is affected by a number of dynamic processes occurring in the soil and soil solution (Fig. 12.1). Usually agronomists

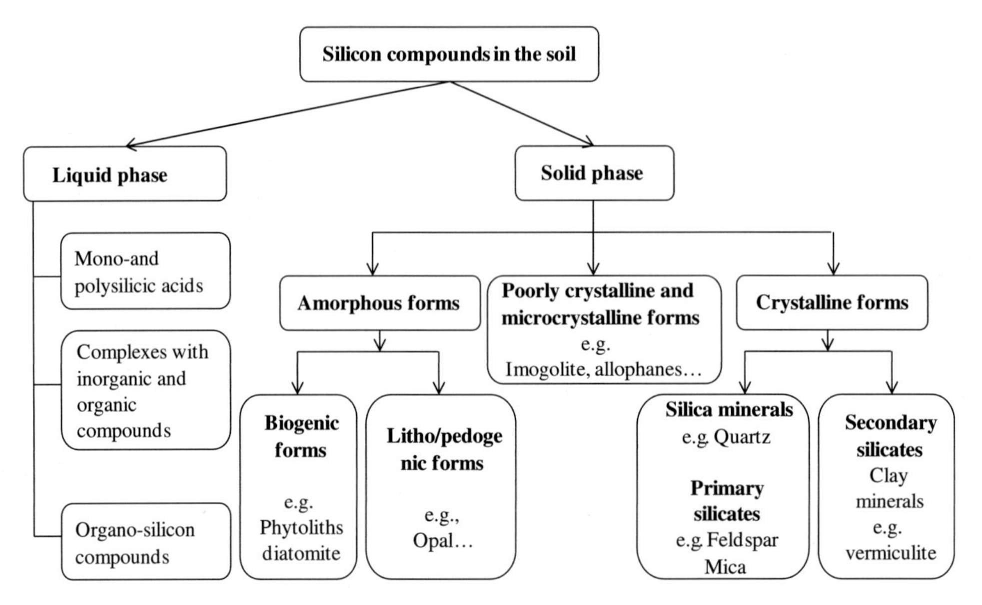

FIGURE 12.2 Classification of Si compounds in soil. Source: *Modified from Matichenkov, V.V., Bocharnikova, E. A., 2001. The relationship between silicon and soil physical and chemical properties. In: Datnoff, L.E., Snyder, G.H., Korndörfer, G.H. (Eds.), Silicon in Agriculture. Studies in Plant Science. Elsevier, pp. 209–219; Cornelis, J.T., Titeux, H., Ranger, J., Delvaux, B., 2011. Identification and distribution of the readily soluble silicon pool in a temperate forest soil below three distinct tree species. Plant Soil 342, 369–378.*

define bioavailable silica by the easily-leached silica extracted with an acetate buffered at pH = 4 (Ma and Takahashi, 2002). However, this procedure and others defined to assess the nutrient need of crop species do not dissolve amorphous biogenic silica (Savant et al., 1997), while it has been shown that phytoliths are a pool of Si that should not be neglected, and the amorphous silicon is the important pool of bioavailable silicon. Amorphous silica consists of both ASi from plants and other inorganic ASi forms present in the soil (Sauer et al., 2006). Both under a tropical and a temperate forest, phytoliths have been considered the first pool of plant available Si (Alexandré et al., 1997). The phytoliths are mainly comprised of about 92 wt% Si and 6 wt% water with trace amounts of carbon, Al, and Fe (Meunier et al., 1999). Isotopic studies on a bamboo forest also showed that Si in the soil solution comes mainly from phytoliths dissolution (Ding et al., 2008). Indeed, Fraysse et al. (2006, 2009) have shown that phytoliths dissolve faster than other silicates at pH > 4 and solubility of phytoliths is about 17 times higher than that of quartz. Therefore, in natural ecosystems, soil phytoliths can be a major source of bioavailable Si for plants (Farmer et al., 2005).

Man ignored exogenous use of Si with the belief that the soil itself can maintain Si supply to plants, but most recently Guntzer et al. (2012) showed that continuous cropping and removal of wheat straw reduced the bioavailable Si from the soil. Therefore, there is need of external plant available Si amendments to maintain higher level of bioavailable Si in soils. The practice of Si amendment on rice and sugarcane is already well developed in different parts of the world (Korndörfer and Lepsch, 2001). However, due to low solubility of Si, there is a practical limit to increase silicic acid bioavailability in the field. Therefore, there is need to use non-toxic sources of Si with higher capacity to release bioavailable Si.

12.2.3 Effect of Si on Plants

All plants could absorb Si from the growth medium; its biochemical and physiological role in plants is clearly defined, especially under abiotic and biotic stresses (Guntzer et al., 2012; Adrees et al., 2015). Silicon has been shown to enhance disease resistance, water-use-efficiency and photosynthesis, and remediating nutrient imbalances in plants (Liang et al., 2007). Direct and indirect positive impacts of Si on growth and development have been seen in many plant species (Liang et al., 2007). Plant roots uptake Si and transport it to other plant parts via transpiration streams. Its accumulation results in formation of a hurdle for pathogens' attacks by expanding polymerization in the intercellular spaces and beneath the cuticles. Furthermore, these completely dissolvable Si generate different metabolic pathways. According to some analysis though Si supply, natural protective shield toward insects attack; however, researchers, policymakers and farmers did not consider application of Si for insect control. In plants, Si withstands lodging and enhances power of plants and keeps them turgid. According to the scientists, the firmness of the cell wall in plants developed because of the connection of Si to the cell wall (Bakhat et al., 2018).

By application of Si, clear impacts on grain and straw were seen, but roots did not show any special impact. Si applications enhanced the iron plaque in various genotypes but reduced the Sb concentration in grain, husk, straw, and roots. Also, it lessens the inorganic Sb while amplified the trimethylantimony to some extent (Zhang et al., 2017). Total and soluble Cd in rice can be lessened by applying treatments of Si. Intensity and distribution of the Si–Cd deposition depend on the characteristics of Si-rich amendments. The transfer of Cd into stems and leaves and the vigor of Cd in the soil mainly depends on the concentration of monosilicic acid in the paddy soil rice system. Si-agro had superior effect on aboveground biomass of rice and hence reduces the leaf Cd to a higher extent (Ji et al., 2017). Si in Cd stressed plant visibly enhanced the growth and total protein and stability of membrane of disturbed plant and it signifies that Si have crucial roles in Cd detoxification in plants. Moreover, in peas, the Fe transporter (RIT1) caused the regulation in shoots by applying Si with Cd treated conditions. By observing the physiological observations, the improvement of Cd toxicity in peas can be linked with Cd sequestration in roots and decreased Cd transference in shoots via ruling of Fe. Furthermore, increased antioxidant activity along with elevated S-metabolites (cysteine, methionine, glutathione) implies the active involvement of ROS scavenging and play an important role in Si-mediated reduction of Cd toxicity in plants (Rahman et al., 2017).

12.3 BIOCHAR

Biochar is a gorgeous and latest ameliorating technique for degraded and contaminated agricultural soils (Rizwan et al., 2016). The preparation of biochar from feedstock natural organic waste matter, such as manure, crop residues, or wood chips is pyrolysed under synthetic environmental conditions. The biochar is produced at limited oxygen and low temperature, a blackish material prepared by solid organic feedstock under limited oxygen supply at different temperature. Its characteristics and physiochemical properties depend on the combustible biomass and pyrolysis temperature, and these properties, such as broader surface area, higher pH, carbon and nitrogen ratio (Kim et al., 2015; Puga et al., 2015). Biochar is the latest soil remediation agent, practicing in agriculture sectors for improving soils fertility, enhancing crops productivity, and making sure environment sustainability through C sequestration (Abiven et al., 2015), where, C sequestered in soil via biochar remains for 100–1000 years (Lehmann et al., 2006).

In the past, numbers of conventional methods were adopted for soil protection and remediation such as excavating, washing of soil, and land filling, but all these techniques are known as unsuitable and not economically practicable (Houben et al., 2013). The most feasible alternative techniques include biofertilizers, animal dung, compost, and biochar have been practiced for protection, reduction, and immobilization of Cd in contaminated soils. Among these alternative soil amendments methods, the using of biochar is the latest and innovative idea for soil metal remediation, metal immobilization, best sorbent of organic and inorganic pollutants, carbon sequestration, and climate change mitigation, due to its unique structural properties, such as higher surface area (Ok et al., 2015). Biochar has two main mechanisms for promoting plant growth, such as direct and indirect actions. Direct action mechanism of biochar, when added in soil environment, directly releases minerals and nutrients for plant growth promotion, like Ca^{+}, Mg, and NPK, while indirect action mechanisms of biochar stabilize physiological, chemical, and biological properties of soils (Sohi et al., 2010). Biochar addition in degraded, contaminated, and nutrient deficient soils can play an excellent role rather than in well-structured fertile soils (Abiven et al., 2015). Biochar acts as an efficient adsorbent for both organic and inorganic nature of contaminants in polluted soils (Uchimiya and Bannon, 2013; Rizwan et al., 2016).

12.3.1 Sources of Feedstock for Biochar

Biochar is prepared from organic waste materials under anaerobic conditions at various pyrolysis temperatures (Sohi et al., 2010). Agricultural, forestry, domestic, and animal wastes are converted to stable solid carbon (biochar); in this way carbon can be sequestered into soils from environment for mitigating global warming (Liu et al., 2011; Dong et al., 2013). A highly porous black carbon (biochar) obtained from organic waste materials is also an efficient carbon confiscation technique (Spokas et al., 2012) for agriculture and environmental management. The charcoal resemblance material could be used as an efficient amendment for reclamation of contaminated and degraded soils (Sohi et al., 2010). Pyrolysis temperature is primarily responsible for characterization of the stable carbonaceous material, such as higher volatile biochar derived at temperatures less than 500°C that is an efficient supporter of plant growth, and while with broader surface area higher, adsorption capacity was derived at temperatures greater than 500°C (Yi et al., 2015). Biochar production at low pyrolysis temperature ($<$700°C) from various feedstocks could be potentially used for heavy metals (HMs) immobilization, carbon sequestration, and soil fertilization (Cao and Harris, 2010). The production of biochar at higher pyrolysis temperatures (at $\geq$ 700°C) is more efficient to limit leachability and phytoavailability than prepared at lower pyrolysis temperatures (at $\leq$ 300°C) (Ahmad et al., 2016). Where the biochar prepared at 450°C could more efficient to limit the Cd uptake in plants rather than at 400, 500, and 550°C temperatures from the same feedstock, therefore, the bioavailability of metals to plants, depends on biochar characteristics (Al-Wabel et al., 2015; Puga et al., 2015).

12.3.2 Characterization of Biochar

Physiochemical properties and characteristics of biochar played important roles in immobilization of Cd in agricultural soils. The biochar properties were highly dependent on feedstock materials and pyrolysis temperatures (Wu et al., 2016). Structural properties and chemical characteristics of biochar were highly affected by pyrolysis conditions, such as production of biochar with various sorption capacities for organic and inorganic pollutants, surface area, and acidic groups, cation exchange capacity (CEC), total organic carbon (TOC), and other chemical parameters (Ding et al., 2016; Rajapaksha et al., 2016). Biochar can enhance the microbial activities and surveillance, such as mycorrhizae fungi in agricultural soil, which can develop soil conditions, because microorganisms could be efficiently reproduced, colonized, and raised in the porous structure of biochar (Ahmad et al., 2016). Physical properties of soil could be affected by the plant characteristics, such as growth, development, respiration rate, nutrients, and water uptake. These physiological properties of biochar are highly dependent on the organic waste material feedstock and other pyrolysis parameters, including kiln temperature and anaerobic conditions (Li et al., 2016; Shen et al., 2016). Biochar could be produced with different characteristics, but it would be recognized by large surface area, micro porous structure, activated functional groups, and higher pH, etc. (Wu et al., 2012; Rizwan et al., 2016). Commonly, the bioavailability of Cd depends on the primary factors of soil pH and redox potential (Uchimiya et al., 2010). Hence, mostly biochar materials have higher affinity for Cd adsorption (Yuan et al., 2011; Inyang et al., 2016).

12.3.3 Benefits of Biochar in Agricultural Practices

Many benefits of biochar practice for soil remediation have been reported, such as increased carbon sequestration trends, soil fertility, biomass production, crop yield, and minimize emission of greenhouse gases (Mukherjee and Lal, 2013). Biochar, also renowned as biomass-derived black carbon, is becoming a valuable amendment for sustainable agriculture (Barrow, 2012). It is a recalcitrant carbonaceous material derived from bio-waste matter combusted under anaerobic conditions, which is an economical and eco-friendly method of carbon sequestration (Lorenz and Lal, 2014; Jeffery et al., 2015). Soil physiological properties have been affected by using biochar with various characteristics and application rates for soil amendments (Herath et al., 2013). Biochar can improve physicochemical and biological properties of cultivated soils. Three major role of biochar amendments are: first, management of physiological properties of soil, including bulk density, soil moisture contents, porosity, water holding capacity (WHC) and aggregated soil stability; second, improvements in chemical parameters of agro soils like pH, EC, and CEC; and, third, stabilizing biotic properties involving enzyme promotion, microbial population, and biomass in rhizosphere (Sohi et al., 2010). Biochar amendment can enhance the production of biomass and plant yield (Jeffery et al., 2011). Biochar is an efficient carbon sequestration and GHG reduction strategy; it may remain in soils for a long time (even thousands of years) due to its higher recalcitrant contents of carbon, which improve soil fertility (Sohi et al., 2010).

Biochar is a novel technology for agriculture productivity, which is commonly known by black gold and recalcitrant nature due to high degree of aromaticity (aromatic or aryl functional group) (Ok et al., 2015; Rizwan et al., 2016). Biochar is also a unique soil amendment that bound and immobilized the Cd contents in the soil, promotes plants growth and reduces the environmental issues (Younis et al., 2016). The higher porous structure and functional groups of biochar could play a main role in Cd adsorption and reduce the risk of food chain contamination (Abbas et al., 2017). The addition of biochar can potentially add the total organic carbon contents in soils: improve soil fertility and agricultural productivity (Rizwan et al., 2016). The most prominent benefit of biochar is longevity; it can remain in the soil for the years and reduce the demand of repeated addition of soil amendments and minimize the possibility of new contaminants from intensive use of synthetic soil amendments, like phosphate fertilizers. Biochar is a unique weapon to combat against climate change and global warming via sequestration of atmospheric CO_2 (Rizwan et al., 2016). The efficient way of reduction the mobility and bioavailability of metals such as Cd, Pb, Cu, and Ni in contaminated cultivable sites through using biochar prepared from different organic waste stock (Kim et al., 2015; Puga et al., 2015).

According to chemical extractions and extended X-ray absorption fine structure spectroscopy analysis, the metal concentration decreased in the gains of plants by addition of biochar for soils remediation was found to be in a more stabilized form (Rizwan et al., 2016): soil aggregate formations, stabilization, enhanced organic matter, and increased microbial surveillance by adding organic amendments. Biochar amendment in soil improved microbial respiration and habitation and decreased bulk density of soil with poor organic content (Herath et al., 2013). Biochar has been recommended as a paramount remediation of organic and inorganic soil pollutants (Beesley and Marmiroli, 2011; Ahmad et al., 2016). Such potential toxic effects of Cd to humans health could be minimized by using carbon-rich material (biochar) for amendment of agricultural soils (Komkiene and Baltrenaite, 2016). The bioavailability and mobility of metal ions has been reduced by the application of biochar for soil remediation (Rajapaksha et al., 2016). It has been reported that the phytotoxicity of metal in maize plants was reduced and the biomass and yield was increased by applying of biochar in Cu and As contaminated soil (Brennan et al., 2014). The biochar application increased wheat biomass and yield under control conditions as well as in Cd-contamination (Abbas et al., 2017).

Biochar is considered a direct source of mineral nutrients because it is produced in enclosed combustion of nutrient rich biomass. It can also improve soil nutrients' bioavailability through incorporating with soil pH, cations exchange, redox potential, and soils biotic commotion (Rizwan et al., 2016). The recent study revealed that the application of wheat straw biochar, along with fertilizer, increased yields of wheat grain about 20%–30% as compared with fertilizer alone (Alburquerque et al., 2013). The enclosed burning of abundantly available crop residual straw would be the best solution of GHG mitigation and carbon sequestration by converting into biochar. Biochar has duel beneficial properties for soil and as well as environmental health. The addition of rice straw biochar can enhance crop production and reduce Cd availability (Abbas et al., 2018). Long term application of rice straw biochar in a wheat rice cropping system could obtain long term benefits, such as field sustainability, atmospheric carbon reduction, and decrease in the rate of fertilizer applications (Rizwan et al., 2016). Therefore, the appropriate use of biochar could professionally curtail environmental issues.

12.4 BIOCHAR IS A POTENTIAL SOURCE OF BIOAVAILABLE SI

Biochar is gradually documented as environmentally friendly and one of the cheapest remediation agents for contaminated soil. On the other hand, Si is considered a potential beneficial element for crops, especially under abiotic stress. However, potential bioavailable sources of Si are lacking. Plant available Si rich amendments may be used as a source of Si (Abbas et al., 2017; Teasley et al., 2017). Plant-derived biochar could be a potential source of plant-available Si (Xu and Chen, 2015; Abbas et al., 2017). In literature, the parts of Si in biochar and connections between Si and carbon have been ignored, as well as morphology, transformation, and dissolution of Si and their relation with carbon have been poorly addressed. The form of Si in biochar depends upon the producing temperature. At low producing temperature (≤300°C), Si in biochar mostly exists in amorphous form, whereas high producing temperature converts Si from amorphous to crystalline form (Guo and Chen, 2014; Jindo et al., 2014). Xiao et al. (2014) carried out a study to understand the morphology, transformation of carbon, and Si in rice straw biochar prepared at a range of pyrolytic temperature. Si-rich biochar obtained from rice straw was prepared under 150 − 700°C. Particles in biochar were monitored through Fourier-transform infrared spectroscopy, X-rat diffract meter, and scanning electron microscopy. By increasing pyrolytic temperature, Si stored and its speciation transformed from amorphous to crystalline substance, although the organic matter changed from aliphatic to aromatic. Amorphous carbon and amorphous Si are found in rice straw biochar. At medium pyrolysis temperatures (250 − 350°C), a strong cracking of carbon components occurred, and, thus, the Si located in the inside tissue was exposed. The biochar became condensed at high pyrolysis temperatures (500 − 700°C), due to the aromatization of carbon and crystallization of Si. The Si dissolution kinetics suggests that high Si biochar could serve as a novel slow release source of biologically available Si in low Si agricultural soils (Xiao et al., 2014).

Soil type, feedstock pretreatment, as well as source of feedstock may enhance the release of Si from biochar. For example, pretreatment of biochar with KOH enhanced the plant available Si as compared to the respective untreated biochars. Furthermore, biochar from rice straw has higher plant available Si contents than other feedstocks studied, such as miscanthus, sugarcane harvest residues, and switchgrass (Wang et al., 2018a). It has been reported that biochar obtained from Si-rich feedstock (rice husk and straw) increased Si dissolution in soil whereas the response was higher in high Si soil than low Si soil. On the other hand, biochar obtained from Si-deficient feedstock (wood saw dust and orange peel) decreased the Si dissolution than control. The authors concluded that Si-rich biochar may be a source of Si whereas Si-deficient biochar may serve as a sink of Si in soil (Wang et al., 2018b). Houben et al. (2014) demonstrated that biochar produced from Miscanthus

(600°C for 30 min) increased the kinetic release of plant available Si than the biochar produced from other feedstock.

Studies also demonstrated that biochar application enhanced the Si concentration in plants. For example, it was reported that rice husk biochar increased the Si concentrations in shoots, leaves, and roots of alfalfa in a dose-additive manner (Ibrahim et al., 2016). Furthermore, rice straw biochar increased the Si concentration in wheat shoots more than the control (Abbas et al., 2017). The application of rice straw biochar enhanced the soluble Si in the soil, soil available Si, and Si concentrations in wheat root tips (Qian et al., 2016). Rice husk char application increased Si concentrations in rice compared to the control (Limmer et al., 2018). Field application of wheat straw biochar increased the Si contents in rice shoots more than the control (Liu et al., 2014). Overall, biochar derived from wheat and rice straw, as well as sugarcane residues, has more importance because it may be a cost effective and environmentally friendly source of Si.

12.5 CONCLUSION AND PERSPECTIVES

Biochar produced from Si-accumulating crops, such as wheat, rice, and sugarcane residues might be a good source of plant-available Si. The release of bioavailable Si depends upon the producing temperature of biochar. The use of biochar prepared from Si-rich feedstocks could be a practical strategy to alleviate Si deficiency in soils and ultimately in crops. Overall, Si-rich biochar would be good source of organic fertilizer for bioavailable Si in the soil. However, Si in biochar and the connections between Si and carbon have been ignored. Further studies are worth investigating regarding the morphology, transformation, and dissolution about Si and their relation with carbon. More studies are required to understand the relation between Si and carbon in biochar.

Acknowledgments

Financial support from Government College University, Faisalabad, is greatly acknowledged.

References

Abbas, T., Rizwan, M., Ali, S., Rehman, M.Z., Qayyum, M.F., Abbas, F., et al., 2017. Effect of biochar on cadmium bioavailability and uptake in wheat (*Triticum aestivum* L.) grown in a soil with aged contamination. Ecotoxicol. Environ. Saf. 140, 37–47.

Abbas, T., Rizwan, M., Ali, S., Adrees, M., Mahmood, A., Rehman, M.Z., et al., 2018. Biochar application increased the growth and yield and reduced cadmium in drought stressed wheat grown in an aged contaminated soil. Ecotoxicol. Environ. Saf. 148, 825–833.

Abiven, S., Hund, A., Martinsen, V., Cornelissen, G., 2015. Biochar amendment increases maize root surface areas and branching: a shovelomics study in Zambia. Plant Soil 395, 45–55.

Adrees, M., Ali, S., Rizwan, M., Rehman, M.Z., Ibrahim, M., Abbas, F., et al., 2015. Mechanisms of silicon-mediated alleviation of heavy metal toxicity in plants: a review. Ecotoxicol. Environ. Saf. 119, 186–197.

Ahmad, M., Ok, Y.S., Rajapaksha, A.U., Lim, J.E., Kim, B.Y., Ahn, J.H., et al., 2016. Lead and copper immobilization in a shooting range soil using soybean stover-and pine needle-derived biochars: chemical, microbial and spectroscopic assessments. J. Hazard. Mater. 301, 179–186.

Alburquerque, J.A., Salazar, P., Barrón, V., Torrent, J., del Campillo, M.D.C., Gallardo, A., et al., 2013. Enhanced wheat yield by biochar addition under different mineral fertilization levels. Agron. Sustain. Dev. 33, 475–484.

Alexandré, A., Meunier, J.D., Lezine, A.M., Vincens, A., Schwartz, D., 1997. Phytoliths: indicators of grassland dynamics during the late Holocene in intertropical Africa. Palaeogeogr. Palaeoclimatol. Palaeoecol. 136, 213–229.

Al-Wabel, M., Usman, A.R.A., El-Naggar, A.H., Aly, A.A., Ibrahim, H.M., Elmaghraby, S., et al., 2015. Conocarpus biochar as a soil amendment for reducing heavy metals availability and uptake by maize plants. Saudi J. Biol. Sci. 22, 503–511.

Bakhat, H.F., Bibi, N., Zia, Z., Abbas, S., Hammad, H.M., Fahad, S., et al., 2018. Silicon mitigates biotic stresses in crop plants: a review. Crop. Prot. 104, 21–34.

Barrow, C.J., 2012. Biochar: potential for countering land degradation and for improving agriculture. Appl. Geogr. 34, 21–28.

Beesley, L., Marmiroli, M., 2011. The immobilisation and retention of soluble arsenic, cadmium and zinc by biochar. Environ. Pollut. 159, 474–480.

Berthelsen, S., Noble, A.D., Garside, A.L., 2001. Silicon research down under: past, present and future. In: Datnoff, L.E., Snyder, G.H., Korndorfer, G.H. (Eds.), Silicon Deposition in Higher Plants. Silicon in Agriculture. Elsevier Science, pp. 241–255.

Bian, R., Joseph, S., Cui, L., Pan, G., Li, L., Liu, X., et al., 2014. A three-year experiment confirms continuous immobilization of cadmium and lead in contaminated paddy field with biochar amendment. J. Hazard. Mater. 272, 121–128.

Brennan, A., Jiménez, E.M., Puschenreiter, M., Alburquerque, J.A., Switzer, C., 2014. Effects of biochar amendment on root traits and contaminant availability of maize plants in a copper and arsenic impacted soil. Plant Soil 379, 351–360.

Cao, X., Harris, W., 2010. Properties of dairy-manure-derived biochar pertinent to its potential use in remediation. Biores. Technol. 101, 5222–5228.

Cornelis, J.T., Titeux, H., Ranger, J., Delvaux, B., 2011. Identification and distribution of the readily soluble silicon pool in a temperate forest soil below three distinct tree species. Plant Soil 342, 369–378.

Ding, T.P., Ma, G.R., Shui, M.X., Wan, D.F., Li, R.H., 2005. Silicon isotope study on rice plants from the Zhejiang province, China. Chem. Geol. 218, 41–50.

Ding, T.P., Zhou, J.X., Wan, D.F., Chen, Z.Y., Zhang, F., 2008. Silicon isoope fractionation in bamboo and its significance to the biogeochemical cycle of cilicon. Geochim. Cosmochim. Acta 72, 1381–1395.

Ding, Z., Wan, Y., Hu, X., Wang, S., Zimmerman, A.R., Gao, B., 2016. Sorption of lead and methylene blue onto hickory biochars from different pyrolysis temperatures: importance of physicochemical properties. J. Ind. Eng. Chem. 37, 261–267.

Dong, D., Yang, M., Wang, C., Wang, H., Li, Y., Luo, J., et al., 2013. Responses of methane emissions and rice yield to applications of biochar and straw in a paddy field. J. Soils Sedim. 13, 1450–1460.

Epstein, E., 1994. The anomaly of silicon in plant biology. Proc. Natl. Acad. Sci. U.S.A. 91, 11–17.

Epstein, E., 1999. Silicon. Annu. Rev. Plant Physiol. 50, 641–664.

Farmer, V.C., Delbos, E., Miller, J.D., 2005. The role of phytolith formation and dissolution in controlling concentrations of silica in soil solutions and streams. Geoderma 127, 71–79.

Feng, J., Shi, Q., Wang, X., Wei, M., Yang, F., Xu, H., 2010. Silicon supplementation ameliorated the inhibition of photosynthesis and nitrate metabolism by cadmium (Cd) toxicity in *Cucumis sativus* L. Sci. Hortic. 123, 521–530.

Fraysse, F., Pokrovsky, O.S., Schott, J., Meunier, J.D., 2006. Surface properties, solubility and dissolution kinetics of bamboo phytoliths. Geochim. Cosmochim. Acta 70, 1939–1951.

Fraysse, F., Pokrovsky, O.S., Schott, J., Meunier, J.D., 2009. Surface chemistry and reactivity of plant phytoliths in aqueous solutions. Chem. Geol. 258, 197–206.

Guntzer, F., Keller, C., Poulton, P.R., McGrath, S.P., Meunier, J.D., 2012. Long term removal of wheat straw decreases soil amorphous silica at Broadbalk, Rothamsted. Plant Soil 352, 173–184.

Guo, J., Chen, B., 2014. Insights on the molecular mechanism for the recalcitrance of biochars: interactive effects of carbon and silicon components. Environ. Sci. Technol. 48, 9103–9112.

Herath, H.M.S.K., Camps-Arbestain, M., Hedley, M., 2013. Effect of biochar on soil physical properties in two contrasting soils: an Alfisol and an Andisol. Geoderma. 209, 188–197.

Hodson, M.J., White, P.J., Mead, A., Broadley, M.R., 2005. Phylogenetic variation in the silicon composition of plants. Ann. Bot. 96, 1027–1046.

Houben, D., Evrard, L., Sonnet, P., 2013. Beneficial effects of biochar application to contaminated soils on the bioavailability of Cd, Pb and Zn and the biomass production of rapeseed (*Brassica napus* L.). Biomass Bioenergy 57, 196–204.

Houben, D., Sonnet, P., Cornelis, J.T., 2014. Biochar from Miscanthus: a potential silicon fertilizer. Plant Soil 374, 871–882.

Ibrahim, M., Khan, S., Hao, X., Li, G., 2016. Biochar effects on metal bioaccumulation and arsenic speciation in alfalfa (*Medicago sativa* L.) grown in contaminated soil. Int. J. Environ. Sci. Technol. 13, 2467–2474.

Inyang, M.I., Gao, B., Yao, Y., Xue, Y., Zimmerman, A., Mosa, A., et al., 2016. A review of biochar as a low-cost adsorbent for aqueous heavy metal removal. Crit. Rev. Environ. Sci. Technol. 46, 406–433.

Jeffery, S., Verheijen, F.G., Van Der Velde, M., Bastos, A.C., 2011. A quantitative review of the effects of biochar application to soils on crop productivity using meta-analysis. Agri. Ecosyst. Environ 144, 175–187.

Jeffery, S., Bezemer, T.M., Cornelissen, G., Kuyper, T.W., Lehmann, J., Mommer, L., et al., 2015. The way forward in biochar research: targeting trade-offs between the potential wins. GCB Bioenergy 7, 1–13.

Ji, X., Liu, S., Juan, H., Bocharnikova, E.A., Matichenkov, V.V., 2017. Effect of silicon fertilizers on cadmium in rice (*Oryza sativa*) tissue at tillering stage. Environ. Sci. Pollut. Res. 24, 10740–10748.

Jindo, K., Mizumoto, H., Sawada, Y., Sanchez-Monedero, M.A., Sonoki, T., 2014. Physical and chemical characterization of biochars derived from different agricultural residues. Biogeoscience 11, 6613–6621.

Kim, H.S., Kim, K.R., Kim, H.J., Yoon, J.H., Yang, J.E., Ok, Y.S., et al., 2015. Effect of biochar on heavy metal immobilization and uptake by lettuce (*Lactuca sativa* L.) in agricultural soil. Environ. Earth Sci. 74, 1249–1259.

Knight, C.T.G., Kinrade, S.D., 2001. A primer on the aqueous chemistry of silicon. In: Datnoff, L.E., Snyder, G.H., Korndörfer, G.H. (Eds.), Silicon in Agriculture. Elsevier, Amsterdam, pp. 57–84.

Komkiene, J., Baltrenaite, E., 2016. Biochar as adsorbent for removal of heavy metal ions [cadmium (II), copper (II), lead (II), zinc (II)] from aqueous phase. Int. J. Environ. Sci. Technol. 13, 471–482.

Korndörfer, G.H., Lepsch, I., 2001. Effect of silicon on plant growth and crop yield. In: Datnoff, L.E., Snyder, G.H., Korndörfer, G.H. (Eds.), Silicon in Agriculture. Studies in Plant Science. Elsevier, pp. 133–147.

Lehmann, J., Gaunt, M., Rondon, J., 2006. Bio-char sequestration in terrestrial ecosystems – a review. Mitig. Adapt. Strat. Global Change 11, 403–427.

Li, L.F., Ai, S.Y., Wang, Y.H., Tang, M.D., Li, Y.C., 2016. In situ field-scale remediation of low Cd-contaminated paddy soil using soil amendments. Water Air Soil Pollut. 227, 342.

Liang, Y., Sun, W., Zhu, Y.G., Christie, P., 2007. Mechanisms of silicon-mediated alleviation of abiotic stresses in higher plants: a review. Environ. Pollut. 147, 422–428.

Limmer, M.A., Mann, J., Amaral, D.C., Vargas, R., Seyfferth, A.L., 2018. Silicon-rich amendments in rice paddies: effects on arsenic uptake and biogeochemistry. Sci. Total Environ. 624, 1360–1368.

Liu, L., Tan, Z., Zhang, L., Huang, Q., 2017. Influence of pyrolysis conditions on nitrogen speciation in a biochar 'preparation-application'process. J. Energy Inst. . Available from: http://dx.doi.org/10.1016/j.joei.2017.09.004.

Liu, X., Li, L., Bian, R., Chen, D., Qu, J., Wanjiru Kibue, G., et al., 2014. Effect of biochar amendment on soil-silicon availability and rice uptake. J. Plant Nutr. Soil Sci. 177, 91–96.

Liu, Y., Yang, M., Wu, Y., Wang, H., Chen, Y., Wu, W., 2011. Reducing CH_4 and CO_2 emissions from waterlogged paddy soil with biochar. J. Soils Sedim. 11, 930–939.

Lorenz, K., Lal, R., 2014. Biochar application to soil for climate change mitigation by soil organic carbon sequestration. J. Plant. Nutr. Soil Sci. 177, 651–670.

Ma, J.F., Takahashi, E., 2002. Soil, Fertiliser, and Plant Silicon Research in Japan. Elsevier.

Ma, J.F., Yamaji, N., 2006. Silicon uptake and accumulation in higher plants. Trends Plant Sci. 11, 392–397.

Ma, J.F., Yamaji, N., 2008. Functions and transport of silicon in plants. Cell. Mol. Life Sci. 65, 3049–3057.

Ma, J.F., Goto, S., Tamai, K., Ichii, M., 2001. Role of root hairs and lateral roots in silicon uptake by rice. Plant Physiol. 127, 1773–1780.

Matichenkov, V.V., Bocharnikova, E.A., 2001. The relationship between silicon and soil physical and chemical properties. In: Datnoff, L.E., Snyder, G.H., Korndörfer, G.H. (Eds.), Silicon in Agriculture. Studies in Plant Science. Elsevier, pp. 209–219.

Mayland, H.F., Wright, J.L., Sojka, R.E., 1991. Silicon accumulation and water uptake by wheat. Plant Soil 137, 191–199.

Meunier, J.D., Colin, F., Alarcon, C., 1999. Biogenic silica storage in soils. Geology 27, 835–838.

Meunier, J.D., Alexandre, A., Colin, F., Braun, J.J., 2001. Intérêt de l'étude du cycle biogéochimique du silicium pour interpréter la dynamique des sols tropicaux. Bull. De La Soc. Geol. De France 172, 533–538.

Mukherjee, A., Lal, R., 2013. Biochar impacts on soil physical properties and greenhouse gas emissions. Agronomy 3, 313–339.

Nowakowski, W., Nowakowska, J., 1997. Silicon and copper interaction in the growth of spring wheat seedlings? Biol. Plant. 39, 463–466.

Ok, Y.S., Chang, S.X., Gao, B., Chung, H.J., 2015. SMART biochar technology—a shifting paradigm towards advanced materials and healthcare research. Environ. Technol. Innov. 4, 206–209.

Puga, A.P., Abreu, C.A., Melo, L.C.A., Beesley, L., 2015. Biochar application to a contaminated soil reduces the availability and plant uptake of zinc, lead and cadmium. J. Environ. Manag. 159, 86–93.

Qian, L., Chen, B., Chen, M., 2016. Novel alleviation mechanisms of aluminum phytotoxicity via released biosilicon from rice straw-derived biochars. Sci. Rep. 6, 29346.

Rafi, M.M., Epstein, E., 1999. Silicon absorption by wheat (*Triticum aestivum* L.). Plant Soil 211, 223–230.

Rahman, M.F., Ghosal, A., Alam, M.F., Kabir, A.H., 2017. Remediation of cadmium toxicity in field peas (*Pisum sativum* L.) through exogenous silicon. Ecotoxicol. Environ. Saf. 135, 165–172.

Rajapaksha, A.U., Chen, S.S., Tsang, D.C.W., Zhang, M., Vithanage, M., Mandal, S., et al., 2016. Engineered/designer biochar for contaminant removal/immobilization from soil and water: potential and implication of biochar modification. Chemosphere 148, 6–291.

Rizwan, M., Meunier, J.D., Miche, H., Keller, C., 2012. Effect of silicon on reducing cadmium toxicity in durum wheat (*Triticum turgidum* L. cv. Claudio W.) grown in a soil with aged contamination. J. Hazard. Mater. 209–210, 326–334.

Rizwan, M., Ali, S., Adrees, M., Rizvi, H., Zia-ur-Rehman, M., Hannan, F., et al., 2016. Cadmium stress in rice: toxic effects, tolerance mechanisms, and management: a critical review. Environ. Sci. Pollut. Res. 23, 17859–17879.

Sauer, D., Saccone, L., Conley, D.J., Herrmann, L., Sommer, M., 2006. Review of methodologies for extracting plant-available and amorphous Si from soils and aquatic sediments. Biogeochemistry 80, 89–108.

Savant, N.K., Datnoff, L.E., Snyder, G.H., 1997. Depletion of plant available silicon in soils: a possible cause of declining rice yields. Commun. Soil Sci. Plant Anal. 28, 1245–1252.

Shen, X., Huang, D.Y., Ren, X.F., Zhu, H.H., Wang, S., Xu, C., et al., 2016. Phytoavailability of Cd and Pb in crop straw biochar-amended soil is related to the heavy metal content of both biochar and soil. J. Environ. Manage. 168, 245–251.

Sohi, S.P., Krull, E., Lopez-Capel, E., Bol, R., 2010. A review of biochar and its use and function in soil. Adv. Agron. 105, 47–82.

Sommer, M., Kaczorek, D., Kuzyakov, Y., Breuer, J., 2006. Silicon pools and fluxes in soils and landscapes: a review. J. Plant Nutr. Soil Sci. 169, 310–329.

Spokas, K.A., Cantrell, K.B., Novak, J.M., Archer, D.W., Ippolito, J.A., Collins, H.P., et al., 2012. Biochar: a synthesis of its agronomic impact beyond carbon sequestration. J. Environ. Qual. 41, 973–989.

Teasley, W.A., Limmer, M.A., Seyfferth, A.L., 2017. How rice (*Oryza sativa* L.) responds to elevated As under different Si-rich soil amendments. Environ. Sci. Technol. 51, 10335–10343.

Uchimiya, M., Bannon, D.I., 2013. Solubility of lead and copper in biochar-amended small arms range soils: influence of soil organic carbon and pH. J. Agric. Food Chem. 61, 7679–7688.

Uchimiya, M., Lima, I.M., Klasson, K.T., Wartelle, L.H., 2010. Contaminant immobilization and nutrient release by biochar soil amendment: roles of natural organic matter. Chemosphere 80, 935–940.

Wang, M., Wang, J.J., Wang, X., 2018a. Effect of KOH-enhanced biochar on increasing soil plant-available silicon. Geoderma 321, 22–31.

Wang, Y., Xiao, X., Chen, B., 2018b. Biochar impacts on soil silicon dissolution kinetics and their interaction mechanisms. Sci. Rep. 8 (1), 1–11.

Wu, M., Han, X., Zhong, T., Yuan, M., Wu, W., 2016. Soil organic carbon content affects the stability of biochar in paddy soil. Agric. Ecosyst. Environ. 223, 59–66.

Wu, W., Yang, M., Feng, Q., McGrouther, K., Wang, H., Lu, H., et al., 2012. Chemical characterization of rice straw-derived biochar for soil amendment. Biomass Bioenergy 47, 268–276.

Xiao, X., Chen, B., Zhu, L., 2014. Transformation, morphology, and dissolution of silicon and carbon in rice straw-derived biochars under different pyrolytic temperatures. Environ. Sci. Technol. 48, 3411–3419.

Xu, Y., Chen, B., 2015. Organic carbon and inorganic silicon speciation in rice-bran-derived biochars affect its capacity to adsorb cadmium in solution. J. Soils Sedim. 15, 60–70.

Yi, P., Pignatello, J.J., Uchimiya, M., White, J.C., 2015. Heteroaggregation of cerium oxide nanoparticles and nanoparticles of pyrolyzed biomass. Environ. Sci. Technol. 49, 13294–13303.

Younis, U., Malik, S.A., Rizwan, M., Qayyum, M.F., Ok, Y.S., Shah, M.H.R., et al., 2016. Biochar enhances the cadmium tolerance in spinach (*Spinacia oleracea*) through modification of Cd uptake and physiological and biochemical attributes. Environ. Sci. Pollut. Res. 23, 21385–21394.

Yuan, J.H., Xu, R.K., Zhang, H., 2011. The forms of alkalis in the biochar produced from crop residues at different temperatures. Biores. Technol. 102, 3488–3497.

Zhang, L., Yang, Q., Wang, S., Li, W., Jiang, S., Liu, Y., 2017. Influence of silicon treatment on antimony uptake and translocation in rice genotypes with different radial oxygen loss. Ecotoxicol. Environ. Saf. 144, 572–577.

CHAPTER

13

Sludge-Derived Biochar and Its Application in Soil Fixation

Weihua Zhang[1] *and Daniel C.W. Tsang*[2]

[1]School of Environmental Science and Engineering, Sun Yat-sen University, Guangzhou, P.R. China [2]Department of Civil and Environmental Engineering, The Hong Kong Polytechnic University, Hung Hom, Kowloon, Hong Kong, P.R. China

13.1 SEWAGE SLUDGE PRODUCTION AND DISPOSAL IN CHINA

With the industrialization and urbanization in China, more and more municipal sewage treatment plants have been established, and thus large amount of sludge, as a by-product of sewage treatment processes, has been produced. Total sludge production in China increased 13% annually from 2007 to 2013, and 6.25 million tons of dry sludge was produced in 2013; this number is still quickly increasing (Yang et al., 2015).

Sludge has many toxic substances, such as pathogens, heavy metals, and some organic contaminants, which can cause serious environment pollution if there is no further treatment or disposal. Therefore, the problem of treating and disposing of such a large amount of sewage sludge is now a particular concern to local authorities and environmental engineers. Many approaches have been applied in China, including landfill, land application, incineration, and producing building materials. Based on the data given by NBSC (2013), the most commonly used approach was landfill, which accounted for 13.4%, followed by land application (2.4%), incineration (0.36%), and building materials (0.24%). However, more than 80% of sludge was not properly treated and disposed and may cause serious secondary pollution. Especially the potential release of toxic heavy metals from the unprocessed or improperly treated/disposed sludge may cause ecological risks there (Singh and Agrawal, 2010), as approximately 50%–80% of heavy metal contents in sewage is fixed into sludge during sewage treatment and thus the sewage sludge often contains chromium, lead, copper, nickel, and other metals with mean concentrations approximately in the range of 0.1%–0.3% (w/w). Table 13.1 summarizes the average heavy metal contents

Biochar from Biomass and Waste
DOI: https://doi.org/10.1016/B978-0-12-811729-3.00013-3

TABLE 13.1 Heavy Metal Contents in Sewage Sludge in China

Heavy Metals	Contents in Sludge (mg kg^{-1} dry mass)		
As	20.2[a]	NA	3.15–11.70[c]
Cd	1.97[a]	3.03[b]	0.31–6.16[c]
Cr	93.1[a]	261.15[b]	7.86–200[c]
Cu	218.8[a]	338.98[b]	14.48–239.93[c]
Hg	2.14[a]	NA	NA
Ni	48.7[a]	87.8[b]	NA
Pb	72.3[a]	164.09[b]	6.10–121[c]
Zn	1058[a]	789.82[b]	NA
Data from	Yang et al. (2014)	Wu et al. (2012)	Yao et al. (2010)

[a]*The average of sludge samples collected from 107 STP.*
[b]*The average of sludge samples collected from 44 STP.*
[c]*Values of sludge samples collected from 16 STP.*
NA: not available.

in sewage sludge in China. However, most sludge treatment disposal approaches, such as incineration, land application, and composting may destroy the pathogens, and degrade organic contaminants, but seem not to cause much change on the metal species. Therefore, pyrolysis that can stabilize toxic metals in sewage sludge are thus catching more and more attention (Hossain et al., 2010; Zhang et al., 2013; Jin et al., 2016).

13.2 PYROLYSIS OF SEWAGE SLUDGE AND THE ENVIRONMENTAL SAFETY OF HEAVY METALS IN SLUDGE-DERIVED BIOCHARS

13.2.1 Pyrolysis of Sewage Sludge Under Various Conditions

Pyrolysis refers to a thermal decomposition of materials at elevated temperatures in an inert atmosphere. The production of the pyrolysis of sewage sludge includes the sludge-derived biochar (SDBC) and bio-gas. Many research studies indicate that the pyrolysis of sewage sludge is a self-sustaining process, as the energy derived from the combustion of the pyrolyzed bio-gas is comparable to that required for the pyrolysis (Hossain et al., 2010; Fonts et al., 2012; Jorge and Dinis, 2013; Mills, 2015).

During pyrolysis, both temperature and residence time are reported to affect the yield and physico-chemical properties of the formed biochar considerably. In general, four distinct categories of biochar generally exist under different temperatures (Verheijen et al., 2009): (1) below 250°C transition chars, where the crystalline character of the precursor materials is preserved; (2) between 250 and 350ºC amorphous chars, where cellulose is thermally degraded as volatiles, remaining a rigid randomly mixed C matrix; (3) 330–600°C composite chars, where the poorly ordered poly-aromatic grapheme stacks grow laterally, embedded in amorphous phases, and eventually coalesce; and (4) above 600°C turbostratic chars, where disordered graphitic crystallites dominate and most of the remaining non-C atoms are removed.

Table 13.2 summarizes the reported biochar yield (%, by mass) under various pyrolysis temperatures and periods from municipal sewage sludge with different sewage treatment and digestion modes. It was found the biochar yields ranged widely from 27% to 81.6%.

TABLE 13.2 The Reported Biochar Yield and BET Surface Area Under Different Pyrolysis Conditions

Time (h)	Temp. (°C)	Biochar Yield (%, m/m)	BET Surface Area ($m^2 g^{-1}$)	Ash in Sludge (%)	Sewage Treatment	Sludge Digestion	References
1	400	60.6	5.49	41.3	Triple oxidation ditch process	None	Jin et al. (2016)
	450	59.6	7.21				
	500	59.0	7.73				
	550	58.7	8.45				
	600	53.1	5.99				
1	400	77	14.1	62.7	Oxidation ditch without the primary settlement	None	Zhang et al. (2013)
	500	74.1	17.2				
	600	72.1	13.2				
2	300	81.6	7.90				
	400	76.1	23.7				
	500	73.6	20.1				
	600	71.3	16.0				
1	500	45.1–54.3	7.1–19.4	55.8–58.5	Mechanical biological treatment	Anaerobic digestion	Zielinska et al. (2015)
	600	43.2–51.3	2.8–7.7				
	700	40.2–49.5	1.4–23.7				
0.5	300	62.5	4.0	25.9	Activated sludge	Anaerobic digestion	Agrafioti et al. (2013)
	400	28.5					
	500	27.3	18.0				
1	300	58.1					
	400	25.5					
	500	27					
1.5	300	64.2					
	400	27.5					
	500	31					

The sludge with an anaerobic digestion often resulted in a lower biochar yield (53.1%–81.6% vs. 27%–64.2%). Most of them were much higher than those from the agricultural wastes, which are typically 25%–35% (Boudrahem et al., 2009). Such a disparity could likely be attributed to their different mineral contents. The municipally activated sludge itself also contains massive biological flocs, which are composed of bacteria and protozoa, and account for about 30%–80% of the sludge dry mass, depending on the wastewater treatment process and source, and can also likewise be pyrolyzed into carbonaceous residue (Smith et al., 2009). However, the sludge often owns 25.9%–62.7% of ash contents (Jin et al., 2016; Zhang et al., 2013), while the ash content in most typical biomass is often much lower: 0.43%–1.82% in wood chips (Tsang et al., 2007) and 4%–19% in straw (Chen et al., 2011).

These studies all showed that an increase in the heating temperature always resulted in less biochar yield, likely due to the loss of volatile compounds. However, such a decrement in biochar yield progressively diminished when the temperature rose to 400°C, likely because mass cellulose in sludge is thermally degraded at 250–350°C as volatiles, the amorphous C matrix remaining (Verheijen et al., 2009). Although a higher temperature often induces the transformation from alkyl and O-alkyl C to aryl C, and from amorphous C phase to polyaromatic graphene sheets (Verheijen et al., 2009), the total mass seems unchanged. Therefore, the further mass loss at a higher temperature seems insubstantial.

Table 13.2 also lists the Brunauer–Emmett–Teller (BET) specific surface areas of all biochar pyrolyzed under various temperatures and periods from different municipal sewage sludge. It was found that the BET surface areas ranged from several to two dozen $m^2\,g^{-1}$. These values were lower than most commercial adsorbents, such as activated carbons and activated aluminum oxides or silicates, but comparable to some natural mineral adsorbents, such as zeolite (Sakintuna et al., 2003), or Italian red soil (Papini et al., 2004), and the biochar from the pig or cow mature (Kołodynska et al., 2012). In addition, 400°C for 2 h, 450°C for 1 h or 500°C for 0.5 h was found to always result in a high BET surface area.

13.2.2 Environmental Safety of Heavy Metals in Sludge-Derived Biochars

Table 13.3 lists the reported total and extractable metal contents in sludge-derived biochars under different pyrolysis conditions. It was found that the total concentrations of Cd, Cr, Cu, Pb, Ni, and Zn in the biochars were basically higher than, or even double of those in the raw sewage sludge. It can be ascribed to the lower loss in mass of the heavy metals compared to the loss in organic compounds during pyrolysis, resulting in the enrichment of heavy metals in the biochar matrix. In addition, the increased pyrolysis temperature led to the heavy metal accumulation in the biochar, as a result of the greater loss of coal mass (mainly composed of C, N, O) during pyrolysis. Overall, the contents of heavy metals in the sludge and biochars generally followed in the order as follows. Zn > Cu > Cr > Ni > Pb > Cd. The high Zn contents in the sewage sludge can be contributed to the widespread use of galvanized pipeline in municipal water supply system. The high Cu levels may be due to the high proportion of trade effluent contaminated with a range of chemicals, including heavy metal rinses.

TABLE 13.3 The Total and Extractable Metal Contents in Sludge Derived Biochar

Heavy Metals	Total Contents (mg kg^{-1})	Extractable Contents (mg kg^{-1})	Pyrolysis Conditions	References
Cr	449	22.4	Raw sludge	Jin et al. (2016) (acid extractable in BCR)
Cu	1218	138		
Ni	112	31.5		
Pb	95.1	ND		
Zn	2580	694		
Cr	665	19.42	400°C for 1 h	
Cu	1551	7.19		
Ni	147	2.2		
Pb	85	ND		
Zn	2572	179.93		
Cr	948	14.8	450°C for 1 h	
Cu	1591	12.7		
Ni	177	2.71		
Pb	92	ND		
Zn	2727	121.9		
Cr	1065	18.4	500°C for 1 h	
Cu	1674	17.7		
Ni	187	2.47		
Pb	99	ND		
Zn	2822	156		
Cr	1318	18.6	550°C for 1 h	
Cu	1640	20.7		
Ni	231	2.19		
Pb	109	ND		
Zn	3043	155		
Cr	1374	19.2	600°C for 1 h	
Cu	1697	33.3		
Ni	219	3.99		
Pb	111	ND		
Zn	3368	217		

(*Continued*)

TABLE 13.3 (Continued)

Heavy Metals	Total Contents ($mg\ kg^{-1}$)	Extractable Contents ($mg\ kg^{-1}$)	Pyrolysis Conditions	References
Cd	2.28–5.26	0.28–3.20	Raw sludge	Lu et al. (2013) (DTPA-extractable)
Cu	401–611	63.7–265		
Pb	136–224	21.5–55.6		
Zn	629–1238	324–696		
Cd	3.30–7.45	ND	300°C for 2 h	
Cu	480–1034	1.28–2.07		
Pb	242–350	ND		
Zn	849–1909	1.30–8.57		
Cd	3.76–9.82	ND	400°C for 2 h	
Cu	549–1198	9.2–17.6		
Pb	194–438	1.30–10.4		
Zn	911–2104	15.0–76.5		
Cd	4.25–8.85	ND	500°C for 2 h	
Cu	565–1127	14.7–23.0		
Pb	211–506	2.46–9.50		
Zn	1014–2309	18.0–92.6		

Among the total contents of heavy metals, only part can usually be mobile and bioavailable to enter the food chain, to result in the ecologic risk. Therefore, to access the ecological safety of these heavy metals in biochar when they are applied in environment, the sequential extraction procedure proposed by the Community Bureau of Reference (BCR) or the extraction procedures with diethylenetriaminepentaacetic acids (DTPA) were often employed in many previous studies, and reported herein in Table 13.3.

As listed in Table 13.3, the pyrolysis process significantly decreases the acid-extractable and DTPA extractable metals in the sludge, especially to a great extent for the cationic metals, such as Cu, Cd, Pb, Ni, and Zn. As reported by Lu et al. (2013), the DTPA-extractable metal contents in the biochars only accounted for 0%–3%, 0%–4%, 0%–9% of the total Cu, Pb, and Zn levels, respectively. It may be the primary reason that metal oxides and mineral residues as ash content fixed in the biochars rendered the pH more alkaline, which is widely reported to favor the stabilization of cationic metals.

However, a higher pyrolysis temperature seems to disfavor the immobilization of heavy metals in the biochar. There are several reasons for the reduction of extractable metals at low temperature, i.e., 300°C for 2 h, or 400/450°C for 1 h. First, at the low time such as 300°C, the biochar surface is rich in functional groups (such as carboxyl, hydroxyl, amine, and amides (Zhang et al., 2013), which can contribute to the formation of organo-metallic complex or ligands for binding metals in the biochar particles. In addition, the available phosphorus content was the highest at the temperature of 300°C, and thus led to the formation of metal phosphate precipitates. Both mechanisms can lead to a decrease of extractable metal contents in the sludge derived biochar.

Therefore, the sludge pyrolysis may accumulate the heavy metals, but as well result in a significant decline in metal mobility and bioavailability, leading to a very low ecological risk of sludge exposed to the environment. However, it is still recommended to monitor the impacts of biochar on soil, water, and plants under field conditions.

13.3 ADSORPTION OF CONTAMINANTS IN SLUDGE-DERIVED BIOCHARS

As discussed above, to obtain the ecologically acceptable SDBC, the pyrolysis temperature should not be more than 600°C. At these temperatures, the SDBS cannot be fully carbonized, so both the carbonized and non-carbonized phases of organic components coexist, and thus generally represent different sorption mechanisms of various contaminants, such as inner-sphere complex and organic partitioning. In addition, unlike most biochar from bio-solids, i.e., plant residual, the SDBCs have a high mineral content, and therefore it is not possible to ignore the roles of the minerals, which also can adsorb some contaminants by electrostatic ion exchange, specific sorption and surface precipitation.

Therefore, different from most traditional sorbents, such as activated carbons or mineral oxides, SDBC are carbon-mineral mixed adsorbents with abundant organic functional groups and mineral oxides as well as a considerable alkalinity, owning considerable sorption capacity for diverse range of contaminants, such as cationic and oxyanionic metal species as well as organic contaminants.

13.3.1 Cationic Metals

Table 13.4 summarizes the cationic metal sorption capacity of SDBC. Lu et al. (2012) reported that SDBC owned the Pb(II) sorption capacity at the initial pH 2.0, 3.0, 4.0, and 5.0, were 16.1, 20.1, 24.8, and 30.9 mg g^{-1}, respectively. Although SDBC had a lower Pb sorption capacity than some commercial adsorbents (i.e., activated alumina and thermo-modified silica) at moderately acidic or neutral conditions such as pH 5.0, its capacity at the acidic pH was comparable to them, likely due to its own great alkalinity. Compared with other biochar from biomass, the Pb sorption capacity of SDBC seems much higher (Inyang et al., 2012). Ho et al. (2017) even reported as high as 51.2 mg g^{-1} of Pb sorption capacity of the biochar from an anaerobic digestion sludge.

The Pb(II) sorption mechanisms were investigated by Lu et al. (2012), who concluded that the mechanisms involved (1) electrostatic ion exchange with K^+ and Na^+, (2) co-precipitation and inner-sphere complexation with complexed organic matter and mineral oxides of SDBC, surface complexation with free carboxyl and hydroxyl functional groups, inner-sphere complexation with free hydroxyl groups of mineral oxides, and other surface precipitation. Among them, the contribution of irreversible precipitation (co-precipitation) and inner-sphere complexation was dominant. Therefore, the sorbed Pb on SDBC is rather

TABLE 13.4 Cationic Metal Sorption Capacity of SDBC

Cationic Metals	Sorption Capacity (mg kg^{-1})	Biochar Pyrolysis Condition	pH in Solution	References
Pb	16.1	550°C for 2 h	2	Lu et al. (2012)
	20.1		3	
	24.8		4	
	30.9		5	
Pb	12.6	300°C 2 h	5	Zhang et al. (2013)
	15.3	400°C 1 h		
	18.2	400°C 2 h		
	16.6	500°C 1 h		
	15.4	500°C 2 h		
	12.4	600°C 1 h		
	12.8	600°C 1 h		
Pb	51.2	600°C for 1.5 h from anaerobic digestion sludge	6.0	Ho et al. (2017)
Cu	21.2		5.5	
Pb	47		5.5	
Cd	16.2		5.5	
Ni	22.5		5.5	
Zn	18.1		5.5	

stable. Similarly, Ho et al. (2017) also contributed the high-efficient Pb removal from solution by SDBC to the precipitation with minerals, especially for phosphates, which accounts for 53.5% of the overall Pb removal.

Zhang et al. (2013) discussed the Pb sorption on the biochar pyrolyzed under different conditions. They found that the biochar samples prepared under 400°C for 2 h, which owned the highest BET surface area, thus had the highest Pb(II) sorption capacity. It shows that the surface area may be the key factor to dominate the Pb sorption capacity. However, such an influence seem not always positively correlated. A biochar sample pyrolyzed at lower temperature with a lower surface area may own a higher Pb sorption than those pyrolyzed at 600°C. It is likely because the low temperature may lead to the incomplete formation of the meso- or micro-pores of SDBC, and thus a low surface area, but maintains the bulk of organic functional groups, which are considered as the main Pb sorption sites.

Besides Pb, SDBC also can sorb considerably other cationic metals. As reported by Ho et al. (2017), the sorbed Cu, Pb, Cd, Ni, and Zn amounts on the biochar pyrolyzed at 600°C for 1.5 h from anaerobic digestion sludge was 21.2, 47.0, 16.2, 22.5, 18.1, and 8.3 mg g^{-1}, respectively, when the 20 mL of 100 mg L^{-1} cationic metal solution was mixed with 0.04 g biochar for 24 h. The sorption capacity of SDBC for the heavy metals followed the order of Pb > Ni > Cu > Ni > Cd > Zn at the fixed initial mass concentrations of these metals (Ho et al., 2017; Zhang et al., 2017). The sequence seems different from the order obtained for other biochar form biomass (Inyang et al., 2012). The distinct sorption mechanisms involved for metal sorption on SDBC may be the behind reasons for such a difference.

13.3.2 Oxyanionic Metals

In addition to cationic metals, the SDBC was also found to immobilize some oxyanionic metals, such as highly toxic Cr(VI) and As(III).

As reported by Zhang et al. (2013), SDBC prepared at 400°C for 2 h sorption showed the greatest sorption capacity at the same equilibrium concentrations in the solution compared with the SDBC pyrolyzed at other conditions. Under the same pyrolysis temperature, SDBC with a longer residence time seems to own a stronger sorption bond between the Cr(VI) species and biochar surfaces.

In addition, as reported by Zhang et al. (2013), the Cr(VI) sorption was highly pH-dependent at pH 2.0–5.0, and an acidic environment favor the Cr(VI) sorption. It is consistent with the previous studies (Demirbas et al., 2004; Vinod et al., 2010; Dong et al., 2011). It can be primarily ascribed to a favorable reduction of Cr(VI) into Cr(III) by the biochar at the acidic environment, as X-ray photoelectron spectra on the SDBC surface with the sorbed Cr illustrated the coexistence of Cr(III) at pH 2 and 2.5. That phenomenon can be explained by the following two mechanisms: (1) a low pH results in a high redox potential of Cr(VI)/Cr(III) couple; and (2) a decreasing pH facilitates to form complexes on the SDBCs surface, which was the prerequisite for subsequent electron transfer between the redox pair (Zhou et al., 2015). Thiol, phenolic, carboxylic, methoxy, and carbonyl functional groups on SBDC may be effective electron-donor groups for Cr(VI) reduction (Dupont and Guillon, 2003; Suksabye et al., 2009). In addition, the coexisting humic substance in the solution was also reported to increase the amount of the sorbed Cr on the SDBC surface (Zhou et al., 2015), since they also favor the Cr(VI) reduction to Cr(III).

Compared with cationic metals and Cr(VI), the As(III) sorption capacity on the SDBC was much less. Zhang et al. (2015a) reported the As(III) Langmuir sorption capacity of SDBC pyrolyzed at 400–500°C for 1–2 h was only 3.08–6.04 mg g^{-1}. However, the values were comparable to and even more than minerals (i.e., red mud 0.46–0.66 mg g^{-1} Altundoğan et al., 2000; activated alumina grains 0.77–3.48 mg g^{-1}, Lin and Wu, 2001), and biochar from biomass (i.e., activated carbons prepared from olive pulp and stones 0.21–1.39 mg g^{-1}, Budinova et al., 2006).

In general, the larger BET surface area and finer particle size of the SDBC in part contributed to the greater As(III) sorption capacity and affinity the high As(III) sorption affinity and capacity (Zhang et al., 2015a). The physicochemical characteristics (e.g., surface site density, amorphous Fe oxides, soluble Ca minerals, C content, and morphology) of sorbents could also affect the surface complexation and (co-)precipitation of As(III) (Oh et al., 2012; Tsang et al., 2014).

The occurrence of ligand exchange between As(III) and surface hydroxyl groups of the SDBC is considered the main mechanism for As(III) sorption at the initial stage

(Zhang et al. 2015a). When the As(III)-sorbed SDBC was exposed in the air, the immobilized As(III) can be oxidized to less toxic As(V) by the co-oxidation of As(III) and Fe(II) in SDBC by O_2 in ambient air, as concluded by Katsoyiannis et al. (2008) and Wang et al. (2013).

13.3.3 Organic Contaminants

SDBC is also a good sorbent to immobilize organic contaminants, such as atrazine in water. When the equilibrium concentration in solution was $10\,mg\,L^{-1}$, atrazine sorption capacity was $3.83-17.6\,mg\,g^{-1}$ (Zhang et al., 2015b), which was comparable to values for the soils (Dehghani et al., 2005; Báez et al., 2013), but was less than the traditional biochar from biomass (Zheng et al., 2010; Gupta et al., 2011; Rambabu et al., 2012). This can be explained by the fact that the SDBCs had relative low surface areas and limited organic carbon (only approximately 10% of dry mass).

The atrazine sorption isotherms on SDBCs fit well with the Freundlich equation, a heating temperature at 400°C for a duration of 2 h was found to obtain the higher atrazine sorption, as reported by Zhang et al. (2015b). It is well known that the H-bond, electrostatic attraction between the hydrogen atom in these polar groups and the highly electronegative heterocyclic ring in atrazine (electron donor), and organic partitioning plays an important role in the atrazine sorption processes. However, atrazine sorption on SDBC may mainly involve site-specific process, i.e., H-bonding, especially under the low atrazine loading (Zhang et al., 2015b), but hydrophobic force with silanol surface of minerals (Báez et al., 2013) or amorphous black carbons (Lima et al., 2010) at a high atrazine loading.

In general, a deeper carbonization of organic compounds with an increase in temperature (Hao et al., 2013) may lead to more site-specific interactions during the adsorption process (Chen and Chen, 2009). However, a temperature higher than 400°C significantly destroys the primary active sites of virgin organic components, such as carboxylic and phenolic moieties, which are essential components for H-bonding formation for atrazine adsorption under the low loading (Wang et al., 2011).

The coexisting humic acids in aqueous solution decreased and slowed the atrazine sorption especially under a higher atrazine loading (Zhou et al., 2015). This can be partially ascribed to the increase in the hydrophilicity of the SDBC surface induced by the sorbed humic acids. In turn, the repulsion with the hydrophobic atrazine is enhanced, thus disfavoring its sorption by the hydrophobic force. However, at a lower atrazine loading, the sorbed humic acids can strengthen the atrazine sorption binding by providing more oxygen-containing sites for H-bonding (Zhou et al., 2015), due to the less contribution of hydrophobic force to the overall atrazine sorption.

13.4 METAL STABILIZATION IN SOILS BY SLUDGE-DERIVED BIOCHARS

As discussed above, the potential risks related to metal(loid)s leaching from the SDBC itself are considered acceptable and much lower than the direct application of sewage sludge (Hwang et al., 2007). Moreover, SDBC has a substantial sorption capacity for cationic and oxyanionic metals in the aqueous phase, and the main involved mechanisms are irreversible inner-complex with active mineral oxides or hydroxides and organic functional groups chemical oxidation/reduction, as well as surface precipitation or co-precipitation. Therefore, SDBC seems a promising soil amendment agent to be applied in contaminated sites.

Many previous studies (Fang et al., 2016) showed that the applied SDBCs obviously reduced the leachable levels of Pb(II), Ni(II), Cu(II), and Cd(II) in soils, companied with an increase in soil pH, although the incremental pH was not alkaline enough to form surface precipitates or co-precipitates of cationic metal species. Therefore, the direct sorption of liable metal species on the SDBC surface may be the main mechanism for its immobilization in soils. Their performance for Pb(II) immobilization was found better in the acidic soil with lower extractable Al(III) content. In the soil with high extractable Al(III) content, the mineral-rich SDBC is greater priority over the biochar from the biomass (Yang et al., 2015).

However, if the SDBC is applied as a separate layer in the soil matrix, the concentrated cationic metals locally raised the peak levels of metal release under continuous acid leaching. It may bring a new ecologic risk for local environment (Fang et al., 2016).

However, the added SDBC seem not substantially to decrease the leachable Zn(II) levels (Fang et al., 2016; Yang et al., 2015). Although the SDBC have some Zn(II) sorption capacity, the competition of other cationic metals, such as Pb(II) and Ni(II), which SDBC prefer to sorb rather than Zn(II), and intrinsic Zn(II) release from SDBC itself led to the trivial change of final extractable Zn(II) level induced by added SDBC.

In addition, SDBC could effectively stabilize oxyanionic Cr(VI) and As(III), respectively, which are often favored by elevated soil temperature and longer application period. At the initial stage, the immobilized Cr(VI) and As(III) are not stable enough, so periodic temperature decrease from 45 to 4°C resulted in their release, as reported by Zhou et al. (2015). A longer application can facilitate their stabilization in the soil. Cationic metals can be strongly bound and even form stable precipitates. With increasing time, Cr(VI) can be reduced to Cr(III), while immobilized As(III) was co-oxidized to As(V).

Yet, a long-term monitoring and investigation may be required to evaluate the potential ecological risks and bioavailability/toxicity of these metal species in the SDBC-amended soils.

13.5 AGEING OF SLUDGE-DERIVED BIOCHARS IN THE ENVIRONMENT

Ageing is a common phenomenon during biochar storage and its soil application. When SDBC is exposed in the air, the surface oxidation with O_2, acidification by CO_2, bring an increase in acidity, cation exchange capacity, and carboxyl groups on the SDBC surface; it also causes a decrease in alkalinity and Fe(III) species. Therefore, SDBC exposed in the air for more than 30 days was found to favor the Pb(II) and As(III) sorption (Wang et al., 2017) because of higher density of available oxygen-containing groups. The Cr(VI) sorption was found to be compromised by the atmospheric exposure, because some reducing agents for Cr(VI) reduction was consumed there. Higher temperatures accelerate the above-mentioned ageing effect.

Similarly, the continuous leaching of SDBC also gradually decreases the density of basic functional groups and increased the density of carboxyl groups as well as cation exchange capacity on the SDBC surface, as reported by Feng et al. (2018). It can be attributed to surface dissolution as well as acidification, and oxidation process by the leaching process. Continuous leaching was also found to increase Pb(II), Cr(VI), and As(III) sorption capacity of the SDBC (Feng et al., 2018), probably because the increase in carboxyl groups promotes inner-sphere complexation and Fe oxidation.

Yet, when the SDBC is applied in contaminated soils, its performance would be affected by both the ageing of SDBC itself under varying redox conditions or intermittent leaching process and long-term interactions among soil components, such as soil colloids and solution, metal species, and SDBC. The good understanding of those processes requires further investigation.

13.6 CONCLUSIONS

Pyrolysis of sewage sludge into biochar is a particularly promising disposal approach for municipal sewage sludge because the pyrolysis process can effectively stabilize heavy metals in sludge to reduce the ecological risk, as well as the by-product, SDBC can provide considerable active sites for metal complexation and alkalinity for precipitation. So, the SDBC is considered as an efficient sorbent for immobilization of cationic Pb(II), oxyanionic Cr(VI) and As(III) species, as well as atrazine. The pyrolysis conditions can influence the yield and properties of the produced SDBC. The SDBC prepared at 400°C for 2 h or 500°C for 1 or 1.5 h seem to own an excellent sorption capacity for contaminants with high BET surface area and a low extractable metal level in itself, as many organic functional groups and phosphorous components still remain.

When SDBC is applied in the metal contaminated soil, it can effectively reduce the mobility of the cationic metals, such as Pb(II), Cu(II), Ni(II), and Cd(II), as well as the oxyanionic Cr(V) and As(III). The long contact time favors their stabilization. When SDBC is exposed to the air and flushed by soil solution, surface oxidation and acidification occur to change the properties of SDBC, influencing the metal stabilization by SDBC. Further investigation on the long-term interactions among soil components, metal species, and SDBC should be performed.

References

Agrafioti, E., Bouras, G., Kalderis, D., Diamadopoulos, E., 2013. Biochar production by sewage sludge pyrolysis. J. Anal. Appl. Pyrol. 101, 72–78.

Altundoğan, H.S., Altundoğan, S., Tumen, F., Bildik, M., 2000. Arsenic removal from aqueous solutions by adsorption on red mud. Waste Manage. 20, 761–767.

Báez, M., Fuentes, E., Espinoza, J., 2013. Characterization of the atrazine sorption process on Aandisol and Ultisol volcanic ash-derived soils: kinetic parameters and the contribution of humic fractions. J. Agric. Food Chem. 61, 6150–6160.

Boudrahem, F., Aissani-Benissad, F., Aıt-Amar, H., 2009. Batch sorption dynamics and equilibrium for the removal of lead ions from aqueous phase using activated carbon developed from coffee residue activated with zinc chloride. J. Environ. Manage. 90, 3031–3039.

Budinova, T., Petrov, N., Razvigorova, M., Parra, J., Galiatsatou, P., 2006. Removal of arsenic(III) from aqueous solution by activated carbons prepared from solvent extracted olive pulp and olive stones. Ind. Eng. Chem. Res. 45, 1896–1901. Available from: https://doi.org/10.1021/ie051217a.

Chen, B.L., Chen, Z.M., 2009. Sorption of naphthalene and 1-naphthol by biochars of orange peels with different pyrolytic temperatures. Chemosphere 76, 127–133.

Chen, X., Chen, G., Chen, L., Chen, Y., Lehmann, J., McBride, M.B., et al., 2011. Adsorption of copper and zinc by biochars produced from pyrolysis of hardwood and corn straw in aqueous solution. Bioresour. Technol. 102, 8877–8884.

Dehghani, M., Nasseri, S., Amin, S., Naddafi, K., Taghavi, M., Yunosian, M., et al., 2005. Atrazine adsorption desorption behavior in Darehasaluie Kavar corn field soil in fars province of Iran. Iran. J. Environ. Health Sci. Eng. 2, 221–228.

Demirbas, E., Kobya, M., Senturk, E., Ozkan, T., 2004. Adsorption kinetics for theremoval of chromium(VI) from aqueous solutions on the activated carbons prepared from agricultural wastes. Water SA 30, 533–539.

Dong, X., Ma, L.Q., Li, Y., 2011. Characteristics and mechanisms of hexavalent chromium removal by biochar from sugar beet tailing. J. Hazard. Mater. 190, 909–915.

Dupont, L., Guillon, E., 2003. Removal of hexavalent chromium with a lignocellulosic substrate extracted from wheat bran. Environ. Sci. Technol. 37, 4235–4241.

Fang, S., Tsang, D.C.W., Zhou, F.S., Zhang, W.H., Qiu, R.L., 2016. Stabilization of cationic and anionic metal species in contaminated soils using sludge-derived biochar. Chemosphere 149, 263–271.

Feng, M., Zhang, W.H., Wu, X.Y., Jia, Y.M., Jiang, C.X., Wei, H., et al., 2018. Continuous leaching modifies the surface properties and metal(loid) sorption of sludge-derived biochar. Sci. Total Environ. 625, 731–737.

Fonts, I., Gea, G., Azuara, M., Abrego, J., Arauzo, J., 2012. Sewage sludge pyrolysis for liquid production: a review. Renew. Sustain. Energy Rev. 16, 2781–2805.

Gupta, V.K., Gupta, B., Rastogi, A., Agarwal, S., Nayak, A., 2011. Pesticides removal from waste water by activated carbon prepared from waste rubber tire. Water Res. 45, 4047–4055.

Hao, F.H., Zhao, X.C., Ouyang, W., Lin, C.Y., Chen, S.Y., Shan, Y.S., et al., 2013. Molecular structure of corncob-derived biochars and the mechanism of atrazine adsorption. Agronomy J. 105, 773–782.

Ho, S.H., Chen, Y.D., Yang, Z., Nagarajan, D., Chang, J.S., Ren, N., 2017. High-efficiency removal of lead from wastewater by biochar derived from anaerobic digestion sludge. Biores. Technol 246, 142–149.

Hossain, M.K., Strezov, V.K., Chan, Y., Nelson, P., 2010. Agronomic properties of wastewater sludge biochar and bioavailability in production of cherry tomato (*Lycopersicon esculentum*). Chemosphere 78, 1167–1171.

Hwang, I.H., Ouchi, Y., Matsuto, T., 2007. Characteristics of leachate frompyrolysis residue of sewage sludge. Chemosphere 68, 1913–1919.

Inyang, M., Gao, B., Yao, Y., Xue, Y., Zimmerman, A.R., Pullammanappallil, P., et al., 2012. Removal of heavy metals from aqueous solution by biochars derived from anaerobically digested biomass. Bioresour Technol. 110, 50–56. Available from: http://dx.doi.org/10.1016/j.biortech.2012.01.072.

Jin, J., Li, Y., Zhang, J.Y., Wu, S.C., Cao, Y., Liang, P., et al., 2016. Influence of pyrolysis temperature on properties and environmentalsafety of heavy metals in biochars derived from municipal sewagesludge. J. Hazard. Mater. 320, 417–426.

Jorge, F.C., Dinis, M.A.P., 2013. Sewage sludge disposal with energy recovery: a review. Waste Res. Manage. 166, 14–28.

Katsoyiannis, I.A., Ruettimann, T., Hug, S.J., 2008. pH dependence of Fenton reagent generation and As(III) oxidation and removal by corrosion of zero valent iron in aerated water. Environ. Sci. Technol. 242, 7424–7430.

Kołodynska, D., Wneztrzak, R., Leahy, J.J., Hayes, M.H.B., Kwapinski, W., Hubicki, Z., 2012. Kinetic and adsorptive characterization of biochar in metal ions removal. Chem. Eng. J. 197, 295–305.

Lima, D.L., Schneider, R.J., Scherer, H.W., Duarte, A.C., Santos, E.B., Esteves, V.I., 2010. Sorption–desorption behavior of atrazine on soils subjected to different organic long-term amendments. J. Agric. Food Chem. 58, 3101–3106.

Lin, T.F., Wu, J.K., 2001. Adsorption of arsenite and arsenate within activated alumina grains: equilibrium and kinetics. Water Res. 35, 2049–2057.

Lu, H.L., Zhang, W.H., Yang, Y.X., Huang, X.F., Wang, S.Z., Qiu, R.L., 2012. Relative distribution of Pb^{2+} sorption mechanisms by sludge-derived biochar. Water Res. 46, 854–862.

Lu, H.L., Zhang, W.H., Wang, S.Z., Zhuang, L.W., Yang, Y.X., Qiu, R.L., 2013. Characterization of sewage sludge-derived biochars from different feedstocks and pyrolysis temperatures. J. Anal. Appl. Pyrol. 102, 137–143.

Mills, N., 2015. Unlocking the Full Energy Potential of Sewage Sludge. Dissertation, University of Surrey.

NBSC (National Bureau of Statistics of China), 2013. China Statistical Yearbook (2013). China Statistics Press, Beijing, China (in Chinese).

Oh, C., Rhee, S., Oh, M., Park, J., 2012. Removal characteristics of As(III) and As(V) from acidic aqueous solution by steel making slag. J. Hazard. Mater. 213–214, 147–155. Available from: https://doi.org/10.1016/j.jhazmat.2012.01.074.

Papini, M.P., Saurini, T., Bianchi, A., Majone, M., Beccari, M., 2004. Modeling the competitive adsorption of Pb, Cu, Cd, and Ni onto a natural heterogeneous sorbent material (Italian "Red Soil"). Ind. Eng. Chem. Res. 43, 5032–5041.

Rambabu, N., Guzman, C.A., Soltan, J., Himabindu, V., 2012. Adsorption characteristics of atrazine on granulated activated carbon and carbon nanotubes. Chem. Eng. Technol. 35, 272–280.

Sakintuna, B., Aktas, Z., Yürüm, Y., 2003. Synthesis of porous carbon materials by carbonization in natural zeolite nano-channels. Prepr. Pap. Am. Chem. Soc. Div. Fuel Chem. 48, 614–615.

Singh, R.P., Agrawal, M., 2010. Effect of different sewage sludge application on growth and yield of *Vigna radiate* L. field crop: metal uptake by plant. Ecol. Eng. 36, 969–972.

Smith, K.M., Fowler, G.D., Pullket, S., Graham, N.J., 2009. Sewage sludge based adsorbents: a review of their production, properties and use in water treatment applications. Water Res. 43 45, 2569–2594.

Suksabye, P., Nakajima, A., Thiravetyan, P., Baba, Y., Nakbanpote, W., 2009. Mechanism of Cr(VI) adsorption by coir pith studied by ESR and adsorption kinetic. J. Hazard. Mater. 161, 1103–1108.

Tsang, D.C.W., Hu, J., Liu, M.Y., Zhang, W., Lai, K.C.K., Lo, I.M.C., 2007. Activated carbon produced from waste wood pallets: adsorption of three classes of dyes. Water Air Soil Pollut. 184, 141–155.

Tsang, D.C.W., Yip, A.C.K., Olds, W.E., Weber, P.A., 2014. Arsenic and copper stabilisation in a contaminated soil by coal fly ash and green waste compost. Environ. Sci. Pollut. Res. 21, 10194–10204. Available from: https://doi.org/10.1007/s11356-014-3032-3.

Verheijen, F.G.A., Jeffery, S., Bastos, A.C., Van der Velde, M., Diafas, I., 2009. Biochar Application to Soils – A Critical Scientific Review of Effects on Soil Properties, Processes and Functions, EUR 24099 EN. Office for the Official Publications of the European Communities, Luxembourg.

Vinod, V.T.P., Sashidhar, R.B., Sreedhar, B., 2010. Biosorption of nickel and total chromium from aqueous solution by gum kondagogu (*Cochlospermum gossypium*): a carbohydrate biopolymer. J. Hazard. Mater. 178, 8510860.

Wang, H., Feng, M., Zhou, F.S., Huang, X.C., Tsang, D.C.W., Zhang, W.H., 2017. Effects of atmospheric ageing under different temperatures on surface properties of sludge-derived biochar and metal/metalloid stabilization. Chemosphere 184, 176–184.

Wang, X.L., Guo, X.Y., Yang, Y., Tao, S., Xing, B.S., 2011. Sorption mechanisms of phenanthrene, lindane, and atrazine with various humic acid fractions from a single soil sample. Environ. Sci. Technol. 45, 2124–2130.

Wang, Z., Bush, R.T., Liu, J., 2013. Arsenic(III) and iron(II) co-oxidation by oxygen and hydrogen peroxide: divergent reactions in the presence of organic ligands. Chemosphere 93, 1936–1941.

Wu, X., Li, J.Y., Zhang, M., Shi, M.J., 2012. Distribution of Heavy Metal Concentration of Sewage Sludge in China. Biomedical Engineering and Biotechnology (iCBEB), Macau, China, <https://doi.org/10.1109/iCBEB.2012.156>.

Yang, G., Zhang, G., Wang, H., 2015. Current state of sludge production, management, treatment and disposal in China. Water Res. 78, 60–73.

Yang, J., Lei, M., Chen, T., Gao, D., Zheng, G., Guo, G., et al., 2014. Current status and developing trends of the contents of heavy metals in sewage sludges in China. Front. Environ. Sci. Eng. 8, 719–728.

Yao, J., Wang, H., Yu, Y.J., Wang, Q., Wang, R.X., 2010. Pollution status and characteristics of heavy metals in sewage sludge from municipal wastewater treatment plants. Res. Environ. Sci. 23 (6), 696–702.

Zhang, W.H., Mao, S.Y., Chen, H., Huang, L., Qiu, R.L., 2013. Pb(II) and Cr(VI) sorption by biochars pyrolyzed from the municipal wastewater sludge under different heating conditions. Bioresour. Technol. 147, 545–552.

Zhang, W.H., Zheng, J., Zheng, P., Tsang, D., Qiu, R.L., 2015a. Sludge-derived biochar for As(III) removal/immobilization: effects of solution chemistry on sorption behavior. J. Environ. Qual. 44 (4), 1119–1126.

Zhang, W.H., Zheng, P., Zheng, J., Qiu, R.L., 2015b. Atrazine immobilized on sludge derived biochar and the interactive influence of coexisting Pb(II) or Cr(VI) ions. Chemosphere 134, 438–445.

Zhang, W.H., Huang, X.C., Rees, F., Tsang, D.C.W., Qiu, R.L., Wang, H., 2017. Metal immobilization by sludge-derived biochar: roles of mineral oxides and carbonized organic compartment. Environ. Geochem. Health 39 (2), 379–389.

Zheng, W., Guo, M., Chow, T., Bennett, D.N., Rajagopalan, N., 2010. Sorption properties of greenwaste biochar for two triazine pesticides. J. Hazard. Mater. 181, 121–126.
Zhou, F.S., Wang, H., Fang, S.E., Zhang, W.H., Qiu, R.L., 2015. Pb(II), Cr(VI) and atrazine sorption behavior on sludge-derived biochar: role of humic acids. Environ. Sci. Pollut. Res. 22 (20), 16031–16039.
Zielinska, A., Oleszczuk, P., Charmas, B., Skubiszewska-zi̧eba, J., Pasieczna-Patkowska, S., 2015. Effect of sewage sludge properties on the biochar characteristic. J. Anal. Appl. Pyrol. 112, 201–213.

Further Reading

Yang, Y.X., Zhang, W.H., Qiu, H., Tsang, D.C.W., Morel, J.L., Qiu, R.L., 2016. Effect of coexisting Al(III) ions on Pb(II) sorption on biochars: role of pH buffer and competition. Chemosphere 161, 438–444.

CHAPTER

14

Biochar as an (Im)mobilizing Agent for the Potentially Toxic Elements in Contaminated Soils

Sabry M. Shaheen[1,2], *Ali El-Naggar*[3,4], *Jianxu Wang*[2,5], *Noha E.E. Hassan*[6], *Nabeel Khan Niazi*[7,8], *Hailong Wang*[9,10], *Daniel C.W. Tsang*[11], *Yong Sik Ok*[4], *Nanthi Bolan*[12] *and Jörg Rinklebe*[2]

[1]Department of Soil and Water Sciences, University of Kafrelsheikh, Faculty of Agriculture, Kafr El-Sheikh, Egypt [2]University of Wuppertal, School of Architecture and Civil Engineering, Institute of Foundation Engineering, Water- and Waste-Management, Laboratory of Soil- and Groundwater-Management, Wuppertal, Germany [3]Department of Soil Sciences, Faculty of Agriculture, Ain Shams University, Cairo, Egypt [4]Korea Biochar Research Center, O-Jeong Eco-Resilience Institute (OJERI) & Division of Environmental Science and Ecological Engineering, Korea University, Seoul, Republic of Korea [5]State Key Laboratory of Environmental Geochemistry, Institute of Geochemistry, Chinese Academy of Sciences, Guiyang, P.R. China [6]Agricultural Extension and Rural Development Research Institute, Agricultural Researcher Center, Giza, Egypt [7]Institute of Soil and Environmental Sciences, University of Agriculture Faisalabad, Faisalabad, Pakistan [8]Southern Cross GeoScience, Southern Cross University, Lismore, NSW, Australia [9]School of Environment and Chemical Engineering, Foshan University, Foshan, Guangdong, P.R. China [10]Key Laboratory of Soil Contamination Bioremediation of Zhejiang Province, Zhejiang A & F University, Hangzhou, Zhejiang, P.R. China [11]Department of Civil and Environmental Engineering, The Hong Kong Polytechnic University, Hung Hom, Kowloon, Hong Kong, P.R. China [12]Global Centre for Environmental Remediation (GCER), Faculty of Science, The University of Newcastle, Callaghan, NSW, Australia

Biochar from Biomass and Waste
DOI: https://doi.org/10.1016/B978-0-12-811729-3.00014-5

14.1 INTRODUCTION

Soil pollution with potentially toxic elements (PTE) has significantly increased over the last decades, which represents a health risk to the humans, animals, and plants (Antoniadis et al., 2017a,b). Therefore, remediation of PTE-contaminated soils has received an increasing attention amongst the scientific community and regulating agencies. There are different approaches effectively to remediate the soils contaminated with PTEs. Recently, "green remediation" technologies are increasingly being used for sustainable remediation of contaminated sites (Gill, 2008; Tomasevic et al., 2013; Shaheen and Rinklebe, 2015). Remediation of PTE-contaminated soils using (im)mobilization technique has many advantages, such as its simplicity and high effectiveness, in situ applicability, and low cost (Bolan et al., 2014; Shaheen et al., 2015a,b, 2017).

The safe recycling of environmental wastes as soil amendments and immobilizing agents for PTE (waste for waste treatment) is an important issue for an adequate environmental management of contaminated soils. Several types of immobilizing/mobilizing agents, such as coal fly ash, brick dust, clay minerals, activated carbon, liming materials, and nano-materials, have been used to immobilize PTE in contaminated soils. However, the majority of these amendments have some demerits, including difficulties in recyclability, and relatively high operational cost (Shaheen and Rinklebe, 2015). Therefore, there is an increasing demand for new, applicable, and cheap waste-derived (im)mobilizing agents for the remediation of PTE-contaminated soils (Ahmad et al., 2014; Bolan et al., 2014; Wang et al., 2018a,b). These materials should be abundant in nature, locally available, originate from renewable sources, and potentially recyclable or renewable (Bolan et al., 2014; Rinklebe and Shaheen, 2015). Among different types of these materials, biomass feedstocks are readily available resources for recycling as (im)mobilizing agents and applicable amendments (Shaheen et al., 2015a,b; Lahori et al., 2017). The efficiency of raw biomass feedstocks in (im)mobilization of PTE in soils has been investigated; however, the efficiency of raw biomass materials is much lower than the those that are pyrolyzed (biochar; BC), possibly because pyrolysis process might improve their physical and chemical properties, as well as surface activities of raw materials for retention of PTE (Wang et al., 2018a).

Research on biochars as low-cost and eco-friendly material for the treatment of PTE-contaminated soils has received significant attention in the last few years (Lahori et al., 2017; Wang et al., 2018a). Biochar has exhibited a greater potential for treatment of contaminated soils than the other low-cost sorbents (e.g., brick dust and inorganic liming materials) as it contains micro and/or meso-porous structures, a variety of surface functional groups and inorganic mineral species (Ahmad et al., 2014; Bian et al., 2014; Chen et al., 2016). However, biochars can cause contradictory effects on the element mobilization and phytoavailability depending on the type of element, the source of feedstock, and soil conditions. Therefore, the aim of this chapter is to present the contrasting effects of different types of biochars on the mobilization and phytoavailability of different PTE in different soils.

14.2 BIOCHAR AS AN IMMOBILIZING AGENT FOR POTENTIALLY TOXIC ELEMENTS IN CONTAMINATED SOILS

14.2.1 Reducing Mobility and Phytoavailability of Potentially Toxic Elements in Soils Using Biochar

Recent studies (e.g., Ahmad et al., 2014; Yu et al., 2017; Rehman et al., 2018; Nie et al., 2018; Qi et al., 2018; Wang et al., 2018a) have reported that different types of biochars can be used for immobilization and reducing the bioavailability of PTE in contaminated soils (Table 14.1). Biochars showed high potential for Cd and Pb immobilization in soils

TABLE 14.1 Biochar Potential for (Im)mobilization of Potentially Toxic Elements in Contaminated Soils

Biochar Feedstock	Pyrolysis Temperature	Application Rate/Type of Experiment	Soil Type (Texture/ Classification)	Element	Impact	References
Sewage sludge	500–550°C	5% and 10% (w/w) Pot experiment	Acidic soil	As, Cd, Cr, Co, Cu, Ni, Pb, and Zn	Immobilization of As, Cr, Co, Ni, and Pb in the soil, due to rise in soil pH Mobilization of Cu, Zn, and Cd in the soil, due to high available concentrations in biochar	Khan et al. (2013)
Broiler litter	350–650°C	2–20wt% Incubation experiment	Acidic soil	Pb, Cu, and Sb	Immobilization of Pb and Cu Mobilization of Sb	Uchimiya et al. (2012)
Oak wood	400°C	5wt% Incubation and seed germination test	Acidic soil	Pb	Immobilization of Pb in the soil, due to rise in soil pH and adsorption onto biochar surfaces	Ahmad et al. (2012)
Cottonseed hulls	200–800°C	Stimulation of contaminant leaching by percolating rain water	Norfolk loamy sand—acidic soil	Ni, Cu, Pb, and Cd	Immobilization of PTE in the soil, due to the surface complexation with functional groups on biochar surfaces	Uchimiya et al. (2011)
Chicken manure and Green waste	550°C	5% (w/w) Plant growth experiment	Acidic soil	Cd, Cu, and Pb	Immobilization of PTE in the soil, due to their stabilization into the organic bound fraction	Park et al. (2011)
Eucalyptus saligna wood	450°C	0%, 1%, 2% Pot experiment	Sandy clay loam—neutral	Ni	Immobilization of Ni by 22% and 33% with application rate of 1% and 2%, respectively, due to the rise in soil pH and CEC	Rehman et al. (2016)

(Continued)

TABLE 14.1 (Continued)

Biochar Feedstock	Pyrolysis Temperature	Application Rate/Type of Experiment	Soil Type (Texture/ Classification)	Element	Impact	References
Mix of Shell limestone, perlite, organic substrates	–	1% Greenhouse pot experiment	Silty soil- weakly acidic	Ni and Zn	The biochar decreased significantly the water soluble Ni and Zn in soil, and their concentrations in plant	Shaheen et al. (2015a)
Woody biomass	900°C	0%, 1%, 2.5%, and 5% Pot experiment	Serpentine soil	Cr, Ni, and Mn	Immobilization of Cr, Ni, and Mn in soil. Bioaccumulation of toxic elements decreased in tomato plants, especially with biochar rate of 5%. Bioavailable Cr, Ni, and Mn decreased by 99%, 61%, and 42%, respectively, with 5% application rate	Herath et al. (2015)
Chicken litter Wood shaving	550°C 650°C	5% Incubation experiment	Entisol, andisol, vertisol, inceptisol	Cd	Neither biochar affected the Cd bioavailability in Andisol and Inceptisol, due to the high sorption capacity of both soils. Chicken litter biochar reduced the Cd bioavailability in soils with lower sorption capacity (Entisol and Vertisol soils) with up to 50%	Qi et al. (2018)
Sugarcane bagasse	450°C	0, 1.5, 2.25, and 3.0 t ha^{-1} Field experiment	Sandy loam	Cd, Cu, and Pb	Bioavailability of PTE decreased with increasing rate of biochar. Biochar promoted the transformation of the elements from labile fractions to more stable fractions	Nie et al. (2018)
Barley straw, Tomato waste, Chicken manure, Duck manure, Swine manure	500°C	0%, 2.5%, and 5.0% (w/w) Greenhouse experiment	Sandy clay loam	Cd	Immobilization of Cd was more pronounced by the tomato waste and chicken manure derived biochars by 34% and 76%, respectively	Khan et al. (2017)

(*Continued*)

TABLE 14.1 (Continued)

Biochar Feedstock	Pyrolysis Temperature	Application Rate/Type of Experiment	Soil Type (Texture/ Classification)	Element	Impact	References
Chicken manure	500°C	0%, 5%, and 10% (w/w) Incubation experiment	Sandy	Cu	Immobilization of Cu in soil. The Cu concentration in the organic-bound and residual fractions increased in the biochar amended soil as compared to control	Meier et al. (2017)
Bamboo Rice straw	500°C	0%, 1%, and 5% Incubation experiment	Sandy loam	Cd, Cu, Pb, and Zn	Immobilization of different PTE with both types of biochar, due to the increased pH, electrical conductivity, and CEC in biochar amended soils	Yang et al. (2016)
Tobacco stalk and Dead pigs	450–650°C	0%, 1%, 2.5%, and 5% Plant growth experiment	Clay loam	Cd and Zn	Immobilization of Cd and Zn with both biochars, especially with higher application rates, due to the rise in soil pH	Yang et al. (2017)
Rice hull	500°C	5% Incubation, then microcosm experiment under dynamic redox conditions	Loamy sand	Cd, Cu, Ni, and Zn	The biochar increased the Cd, Cu, Ni, and Zn concentrations in the dissolved phase under flooded-oxic conditions. The biochar increased the phytoavailability and potential mobility of those elements under oxic conditions as well	El-Naggar et al. (2018)
Soybean stover Pine needles	300°C 700°C	5% Incubation experiment	Sandy loam	Pb and Cu	Immobilization of Pb and Cu was more pronounced by soybean stover biochar than pine needles biochar	Ahmad et al. (2016)
Ramie stick Rice straw	350–550°C	1% (w/w) Incubation experiment	Clayey soil	Cd and Pb	Immobilization of Cd and Pb in soil with both biochars, due to the rise in soil pH	Shen et al. (2016)

(Qin et al., 2018). For example, Khan et al. (2017) used biochars produced from barley straw, tomato green waste, chicken manure, duck manure, and swine manure and found that biochars from tomato green waste and chicken manure were more effective than the other biochars in reducing Cd mobilization and its Cd uptake by pakchoi cultivars. The significant impact of oil mallet and wheat chaff biochars on immobilizing Cd in soil was also reported by Zhang et al. (2013). Also, Tan et al. (2015) investigated the impact of bamboo (BB), coconut shells, pine wood shavings, and sugarcane bagasse derived biochars on Cd mobilization and found that the BB biochar was more effective for Cd immobilization than the others.

Recently, Qi et al. (2018) found that chicken litter biochar reduced Cd bioavailability by 50% as compared to the control in low sorption capacity soils. Also, Rehman et al. (2018) reported that application of rice straw biochar in combination with farmyard manure (0.05% each) decreased the phytoavailability and uptake of Cd by wheat and rice grains. Similar to Cd, biochars were also effective for immobilization of Pb and other PTE. In this respect, Bian et al. (2014) indicated that wheat straw biochar significantly reduced the mobility and plant uptake of Cd and Pb during a 3-year field experiment using rice plant (Fig. 14.1). Similarly, Shaheen and Rinklebe (2015) investigated the impact of biochar on the mobilization, fractionation, and plant uptake of Cd and Pb in a contaminated floodplain soil. They found that biochar improved plant growth of rapeseed, decreased mobile

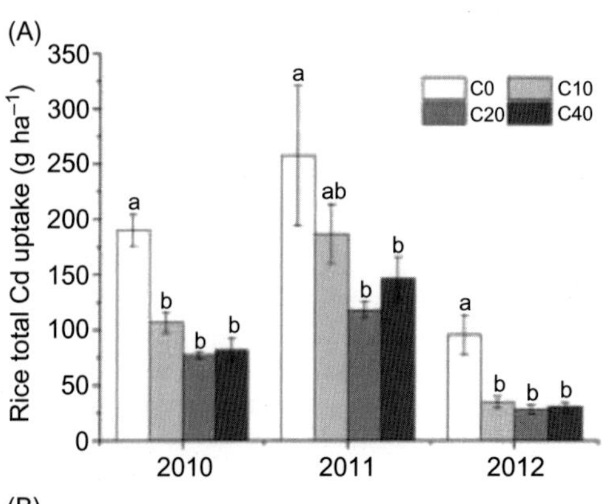

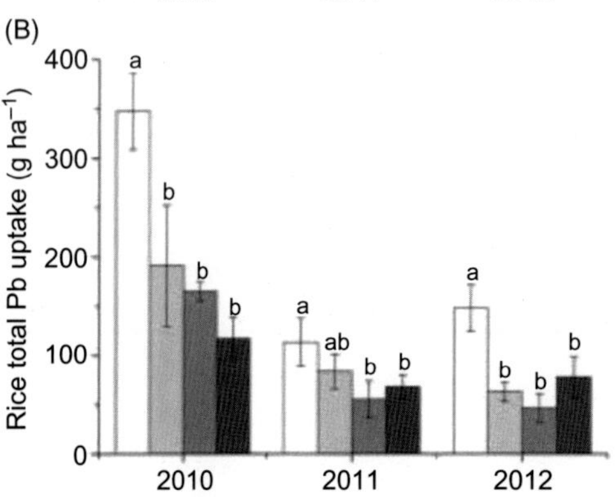

FIGURE 14.1 Total Cd (A) and Pb (B) uptake by rice across 3 years field experiment (mean ± S.D., $n = 3$). The bar above the block represents the standard deviation of three replicates; different letters above the blocks indicate significant differences ($P < 0.05$) between the biochar treatments within single year. Source: *Reproduced from Bian, R., Joseph, S., Cui, L., Pan, G., Li, L., Liu, X., et al., 2014. A three-year experiment confirms continuous immobilization of cadmium and lead in contaminated paddy field with biochar amendment. J. Hazard. Mater. 272, 121–128 with permission from the publisher.*

fraction of Cd by 18% and Pb by 41%, and decreased plant tissue concentrations of Cd by 23% and Pb by 63% as compared to the untreated soil. Likewise, Moon et al. (2013) found that soybean stover-derived biochar is a useful immobilizing agent for Pb in contaminated soil.

Recently, Qin et al. (2018) evaluated the effect of biochars derived from BB and pig (PB) on the leachability of Cd, and Pb in two soils with different organic carbon content. They found that application of PB to the low organic carbon soil significantly reduced the leaching loss by up to 38% for Cd, and 71% for Pb, which might be related to the higher specific surface area, surface alkalinity, pH, and mineral contents of PB than those of BB.

Yang et al. (2016) found that rice straw and BB biochar, in particular, rice straw biochar, decreased significantly the bioavailability of Cd, Cu, Pb, and Zn in soil. Similarly, Yang et al. (2017) reported that tobacco stalks and dead PBs' biochars, the earlier particularly, decreased the $CaCl_2$-extractable Cd and Zn. Prapagdee et al. (2014) demonstrated that cassava stem biochar decreased the bioavailability and uptake of Zn by green bean plants. The effective reducing of the bioavailability and uptake of Cd, Pb, and Zn by jack bean plants in mine polluted soils was reported by Puga et al. (2015).

In a field trial over four rice seasons, Chen et al. (2016) found that biochar application with 20 and 40 t ha^{-1} reduced Cd and Zn mobility significantly as compared to the untreated soil (Fig. 14.2). Biochar reduced rice grain Cd by a maximum of 61%, 86%, and 57% over the last three seasons. The translocation of Cd from rice roots to shoots was reduced from 20% to 80% by biochar. Zinc accumulation in the rice grains was not decreased by biochar application, although available soil Zn was sharply reduced (35%–91%).

Sewage sludge-derived biochar has been used for improving soil properties and reducing the solubility and bioavailability of PTE in agricultural soils. For example, Khan et al. (2013) investigated the influence of sewage sludge-derived biochar upon biomass yield and bioaccumulation of various metal(loid)s and found that the biochar increased soil pH and decreased bioavailability and bioaccumulation of As, Cr, Co, Ni, and Pb (but not Cd, Cu, and Zn). Also, Méndez et al. (2012) found that sewage sludge derived biochar decreased the mobile forms of Cu, Ni, Zn, Cd, and Pb and their plant-available forms.

In addition, Zhu et al. (2015) reported that wine lees-derived BC increased soil pH and reduced soil exchangeable Cr, Ni, Cu, Pb, Zn, and Cd and their content in the plant roots, stems, leaves, rice husk, and rice grains as compared to the control. The significant impact of wood-derived biochar in reducing bioaccumulation of Cr and Ni in tomato plants was reported by Herath et al. (2015). The effective reducing of mobile Cd and Cu by hardwood-derived biochar (57.9% for Cd and 63.8% for Cu as compared to the untreated soil) was documented in a long term (3 years of incubation) by Li et al. (2017). Similar positive impact of soybean biochar on (im)mobilization of Pb and Cu was observed by Ahmad et al. (2016).

The impact of biochar on the mobilization, fractionation, and plant uptake of Ni and Zn in a contaminated floodplain soil was studied by Shaheen et al. (2015a). They found that biochar decreased the water soluble Ni by 99% and Zn by 56%, decreased the soluble + exchangeable fraction of Ni by 25% and Zn by 17%, and decreased plant tissue concentrations of Ni by 31% and Zn by 17% compared to untreated soil. Similarly, Rehman et al. (2016) reported that biochar addition to soil stabilized Ni in the soil and

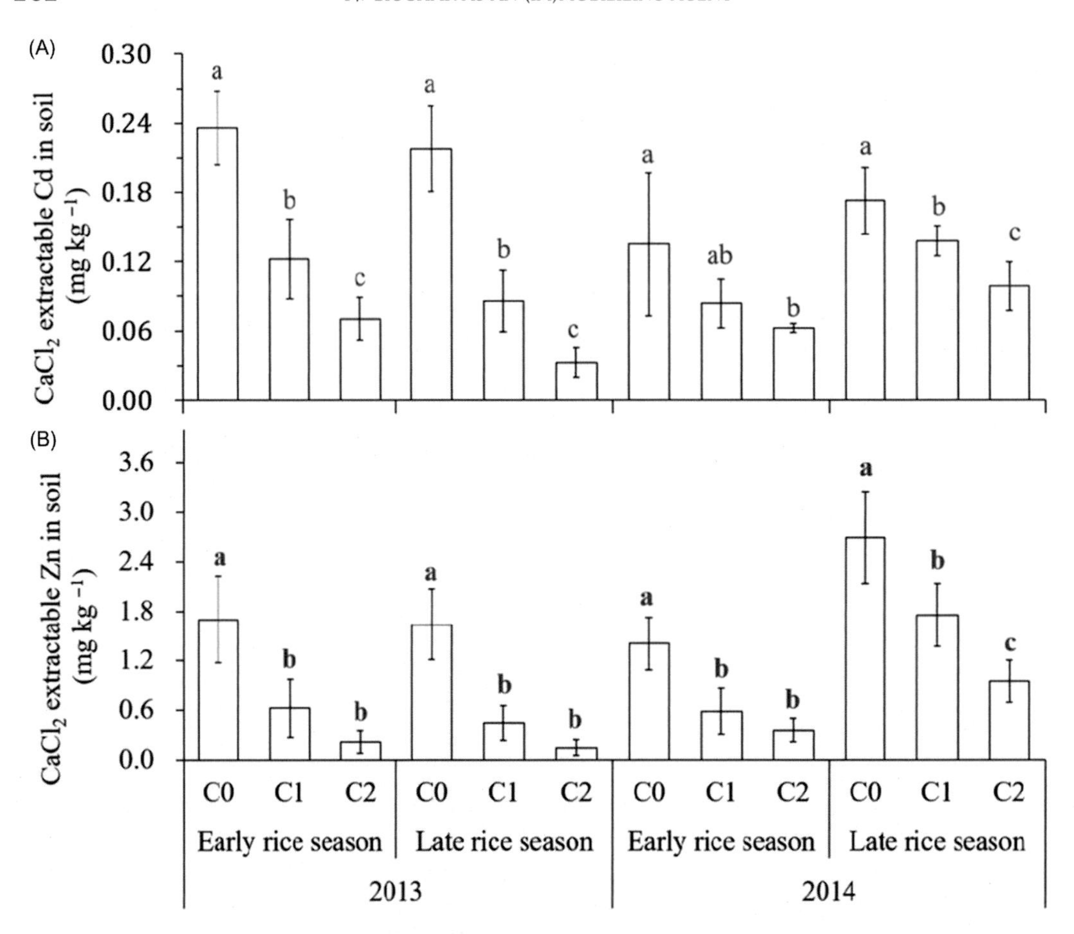

FIGURE 14.2 Effects of biochar on $CaCl_2$ extractable Cd and Zn concentrations (means ± S.D., $n = 6$). C0, C1, and C2 indicate the biochar application rates of 0, 20, and 40 t ha^{-1}. The different lower case letters indicate a significant difference between biochar treatments with the same cultivar in each season ($P < 0.05$). Source: *Reproduced from Chen, D., Guo, H., Li, R., Li, L., Pan, G., Chang, A., et al., 2016. Low uptake affinity cultivars with biochar to tackle Cd-tainted rice—a field study over four rice seasons in Hunan, China. Sci. Total Environ. 541, 1489–1498 with permission from the publisher.*

alleviated Ni toxicity by decreasing its uptake by maize seedlings. In another study by Turan et al. (2017), the application of biochar to Ni-contaminated soil reduced the Ni phytoavailability for sunflower seed up to 17% as compared to the un-treated soil. In a multi-contaminant soil, the study of Beesley et al. (2010) showed that biochar reduced significantly the mobility of Cd and Zn and their pore water concentrations during 60 days of field study. In dynamic redox conditions, concentrations of dissolved Al, As, Cd, Cu, Ni, and Zn were significantly lower in a biochar-treated contaminated floodplain soil than in the untreated soil (Rinklebe et al., 2016; Fig. 14.3).

Biochar also showed a promising performance in remediation of Hg contaminated soils (Wang et al., 2017). Addition of malt spent rootlets derived biochar at rates of

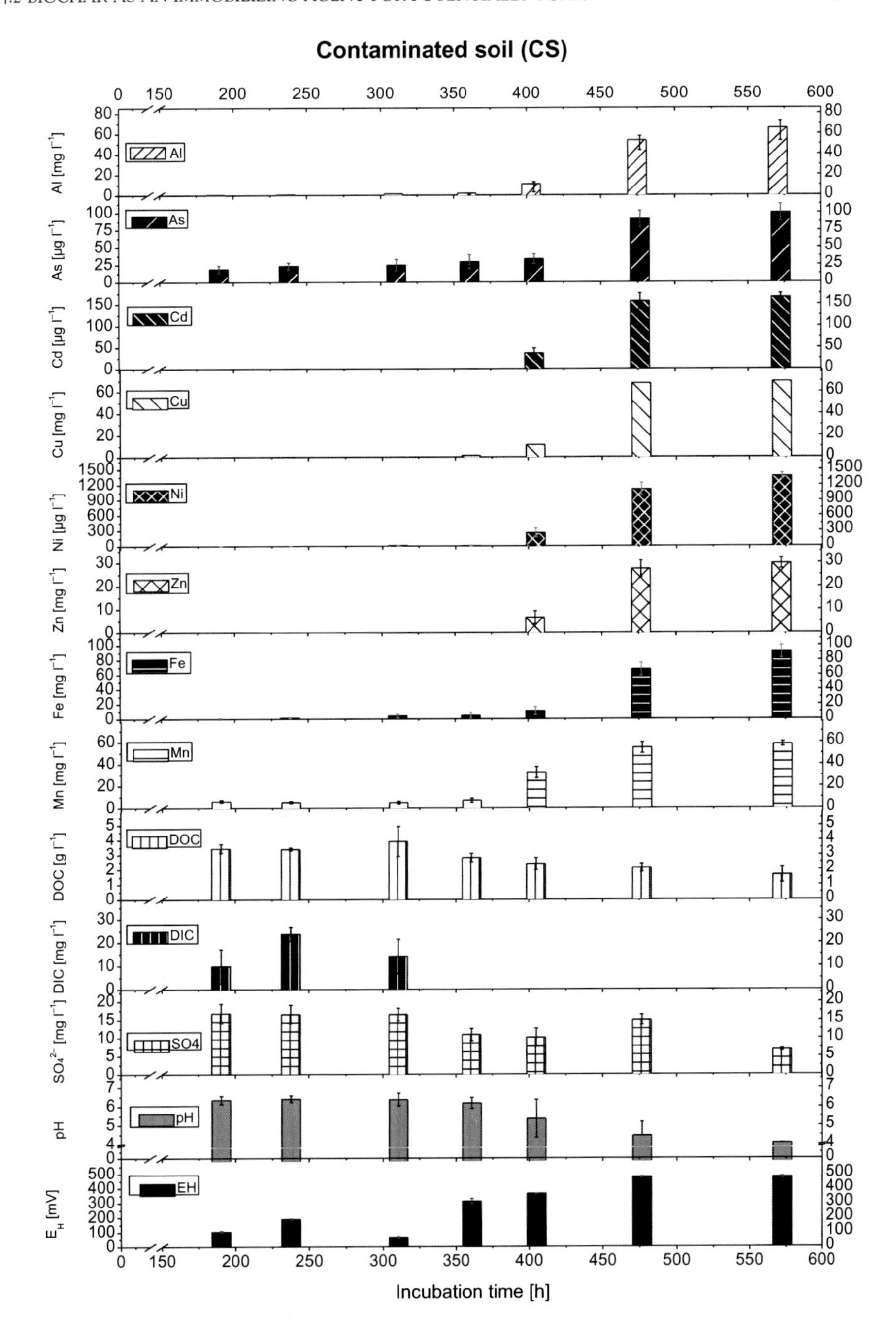

FIGURE 14.3 Impact of biochar on release dynamics of dissolved As, Cd, Cu, Ni, Zn, Fe, Mn, DOC, dissolved inorganic carbon (DIC), SO_4^{2-}, and pH in a contaminated floodplain soil under pre-definite Eh-conditions. Columns represent mean and whiskers represent standard deviation of three replicates using biogeochemical microcosm systems. Source: *Reproduced from Rinklebe, J., Shaheen, S.M., Frohne, T., 2016. Amendment of biochar reduces the release of toxic elements under dynamic redox conditions in a contaminated floodplain soil. Chemosphere 142, 41–47 with permission from the publisher.*

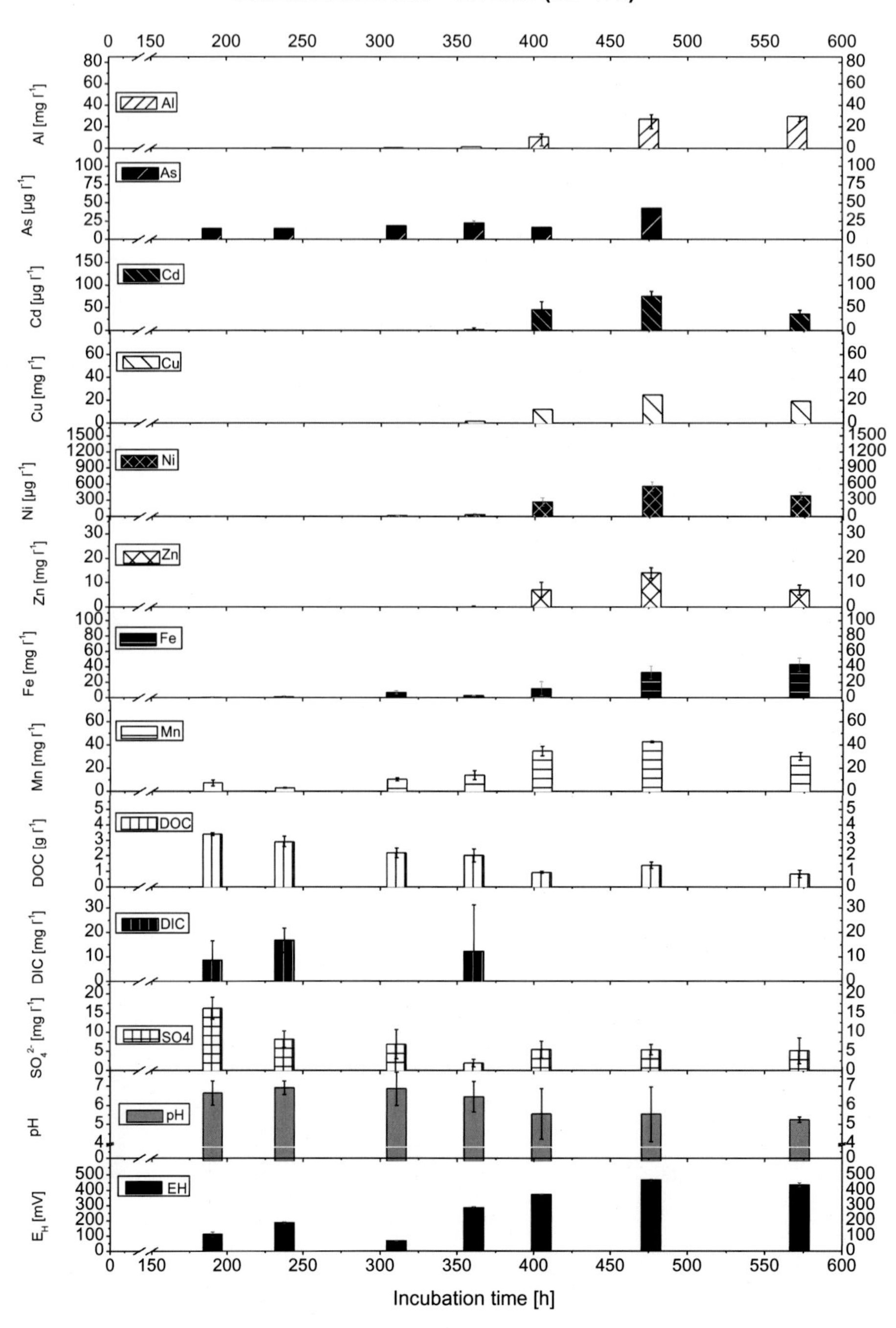

FIGURE 14.3 (Continued)

0.3–1.0 g L^{-1} to the Hg-polluted solution (pH 5.0) under 25°C decreased the concentration of Hg by 71%–100% (Boutsika et al., 2014).

Kong et al. (2011) reported that application of soybean stalk-based biochar at a rate of 0.33 g L^{-1} to a solution containing 2 μg Hg^{2+} L^{-1} under 25°C decreased the concentration of Hg by 86.4%. The removal of Hg by biochar depends upon the biochar properties and its production condition such as pyrolysis temperature. For example, Liu et al.(2016) found that the concentration of Hg was decreased by 90% when the solution was added with biochar produced under higher temperature (600–700°C), while that was decreased by 40%–90% when the solution was added with biochar pyrolyzed under lower temperature (300°C).

The immobilization of Hg^{2+} by biochar was attributed mainly to the interactions between Hg^{2+} and functional groups on biochar. Liu et al. (2016) found that the Hg was complexed predominately by sulfur, oxygen, and chloride in biochars as demonstrated by synchrotron based-X-ray absorption spectroscopy results. The chemical modification of biochar might enhance the ability of biochar to remove Hg. Tan et al. (2016) investigated the modification of corn straw based-biochar using Na_2S, KOH and activated carbon on Hg removal from solution. Results showed that the sorption capacity of Na_2S, KOH and activated carbon-modified biochar for Hg increased by 77%, 32%, and 42% respectively compared to the non-modified biochar, and S-modified biochar was more efficient than the KOH and activated carbon-modified biochar. O'Connor et al. (2018) also reported that the modification of biochar by sulfur could increase the adsorption of Hg^{2+} by biochar by −73% as compared to the non-modified biochar. In addition, application of this S-modified biochar decreased the concentration of Hg in toxicity characteristic leaching procedure (TCLP) leachates by 95.4%–99.3% in Hg-contaminated soil compared to untreated soil (Gilmour et al., 2018; O'Connor et al., 2018). The removal of Hg by S-modified biochar is attributed to the formation of strong Hg-S coordination in biochar (Liu et al., 2018).

Exploiting biochar to treat Hg-contaminated soils/sediments was reported. For example, Gomez-Eyles et al. (2013) reported that the pine dust-phragmites derived biochar was effective in the sorption of methyl mercury (MeHg) from sediment, and the concentration of MeHg in pore water was decreased by 57%–92%. Bussan et al. (2016) found that the methylation rate of ^{200}Hg was reduced by 88% when biochar was added in sediments. On the other hand, Shu et al. (2016a) found that application of biochar increased the total concentration of MeHg in the soil. However, Shu et al. (2016b) explained that biochar reduced the phytoavailability of MeHg and its accumulation in rice plants. A Potential pathway of decreasing the total MeHg levels in soils is that adding biochar assisted by nitrate, which decreases the concentration of MeHg by 34% as compared to the single biochar treatment (Zhang et al., 2018). Also, biochar was exploited to remove the gas element gaseous Hg (Hg^0). Moreover, it was usually modified by chloride to increase its activated C–Cl groups to enhance the adsorption capacity of biochar for Hg (Hg^0) (Li et al., 2015; Wang et al., 2018a,b).

Chromite (Cr(III)) and chromate (Cr(VI)) are the major two oxidation states of Cr in contaminated soils. The reduction of Cr(VI) to Cr(III) induced by biochar helps to reduce its bioavailability and toxicity in soils (Choppala et al., 2016). Therefore, biochar with high amounts of functional groups can enhance Cr(VI) immobilization via reduction of toxic

Cr(VI) to less toxic Cr(III) and as such facilitates Cr(VI) remediation in soil (Park et al., 2011; Choppala et al., 2012, 2016; Shaheen et al., 2018).

The above studies revealed that immobilization of PTE with the use of different types of biochars reduced the bioavailability and uptake of these pollutants by plants.

14.2.2 Immobilization Mechanisms of Potentially Toxic Elements by Biochar

As mentioned above, biochar can adsorb PTE in soils (Beesley et al., 2010). The PTE immobilization ability of biochar in soil pore water might be attributed to their surface properties (Zhen et al., 2008). For instance, the abundance of negatively charged surface sites on biochars due to presence of functional groups such as carboxyl, hydroxyl and phenolic ($-COO^-$ and $-OH^-$) could bind cationic PTE and thus reduce their mobility in soils (Houben et al., 2013; Yin et al., 2015; Zama et al., 2017; Shaheen et al., 2018).

A number of studies showed that biochar is able to reduce the release of PTE by sorption of these elements as a result of its active components, specific micro-pore structure, and its functional active groups (Uchimiya et al., 2011; Qian et al., 2015, 2016; Xu et al., 2016). Zhao et al. (2015) also suggested that the sorption of Pb, Cu, and Cd by different biochars may be affected by its pore structure and its surface functional groups. Increasing soil pH using ash-rich biochars promote its ability to bind Cu as a result of activation of surface functional groups on soil particles (Lima et al., 2010; Mukherjee et al., 2011; Tong et al., 2011; Zhang et al., 2015). Bian et al. (2014) found that Cd and Pb may be bonded with the mineral phases of Al, Fe and P on the contaminated biochar particle and/or also might occur due to cation exchange on the porous carbon structure. Xu et al.(2016) reported that the sorption of Hg by hickory chips derived biochar was due to the complexation of Hg with C=C and C=O via Hg-π binding, while Hg was complexed with phenolic hydroxyl (CO^-) and carboxylic (COO^-) groups in bagasse derived biochar (Xu et al., 2016). Dong et al. (2015) reported that the predominate functional groups in Brazilian pepper-derived biochar that bond with Hg were phenolic hydroxyl and carboxylic groups, and the reactions among those functional groups and Hg were affected by the pyrolysis temperature for biochar. For example, about 23% − 31% and 77% − 69% of the complexed Hg was associated with carboxylic and phenolic hydroxyl groups on the biochar produced at 300−450°C respectively, while about 91% of the sorbed Hg was associated with a graphite-like domain on an aromatic structure on the biochar produced at 600°C (Dong et al., 2015).

Biochars can adsorb both negatively and positively charged species, including $AsOx^{n-}$ and $CrOx^{n-}$ (Solaiman and Anawar, 2015). As compared to other PTE, Cr shows different behavior with biochar. The Cr species govern its sorption/immobilization mechanisms by biochar (Li et al., 2017; Mandal et al., 2017; Shaheen et al., 2018). Reduction of Cr(VI) to Cr(III) followed by Cr(III) complexation is a major immobilization mechanism for Cr(VI) by biosorbents, including biochar. The pH-dependent sorption of Cr(VI) in biochar shows different behavior as compared to other PTEs, where its sorption capacity increases under acidic conditions. This behavior might be explained by the highly protonated biochar surface at low pH value, which favors the formation of ion-pair interaction mechanism between chromate (CrO_4^{2-}) ion and the positively charged functional groups

Concomitant reduction and immobilization of chromium by biochar

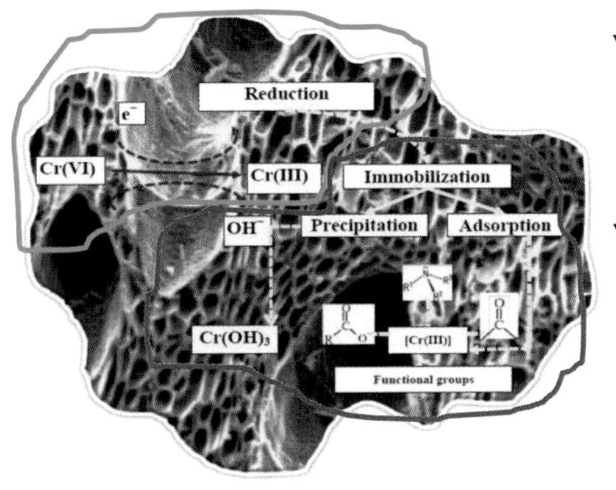

- ✓ Carbon amendments enhanced Cr(VI) reduction to Cr(III) species through supply of electrons.
- ✓ The Cr(III) species subsequently gets immobilized through adsorption and precipitation reactions, thereby decreasing its mobility and bioavailability.

FIGURE 14.4 A schematic diagram of the impact of biochar on Cr in soil (biochar-mediated interactions with Cr in soil).

(Abdel-Fattah et al., 2015; Shaheen et al., 2018). Also, the presence of electron donor functional groups (−C−OH, C−O, C−O−R) could promote sorption of metal anions such as Cr(VI) on biochar by sorption coupled reduction of Cr(VI) to Cr(III) in order to enable surface sorption (Shaheen et al., 2018). The oxygen-containing compounds in biochars may be responsible for the reducing Cr(VI) to Cr(III) and its subsequent sorption by biochar (Hsu et al., 2009; Choppala et al., 2012; Mohan et al., 2014). The reduction mechanisms can occur in the aqueous phase (Wang et al., 2015), or on solid biochar surface (Agrafioti et al., 2014). These two processes, that is, reduction followed by sorption, are presented in Fig. 14.4. The above-mentioned studies demonstrate that application of biochar in redox-sensitive media such as aquatic environments, including wetlands, might be an efficient tool to reduce and detoxify PTE such as Cr(VI) (Yuan et al., 2018).

In conclusion, different spectroscopic techniques have been used to depict sorption/immobilization of PTE on biochar (Zama et al., 2017; Shaheen et al., 2018). The studies suggested an element specific sorption of PTE by biochar. Moreover, the properties of wood biochar (e.g., organic functional groups, mineral content, ionic content, and π-electrons) have a direct influence on sorption/immobilization mechanisms of PTE (Shaheen et al., 2018).

14.3 BIOCHAR AS A MOBILIZING AGENT FOR POTENTIALLY TOXIC ELEMENTS IN CONTAMINATED SOILS: MOBILIZATION MECHANISMS

In this section, we briefly describe some studies indicating that biochar may increase the solubility, mobility, and phytoavailability of PTE in soils. For example, Prapagdee et al. (2014) demonstrated that cassava biochar increased Cd uptake by green beans. Khan et al. (2013) found that sewage sludge-derived biochar increased the bioavailability of Cd, Cu, and Zn. Beesley and Dickinson (2011) reported that biochar increased the concentration of Cu in the pore water more than 30-fold, and they attributed that to the increase of dissolved organic carbon (DOC) in the biochar-treated soil as compared to that which was untreated. Similarly, Park et al. (2011) found that addition of biochar leads to an increase in DOC and thus to Cu mobilization. Beesley et al. (2010) also found that biochar increased the pore water concentrations of As and Cu. In addition, Rinklebe and Shaheen (2015) assessed the biochar-induced mobilization and phytoavailability of Cu in a contaminated floodplain soil. Application of biochar increased the rapeseed biomass production compared to the control. Application of biochar increased the water soluble Cu by 54% as compared to the control. Incorporation of the soil with biochar increased Cu plant tissue concentration and uptake by rapeseed by approximately 160% as compared to the control. Their results (Rinklebe and Shaheen, 2015) suggested that under the high concentrations of total and mobile Cu, application of biochar with the 1% application rate is low for Cu immobilization, and this low rate enhanced the uptake of Cu by rapeseed, which might be useful for increasing Cu phytoremediation and removal from this highly contaminated soil using bioenergy crops like rapeseed. Recently, El-Naggar et al. (2018) found that rice husk biochar increased the dissolved concentration of Cd, Ni, Zn, and in particular Cu under oxic conditions in the treated soil as compared to the un-treated soil (Fig. 14.5).

Application of biochar to As contaminated soils affect its mobilization and bioavailability. The effect of biochar on the mobilization of As depends on As species (Bolan et al., 2014; Choppala et al., 2016). In this respect, Yin and Zhu (2016) investigated the effect of biochars (2% and 5%) derived from water hyacinth and rice straw on immobilization of As in paddy soil. Both biochars with the two rates increased the mobilization of As compared to no-biochar control. Ahmad et al. (2016) found that biochars derived from soybean stover and pine needles increased significantly the mobility of As. Zheng et al. (2012) investigated how BCs produced from different parts of rice plants (straw, husk, and bran) can influence the solubility of As in a historically multi-metal contaminated soil and found that biochar amendments significantly increased the pore water concentrations of As. Beesley and Dickinson (2011) reported that As concentrations in pore water increased after biochar application to soil, which is closely related to the increase in pH induced by biochar. Beesley et al. (2013) determined the impact of residues of orchard prune biochar on the pore water concentrations of As and they exhibited a greater As concentrations in pore water (500–2000 mg L^{-1}) following the biochar application.

Recently, a soil contaminated with As (2047 mg kg^{-1}) was pre-incubated for 105 days with pine biochar under dynamic redox conditions (Beiyuan et al., 2017). The biochar increased the mobility of As at moderate reducing conditions, while enhanced the phytoavailability of As under oxidizing condition. Other studies (e.g., Beesley et al., 2014) revealed that biochar application to As contaminated soils may enhance As bioavailability, which may enhance phytoremediation of As contaminated soils (Beesley et al., 2014; Choppala et al., 2016). Biochar could promote the reduction of As(V) to As(III), thereby increasing its bioavailability in soils (Bolan et al., 2014; Choppala et al., 2016). A brief illustration how BC interferes with soil As is shown in Fig. 14.6.

Beesley et al. (2010) explained the increase of As mobility in the biochar treated soil for the increase in DOC and soil pH for As mobilization. This hypothesis was further supported by Yin and Zhu (2016), who found that biochar increased soil pH and thus increased the

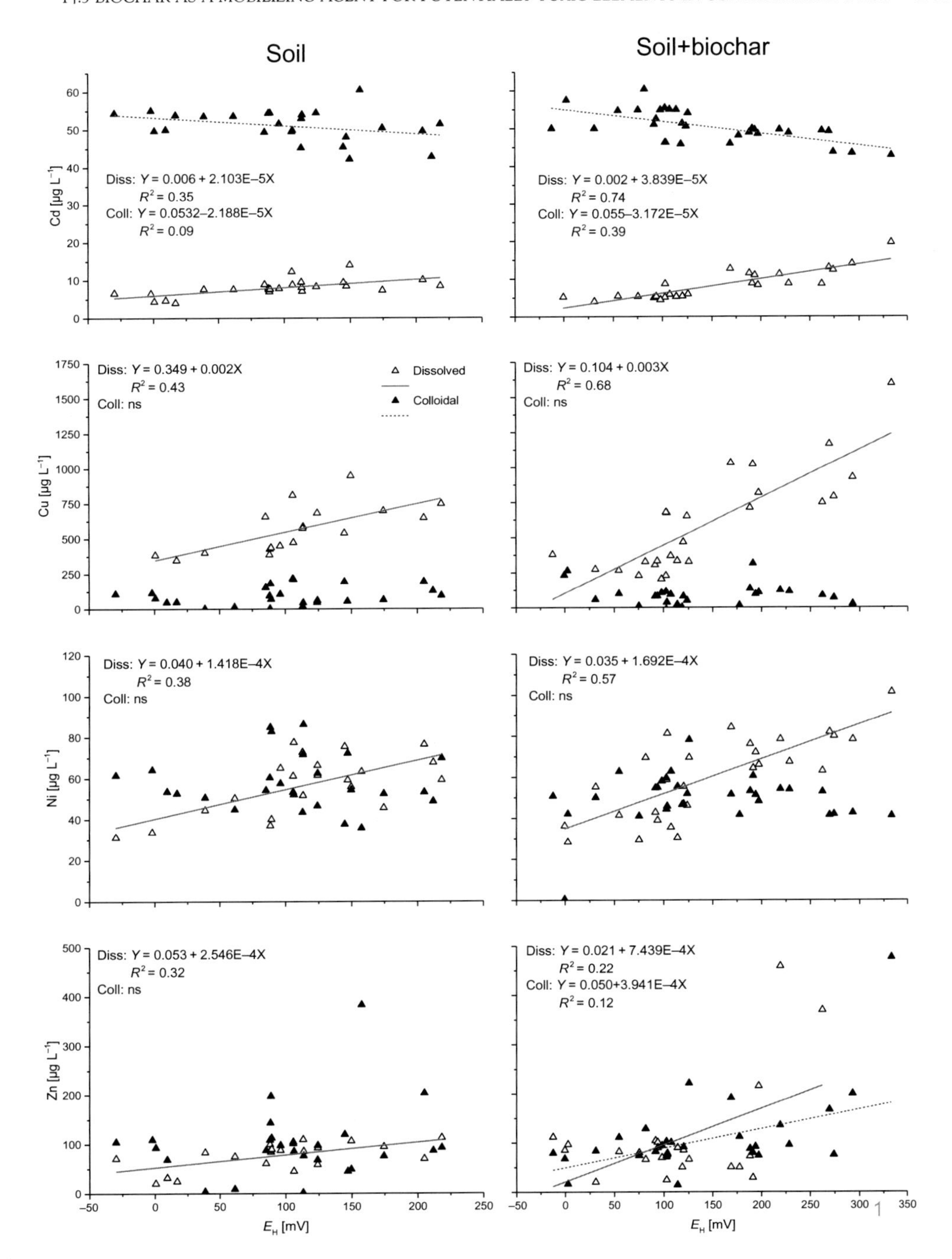

FIGURE 14.5 Relationships between E_H and Cd, Cu, Ni, and Zn in the dissolved and colloidal phases of the non-treated soil and the biochar-treated soil. Source: *Reproduced from El-Naggar, A., Shaheen, S.M., Ok, Y.S., Rinklebe, J., 2018. Biochar affects the dissolved and colloidal concentrations of Cd, Cu, Ni, and Zn and their phytoavailability and potential mobility in a mining soil under dynamic redox-conditions. Sci. Total Environ. 624, 1059–1071 with permission from the publisher.*

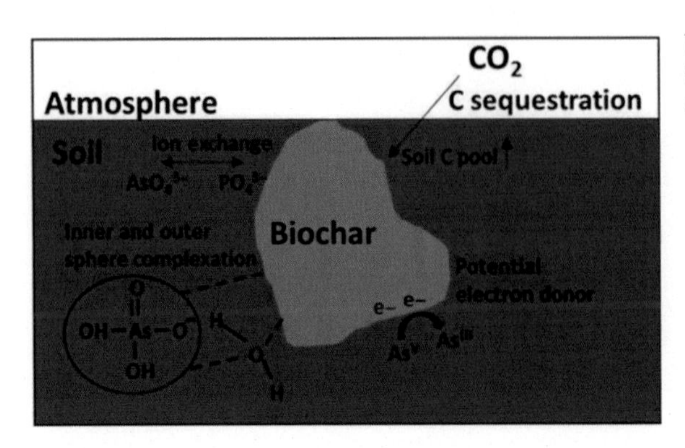

FIGURE 14.6 A schematic diagram of the impact of biochar on arsenic in soil (biochar-mediated interactions with As in soil).

mobilization of As up to 16.7% as compared to the untreated soil. Increasing mobility of As in biochar treated soils may also result from the competition of OH^- with $HAsO_4^{2-}$ for binding sites under alkaline conditions (Inyang et al., 2010; Yin and Zhu, 2016).

14.4 CONCLUSIONS

Applying biochars to PTE contaminated soils can have contradictory effects on the mobilization and phytoavailability of a specific PTE, depending on the soil and biochar properties and the nature and speciation of PTE. Thus, application of biochars to PTE contaminated soils may affect the bio-availability of soil PTE either by immobilizing or mobilizing these elements. However, those impacts are element specific and also depend on the biomass feedstock, application rate, and soil type. For example, biochar application to soils immobilized many PTE as a result of increasing soil pH, increasing soil surface area and sorption capacity, and by its sorption of these elements. However, immobilization of PTE by biochar is an element specific, and also affected by the properties of biochar (e.g., organic functional groups, mineral content, ionic content, and π-electrons), which have a direct influence on sorption/immobilization mechanisms of PTE.

Biochar may increase the mobilization of Cu as a result of the increase of DOC. Biochar application increases the solubility and mobilization of As. The increase of As mobility in the biochar treated soil might be explained by the increase in DOC and soil pH and thus the competition of OH^- with $HAsO_4^{2-}$ for binding sites under alkaline conditions. Also, biochar could promote the reduction of As(V) to As(III), thereby increasing its bioavailability in soils. Increasing the solubility of PTE such as Cu and As using biochar might be useful for increasing phytoextraction and removal of these elements from this highly contaminated soil using bioenergy crops.

Acknowledgments

We thank the German Alexander von Humboldt Foundation for financial support of the experienced researcher's fellowships of Prof. Dr. Shaheen (Ref. 3.4-EGY-1185373-GF-E) at the University of Wuppertal, Germany. Also, we would like to express our thanks to the authors and the publishers for their permission to reuse data from publications which are mentioned in the text.

References

Abdel-Fattah, T.M., Mahmoud, M.E., Ahmed, S.B., Huff, M.D., Lee, J.W., Kumar, S., 2015. Biochar from woody biomass for removing metal contaminants and carbon sequestration. J. Ind. Eng. Chem. 22, 103–109.

Agrafioti, E., Kalderis, D., Diamadopoulos, E., 2014. Ca and Fe modified biochars as adsorbents of arsenic and chromium in aqueous solutions. J. Environ. Manage. 146, 444–450.

Ahmad, M., Lee, S.S., Yang, J.E., Ro, H.M., Lee, Y.H., Ok, Y.S., 2012. Effects of soil dilution and amendments (mussel shell, cow bone, and biochar) on Pb availability and phytotoxicity in military shooting range soil. Ecotoxicol. Environ. Saf. 79, 225–231.

Ahmad, M., Rajapaksha, A.U., Lim, J.E., Zhang, M., Bolan, N., Mohan, D., et al., 2014. Biochar as a sorbent for contaminant management in soil and water: a review. Chemosphere 99, 19–33.

Ahmad, M., Ok, Y.S., Rajapaksha, A.U., Lim, J.E., Kim, B.Y., Ahn, J.H., et al., 2016. Lead and copper immobilization in a shooting range soil using soybean stover-and pine needle-derived biochars: chemical, microbial and spectroscopic assessments. J. Hazard. Mater. 301, 179–186.

Antoniadis, V., Shaheen, S.M., Boersch, J., Frohne, T., Du Laing, G., Rinklebe, J., 2017a. Bioavailability and risk assessment of potentially toxic elements in garden edible vegetables and soils around a highly contaminated former mining area in Germany. J. Environ. Manage. 186, 192–200.

Antoniadis, V., Levizou, E., Shaheen, S.M., Ok, Y.S., Sebastian, A., Baum, C., et al., 2017b. Trace elements in the soil-plant interface: phytoavailability, translocation, and phytoremediation–a review. Earth Sci. Rev. 171, 621–645.

Beesley, L., Dickinson, N., 2011. Carbon and trace element fluxes in the pore water of an urban soil following greenwaste compost, woody and biochar amendments, inoculated with the earthworm *Lumbricus terrestris*. Soil Biol. Biochem. 43 (1), 188–196.

Beesley, L., Moreno-Jiménez, E., Gomez-Eyles, J.L., 2010. Effects of biochar and greenwaste compost amendments on mobility, bioavailability and toxicity of inorganic and organic contaminants in a multi-element polluted soil. Environ. Pollut. 158 (6), 2282–2287.

Beesley, L., Marmiroli, M., Pagano, L., Pigoni, V., Fellet, G., Fresno, T., et al., 2013. Biochar addition to an arsenic contaminated soil increases arsenic concentrations in the pore water but reduces uptake to tomato plants (*Solanum lycopersicum* L.). Sci. Total Environ. 454, 598–603.

Beesley, L., Inneh, O.S., Norton, G.J., Moreno-Jimenez, E., Pardo, T., Clemente, R., et al., 2014. Assessing the influence of compost and biochar amendments on the mobility and toxicity of metals and arsenic in a naturally contaminated mine soil. Environ. Pollut. 186, 195–202.

Beiyuan, J., Awad, Y.M., Beckers, F., Tsang, D.C., Ok, Y.S., Rinklebe, J., 2017. Mobility and phytoavailability of As and Pb in a contaminated soil using pine sawdust biochar under systematic change of redox conditions. Chemosphere 178, 110–118.

Bian, R., Joseph, S., Cui, L., Pan, G., Li, L., Liu, X., et al., 2014. A three-year experiment confirms continuous immobilization of cadmium and lead in contaminated paddy field with biochar amendment. J. Hazard. Mater. 272, 121–128.

Bolan, N., Kunhikrishnan, A., Thangarajan, R., Kumpiene, J., Park, J., Makino, T., et al., 2014. Remediation of heavy metal (loid) s contaminated soils–to mobilize or to immobilize? J. Hazard. Mater. 266, 141–166.

Boutsika, L.G., Karapanagioti, H.K., Manariotis, I.D., 2014. Aqueous mercury sorption by biochar from malt spent rootlets. Water Air Soil Pollut. 225 (1), 1805.

Bussan, D.D., Sessums, R.F., Cizdziel, J.V., 2016. Activated carbon and biochar reduce mercury methylation potentials in aquatic sediments. Bull. Environ. Contam. Toxicol. 96 (4), 536–539.

Chen, D., Guo, H., Li, R., Li, L., Pan, G., Chang, A., et al., 2016. Low uptake affinity cultivars with biochar to tackle Cd-tainted rice—a field study over four rice seasons in Hunan, China. Sci. Total Environ. 541, 1489–1498.

Choppala, G., Bolan, N., Kunhikrishnan, A., Bush, R., 2016. Differential effect of biochar upon reduction-induced mobility and bioavailability of arsenate and chromate. Chemosphere 144, 374–381.

Choppala, G.K., Bolan, N.S., Megharaj, M., Chen, Z., Naidu, R., 2012. The influence of biochar and black carbon on reduction and bioavailability of chromate in soils. J. Environ. Qual. 41 (4), 1175–1184.

Dong, D., Feng, Q., McGrouther, K., Yang, M., Wang, H., Wu, W., 2015. Effects of biochar amendment on rice growth and nitrogen retention in a waterlogged paddy field. J. Soils Sediments 15 (1), 153–162.

El-Naggar, A., Shaheen, S.M., Ok, Y.S., Rinklebe, J., 2018. Biochar affects the dissolved and colloidal concentrations of Cd, Cu, Ni, and Zn and their phytoavailability and potential mobility in a mining soil under dynamic redox-conditions. Sci. Total Environ. 624, 1059–1071.

Gill, M.D., 2008. Green Remediation: An EPA Perspective. EPA Region 9, San Francisco NASA/C3P. Int'l Workshop on P2 and Sustainable Development San Diego, CA–November 19, 2008.

Gilmour, C., Bell, T., Soren, A., Riedel, G., Riedel, G., Kopec, D., et al., 2018. Activated carbon thin-layer placement as an in situ mercury remediation tool in a Penobscot River salt marsh. Sci. Total Environ. 621, 839–848.

Gomez-Eyles, J.L., Yupanqui, C., Beckingham, B., Riedel, G., Gilmour, C., Ghosh, U., 2013. Evaluation of biochars and activated carbons for in situ remediation of sediments impacted with organics, mercury, and methylmercury. Environ. Sci. Technol. 47 (23), 13721–13729.

Herath, I., Kumarathilaka, P., Navaratne, A., Rajakaruna, N., Vithanage, M., 2015. Immobilization and phytotoxicity reduction of heavy metals in serpentine soil using biochar. J. Soils Sediments 15 (1), 126–138.

Houben, D., Evrard, L., Sonnet, P., 2013. Beneficial effects of biochar application to contaminated soils on the bioavailability of Cd, Pb and Zn and the biomass production of rapeseed (*Brassica napus* L.). Biomass Bioenergy 57, 196–204.

Hsu, N.H., Wang, S.L., Lin, Y.C., Sheng, G.D., Lee, J.F., 2009. Reduction of Cr(VI) by crop-residue-derived black carbon. Environ. Sci. Technol. 43 (23), 8801–8806.

Inyang, M., Gao, B., Pullammanappallil, P., Ding, W., Zimmerman, A.R., 2010. Biochar from anaerobically digested sugarcane bagasse. Bioresour. Technol. 101 (22), 8868–8872.

Khan, K.Y., Ali, B., Cui, X., Feng, Y., Yang, X., Stoffella, P.J., 2017. Impact of different feedstocks derived biochar amendment with cadmium low uptake affinity cultivar of pak choi (*Brassica rapa* ssb. chinensis L.) on phytoavoidation of Cd to reduce potential dietary toxicity. Ecotoxicol. Environ. Saf. 141, 129–138.

Khan, S., Chao, C., Waqas, M., Arp, H.P.H., Zhu, Y.G., 2013. Sewage sludge biochar influence upon rice (*Oryza sativa* L.) yield, metal bioaccumulation and greenhouse gas emissions from acidic paddy soil. Environ. Sci. Technol. 47 (15), 8624–8632.

Kong, H., He, J., Gao, Y., Wu, H., Zhu, X., 2011. Cosorption of phenanthrene and mercury (II) from aqueous solution by soybean stalk-based biochar. J. Agric. Food Chem. 59 (22), 12116–12123.

Lahori, A.H., Zhanyu, G.U.O., Zengqiang, Z.H.A.N.G., Ronghua, L.I., mahar, A., Awasthi, M.K., et al., 2017. Use of biochar as an amendment for remediation of heavy metal-contaminated soils: prospects and challenges. Pedosphere 27 (6), 991–1014.

Li, G., Shen, B., Li, F., Tian, L., Singh, S., Wang, F., 2015. Elemental mercury removal using biochar pyrolyzed from municipal solid waste. Fuel Process. Technol. 133, 43–50.

Li, H., Dong, X., da Silva, E.B., de Oliveira, L.M., Chen, Y., Ma, L.Q., 2017. Mechanisms of metal sorption by biochars: biochar characteristics and modifications. Chemosphere 178, 466–478.

Lima, I.M., Boateng, A.A., Klasson, K.T., 2010. Physicochemical and adsorptive properties of fast pyrolysis biochars and their steam activated counterparts. J. Chem. Technol. Biotechnol. 85 (11), 1515–1521.

Liu, P., Ptacek, C.J., Blowes, D.W., Landis, R.C., 2016. Mechanisms of mercury removal by biochars produced from different feedstocks determined using X-ray absorption spectroscopy. J. Hazard. Mater. 308, 233–242.

Liu, P., Ptacek, C.J., Elena, K.M., Blowes, D.W., Gould, W.D., Finfrock, Y.Z., et al., 2018. Evaluation of mercury stabilization mechanisms by sulfurized biochars determined using X-ray absorption spectroscopy. J. Hazard. Mater. 347, 114–122.

Mandal, S., Sarkar, B., Bolan, N., Ok, Y.S., Naidu, R., 2017. Enhancement of chromate reduction in soils by surface modified biochar. J. Environ. Manage. 186, 277–284.

Meier, S., Curaqueo, G., Khan, N., Bolan, N., Cea, M., Eugenia, G.M., et al., 2017. Chicken-manure-derived biochar reduced bioavailability of copper in a contaminated soil. J. Soils Sediments 17 (3), 741–750.

Méndez, A., Gómez, A., Paz-Ferreiro, J., Gascó, G., 2012. Effects of sewage sludge biochar on plant metal availability after application to a Mediterranean soil. Chemosphere 89 (11), 1354–1359.

Mohan, D., Sarswat, A., Ok, Y.S., Pittman Jr, C.U., 2014. Organic and inorganic contaminants removal from water with biochar, a renewable, low cost and sustainable adsorbent—a critical review. Bioresour. Technol. 160, 191–202.

Moon, D.H., Park, J.W., Chang, Y.Y., Ok, Y.S., Lee, S.S., Ahmad, M., et al., 2013. Immobilization of lead in contaminated firing range soil using biochar. Environ. Sci. Pollut. Res. 20 (12), 8464–8471.

Mukherjee, A., Zimmerman, A.R., Harris, W., 2011. Surface chemistry variations among a series of laboratory-produced biochars. Geoderma 163 (3–4), 247–255.

Nie, C., Yang, X., Niazi, N.K., Xu, X., Wen, Y., Rinklebe, J., et al., 2018. Impact of sugarcane bagasse-derived biochar on heavy metal availability and microbial activity: a field study. Chemosphere 200, 274–282.

O'Connor, D., Peng, T., Li, G., Wang, S., Duan, L., Mulder, J., et al., 2018. Sulfur-modified rice husk biochar: a green method for the remediation of mercury contaminated soil. Sci. Total Environ. 621, 819–826.

Park, J.H., Choppala, G.K., Bolan, N.S., Chung, J.W., Chuasavathi, T., 2011. Biochar reduces the bioavailability and phytotoxicity of heavy metals. Plant Soil 348 (1–2), 439.

Prapagdee, S., Piyatiratitivorakul, S., Petsom, A., Tawinteung, N., 2014. Application of biochar for enhancing cadmium and zinc phytostabilization in *Vigna radiata* L. cultivation. Water Air Soil Pollut. 225 (12), 2233.

Puga, A.P., Abreu, C.A., Melo, L.C.A., Beesley, L., 2015. Biochar application to a contaminated soil reduces the availability and plant uptake of zinc, lead and cadmium. J. Environ. Manage. 159, 86–93.

Qi, F., Lamb, D., Naidu, R., Bolan, N.S., Yan, Y., Ok, Y.S., et al., 2018. Cadmium solubility and bioavailability in soils amended with acidic and neutral biochar. Sci. Total Environ. 610, 1457–1466.

Qian, L., Chen, M., Chen, B., 2015. Competitive adsorption of cadmium and aluminum onto fresh and oxidized biochars during aging processes. J. Soils Sediments 15 (5), 1130–1138.

Qian, L., Zhang, W., Yan, J., Han, L., Gao, W., Liu, R., et al., 2016. Effective removal of heavy metal by biochar colloids under different pyrolysis temperatures. Bioresour. Technol. 206, 217–224.

Qin, P., Wang, H., Yang, X., He, L., Müller, K., Shaheen, S.M., et al., 2018. Bamboo-and pig-derived biochars reduce leaching losses of dibutyl phthalate, cadmium, and lead from co-contaminated soils. Chemosphere 198, 450–459.

Rehman, M.Z., Rizwan, M., Khalid, H., Ali, S., Naeem, A., Yousaf, B., et al., 2018. Farmyard manure alone and combined with immobilizing amendments reduced cadmium accumulation in wheat and rice grains grown in field irrigated with raw effluents. Chemosphere 199, 468–476.

Rehman, M.Z.U., Rizwan, M., Ali, S., Fatima, N., Yousaf, B., Naeem, A., et al., 2016. Contrasting effects of biochar, compost and farm manure on alleviation of nickel toxicity in maize (*Zea mays* L.) in relation to plant growth, photosynthesis and metal uptake. Ecotoxicol. Environ. Saf. 133, 218–225.

Rinklebe, J., Shaheen, S.M., 2015. Miscellaneous additives can enhance plant uptake and affect geochemical fractions of copper in a heavily polluted riparian grassland soil. Ecotox. Environ. Safe. 119, 58–65.

Rinklebe, J., Shaheen, S.M., Frohne, T., 2016. Amendment of biochar reduces the release of toxic elements under dynamic redox conditions in a contaminated floodplain soil. Chemosphere 142, 41–47.

Shaheen, S.M., Rinklebe, J., 2015. Impact of emerging and low cost alternative amendments on the (im) mobilization and phytoavailability of Cd and Pb in a contaminated floodplain soil. Ecol. Eng. 74, 319–326.

Shaheen, S.M., Rinklebe, J., Selim, M.H., 2015a. Impact of various amendments on immobilization and phytoavailability of nickel and zinc in a contaminated floodplain soil. Int. J. Environ. Sci. Technol. 12 (9), 2765–2776.

Shaheen, S.M., Tsadilas, C.D., Rinklebe, J., 2015b. Immobilization of soil copper using organic and inorganic amendments. J. Plant Nutr. Soil Sci. 178 (1), 112–117.

Shaheen, S.M., Shams, M.S., Khalifa, M.R., El-Daly, M.A., Rinklebe, J., 2017. Various soil amendments and wastes affect the (im)mobilization and phytoavailability of potentially toxic elements in a sewage effluent irrigated sandy soil. Ecotoxicol. Environ. Saf. 142, 375–387.

Shaheen, S.M., Niazi, N.K., Hassan, N.E.E., Bibi, I., Wang, H., Tsang, D.C., et al., 2018. Wood-based biochar for removal of potentially toxic elements (PTEs) in water and wastewater: a critical review. Int. Mater. Rev. Available from: https://doi.org/10.1080/09506608.2018.1473096.

Shen, Z., McMillan, O., Jin, F., Al-Tabbaa, A., 2016. Salisbury biochar did not affect the mobility or speciation of lead in kaolin in a short-term laboratory study. J. Hazard. Mater. 316, 214–220.

Shu, R., Dang, F., Zhong, H., 2016a. Effects of incorporating differently-treated rice straw on phytoavailability of methylmercury in soil. Chemosphere 145, 457–463.

Shu, R., Wang, Y., Zhong, H., 2016b. Biochar amendment reduced methylmercury accumulation in rice plants. J. Hazard. Mater. 313, 1–8.

Solaiman, Z.M., Anawar, H.M., 2015. Application of biochars for soil constraints: challenges and solutions. Pedosphere 25, 631–638.

Tan, G., Sun, W., Xu, Y., Wang, H., Xu, N., 2016. Sorption of mercury (II) and atrazine by biochar, modified biochars and biochar based activated carbon in aqueous solution. Bioresour. Technol. 211, 727–735.

Tan, X.F., Liu, Y.G., Gu, Y.L., Zeng, G.M., Wang, X., Hu, X.J., et al., 2015. Immobilization of Cd(II) in acid soil amended with different biochars with a long term of incubation. Environ. Sci. Pollut. Res. 22, 12597–12604.

Tomasevic, D.D., Dalmacija, M.B., Prica, M.D., Dalmacija, B.D., Kerkez, D.V., Bečelić-Tomin, M.R., et al., 2013. Use of fly ash for remediation of metals polluted sediment–green remediation. Chemosphere 92 (11), 1490–1497.

Tong, X.J., Li, J.Y., Yuan, J.H., Xu, R.K., 2011. Adsorption of Cu (II) by biochars generated from three crop straws. Chem. Eng. J. 172 (2–3), 828–834.

Turan, V., Ramzani, P.M.A., Ali, Q., Abbas, F., Iqbal, M., Irum, A., et al., 2017. Alleviation of nickel toxicity and an improvement in zinc bioavailability in sunflower seed with chitosan and biochar application in pH adjusted nickel contaminated soil. Arch. Agron. Soil Sci. 64, 1–15.

Uchimiya, M., Chang, S., Klasson, K.T., 2011. Screening biochars for heavy metal retention in soil: role of oxygen functional groups. J. Hazard. Mater. 190 (1–3), 432–441.

Uchimiya, M., Bannon, D.I., Wartelle, L.H., Lima, I.M., Klasson, K.T., 2012. Lead retention by broiler litter biochars in small arms range soil: impact of pyrolysis temperature. J. Agric. food chem. 60, 5035–5044.

Wang, H., Yang, X., He, L., Kouping Lu, K., et al., 2018a. Using biochar for remediation of contaminated soils. In: Luo, Y., Tu, C. (Eds.), Twenty Years of Research and Development on Soil Pollution and Remediation in China. Springer. Available from: http://dx.doi.org/10.1007/978-981-10-6029-8_47.

Wang, J., Xia, J., Feng, X., 2017. Screening of chelating ligands to enhance mercury accumulation from historically mercury-contaminated soils for phytoextraction. J. Environ. Manage. 186, 233–239.

Wang, S., Gao, B., Li, Y., Mosa, A., Zimmerman, A.R., Ma, L.Q., et al., 2015. Manganese oxide-modified biochars: preparation, characterization, and sorption of arsenate and lead. Bioresour. Technol. 181, 13–17.

Wang, T., Liu, J., Zhang, Y., Zhang, H., Chen, W.Y., Norris, P., et al., 2018b. Use of a non-thermal plasma technique to increase the number of chlorine active sites on biochar for improved mercury removal. Chem. Eng. J. 331, 536–544.

Xu, X., Schierz, A., Xu, N., Cao, X., 2016. Comparison of the characteristics and mechanisms of Hg(II) sorption by biochars and activated carbon. J. Colloid Interface Sci. 463, 55–60.

Yang, X., Liu, J., McGrouther, K., Huang, H., Lu, K., Guo, X., et al., 2016. Effect of biochar on the extractability of heavy metals (Cd, Cu, Pb, and Zn) and enzyme activity in soil. Environ. Sci. Pollut. Res. 23 (2), 974–984.

Yang, X., Lu, K., McGrouther, K., Che, L., Hu, G., Wang, Q., et al., 2017. Bioavailability of Cd and Zn in soils treated with biochars derived from tobacco stalk and dead pigs. J. Soils Sediments 17 (3), 751–762.

Yin, H., Zhu, J., 2016. In situ remediation of metal contaminated lake sediment using naturally occurring, calcium-rich clay mineral-based low-cost amendment. Chem. Eng. J. 285, 112–120.

Yin, R.S., Feng, X.B., Chen, B.W., Zhang, J.J., Wang, W.X., Li, X.D., 2015. Identifying the sources and processes of mercury in subtropical estuarine and ocean sediments using hg isotopic composition. Environ. Sci. Technol. 49, 1347– – 1355.

Yu, Z., Qiu, W., Wang, F., Lei, M., Wang, D., Song, Z., 2017. Effects of manganese oxide-modified biochar composites on arsenic speciation and accumulation in an indica rice (*Oryza sativa* L.) cultivar. Chemosphere 168, 341–349.

Yuan, Y., Bolan, N., Prévoteau, A., Vithanage, M., Biswas, J.K., Ok, Y.S., et al., 2018. Applications of biochar in redox-mediated reactions. Bioresour. Technol. 246, 271–281.

Zama, E.F., Zhu, Y.G., Reid, B.J., Sun, G.X., 2017. The role of biochar properties in influencing the sorption and desorption of Pb(II), Cd(II) and As(III) in aqueous solution. J. Cleaner Prod. 148, 127– – 136.

Zhang, F., Wang, X., Yin, D., Peng, B., Tan, C., Liu, Y., et al., 2015. Efficiency and mechanisms of Cd removal from aqueous solution by biochar derived from water hyacinth (*Eichornia crassipes*). J. Environ. Manage. 153, 68–73.

Zhang, Y., Liu, Y.R., Lei, P., Wang, Y.J., Zhong, H., 2018. Biochar and nitrate reduce risk of methylmercury in soils under straw amendment. Sci. Total Environ. 619, 384–390.

Zhang, Z., Solaiman, Z.M., Meney, K., Murphy, D.V., Rengel, Z., 2013. Biochars immobilize soil cadmium, but do not improve growth of emergent wetland species *Juncus subsecundus* in cadmium-contaminated soil. J. Soils Sediments 13 (1), 140– – 151.

Zhao, F.J., Ma, Y.B., Zhu, Y.G., Tang, Z., McGrath, S.P., 2015. Soil contamination in China: current status and mitigation strategies. Environ. Sci. Technol. 49, 750–759.

Zhen, Y.H., Cheng, Y.J., Pan, G.X., Li, L.Q., 2008. Cd, Zn and Se content of the polished rice samples from some Chinese open markets and their relevance to food safety. J. Saf. Environ. 1, 34–36.

Zheng, R.L., Cai, C., Liang, J.H., Huang, Q., Chen, Z., Huang, Y.Z., et al., 2012. The effects of biochars from rice residue on the formation of iron plaque and the accumulation of Cd, Zn, Pb, As in rice (*Oryza sativa* L.) seedlings. Chemosphere 89 (7), 856–862.

Zhu, Q.L., Wu, J., Wang, L.L., Yang, G., Zhang, X.H., 2015. Effect of biochar on heavy metal speciation of paddy soil. Water Air Soil Pollut. 226, 429–436.

CHAPTER

15

Hydrothermal Carbonization for Hydrochar Production and Its Application

Shicheng Zhang[1,2], Xiangdong Zhu[1,2], Shaojie Zhou[1,2], Hua Shang[1,2], Jiewen Luo[1,2] and Daniel C.W. Tsang[3]

[1]Shanghai Key Laboratory of Atmospheric Particle Pollution and Prevention (LAP3), Department of Environmental Science and Engineering, Fudan University, Shanghai, P.R. China [2]Shanghai Institute of Pollution Control and Ecological Security, Shanghai, P.R. China [3]Department of Civil and Environmental Engineering, The Hong Kong Polytechnic University, Hung Hom, Kowloon, Hong Kong, P.R. China

15.1 INTRODUCTION

Recently, hydrochar from wet chemical reaction of renewable biomass has distinct properties compared with pyrolysis char, such as abundant functional groups, moderate carbonization degree, and weak acidity and porosity. Therefore, more and more attention has been paid to hydrochar preparation for properties optimization, assessment of environmental benefits, activation for environmental mediation, and energy storage. This section will comprehensively review recent research progress on hydrochar properties and soil modification benefits affected by its production conditions, and hydrochar activation route for its advanced application on environmental remediation and energy storage.

15.2 PRODUCTION OF HYDROCHAR

Hydrochar with a rich oxygenated functional group is a high value-added carbonaceous material derived from dry/wet biomass by hydrothermal carbonization (HTC), which is a

Biochar from Biomass and Waste
DOI: https://doi.org/10.1016/B978-0-12-811729-3.00015-7

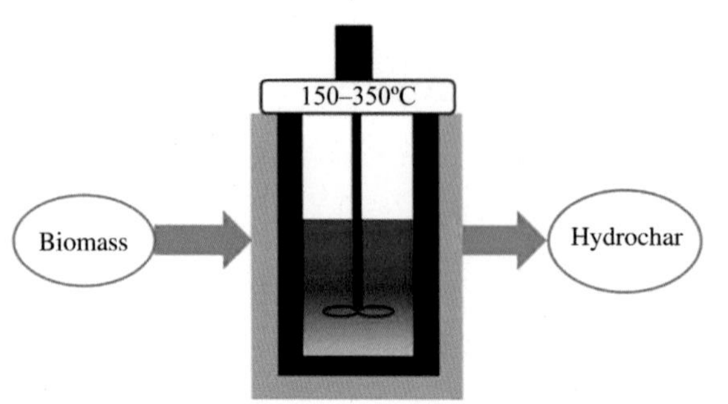

FIGURE 15.1 Hydrothermal carbonization of biomass for hydrochar production.

thermo-chemical conversion process using water as a solvent reaction medium, at the temperature of 150–350°C and autogenously pressure (Falco et al., 2012; Parshetti et al., 2013; Liu et al., 2013b, 2010; Sevilla et al., 2011a; Titirici et al., 2012; Zhu et al., 2015a,b,c). Usually the reaction pressure is not controlled in the process and is autogenic with the saturation vapor pressure of water corresponding to the reaction temperature. At high temperatures, water with high ionization constant can facilitate hydrolysis and cleavage of lignocellulosic biomass (Hatcher, 1997; Masselter et al., 1995; Bobleter, 1994); water is responsible for hydrolysis of organics, which can further be catalyzed by acids or bases (Titirici et al., 2012, 2007; Bobleter, 1994; Libra et al., 2011). The typical procedure for hydrochar production can be depicted in Fig. 15.1.

15.2.1 Influence of Feedstock

Feedstock type plays a significant role in hydrochar composition and performance. Cellulose, hemicellulose, and lignin are the three main components in common lignocellulosic biomass. Different biomass has different cellulose, hemicellulose, and lignin weight ratios, and there are big differences among the structures of these components. Falco et al. (2011) compared the structural differences among glucose-, cellulose-, and lignocellulosic biomass-derived HTC materials. Whereas the cellulose signals of glucose- and cellulose-based hydrochars were lost at 200°C, they were present in the spectrum of rye straw-based material up to 200°C. This difference was attributed to the presence of lignin in rye straw, which might stabilize the cellulose and prevent the disruption of its crystalline structure at lower temperatures. Kang et al. (2012) investigated the chemical and structural properties of hydrochars from lignin, cellulose, D-xylose, and wood meal. The hydrochar yield is between 45% and 60%, and the yield trend of the feedstock is lignin > wood meal > cellulose > D-xylose. The hydrochars seem stable below 300°C, and aromatic structure is formed in all of these hydrochars. The C content, C recovery, energy recovery, ratio of C/O, and ratio of C/H in all of these hydrochars are among 63%–75%, 80%–87%, 78%–89%, 2.3–4.1, and 12–15, respectively. The higher heating value (HHV) of the hydrochars is among 24–30 MJ kg^{-1}, with an increase of 45%–91% compared with the corresponding feedstock. The carbonization mechanism is proposed, and furfural is found to be an important intermediate product during D-xylose hydrochar production, while lignin HTC products are made of polyaromatic and phenolic structures.

In addition, hydrolysis of cellulose chains produces different oligomers (cellobiose, cellohexaose, cellopentaose, cellotetraose, and cellotriose) and glucose. These oligomers undergo complex reaction mechanisms to form polymers and hydrochar skeletons upon their supersaturation. Sevilla and Fuertes (2009a) have investigated extensively the formation of hydrochar particles derived from sucrose and starch. Sucrose undergoes hydrolysis and leads to the formation of glucose and fructose, whereas starch forms maltose (oligosaccharide), glucose, and fructose (from the isomerization of glucose). At the same time, the decomposition of glucose and fructose leads to the formation of organic acids (e.g., acetic, lactic, propenoic, levulinic, and formic acids) (Titirici et al., 2012; Girisuta et al., 2006; Amarasekara and Ebede, 2009; Antal et al., 1990). Reduction in pH due to the release of organic acids catalyzes the hydrolysis (Titirici et al., 2012; Funke and Ziegler, 2010) of oligosaccharides to monosaccharides, which further undergo dehydration and fragmentation (ring opening and C–C bond breaking) resulting in soluble products. On the other hand, the glucose, fructose, and other products of decomposition present in the solution undergo intermolecular dehydration and aldol condensation, which leads to polymerization. These polymers undergo aromatization, leading to the formation of C=O groups due to dehydration of equatorial hydroxyl groups. The appearance of C=C groups is mainly due to keto-enol tautomerism of the resulting dehydrated species, or due to intramolecular dehydration. Lastly, the formation of aromatic clusters takes place via intermolecular dehydration of aromatic compounds.

15.2.2 Influence of Reaction Temperature

The reaction temperature has a significant influence on the properties of the hydrochar. The chemical composition of the hydrochar tends to be stable with elevating the temperature (Liu et al., 2017). Higher temperature leads to extensive dehydration and an increase in the degree of condensation of the hydrochar. Sevilla and Fuertes (2009b) showed a decrease in the O/C and H/C atomic ratios with an increase in the temperature of HTC from 230 to 250°C. With increasing the HTC temperature, the resultant hydrochar possesses a high degree of aromatization. Moreover, the energy content and thermostability of hydrochar is considerably improved with increasing the reaction temperature. Kang et al. (2012) reported the decrease in ion exchange capacity with an increase in temperature (225, 245, and 265°C) for D-xylose, cellulose, lignin, and lignocellulosic biomass (woodmeal) derived hydrochar. A reduction in oxygenated functional group and OH groups was observed with an increase in temperature for all the starting materials, except lignin. A small increase in oxygenated functional group in the case of lignin was observed from 225 to 245°C due to high temperature requirement for the hydrolysis of lignin (Dinjus et al., 2011). Jain et al. (2015a) investigated the effect of HTC temperature (200, 275, 315, and 350°C) on the formation of oxygenated functional group on coconut shell-based hydrochars. A maximum in oxygenated functional group was observed at 275°C, followed by a decrease at 315 and 350°C because of the higher extent of decomposition of oxygenated functional group and formation of gaseous products at higher temperatures. While the hydrothermal treatment temperature can be tuned for higher oxygenated functional group contents, this content at the same time is also dependent on the type of starting materials, residence time, and substrate concentrations used.

Sevilla et al. demonstrated the effect of temperature on the size of hydrochar microspheres. Cellulose hydrothermally treated at 210°C does not show any sign of microsphere formation. A further increase in temperature to 220, 230, and 250°C results in a progressive increase in the size of microspheres formed (Sevilla and Fuertes, 2009b). The same observation was also reported with glucose (170–230°C) and starch (180–200°C) (Tanaike and Inagaki, 1997), which can be attributed to the higher extent of hydrolysis and extensive polymerization at higher temperatures.

A few studies have reported a decrease in the yield of hydrochar with the increase in temperature (Kang et al., 2012; Xiao et al., 2012). Sevilla and Fuertes (2009a) observed an increase in yield of glucose derived secondary hydrochar microspheres from 1.5% to 36% when the temperature was increased from 170% to 230°C and 5.1% to 25% for starch derived hydrochars when the temperature increased from 180 to 200°C.

15.2.3 Influence of Retention Time

Retention time plays an important role in the extent of reaction and the distribution of different types of products and their quality. Longer retention times can lead to a decrease in yield of hydrochar (He et al., 2013; Hoekman et al., 2011). At shorter retention times, less condensed products (high O/C and H/C atomic ratios) are obtained due to a lesser extent of hydrolysis and polymerization (Sevilla and Fuertes, 2009b). For example, Cao et al. (2013) reported lower H/C ratios in sugar beet hydrochars and marginally decreased O/C ratios in bark hydrochars, with an increase in residence time from 3 to 20 h.

Longer retention time may result in a decrease in oxygenated functional group due to intermolecular dehydration/aldol condensation that converts oxygenated functional group to stable oxygen groups (ether or quinones). However, high biomass concentration at higher residence time may increase the oxygenated functional group content due to the greater extent of reaction and formation of secondary char. Thus, higher residence time can yield high oxygenated functional group content even at relatively (1) lower temperatures because of longer exposure and thus more polymerization of aromatic clusters to hydrochar; however excessive exposure might lead to reduction due to integration of oxygenated functional group into stable groups such as ether, quinone, etc.; (2) higher substrate concentrations because of the higher extent of hydrolysis of more biomass per gram of water; and (3) higher lignin content due to longer exposure; however, the temperature should be enough in order to at least degrade the lignin and its constituents.

Retention time also governs the size of microspheres formed since long residence time leads to excessive polymerization and an increase in size due to the formation of stable oxygen groups from surrounding oxygenated functional groups (Sevilla and Fuertes, 2009a,b).

15.2.4 Influence of Catalyst

Catalysts for biomass decomposition can enhance the reaction rate and tailor the path of reaction for the desirable products. A number of studies have been reported that suitable agents such as acids (acrylic acid and citric acid) (Titirici et al., 2012, 2007; Demir-Cakan et al., 2009), oxidants (hydrogen peroxide) and salts (Fe_2O_3 and $ZnCl_2$) (Cui et al., 2006; Jain et al., 2013, 2014) can be used as a catalyst to promote dehydration. Increased decomposition in the presence of a catalyst is attributed to the increased proton concentration leading to acid catalysis (Titirici et al., 2012, 2007; Funke and Ziegler, 2010). Generally, small amounts of acids catalyze dehydration. Besides, it is also found to be helpful in increasing the oxygenated functional group content in the hydrochar, which promises efficient chemical activation. The typical hydrochar characteristics from various production conditions are concluded in Table 15.1.

TABLE 15.1 Typical Characteristics of Hydrochar Prepared From Different Feedstock (Kang et al., 2012; Liu et al., 2013a)

Feedstock	Reaction Temperature (°C)	Retention Time (h)	C/O	C/H	Ash (%)	HHV
Lignin	225	20	2.34	12.2	1.47	24.41 MJ kg^{-1}
Lignin	245	20	2.59	13.2	1.48	26.93 MJ kg^{-1}
Lignin	265	20	2.90	14.7	1.54	27.41 MJ kg^{-1}
Cellulose	225	20	2.33	13.0	–	25.83 MJ kg^{-1}
Cellulose	245	20	2.75	14.0	–	26.73 MJ kg^{-1}
Cellulose	265	20	3.16	14.3	–	26.99 MJ kg^{-1}
D-Xylose	225	20	2.57	14.7	–	26.69 MJ kg^{-1}
D-Xylose	245	20	2.73	14.8	–	27.64 MJ kg^{-1}
D-Xylose	265	20	3.27	14.7	–	29.71 MJ kg^{-1}
Pine wood meal	225	20	2.71	12.1	1.31	27.12 kJ mol^{-1}
Pine wood meal	245	20	3.08	12.9	1.40	28.39 kJ mol^{-1}
Pine wood meal	265	20	4.14	13.4	1.73	29.57 kJ mol^{-1}
Coconut fiber	220	0.5	2.01	11.83	6.2	24.7 kJ mol^{-1}
Coconut fiber	250	0.5	2.54	12.90	5.0	26.7 kJ mol^{-1}
Coconut fiber	300	0.5	3.62	14.39	4.3	29.4 kJ mol^{-1}
Coconut fiber	350	0.5	3.57	16.23	4.9	28.7 kJ mol^{-1}
Coconut fiber	375	0.5	4.91	18.14	8.6	30.6 kJ mol^{-1}
Eucalyptus leaves	220	0.5	1.99	9.97	7.3	18.9 kJ mol^{-1}
Eucalyptus leaves	250	0.5	2.05	11.39	6.9	25.0 kJ mol^{-1}
Eucalyptus leaves	300	0.5	3.02	11.48	7.1	28.7 kJ mol^{-1}
Eucalyptus leaves	350	0.5	3.45	11.89	9.9	29.4 kJ mol^{-1}
Eucalyptus leaves	375	0.5	3.64	15.01	14.2	28.7 kJ mol^{-1}

15.3 PROPERTIES OF HYDROCHAR

15.3.1 Heating Value

HTC proceeds by reactions including dehydration, decarboxylation, and recondensation and can convert lignocellulosic biomass into a high value solid fuel. Kang et al. (2012) reported that the HHV of the hydrochars is among $24-30$ MJ kg^{-1} at the temperature of 225, 245, and 265°C, with an increase of $45\%-91\%$ compared with the corresponding feedstock (lignin, cellulose, D-xylose, and wood meal). Guo et al. (2015) reported that temperature had the main influence on yield, carbon content, and HHV of hydrochar, and longer residence time was favorable for the dehydration, decarboxylation, and polymerization reactions, which resulted in carbon content increasing and oxygen content decreasing. Liu et al. (2013a) also reported the energy density of hydrochar increased with increasing hydrothermal temperature, with HHVs close to that of lignite. The deoxygenation and dehydration reactions cause hydrochar to have high hydrophobicity and increased carbon content compared to the raw feedstocks. This can be explained as the decrease in the number of low energy H–C and O–C bonds and increase of high energy C–C bond, the energy density of biomass feedstock was improved.

15.3.2 Chemical Properties

Rich oxygenated functional groups are distinguished from other carbon materials. HTC results in efficient hydrolysis and dehydration of biomass and bestows the hydrochar with rich oxygenated functional groups (C–O, C=O, COO–, etc.). So, the hydrochar can be used as adsorbents for heavy metal ion. The product content of oxygenated functional groups largely depends on the processing conditions of HTC and the types of feedstock. Sevilla and Fuertes (2009a) demonstrated that the HTC of saccharides (glucose, sucrose, and starch) can lead to the formation of hydrochar with rich oxygenated functional groups. Zhou et al. (2017) reported that the hydrochar with rich oxygenated functional groups was prepared via the HTC of fresh and dehydrated banana peels.

15.3.3 Microcrystalline Structure and Surface Morphology

Feedstock types and carbonization condition play an important role in the formation of the microcrystalline structure of hydrochar. When the preparation conditions are mild, the hydrochar shows the structure similar to that of the raw material. When the preparation conditions are severe, the structure of the hydrochar is transformed and shows different structures. Sevilla and Fuertes (2009b) reported that the hydrochars from hydrothermal treatment at 210°C exhibit an irregular morphology similar to that of pristine cellulose, the samples obtained at temperatures ≥ 220°C consist mainly of aggregates of microspheres with a diameter in the 2–10 μm range. Guo et al. (2015) reported that longer residence time promoted crystalline cellulose components in lawn grass to form hydrochar at the temperature of 240°C.

The surface morphology of carbonaceous microspheres obtained by HTC is influenced by carbonization conditions and feedstock types. Demir-Cakan et al. (2009) showed that the HTC can convert glucose into amorphous, microporous carbon spheres.

15.4 ENVIRONMENTAL IMPACT OF HYDROCHAR

Due to the change of the natural environment and unreasonable human activities, the problem of soil quality degradation and soil environmental pollution is becoming increasingly serious, which has seriously threatened the sustainable development of agriculture. Hydrochar is increasingly used as a soil amendment in degraded and contaminated soils. Numerous studies show that hydrochar applied to the soil can effectively increase the porosity of the soil and the content of soil organic matter (SOM), and provide a good environment for soil microbial growth and reproduction, thus improving soil quality and increasing crop yield. This section focuses on the impact of hydrochar on soil physical, chemical, and biological properties. On this basis, the impact on crop growth and yield is reviewed, providing theoretical support for the improvement of farmland soil by hydrochar.

15.4.1 Effect of Hydrochar on Soil Physical Characteristics

Soil porosity is the gap between solid particles, which contains water and air. Hydrochar has an important impact on soil porosity. For example, 2% (w/w) hydrochars addition to soil decreased the bulk density of the soil by 0.1 ± 0.0 g cm^{-3} on average, and increased the porosity by $3.4 \pm 1.0\%$. The degree of the increase of the porosity depended significantly on production process and particle size. The increase of the porosity was lower for hydro-200 (hydrothermal carbonized at 200°C) than for hydro-250 (hydrothermal carbonized at 250°C) and pyro-750 (pyrolyzed at 750°C), and significantly higher for the larger particles than for the fine sized chars (Eibisch et al., 2015). Also, Abel et al. (2013) mentioned that a decrease in bulk density and an increase of total porosity of packed soil columns could be observed in significantly greater extent for sandy or loamy soil with hydrochar modification.

Soil moisture is one of the most important material components in agro-ecosystems, which is vital to plants and animals. Soil moisture is usually expressed by the available water capacity (AWC). The addition of hydrochar can increase soil water content by enhancing soil porosity (Abel et al., 2013) and aggregate formation (Rillig et al., 2010) and by changing soil tortuosity; large particles of char can block pores. Moreover, hydrochar particles are known for having more porosity to retain water due to their spherical shape and deformability (Abel et al., 2013).

The AWC is defined as the amount of water held between field capacity and permanent wilting point. Eibisch et al. (2015) investigated the effect of water repellency of hydrochars from two different feedstocks (digestate and woodchips) and at two particle sizes (<0.5 and 0.5–1.0 mm) on soil AWC. Highest increase of AWC occurred for hydrochars from hydrothermal carbonized of digestate material at 200°C with particle size fractions

<0.5 mm. Abel et al. (2013) found that hydrochars have the potential to serve as physical soil conditioners, amending in view of increasing plant available soil water reservoir of coarse-grained, sandy sites with low organic matter content. The majority of the soil samples shows the highest rise of AWC with 2.5 wt% with the addition of hydrochars. However, for the soil with high organic matter content, which initially possesses a high AWC, no increase in AWC was detected.

15.4.2 Effect of Hydrochar on Soil Chemical Characteristics

Soil pH is an important indicator of soil quality and appropriate soil pH is the premise of high-yield crop. On one hand, soil acidification inhibits the plant's root respiration and growth, affecting the root's absorption function. Low pH affects soil nutrient conversion and release, and the effectiveness of nutrients.

The impact of hydrochar on soil pH depended on soil type, with significantly higher pH in the more acidic coarse soil compared to its control. In contrast, the carbonate-containing fine soil amended with hydrochar had lower pH relative to its control but this difference was not significant (Malghani et al., 2015). For instance, George et al. (2012) showed that the application of hydrochar made of spent brewer's yeast led to a significant decrease in soil pH (7.08–6.85) with increasing application rates. However, Rillig et al. (2010) reported that a significant increase in soil pH was observed 1 year after hydrochars' application from pH value of 7.2 to 7.45. Busch et al. (2013) found that the pH value showed a slight increase from 6.29 of pure soil to around 6.5 of soil mixtures with hydrochar at an addition rate of 7% (v/v, 30 Mg ha^{-1}), while the pH of hydrochar was around 3.9. These were likely due to that ash elements in hydrochar such as K, Ca, and Mg are soluble, which can improve the salt-based soil saturation in soil, in order to improve the pH (Warnock et al., 2007; Gaskin et al., 2010). Another possible explanation for the increase in pH is the addition of hydrochar promoted microbial reduction reactions, which leads to an increased microbial activity (Libra et al., 2011; Rillig et al., 2010). This is probably also due to the addition of hydrochar to soil with base cations that leads to better plant growth (Gaskin et al., 2010).

Soil cation exchange capacity (CEC), as an important indicator of soil fertility and buffering capacity, reflects the ability of the soil to absorb and supply exchangeable nutrients. Additionally, the ^{13}C NMR spectra and the SEM-EDS analyses showed that hydrochars are rich in oxygen containing functional groups, such as hydroxyl, ester, aldehyde, nitro, keton, phenolic, and carboxyl groups. Presence of these functional groups on the surface of hydrochar can significantly enhance the CEC (Libra et al., 2011; Vaccari et al., 2011). Glaser et al. (2002) found that adding a small amount of hydrochar can increase the soil CEC, and CEC increased with increasing the amount of charcoal. At present, there is no long-term research on the influence of hydrochar on soil CEC. Therefore, the impact of hydrochar on soil CEC needs to be further investigated.

SOMs are carbon-containing organic compounds in the soil. SOM come from a wide range of sources, including plant, animal, and microbial residues and applied organic fertilizer, in which the higher plants are the main source. A large number items in the literatures reported the promotion effect of hydrochars on composition and conversion of SOM.

CO_2 and CH_4 are main products from composition and conversion of SOM. Kammann et al. (2012) found that CO_2 fluxes from the beet hydrochar soil mixtures were up 20 times larger than those of the control. Also, larger CH_4 emissions than in the control treatments was observed, probably due to anaerobic microsites formed by increased O_2 consumption (Khalil and Baggs, 2005; Mørkved et al., 2006). Andert and Mumme (2015) mentioned that the CO_2 and CH_4 emissions in the soil and hydrochar mixture without fertilizer were considerably elevated compared to char-free control treatments, showing up to 8 and 2.5 times higher gas emissions, respectively.

The mechanisms of CH_4 promotion involved: (1) Hydrochar changed the soil water content, which regulated O_2 and CH_4 diffusion through reduction in soil compaction and improvement of soil aeration (Castro et al., 1994). (2) Increases in the soil pH in acidic soils may also enhance CH_4 uptake rates. (3) In highly nitrogen-loaded ecosystems, inhibition of CH_4 oxidation may occur (Veldkamp et al., 2001). Beneficial changes of CH_4 fluxes between hydrochar-amended soils and the atmosphere clearly leave much room for further research—the issue being resolved.

Soil nutrients can be directly or indirectly absorbed and utilized by plants. Most of the hydrochars are prepared by thermal carbonization of plant biomass, thus retaining most of the nutrients required for plants. Therefore, the addition of hydrochar has a significant impact on soil nutrients. Hydrochar can directly affect soil nitrogen content. Bargmann et al. (2014b) showed that hydrochar amendments significantly decreased both NH_4^+ and NO_3^- concentrations in the soil solution compared to the no char control, mainly due to immobilization of nitrogen. The NH_4^+ concentration had no change when the hydrochar concentration was raised, while increasing hydrochar concentration generally had significant effect on the NO_3^- concentration. Further, hydrochar effects on N concentrations (NO_3^-, NH_4^+) in soils with different N pools (soil N, fertilizer N) was investigated (Bargmann et al., 2014a). In the non-fertilized topsoil, the application of hydrochar resulted in almost disappearance of NO_3^- within the first week. In the fertilized topsoil, hydrochar application significantly decreased the NO_3^- concentration by 32% compared to the fertilized control treatment within the first week.

On the other hand, hydrochar affects soil nitrogen by changing the emission of nitrogen-containing gases (NH_3 and N_2O). According to the study of Kammann et al. (2012), hydrochar soil amendment can lead to higher N_2O emissions compared to controlled soil. Malghani et al. (2013) presented opposite results after hydrochar soil amendment. They observed suppressed N_2O emissions from hydrochar treated arable and forest soil, but another forest soil showed elevated N_2O emissions. Accordingly, individual soil properties are a major factor affecting the hydrochar impact on soil nitrogen.

15.4.3 Effect of Hydrochar on Biological Characteristics

Soil microorganisms are closely related to crop growth and nutrient utilization, attracting most of the attention from researchers. For bacteria, Andert and Mumme (2015) observed that bacterial 16S rRNA gene abundance had no change in response to hydrochar treatments and the relative abundance of bacteria decreased 5–6-fold compared to the char-free control. The similar findings were reported by Harter et al. (2014).

Steinbeiss et al. (2009) investigated that a high condensation grade hydrochar had a lower impact on microbial community composition compared to low condensation hydrochar.

For fungi, Busch et al. (2013) found that hydrochar amendments had a strong avoidance of earthworm and promoted vast growth of saprophytic fungi. Rillig showed that hydrochar had a positive effect on arbuscular mycorrhizal (AM) fungi root colonization at high hydrochar application rates (20%) (Rillig et al., 2010), probably due to changes in the nutrient availability like P, K, etc., and changes in the soil pH (Warnock et al., 2007; Gaskin et al., 2010). However, George et al. (2012) found lessened AM root colonization at even lower application rates (5%, 10%). These contradictory findings may attribute to different HTC conditions and feedstock for hydrochar production (spent brewer's grain and beet root chips, respectively). Steinbeiss et al. (2009) found that feedstock types (i.e., glucose, yeast) had significant effects on soil microbial communities. The addition of glucose derived hydrochars led to an overall microbial biomass decrease, whereas yeast-derived hydrochars had no significant effect.

15.4.4 Effect of Hydrochar on Plant Growth

Busch et al. (2012) reported that the beet-root chip hydrochar showed negative effects in barley germination and growth test. Similar inhibitions of biomass germination, growth, and yield were observed (Busch et al., 2013; Gajić and Koch, 2012; Bargmann et al., 2013). These studies proposed two explanations. One points to the decomposition of volatile or toxic substances over time, lessening the negative impact on seed germination (Busch et al., 2012). Bargmann et al. (2014c) found that hydrochars had only short-term effects on germination and decrease of barley production can be avoided by applying hydrochar at least 4 weeks to sowing. It is likely due to microbial degradation of germination-inhibiting components in hydrochar after weeks. Further, hydrochars' post-treatment, like washing, were used in germination tests with spring barley to decrease phytotoxic effects (Bargmann et al., 2013). And Jandl et al. (2013) proposed that the phytotoxic compounds could also be ethylene prepared by degradation of long C-chain aliphatic compounds.

The second explanation is immobilization of nitrogen resulting in N-limitation of plants (Bargmann et al., 2014a; Nelson et al., 2011). N-immobilization may be attributed to large C/N ratios in hydrochar or microbial immobilization. Gajić found that hydrochars could diminish the negative effects on sugar beet crops' yield with adequate N fertilizer (narrow C/N ratio) (Gajić and Koch, 2012). However, negative effects of addition of freshly-prepared hydrochar on plant growth are not clearly understood yet. More research is urgently needed to identify the inhibiting mechanisms, substances, and possible risks associated with hydrochar.

15.5 MODIFICATION AND APPLICATION OF HYDROCHAR

In the last few years, much attention has been paid to the modification of hydrochar on its environmental remediation and energy storage (Akhtar and Amin, 2011; Zhu et al., 2015d). Table 15.2 concludes the activation conditions and application performances of hydrochar. The hydrochar has a less aromatic structure and thermal recalcitrance, a low

TABLE 15.2 Hydrochar Production and Application for CO_2 Capture, Environmental Remediation and Energy Storage

Feedstock	Activation Conditions	Surface Area ($m^2\ g^{-1}$)	Specific Performance	References
CO_2 CAPTURE				
Alginic acid/ethylenediamine	CO_2	2241	16.20 mmol g^{-1}	Sevilla et al. (2011a)
Potato starch, cellulose, and eucalyptus sawdust	KOH	1260–2850	2.9–4.8 mmol g^{-1}	Gaunt and Lehmann (2008)
Lignocellulosic	KOH	1080–2510	0.66–3.71 mmol g^{-1}	Baronti et al. (2017)
ENVIRONMENTAL REMEDIATION				
Cellulose	–	20	Adsorbent for 1-butyl-3-methyl imidazolium chloride, 0.171 mmol g^{-1}	Pazferreiro et al. (2014)
Swine solids/poultry litter	–	1.9	Adsorb both pyrene or triclosan, estrone, carbamazepine	Elaigwu et al. (2014)
Salix psammophila	–	80.3–316	Tetracycline adsorb	Xue et al. (2012)
S. psammophila	K_2CO_3	1230	Dye removal	Zhang et al. (2016a)
Switchgrass	–	2.9	U^{6+} removal (2.12 mg U g^{-1})	Sun et al. (2011)
Biogas residues	H_3PO_4	–	Removal Pb^{2+} (417.4 mg g^{-1})	Qi et al. (2013)
P. africana shell	–	6.05	45.3 mg g^{-1} for Pb^{2+} and 38.3 mg g^{-1} for Cd^{2+}	Sevilla et al. (2011b)
H2 UPTAKE				
α-D-Glucose, potato starch, cellulose, eucalyptus sawdust, and furfural	KOH	1283–2370	H_2 uptake (6.4 wt%)	Kumar et al. (2011)
Lignin waste	KOH	1157–1924	H_2 uptake (3.2–4.7 wt%)	Tan et al. (2015)
ENERGY STORAGE				
Alginic acid	KOH	1795–2421	Supercapacitor (314 F g^{-1} at current density of 1 A g^{-1})	Zhu et al. (2014a)
Bamboo	KOH	1472	Specific capacitance of 301 F g^{-1} at 0.1 A g^{-1}	Simon and Gogotsi (2008)
Coconut shells	$ZnCl_2$, H_2O_2 and CO_2	1370–2440	Energy density of 7.6 W h kg^{-1} at a power density of 4.5 kW kg^{-1}	Jain et al. (2016)
Commercial glucose	KOH and melamine	75–1107	Specific capacitance reaches 492 F g^{-1} in H_2SO_4 electrolyte and 279 F g^{-1} in KOH electrolyte (at 0.1 A g^{-1})	Sevilla et al. (2016)
Citric acid and urea	KOH	1253–2397	High specific capacitance of 365 F g^{-1} at 0.5 A g^{-1}	Jain et al. (2015b)
Gelatin and citric acid	KOH	828–1620	Specific capacitance reached 272.6 F g^{-1} at 1 A g^{-1}	Gupta et al. (2015)
Chinese hairs	KOH	849–1104	High capacitance of up to 264 F g^{-1} at 0.25 A g^{-1}	Feng et al. (2015)

surface area, and poor porosity, hindering the effective exploitation of hydrochar for environmental and energy storage (Titirici et al., 2012; Libra et al., 2011; Sun et al., 2014; Falco et al., 2013). Thus, some post-activation methods can be applied to increase the surface area, porosity for adsorption. Also, surface functional groups of hydrochar can be enhanced by some modification methods. As a result, hydrochar has become a promising material for environmental remediation and energy storage.

15.5.1 Hydrochar for Climate Change Mitigation

The ever-increasing accumulation of greenhouse gases (GHG), in the atmosphere, especially for CO_2, is significantly responsible for the severely detrimental global warming and concomitant climate changes and other environmental issues (Anderegg et al., 2010). In order to mitigate the effect of GHG on the environment, using porous solids as sorbents for capturing CO_2 by means of pressure, temperature, or vacuum swing adsorption systems is a promising technology to solve this problem (Ma et al., 2014). A large amount of research has been directed toward the biochar from pyrolysis, which can minimize CO_2 emissions to the atmosphere (Titirici and Antonietti, 2010; Gaunt and Lehmann, 2008). HTC can reduce the landfill gas emission by making use of waste biomass and offers an efficient approach for carbon sequestration on a large scale (Sevilla et al., 2011a; Baronti et al., 2017). For example, Sevilla and Fuertes (2011) reported that sustainable porous carbons (PCs), which were prepared from hydrothermally treated polysaccharides (starch and cellulose) or biomass (sawdust) by chemical activation using KOH, exhibited high CO_2 adsorption uptakes. Parshetti et al. (2015) also found that the carbonaceous material prepared from a lignocellulosic feedstock by novel HTC coupled with chemical activation exhibited high performance for CO_2 capture.

However, some studies showed that hydrochar was not suitable for carbon sequestration due to its lesser stability in soil (Eibisch et al., 2013; Naisse et al., 2015). Some researchers noticed that hydrochar decomposed rapidly and stimulated emissions of CH_4 and CO_2 to the atmosphere during field application (Schimmelpfennig et al., 2014; Malghani et al., 2015). Highly microporous activated carbons derived from hydrochar via KOH could be an effective adsorbents for CO_2 capture (Falco et al., 2013). Malghani et al. (2013) found that the addition of hydrochar (1%, w/w) would significantly increase CO_2 emissions in three soils (managed spruce forest, unmanaged deciduous forest, and agriculture), with much of the extra C derived from hydrochar decomposition. Higher emissions of greenhouse gas with the application of hydrochar in the soil are most likely the result of increased microbial activity due to easy degradability of carbon in the hydrochar (Kambo and Dutta, 2015). Therefore, when estimating the carbon sequestration potential of different hydrochars, further in-depth research should be carried out to optimize the application of hydrochar in the soil with high stability, low GHG emissions, and positive effects on the agricultural productivity (Kammann et al., 2012; Parshetti and Balasubramanian, 2014).

15.5.2 Hydrochar for Pollution Control and Remediation

Heavy metals, especially the chromium (Cr), arsenic (As), and lead (Pb), are more problematic and threatening to ecological environment and human beings due to high toxicity, non-biodegradation, and accumulation through the food chain (Burakov et al., 2018). Besides, environmental contamination by organic pollutants (OPs, e.g., pesticide residual, antibiotic, pharmaceuticals, and dyes, etc.) is receiving widespread public attention due to the potential negative ecological effects (Sun et al., 2011). In this situation, various methods have been applied to get efficient removal of heavy metal and OPs from soil and waters, and adsorption is recognized as one of the most available options because of its low cost and high efficiency (Kambo and Dutta, 2015; Pazferreiro et al., 2014; Kumar et al., 2011; Qi et al., 2013).

Among various carbonaceous materials, hydrochar can be as a template for fabrication of metal/carbon composites to adsorb heavy metals and OPs with numerous associated sorption sites. The capability of hydrochars (such as specific surface area, pore size) can be significantly improved by modification. In addition, modified hydrochar can offer numerous associated sorption sites attribute to high concentration of oxygen functional groups and low degrees of aromatic group (Sevilla et al., 2011b; Tang et al., 2016).

For heavy metals, Tang et al. found that 99.5% of Pb^{2+} was removed by Ni/Fe modified hydrochar (prepared by biogas residues) in 1.5 h regardless of initial solution pH (Tang et al., 2016). This material not only acted as an effective adsorbent for Pb^{2+} immobilization, but also played a role of catalyst to generate atomic hydrogen that attacked Pb^{2+} aggressively to form Pb^0. Elaigwu et al. investigated that biochar and hydrochar prepared from *Prosopis africana* shell, respectively, could be used as adsorbents to remove Pb^{2+} and Cd^{2+} ions from aqueous solution. Maximum adsorption capacities for the hydrochar and biochar were 45.3 and 31.3 mg g^{-1} for Pb^{2+} and 38.3 and 29.9 mg g^{-1} for Cd^{2+}, respectively, indicating that hydrochar can be used as effective adsorbents for removal of heavy metal ions from wastewater (Elaigwu et al., 2014). Xue et al. (2012) demonstrated that the H_2O_2-modified hydrochar enhanced Pb^{2+} sorption ability with a sorption capacity of 22.82 mg g^{-1}, which was 20 times more than untreated hydrochar (0.88 mg g^{-1}) and it is comparable to commercially activated carbon. It also mentioned that the model results indicated that the heavy metal removal ability of the modified hydrochar follows the order of $Pb^{2+} > Cu^{2+} > Cd^{2+} > Ni^{2+}$. Thus, application of the adsorption methods with hydrochar derived adsorbent is a promising way for heavy metal pollution.

For OPs, Qi et al. (2013) mentioned that carbonaceous material that is rich in carboxylic groups with low surface area (20 m^2 g^{-1}), was prepared by HTC of cellulose. It exhibited excellent adsorption capacity of 1-butyl-3-methyl-imidazolium chloride. Zhang et al. (2016a) investigated the sorption capacities of hydrochars derived from livestock wastes and it was noted to adsorb effectively both polar pharmaceuticals personal care products and nonpolar pyrene. The result implied that amorphous aromatic C played an important role in the high sorption capacity of hydrochar. Moreover, some researchers found that a novel PC can be prepared from hydrochar via pyrolysis and applied in adsorption of tetracycline (Zhu et al., 2014b). Furthermore, Zhu et al. found that waste hydrochar can be activated and modified to a novel magnetic carbon composite, and this magnetic carbon exhibited a superior malachite green adsorption capacity (476 mg g^{-1}) that is suitable for dye removing (Zhu et al., 2014c). Also, magnetic porous carbon from hydrochar material is an excellent adsorbent for the removal of tetracycline (Zhu et al., 2014a; Liu et al., 2014). Moreover, Zhu et al. found strong linear correlations between hydrochar properties (recalcitrance index, H/C and O/C atomic ratios) and the environmental performances of its derived magnetic carbon composites (adsorption capacity for roxarsone) (Zhu et al., 2015b).

Generally, hydrochar can be used to repair the soil and water polluted by heavy metal or OPs; hence, it has great significance in soil management and environmental pollution remediation.

15.5.3 Hydrochar for Energy Storage

The ever-increasing demand for sustainability in energy has spurred effort worldwide in investing tremendous resources. Electrochemical capacitors (also called supercapacitors), an energy storage device, have received much attention to harvest energy due to its high-power density, long cycle life, quick charge/discharge capability, and reliable safety features (Dr and Dr, 2009; Simon and Gogotsi, 2008; Fan et al., 2015). Carbon-based materials, such as hydrochar, which was low-cost, easily accessible, and environmentally friendly have attracted considerable interest in many energy-related applications (Jain et al., 2016; Reza et al., 2016). However, as a material of supercapacitor, it remains an enormous challenge to improve the energy density. To achieve high capacitance, hydrochar requires proper activation.

Briefly, activation can be done via two methods (Jain et al., 2016): (1) physical (or thermal) activation using CO_2 or steam at 800–900°C; (2) chemical activation using KOH, $ZnCl_2$, H_3PO_4, etc., typically in the range of 450–650°C. For instance, hydrochar can be prepared through HTC of alginic acid and subsequent KOH activation for supercapacitors application. It exhibited high surface area (2421 $m^2\ g^{-1}$) and superior capacitive performance with specific capacitance of 314 $F\ g^{-1}$ at current density of 1 $A\ g^{-1}$ (Fan et al., 2015). A bio-inspired beehive-like hierarchical nanoporous carbon was synthesized by carbonizing the industrial waste of bamboo-based by-product and then activated using KOH at a mass ratio of 1:1. It exhibited remarkable electrochemical performances as a supercapacitor electrode material, such as high specific capacitance of 301 $F\ g^{-1}$ at 0.1 $A\ g^{-1}$, and high power density of 26,000 $Wk\ g^{-1}$ at an energy density of 6.1 $Wh\ kg^{-1}$ (Tian et al., 2015). Sevilla et al. (2016) utilized different activators, such as KOH, K_2CO_3, and $KHCO_3$ to activate the glucose-derived hydrochar, and $KHCO_3$ was considered as a greener and more efficient chemical activation reagent. These carbons treated by H_2O_2 and $ZnCl_2$ have delivered a high capacitance of 246 $F\ g^{-1}$ between 0 and 1 V at 0.25 $A\ g^{-1}$ (Jain et al., 2015b). This enhancement of the charge storage capacity is attributed to creation of a broad distribution in pore size and a larger surface area (Gupta et al., 2015).

In addition, the heteroatom doping (B, N, P, or S) on PCs is significantly important for supercapacitor applications because they were able to undergo redox reactions with the electrolyte, thus increasing their capacity (Chen et al., 2017; Feng et al., 2015; Tan et al., 2015). Zhang et al. (2016b) prepared an oxygen-enriched activated carbon with a layered sedimentary rocks structure using gelatin as precursor, the carbon material showed a capacitance of 272.6 1 $F\ g^{-1}$ at a current density of 1 $A\ g^{-1}$. A user-friendly supercapacitor prepared from biomass, such as glucose and sawdust in the presence of melamine exhibited high specific capacitances in aqueous electrolytes ($>270\ F\ g^{-1}$ in H_2SO_4 and $>190\ F\ g^{-1}$ in Li_2SO_4 at 0.1 $A\ g^{-1}$) (Fuertes and Sevilla, 2015). Si et al. (2013) first used human hair to prepare heteroatom-doped carbon materials, it showed large pseudo capacitance (264 $A\ g^{-1}$ at 0.25 $A\ g^{-1}$) due to the synergistic effect of multi N, O, and S-doped species, especially the positive contribution of sulfur species. However, there are few reports about S-doped carbon materials used as electrodes of electrochemical double layer capacitors. Apart from the existing activation methods, more appropriate and novel treatments for activation should be explored.

Hydrochar also can be modified for gas storage application, and it is an excellent precursor for the preparation of activated carbons. The activated carbon materials were prepared from HTC of various organic feedstocks, exhibiting high hydrogen uptakes (6.4 wt %), and the hydrogen storage density ranges between 12 and 16.4 mmol H_2 m^{-2} (Sevilla et al., 2011b). In Sangchoom's study, lignin-derived hydrochar was used as a precursor for activated carbons (activated by KOH). The material has high surface area (3235 m^2 g^{-1}) and pore volume (1.77 cm^3 g^{-1}), and hydrogen uptake of up to 6.2 wt% (Sangchoom and Mokaya, 2015). As just described, appropriate properties of hydrochar are significantly affected by biomass feedstock type, hydrothermal and activation condition, and the optimal condition should be further studied to improve the performance of supercapacitor and gas storage.

Acknowledgements

This research was funded by the National Key Research and Development Program of China (No. 2017YFC0212205), the National Key Technology Support Program (No. 2015BAD15B06), National Natural Science Foundation of China (No. 21577025), and the International Postdoctoral Exchange Fellowship Program of China supported by Fudan University.

References

Abel, S., Peters, A., Trinks, S., Schonsky, H., Facklam, M., Wessolek, G., 2013. Impact of biochar and hydrochar addition on water retention and water repellency of sandy soil. Geoderma 202–203, 183–191.

Akhtar, J., Amin, N.A.S., 2011. A review on process conditions for optimum bio-oil yield in hydrothermal liquefaction of biomass. Renew. Sust. Energy Rev. 15, 1615–1624.

Amarasekara, A.S., Ebede, C.C., 2009. Zinc chloride mediated degradation of cellulose at 200°C and identification of the products. Bioresour. Technol. 100, 5301–5304.

Anderegg, W.R., Prall, J.W., Harold, J., Schneider, S.H., 2010. Expert credibility in climate change. Proc. Natl. Acad. Sci. U.S.A. 107, 12107–12109.

Andert, J., Mumme, J., 2015. Impact of pyrolysis and hydrothermal biochar on gas-emitting activity of soil microorganisms and bacterial and archaeal community composition. Appl. Soil Ecol. 96, 225–239.

Antal Jr, M.J., Mok, W.S., Richards, G.N., 1990. Mechanism of formation of 5-(hydroxymethyl)-2-furaldehyde from D-fructose and sucrose. Carbohydr. Res. 199, 91–109.

Bargmann, I., Rillig, M., Buss, W., Kruse, A., Kuecke, M., 2013. Hydrochar and biochar effects on germination of spring barley. J. Agron. Crop Sci. 199, 360–373.

Bargmann, I., Rillig, M.C., Kruse, A., Greef, J.M., Kücke, M., 2014a. Effects of hydrochar application on the dynamics of soluble nitrogen in soils and on plant availability. J. Plant Nutr. Soil Sci. 177, 48–58.

Bargmann, I., Martens, R., Rillig, M.C., Kruse, A., Kücke, M., 2014b. Hydrochar amendment promotes microbial immobilization of mineral nitrogen. J. Plant Nutr. Soil Sci. 177, 59–67.

Bargmann, I., Rillig, M.C., Kruse, A., Greef, J.M., Kücke, M., 2014c. Initial and subsequent effects of hydrochar amendment on germination and nitrogen uptake of spring barley. J. Plant Nutr. Soil Sci. 177, 68–74.

Baronti, S., Alberti, G., Genesio, L., Criscuoli, I., Camin, F., Vaccari, F.P., et al., 2017. Hydrochar enhances growth of poplar for bioenergy while marginally contributing to direct soil carbon sequestration. GCB Bioenergy 9, 1618–1626.

Bobleter, O., 1994. Hydrothermal degradation of polymers derived from plants. Prog. Polym. Sci. 19, 797–841.

Burakov, A.E., Galunin, E.V., Burakova, I.V., Kucherova, A.E., Agarwal, S., Tkachev, A.G., et al., 2018. Adsorption of heavy metals on conventional and nanostructured materials for wastewater treatment purposes: a review. Ecotoxicol. Environ. Saf. 148, 702–712.

Busch, D., Kammann, C., Grünhage, L., Müller, C., 2012. Simple biotoxicity tests for evaluation of carbonaceous soil additives: establishment and reproducibility of four test procedures. J. Environ. Qual. 41, 1023–1032.

Busch, D., Stark, A., Kammann, C.I., Glaser, B., 2013. Genotoxic and phytotoxic risk assessment of fresh and treated hydrochar from hydrothermal carbonization compared to biochar from pyrolysis. Ecotoxicol. Environ. Saf. 97, 59–66.

Cao, X., Ro, K.S., Libra, J.A., Kammann, C.I., Lima, I., Berge, N., et al., 2013. Effects of biomass types and carbonization conditions on the chemical characteristics of hydrochars. J. Agric. Food Chem. 61, 9401–9411.

Castro, M.S., Melillo, J.M., Steudler, P.A., Chapman, J.W., 1994. Soil moisture as a predictor of methane uptake by temperate forest soils. Can. J. For. Res. 24, 1805–1810.

Chen, L.F., Lu, Y., Yu, L., Lou, X.W., 2017. Designed formation of hollow particle-based nitrogen-doped carbon nanofibers for high-performance supercapacitors. Energy Environ. Sci. 10, 1777–1783.

Cui, X., Antonietti, M., Yu, S.H., 2006. Structural effects of iron oxide nanoparticles and iron ions on the hydrothermal carbonization of starch and rice carbohydrates. Small 2, 756–759.

Demir-Cakan, R., Baccile, N., Antonietti, M., Titirici, M.-M., 2009. Carboxylate-rich carbonaceous materials via one-step hydrothermal carbonization of glucose in the presence of acrylic acid. Chem. Mater. 21, 484–490.

Dinjus, E., Kruse, A., Troeger, N., 2011. Hydrothermal carbonization–1. Influence of lignin in lignocelluloses. Chem. Eng. Technol. 34, 2037–2043.

Dr, M.S., Dr, A.B.F., 2009. Chemical and structural properties of carbonaceous products obtained by hydrothermal carbonization of saccharides. Chem. Eur. J. 15, 4195–4203.

Eibisch, N., Helfrich, M., Don, A., Mikutta, R., Kruse, A., Ellerbrock, R., et al., 2013. Properties and degradability of hydrothermal carbonization products. J. Environ. Qual. 42, 1565–1573.

Eibisch, N., Durner, W., Bechtold, M., Fuß, R., Mikutta, R., Woche, S.K., et al., 2015. Does water repellency of pyrochars and hydrochars counter their positive effects on soil hydraulic properties? Geoderma 245, 31–39.

Elaigwu, S.E., Rocher, V., Kyriakou, G., Greenway, G.M., 2014. Removal of Pb 2 + and Cd2 + from aqueous solution using chars from pyrolysis and microwave-assisted hydrothermal carbonization of *Prosopis africana* shell. J. Ind. Eng. Chem. 20, 3467–3473.

Falco, C., Baccile, N., Titirici, M.-M., 2011. Morphological and structural differences between glucose, cellulose and lignocellulosic biomass derived hydrothermal carbons. Green Chem. 13, 3273–3281.

Falco, C., Sevilla, M., White, R.J., Rothe, R., Titirici, M.M., 2012. Renewable nitrogen-doped hydrothermal carbons derived from microalgae. ChemSusChem 5, 1834–1840.

Falco, C., Marco-Lozar, J.P., Salinas-Torres, D., Morallon, E., Cazorla-Amorós, D., Titirici, M.-M., et al., 2013. Tailoring the porosity of chemically activated hydrothermal carbons: influence of the precursor and hydrothermal carbonization temperature. Carbon N. Y. 62, 346–355.

Fan, Y., Liu, P.F., Huang, Z.Y., Jiang, T.W., Yao, K.L., Han, R., 2015. Porous hollow carbon spheres for electrode material of supercapacitors and support material of dendritic Pt electrocatalyst. J. Power Sources 280, 30–38.

Feng, G., Shao, G., Qu, J., Lv, S., Li, Y., Wu, M., 2015. Tailoring of porous and nitrogen-rich carbons derived from hydrochar for high-performance supercapacitor electrodes. Electrochim. Acta 155, 201–208.

Fuertes, A.B., Sevilla, M., 2015. Superior capacitive performance of hydrochar-based porous carbons in aqueous electrolytes. ChemSusChem 8, 1049.

Funke, A., Ziegler, F., 2010. Hydrothermal carbonization of biomass: a summary and discussion of chemical mechanisms for process engineering. Biofuels, Bioprod. Biorefin. 4, 160–177.

Gajić, A., Koch, H.-J., 2012. Sugar beet (*Beta vulgaris* L.) growth reduction caused by hydrochar is related to nitrogen supply. J. Environ. Qual. 41, 1067–1075.

Gaskin, J.W., Speir, R.A., Harris, K., Das, K., Lee, R.D., Morris, L.A., et al., 2010. Effect of peanut hull and pine chip biochar on soil nutrients, corn nutrient status, and yield. Agron. J. 102, 623–633.

Gaunt, J.L., Lehmann, J., 2008. Energy balance and emissions associated with biochar sequestration and pyrolysis bioenergy production. Environ. Sci. Technol. 42, 4152–4158.

George, C., Wagner, M., Kücke, M., Rillig, M.C., 2012. Divergent consequences of hydrochar in the plant–soil system: arbuscular mycorrhiza, nodulation, plant growth and soil aggregation effects. Appl. Soil Ecol. 59, 68–72.

Girisuta, B., Janssen, L., Heeres, H., 2006. Green chemicals: a kinetic study on the conversion of glucose to levulinic acid. Chem. Eng. Res. Des. 84, 339–349.

Glaser, B., Lehmann, J., Zech, W., 2002. Ameliorating physical and chemical properties of highly weathered soils in the tropics with charcoal—a review. Biol. Fertil. Soils 35, 219–230.

Guo, S., Dong, X., Liu, K., Yu, H., Zhu, C., 2015. Chemical, energetic, and structural characteristics of hydrothermal carbonization solid products for lawn grass. BioResources 10, 4613–4625.

Gupta, R.K., Dubey, M., Kharel, P., Gu, Z., Fan, Q.H., 2015. Biochar activated by oxygen plasma for supercapacitors. J. Power Sources 274, 1300–1305.

Harter, J., Krause, H.-M., Schuettler, S., Ruser, R., Fromme, M., Scholten, T., et al., 2014. Linking N_2O emissions from biochar-amended soil to the structure and function of the N-cycling microbial community. ISME J. 8, 660–674.

Hatcher, P.G., Clifford, D.J., 1997. , The organic geochemistry of coal: from plant materials to coal. Org. Geochem. 27, 251–274.

He, C., Giannis, A., Wang, J.-Y., 2013. Conversion of sewage sludge to clean solid fuel using hydrothermal carbonization: hydrochar fuel characteristics and combustion behavior. Appl. Energy 111, 257–266.

Hoekman, S.K., Broch, A., Robbins, C., 2011. Hydrothermal carbonization (HTC) of lignocellulosic biomass. Energy Fuel. 25, 1802–1810.

Jain, A., Aravindan, V., Jayaraman, S., Kumar, P.S., Balasubramanian, R., Ramakrishna, S., et al., 2013. Activated carbons derived from coconut shells as high energy density cathode material for Li-ion capacitors. Sci. Rep. 3, 1–6.

Jain, A., Jayaraman, S., Balasubramanian, R., Srinivasan, M., 2014. Hydrothermal pre-treatment for mesoporous carbon synthesis: enhancement of chemical activation. J. Mater. Chem. A 2, 520–528.

Jain, A., Balasubramanian, R., Srinivasan, M.P., 2015a. Tuning hydrochar properties for enhanced mesopore development in activated carbon by hydrothermal carbonization. Microporous Mesoporous Mater. 203, 178–185.

Jain, A., Xu, C., Jayaraman, S., Balasubramanian, R., Lee, J.Y., Srinivasan, M.P., 2015b. Mesoporous activated carbons with enhanced porosity by optimal hydrothermal pre-treatment of biomass for supercapacitor applications. Microporous Mesoporous Mater. 218, 55–61.

Jain, A., Balasubramanian, R., Srinivasan, M.P., 2016. Hydrothermal conversion of biomass waste to activated carbon with high porosity: a review. Chem. Eng. J. 283, 789–805.

Jandl, G., Eckhardt, K.-U., Bargmann, I., Kücke, M., Greef, J.-M., Knicker, H., et al., 2013. Hydrothermal carbonization of biomass residues: mass spectrometric characterization for ecological effects in the soil–plant system. J. Environ. Qual. 42, 199–207.

Kambo, H.S., Dutta, A., 2015. A comparative review of biochar and hydrochar in terms of production, physico-chemical properties and applications. Renewable Sustainable Energy Rev. 45, 359–378.

Kammann, C., Ratering, S., Eckhard, C., Müller, C., 2012. Biochar and hydrochar effects on greenhouse gas (carbon dioxide, nitrous oxide, and methane) fluxes from soils. J. Environ. Qual. 41, 1052–1066.

Kang, S., Li, X., Fan, J., Chang, J., 2012. Characterization of hydrochars produced by hydrothermal carbonization of lignin, cellulose, d-xylose, and wood meal. Ind. Eng. Chem. Res. 51, 9023–9031.

Khalil, M., Baggs, E., 2005. CH_4 oxidation and N_2O emissions at varied soil water-filled pore spaces and headspace CH_4 concentrations. Soil Biol. Biochem. 37, 1785–1794.

Kumar, S., Loganathan, V.A., Gupta, R.B., Barnett, M.O., 2011. An assessment of U(VI) removal from groundwater using biochar produced from hydrothermal carbonization. J. Environ. Manage. 92, 2504–2512.

Libra, J.A., Ro, K.S., Kammann, C., Funke, A., Berge, N.D., Neubauer, Y., et al., 2011. Hydrothermal carbonization of biomass residuals: a comparative review of the chemistry, processes and applications of wet and dry pyrolysis. Biofuels 2, 71–106.

Liu, F., Yu, R., Guo, M., 2017. Hydrothermal carbonization of forestry residues: influence of reaction temperature on holocellulose-derived hydrochar properties. J. Mater. Sci. 52, 1736–1746.

Liu, Y., Zhu, X., Qian, F., Zhang, S., Chen, J., 2014. Magnetic activated carbon prepared from rice straw-derived hydrochar for triclosan removal. RSC Adv. 4, 63620–63626.

Liu, Z., Zhang, F.-S., Wu, J., 2010. Characterization and application of chars produced from pinewood pyrolysis and hydrothermal treatment. Fuel 89, 510–514.

Liu, Z., Quek, A., Kent Hoekman, S., Balasubramanian, R., 2013a. Production of solid biochar fuel from waste biomass by hydrothermal carbonization. Fuel 103, 943–949.

Liu, Z., Quek, A., Parshetti, G., Jain, A., Srinivasan, M., Hoekman, S.K., et al., 2013b. A study of nitrogen conversion and polycyclic aromatic hydrocarbon (PAH) emissions during hydrochar–lignite co-pyrolysis. Appl. Energy 108, 74–81.

Ma, X., Zou, B., Cao, M., Chen, S.-L., Hu, C., 2014. Nitrogen-doped porous carbon monolith as a highly efficient catalyst for CO_2 conversion. J. Mater. Chem. A 2, 18360–18366.

Malghani, S., Gleixner, G., Trumbore, S.E., 2013. Chars produced by slow pyrolysis and hydrothermal carbonization vary in carbon sequestration potential and greenhouse gases emissions. Soil Biol. Biochem. 62, 137–146.

Malghani, S., Jüschke, E., Baumert, J., Thuille, A., Antonietti, M., Trumbore, S., et al., 2015. Carbon sequestration potential of hydrothermal carbonization char (hydrochar) in two contrasting soils; results of a 1-year field study. Biol. Fertil. Soil. 51, 123–134.

Masselter, S., Zemann, A., Bobleter, O., 1995. Analysis of lignin degradation products by capillary electrophoresis. Chromatographia 40, 51–57.

Mørkved, P.T., Dörsch, P., Henriksen, T.M., Bakken, L.R., 2006. N_2O emissions and product ratios of nitrification and denitrification as affected by freezing and thawing. Soil Biol. Biochem. 38, 3411–3420.

Naisse, C., Girardin, C., Lefevre, R., Pozzi, A., Maas, R., Stark, A., et al., 2015. Effect of physical weathering on the carbon sequestration potential of biochars and hydrochars in soil. GCB Bioenergy 7, 488–496.

Nelson, N.O., Agudelo, S.C., Yuan, W., Gan, J., 2011. Nitrogen and phosphorus availability in biochar-amended soils. Soil Sci. 176, 218–226.

Parshetti, G.K., Balasubramanian, R., 2014. Evaluation of hydrothermally carbonized hydrochar in improving energy security and mitigating greenhouse gas emissions. ACS Symp. 1184, 23–48.

Parshetti, G.K., Liu, Z., Jain, A., Srinivasan, M., Balasubramanian, R., 2013. Hydrothermal carbonization of sewage sludge for energy production with coal. Fuel 111, 201–210.

Parshetti, G.K., Chowdhury, S., Balasubramanian, R., 2015. Biomass derived low-cost microporous adsorbents for efficient CO_2 capture. Fuel 148, 246–254.

Pazferreiro, J., Lu, H., Fu, S., Méndez, A., Gascó, G., 2014. Use of phytoremediation and biochar to remediate heavy metal polluted soils: a review. Solid Earth Dis. 5, 2155–2179.

Qi, X.H., Li, L.Y., Tan, T.F., Chen, W.T., Smith Jr, R.L., 2013. Adsorption of 1-butyl-3-methylimidazolium chloride ionic liquid by functional carbon microspheres from hydrothermal carbonization of cellulose. Environ. Sci. Technol. 47, 2792–2798.

Reza, M.T., Yang, X., Coronella, C.J., Lin, H., Hathwaik, U., Shintani, D., et al., 2016. Hydrothermal carbonization (HTC) and pelletization of two arid land plants bagasse for energy densification. ACS Sustain. Chem. Eng. 4, 1106–1114.

Rillig, M.C., Wagner, M., Salem, M., Antunes, P.M., George, C., Ramke, H.-G., et al., 2010. Material derived from hydrothermal carbonization: effects on plant growth and arbuscular mycorrhiza. Appl. Soil Ecol. 45, 238–242.

Sangchoom, W., Mokaya, R., 2015. Valorization of lignin waste: carbons from hydrothermal carbonization of renewable lignin as superior sorbents for CO_2 and hydrogen storage. ACS Sustain. Chem. Eng. 3, 1658–1667.

Schimmelpfennig, S., Müller, C., Grünhage, L., Koch, C., Kammann, C., 2014. Biochar, hydrochar and uncarbonized feedstock application to permanent grassland—effects on greenhouse gas emissions and plant growth. Agric. Ecosyst. Environ. 191, 39–52.

Sevilla, M., Fuertes, A.B., 2009a. Chemical and structural properties of carbonaceous products obtained by hydrothermal carbonization of saccharides. Chem. Eur. J. 15, 4195–4203.

Sevilla, M., Fuertes, A.B., 2009b. The production of carbon materials by hydrothermal carbonization of cellulose. Carbon N. Y. 47, 2281–2289.

Sevilla, M., Fuertes, A.B., 2011. Sustainable porous carbons with a superior performance for CO_2 capture. Energy Environ. Sci. 4, 1765–1771.

Sevilla, M., Macia-Agullo, J.A., Fuertes, A.B., 2011a. Hydrothermal carbonization of biomass as a route for the sequestration of CO_2: chemical and structural properties of the carbonized products. Biomass Bioenergy 35, 3152–3159.

Sevilla, M., Fuertes, A.B., Mokaya, R., 2011b. High density hydrogen storage in superactivated carbons from hydrothermally carbonized renewable organic materials. Energy Environ. Sci. 4, 1400–1410.

Sevilla, M., Fuertes, A.B., Green, A., 2016. Approach to high-performance supercapacitor electrodes: the chemical activation of hydrochar with potassium bicarbonate. ChemSusChem 9, 1880–1888.

Si, W., Zhou, J., Zhang, S., Li, S., Xing, W., Zhuo, S., 2013. Tunable N-doped or dual N, S-doped activated hydrothermal carbons derived from human hair and glucose for supercapacitor applications. Electrochim. Acta 107, 397–405.

Simon, P., Gogotsi, Y., 2008. Materials for electrochemical capacitors. Nat. Mater. 7, 845–854.

Steinbeiss, S., Gleixner, G., Antonietti, M., 2009. Effect of biochar amendment on soil carbon balance and soil microbial activity. Soil Biol. Biochem. 41, 1301–1310.

Sun, K., Ro, K., Guo, M., Novak, J., Mashayekhi, H., Xing, B., 2011. Sorption of bisphenol A, 17 alpha-ethinyl estradiol and phenanthrene on thermally and hydrothermally produced biochars. Bioresour. Technol. 102 (10), 5757–5763.

Sun, Y., Gao, B., Yao, Y., Fang, J., Zhang, M., Zhou, Y., et al., 2014. Effects of feedstock type, production method, and pyrolysis temperature on biochar and hydrochar properties. Chem. Eng. J. 240, 574–578.

Tan, J., Chen, H., Gao, Y., Li, H., 2015. Nitrogen-doped porous carbon derived from citric acid and urea with outstanding supercapacitance performance. Electrochim. Acta 178, 144–152.

Tanaike, O., Inagaki, M., 1997. Ternary intercalation compounds of carbon materials having a low graphitization degree with alkali metals. Carbon N. Y. 35, 831–836.

Tang, Z., Deng, Y., Luo, T., Xu, Y.S., Zhu, N.M., 2016. Enhanced removal of Pb(II) by supported nanoscale Ni/Fe on hydrochar derived from biogas residues. Chem. Eng. J. 292, 224–232.

Tian, W., Gao, Q., Tan, Y., Yang, K., Zhu, L., Yang, C., et al., 2015. Bio-inspired beehive-like hierarchical nanoporous carbon derived from bamboo-based industrial by-product as a high performance supercapacitor electrode material. J. Mater. Chem. A 3, 5656–5664.

Titirici, M.-M., Antonietti, M., 2010. Chemistry and materials options of sustainable carbon materials made by hydrothermal carbonization. Chem. Soc. Rev. 39, 103–116.

Titirici, M.M., Thomas, A., Yu, S.-H., Müller, J.-O., Antonietti, M., 2007. A direct synthesis of mesoporous carbons with bicontinuous pore morphology from crude plant material by hydrothermal carbonization. Chem. Mater. 19, 4205–4212.

Titirici, M.-M., White, R.J., Falco, C., Sevilla, M., 2012. Black perspectives for a green future: hydrothermal carbons for environment protection and energy storage. Energy Environ. Sci. 5, 6796–6822.

Vaccari, F., Baronti, S., Lugato, E., Genesio, L., Castaldi, S., Fornasier, F., et al., 2011. Biochar as a strategy to sequester carbon and increase yield in durum wheat. Eur. J. Agron. 34, 231–238.

Veldkamp, E., Weitz, A.M., Keller, M., 2001. Management effects on methane fluxes in humid tropical pasture soils. Soil Biol. Biochem. 33, 1493–1499.

Warnock, D.D., Lehmann, J., Kuyper, T.W., Rillig, M.C., 2007. Mycorrhizal responses to biochar in soil–concepts and mechanisms. Plant Soil 300, 9–20.

Xiao, L.-P., Shi, Z.-J., Xu, F., Sun, R.-C., 2012. Hydrothermal carbonization of lignocellulosic biomass. Bioresour. Technol. 118, 619–623.

Xue, Y., Gao, B., Yao, Y., Inyang, M., Zhang, M., Zimmerman, A.R., et al., 2012. Hydrogen peroxide modification enhances the ability of biochar (hydrochar) produced from hydrothermal carbonization of peanut hull to remove aqueous heavy metals: batch and column tests. Chem. Eng. J. 200–202, 673–680.

Zhang, D.-L., Zhang, M.-Y., Zhang, C.-H., Sun, Y.-J., Sun, X., Yuan, X.-Z., 2016a. Pyrolysis treatment of chromite ore processing residue by biomass: cellulose pyrolysis and Cr(VI) reduction behavior. Environ. Sci. Technol. 50, 3111–3118.

Zhang, L.L., Li, H.H., Shi, Y.H., Fan, C.Y., Wu, X.L., Wang, H.F., et al., 2016b. A novel layered sedimentary rocks structure of the oxygen-enriched carbon for ultrahigh-rate-performance supercapacitors. ACS Appl. Mater. Interfaces 8, 4233.

Zhou, N., Chen, H., Xi, J., Yao, D., Zhou, Z., Tian, Y., et al., 2017. Biochars with excellent Pb(II) adsorption property produced from fresh and dehydrated banana peels via hydrothermal carbonization. Bioresour. Technol. 232, 204–210.

Zhu, X., Liu, Y., Qian, F., Zhou, C., Zhang, S., Chen, J., 2014a. Preparation of magnetic porous carbon from waste hydrochar by simultaneous activation and magnetization for tetracycline removal. Bioresour. Technol. 154, 209–214.

Zhu, X., Liu, Y., Zhou, C., Luo, G., Zhang, S., Chen, J., 2014b. A novel porous carbon derived from hydrothermal carbon for efficient adsorption of tetracycline. Carbon. 77, 627–636.

Zhu, X., Liu, Y., Qian, F., Zhou, C., Zhang, S., Chen, J., 2015a. Role of hydrochar properties on the porosity of hydrochar-based porous carbon for yheir sustainable application. ACS Sustain. Chem. Eng. 3, 833–840.

Zhu, X., Qian, F., Liu, Y., Zhang, S., Chen, J., 2015b. Environmental performances of hydrochar-derived magnetic carbon composite affected by its carbonaceous precursor. RSC Adv. 5, 60713–60722.

Zhu, X., Liu, Y., Qian, F., Zhang, S., Chen, J., 2015c. Investigation on the physical and chemical properties of hydrochar and its derived pyrolysis char for their potential application: influence of hydrothermal carbonization conditions. Energy Fuels 29, 5222–5230.

Zhu, X.D., Liu, Y.C., Zhou, C., Zhang, S.C., Chen, J.M., 2014c. Novel and high-performance magnetic carbon composite prepared from waste hydrochar for dye removal. ACS Sustain. Chem. Eng. 2, 969–977.

Zhu, Z., Rosendahl, L., Toor, S.S., Yu, D., Chen, G., 2015d. Hydrothermal liquefaction of barley straw to bio-crude oil: effects of reaction temperature and aqueous phase recirculation. Appl. Energy 137, 183–192.

CHAPTER

16

Waste-Derived Biochar for CO_2 Sequestration

Shou-Heng Liu

Department of Environmental Engineering, National Cheng Kung University, Tainan, Taiwan

16.1 INTRODUCTION

The extensive emissions of greenhouse gas (e.g., carbon dioxide (CO_2) and methane (CH_4)) have caused global warming effects that lead to negative results in terms of environmental and economic aspects. The obviously increased concentration of CO_2 in the atmosphere has been blamed for the most possible driver of global climate change. Therefore, it is urgent to reduce the amounts of CO_2 released to the air due to consumption of fossil fuels. There are many routes to decrease the anthropogenic CO_2 emissions, for example, enhancement of energy efficiency, the use of fuels with non/less-carbon contents, and the capture and storage of CO_2 (CCS) (Wilcox, 2012; Wang et al., 2011; Haszeldine, 2009). The Intergovernmental Panel on Climate Change (IPCC) has defined the CCS processes to include the separation of CO_2 from emission sources, transportation of captured CO_2 to a storage site, and long-term isolation from the atmosphere (Metz et al., 2005). Among these processes, the crucial tasks aim to develop cost-effective and less efficiency penalty technologies for separating and purifying CO_2 from flue gases in coal-burned power sectors (Sevilla and Fuertes, 2011; Coromina et al., 2016; Chen et al., 2016). It is proposed that CO_2 adsorption by using solid adsorbents may be a prospective technology that can reduce the cost and the energy penalty as well in comparison to state-of-the-art technologies (Rezaei and Webley, 2010; Sayari et al., 2011; D'Alessandro et al., 2010; Alonso et al., 2017; Oschatz and Antonietti, 2018). In view of this, the preparation of adsorbents from the cheap precursors would further increase the effectiveness of CO_2

Biochar from Biomass and Waste
DOI: https://doi.org/10.1016/B978-0-12-811729-3.00016-9

adsorption process in the practical applications. Biochars can be obtained from abundant natural sources (e.g., agricultural wastes, wild grasses, wood wastes, animal manure, and so on) via a simple carbonization process at elevated temperatures under different gas atmospheres (McCarl et al., 2009; Lehmann and Joseph, 2012; Kinney et al., 2012). Large amounts of the above-mentioned biowastes are asked to pay for their disposal fee. Thus, the valorization of wastes to biochars for the fabrication of CO_2 solid sorbents should be both an economical incentive and environmentally attractive. In this chapter, various bio-sources are reported as raw materials to prepare biochars by using different preparation and modification methods for CO_2 adsorption. In addition, these biochars are tested for CO_2 adsorption performance at low temperatures (25–75°C) and high temperatures (75–150°C).

16.2 TECHNOLOGIES FOR CO_2 CAPTURE

Generally, there are three routes to capture CO_2 from the emissions of fossil fuel combustion systems (Fig. 16.1) (Figueroa et al., 2008): (1) oxy-combustion, in which the pure oxygen is used to initiate the combustion, thus, a mixture of CO_2 and H_2O is formed; (2) pre-combustion, in which the carbon is first removed from the fuel before combustion. This process called partial oxidation, or gasification, can be done by reacting coal with steam and oxygen at high temperature and pressure. The resultant product is a gaseous fuel containing mostly carbon monoxide and hydrogen, known as synthesis gas (syngas),

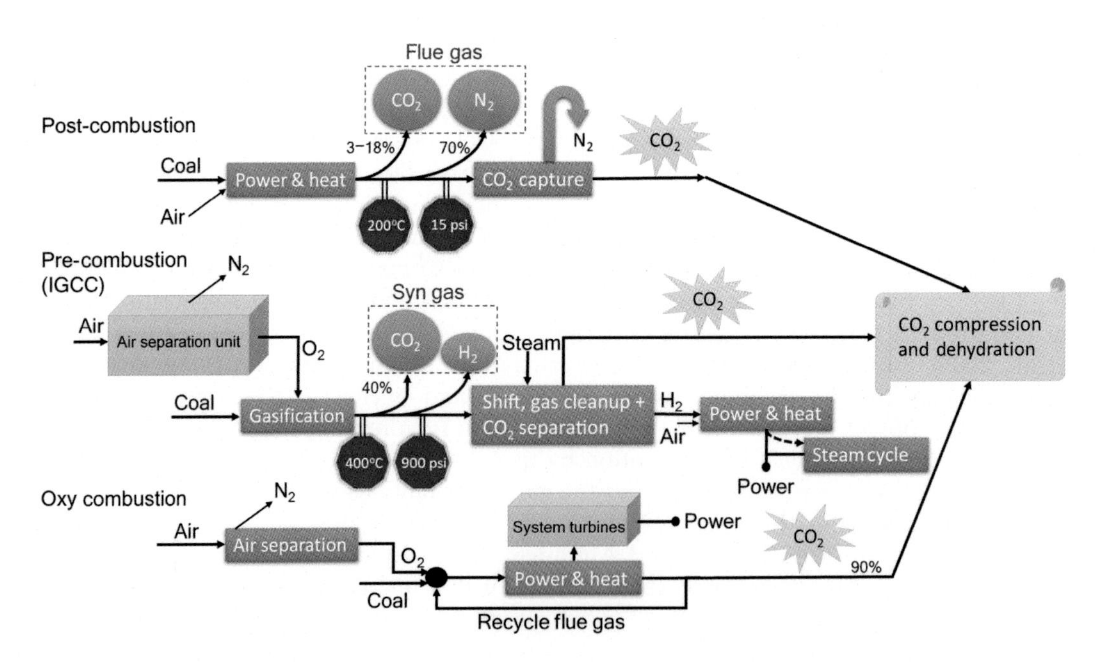

FIGURE 16.1 Technological process for CO_2 capture in the post-combustion, pre-combustion, and oxy-combustion system.

which can be burned to generate electricity in a power plant. (3) Post-combustion, which c. 10–30 vol% of CO_2 is emitted at the end of a pipe from a wet exhaust flue gas after fossil fuels are burned. The critical points to make these processes possible are developing the separation technologies, including O_2/N_2 separation for oxy-combustion, CO_2/H_2 separation for pre-combustion, and CO_2/N_2 separation for post-combustion. Among these separation technologies, CO_2 capture in the post-combustion process is definitely the toughest option because the harsh conditions (a diluted, low-pressure, wet, and hot CO_2/N_2 mixture) need to be treated. However, in terms of its most widely industrial application, it is important to develop the cost-effective method that should be compatible with the pre-existing installations in the coal-fired power plants.

16.3 ADSORPTION

CO_2 capture by solid adsorbents is a process through which the CO_2 adsorbates are attached onto the surface of adsorbents. Typically, adsorption process can be divided into chemical and physical adsorption based on the interaction between adsorbate and adsorbent. If the bonding between CO_2 and the adsorbent surface is weak (i.e., van der Waals and electrostatic forces), the process is attributed to physisorption that can be applied in the pre-combustion process. Most of CO_2 is eliminated from the stream prior to the burning of the fuel under the high pressure and low temperature conditions (García et al., 2011). Therefore, the major features for potential physisorbents include: high surface areas for a superior adsorption capacity and high selectivity for high-purity separation from a mixture gas. Various materials, such as carbon, zeolite, and metal organic-framework adsorbents, have been developed as physisorbents (Hao et al., 2011). In terms of chemisorption, chemical interactions are observed between the adsorbates and the adsorbent surface. During post-combustion processes, CO_2 can be chemically adsorbed at lower concentration and pressures and at higher temperatures. To achieve better adsorption, many nitrogen-containing adsorbents for the adsorption of CO_2 were reported (Pevida et al., 2008a; Maroto-Valer et al., 2005). For power plant applications, the temperature swing adsorption (TSA) systems (Kwon et al., 2011), and rapid adsorption of CO_2 from the flue gas is highly necessary for high efficiencies of the cyclic adsorption/desorption process. In this way, the lower quantities of adsorbents and smaller volumes of adsorption/desorption devices are required. Generally, adsorption performance of adsorbents is varied, depending on various parameters, such as temperature, pressure and CO_2 concentration of flue gas, and properties of adsorbents (i.e., surface area, pore size, and structure of adsorbents). It has been reported (Sjostrom et al., 2011) that a prospective adsorbent has to possess the high selectivity and adsorption capacity of CO_2, fast adsorption and desorption rates, high stability, low attrition rate, low energy requirements for the desorption of CO_2, and should be easy to mass produce.

16.4 WASTE-DERIVED BIOCHARS FOR CO_2 ADSORBENTS

It is estimated that large amounts of solid waste from the coal engineering, agricultural manufacturing, water management, and household zone are produced. As estimated by United Nations Environment Programme (UNEP), about 140 billion metric tons of biomass is produced annually from agriculture all over the world (Wang et al., 2017). However, the present waste treatments (e.g., landfill and incineration) would lead to climate change and water, air, and soil pollution. To reduce the negative impact on environmental issues, extensive research works are concentrating on the development of new strategies toward waste management. Among them, the preparation of high-end products derived from plentiful biomass waste is important because how to decrease the cost of CO_2 capture and the energy penalty in the capturing process is the most critical challenge for reducing CO_2 emissions from flue gases. In addition, due to the varied compositions of biowastes, solid adsorbents with diverse physicochemical properties have been effectively proposed for CO_2 capture in a specific range of temperatures. As can be seen in Tables 16.1 and 16.2, high carbon content wastes (e.g., coal by-products), biowastes or plastic wastes were recycled as precursors to prepare biochar-derived adsorbents for CO_2 capture at low temperatures (25–75°C) and high temperatures (75–150°C). Many groups try to use biomass wastes (e.g., olive stones and almond shells, coconut shells, bagasse, soybeans, and rice husks) for the preparation of biochar adsorbents due to their superior adsorption performance, low ash amount, and low cost (Maroto-Valer et al., 2008; Plaza et al., 2009, 2010, 2011, 2014; Son et al., 2005; Boonpoke et al., 2012; Thote et al., 2010; Arenillas et al., 2005; Olivares-Marín and Maroto-Valer, 2011; Olivares-Marín et al., 2011; Huang et al., 2015, 2014a; Liu and Huang, 2018; Yang et al., 2015; Zhu et al., 2016; Alabadi et al., 2015; Zhang et al., 2014, 2004; Boonpoke et al., 2011). The olive stones were carbonized at 600°C under N_2 flow, followed by activating at 800°C. The resultant biochar adsorbents were found to have a high pH value of 8.7 because of the presence of solid Lewis's base site (Bandosz and Ania, 2006). Also, the surface area and basicity of biochar carbons were varied, depending on the activation and amine grafting process. It can be observed that biochar adsorbents from olive stones possess higher CO_2 adsorption capacity (107 mg g^{-1}) than that prepared from amine-impregnated olive stone's carbon (86 mg g^{-1}) at 25°C. Furthermore, biochars synthesized from olive stones (Plaza et al., 2011) were treated with NH_3 to improve CO_2 capture performance. The adsorbents were impregnated with NH_3 at 800°C for 2 h and their CO_2 adsorption capacity (110 mg g^{-1}) was higher than that of the pristine stone-derived activated carbon (88 mg g^{-1}) in flue gas at 30°C. This can be attributed to the decrease of pore size in the adsorbents by impregnating higher contents of ammonia. The palm kernel shells (Nasri et al., 2014) were also employed as raw materials for preparation of CO_2 adsorption sorbents which were obtained by carbonizing at 700°C for 2 h and then activated at 800°C for 1 h under the atmosphere of CO_2 flow. These palm kernel shell-derived biochars possess high surface areas (167.1 m^2 g^{-1}) and high ratio of micropores volume (89.5%). Therefore, the CO_2 adsorption capacities were reported to be 73.0 (101 kPa) and 170.3 (202 kPa) mg g^{-1} at 30°C. The superior performance of these biochars adsorbents was possibly owing to the formation of micropores via the removal of volatile species from the carbon (Foo and Hameed, 2013). In addition, the contents of fixed

TABLE 16.1 Comparison of CO_2 Capture (Low Temperature) Performance on the Waste-Recovered Adsorbents

Wastes	Synthesis	Modifications	Surface Area ($m^2\ g^{-1}$)	CO_2 Adsorption Capacity ($mg\ g^{-1}$)	Cyclability	Selectivity ($\alpha(CO_2/N_2)$)	Reference
Unburned carbon in fly ash	1. Acid treatments to concentrate the carbon 2. Carbonization under steam flow at 850°C for 30–120 min	Functionalization with MEA, DEA, and MDEA	204–302	68.6 at 30°C	–		Maroto-Valer et al. (2008)
Olive stone and almond shell	1. Pyrolyzed at 600–900°C 2. Activated with CO_2 at 700–1000°C	1. Amination with NH_3 gas at 400–900°C for 2 h 2. Impregnation with PEI 3. Functionalization with amine by EAS	8–1090	117 at 25°C (100% CO_2); 52 at 25°C (15% CO_2)	–	3.5	Plaza et al. (2011, 2010, 2009)
Coconut shell	1. Mix well with 25% of coal tar pitch 2. Carbonization at 800°C and activation with CO_2 at 800°C	Doped with metal (Ca, Co, Cu, Ni) before heat treatment	519 (micropore areas)	98 at 25°C	–		Son et al. (2005)
Bagasse and rice husk	Chemical activation with zinc chloride at 500°C for 1 h	Surface modification with PEI and MEA	923 and 927	76.9 at 30°C (bagasse) 57.13 at 30°C (rice husk)	Averaged CO_2 adsorption capacity of 49.05 (bagasse) and 36.72 (rice husk) at 50°C		Boonpoke et al. (2012)
Soybean	1. Chemical activation with zinc chloride at 600°C for 2 h 2. Physical activation with CO_2 at 600°C for 2.5 h 3. Washing with HCl	–	811	41 at 30°C	CO_2 adsorption capacity of 42 $mg\ g^{-1}$ after four consecutive cycles		Thote et al. (2010)
Sewage sludge	Activated under CO_2 atmosphere at 850°C	Grafting with PEI	127–137	14 at 25°C			Plaza et al. (2014)

(Continued)

TABLE 16.1 (Continued)

Wastes	Synthesis	Modifications	Surface Area ($m^2\ g^{-1}$)	CO_2 Adsorption Capacity ($mg\ g^{-1}$)	Cyclability	Selectivity ($\alpha(CO_2/N_2)$)	Reference
Plastic waste (PET)	Mix the wastes with acridine, carbazole, and urea	Activated with KOH at 500°C for 30 min	472	22–48 at 25°C			Arenillas et al. (2005)
Carpet wastes	Activation with KOH at 600–900°C for 1 h		736–2665	138 at 25°C	CO_2 adsorption capacity can maintain 138 $mg\ g^{-1}$ after five cycles		Olivares-Marín and Maroto-Valer (2011), Olivares-Marín et al. (2011)
Rice straw	Microwave pyrolysis at 100–250 W		122	77 at 20°C	CO_2 adsorption capacity decreased for 8%–10% after eight cycles		Huang et al. (2015)
Coffee grounds	Ammoxidation by melamine at 160°C	Activation with KOH at 600°C for 1 h	990	117.5 at 35°C	CO_2 adsorption capacity decreased to 110 $mg\ g^{-1}$ (35°C) after 10 cycles	74.2	Liu and Huang (2018)
Waste wood	Mix with H_3PO_4 at 450°C	Impregnated with NH_3 at 400°C and 800°C	1889–2079	141.7 at 30°C			Heidari et al. (2014a)
Coconut shell	Carbonized at 500°C for 2 h	Mix with urea in air at 350°C for 2 h; KOH activation at 650°C for 1 h	1535	220 at 25°C	2.5% drop in CO_2 adsorption capacity after five cycles	29	Yang et al. (2015)
Pine-cone	Carbonized at 600°C for 1 h	KOH activation at 600–800°C for 1 h	1170–2110	209 at 25°C	Maintain a stable uptake of 6–7 wt% CO_2 (15% CO_2) during the cyclic tests		Zhu et al. (2016)
Gelatin and starch	Carbonized at 450°C with a ramp temperature 10°C min^{-1}	KOH activation at 700°C for 10 min	714–1957	144.3–167.2 at 25°C	–	98	Alabadi et al. (2015)
Cotton stalk	Carbonization at 500–900°C with N_2 purging	After carbonization, N_2 was replaced by CO_2, NH_3, or CO_2–NH_3	627	99.4 at 20°C			Zhang et al. (2014)

MEA, monoethanolamine; DEA, diethanolamine; MDEA, methyldiethanolamine; PEI, polyethylenimine; EAS, electrophilic aromatic substitution.

TABLE 16.2 Comparison of CO_2 Capture (High Temperature) Performance on the Waste-Recovered Adsorbents

Wastes	Synthesis	Modifications	Surface Area ($m^2 g^{-1}$)	CO_2 Adsorption Capacity ($mg g^{-1}$)	Cyclability	Reference
Unburned carbon in fly ash	Activated under steam flow at 850°C for 0.5–2 h	Impregnation with MEA, DEA, MDEA, and PEI	204–302; 48–1053 (for PEI)	49.8 at 70°C; 43.2 at 100°C; 23.7 at 120°C; 93.6 at 75°C (for PEI)	–	Maroto-Valer et al. (2008), Zhang et al. (2004)
Bagasse and rice husk	Chemical activation with zinc chloride at 500°C for 1 h		923 and 927	24.3 and 6.73 at 75 and 150°C (bagasse); 20.03 and 3.45 at 75 and 150°C (rice husk)		Boonpoke et al. (2011)
Sewage sludge	Activated with CO_2 at 850°C	Grafting with PEI	127–137	25 at 100°C		Plaza et al. (2014)
Plastic waste (PET)	Mix the wastes with acridine, carbazole, and urea	Activated with KOH at 500°C for 30 min	472	12 at 100°C		Arenillas et al. (2005)
Carpet wastes	Activation with KOH at 600–900°C for 1 h		736–2665	31 at 100°C	CO_2 adsorption capacity can maintain 31 $mg g^{-1}$ after five cycles	Olivares-Marín and Maroto-Valer (2011), Olivares-Marín et al. (2011)
Cotton stalk	Carbonization at 500–900°C with N_2 purging	After carbonization, N_2 was replaced by CO_2, NH_3 or CO_2–NH_3	627	39.4 at 20°C		Zhang et al. (2014)

carbon in the biochars may have critical impact on the CO_2 capture. It was reported that the high amounts of fixed carbon content in the adsorbent would increase the inter-particle force between CO_2 and biochars (Rashidi et al., 2013). In other reports, bagasse (Boonpoke et al., 2012) and rice husk (Boonpoke et al., 2011) derived biochars were tested as possible CO_2 adsorbents. They were treated by chemical activation using $ZnCl_2$ at 500°C for 1 h in the nitrogen atmosphere. The surface areas of resultant biochars were in the range of 923–927 $m^2 g^{-1}$. However, the adsorption capacity of CO_2 observed for bagasse-derived biochars (76.9 $mg g^{-1}$) was greater than that of rice husk-derived biochars (57.1 $mg g^{-1}$) at 30°C. There were two possible reasons to explain this result; one is the presence of the higher fixed carbon (84.5 wt%) and lower ash content (1.9 wt%) in bagasse-derived biochars when compared with rice husk-derived biochars. The other is the existence of some minerals which may hinder the CO_2 adsorption performance in rice husk-derived biochars. More importantly, the microporosity in the biochars would be the key parameter in the CO_2 capture process (Maroto-Valer et al., 2005). It was estimated that the best pore size to adsorb gas is five times that of the adsorbate molecule size (Maroto-Valer et al., 2005). So the optimal pore size in the biochars for CO_2 capture should be around 1.0 nm (Boonpoke et al., 2011). Recently, various studies have been focused on wood wastes, which are the probable low-cost biowastes for CO_2 adsorbents (Heidari et al., 2014a,b; Zhu et al., 2014; Creamer et al., 2014). The high carbon amounts in wood wastes and their abundance in the world are the major reasons for their applications (Patnukao and Pavasant, 2008). Eucalyptus wood wastes (Heidari et al., 2014b) were investigated as biochars for CO_2 adsorption. Another study (Timur et al., 2010) showed that the formation of microporosity in biochars can be effected by the variety of biowastes and activation methods. A combination of phosphoric acid, zinc chloride, and potassium hydroxide was used to activate the eucalyptus wood-derived biochars. The results indicated that the activation of biochars by mixture of the phosphoric acid and potassium hydroxide resulted in the formation of adsorbents with higher surface area (2595 $m^2\ g^{-1}$) and volume of microporosity (1.236 $cm^3\ g^{-1}$). Consequently, adsorption capacity of wood wastes treated with activating agents (phosphoric acid and potassium hydroxide) were found to be enhanced (180.4 $mg\ g^{-1}$) at 30°C. The other route (Heidari et al., 2014a) to activate the wood waste-derived biochars was to perform the impregnation of ammonia. The surface area and micropores in the biochar with ammonia impregnation were decreased due to the block of the micropores (Pevida et al., 2008b), resulting in the lower adsorption capacity (141.7 $mg\ g^{-1}$). The hickory wood was carbonized in the temperature range of 300–600°C (Creamer et al., 2014). The results showed that the greater carbonized temperatures the higher surface area in the biochar because of the more micropores formation (Cheng et al., 2008). Also, the formation of nitrogenous groups was induced under high temperature carbonization (600°C). These basic functional groups may cause a strong interaction between the acidic CO_2 molecules (Zhang et al., 2013), thus, enhance the CO_2 capture. The previous investigations indicate that waste-derived biochars are potential candidates for CO_2 adsorbents.

16.5 CONCLUSIONS

In summary, the fabrication of waste-derived biochars for CO_2 capture is reviewed in this chapter. These biochars from wastes with various physicochemical properties, due to their different compositions, were investigated in terms of CO_2 adsorption at low (25–75°C) and high temperatures (75–150°C). The obtained results indicated that these waste-derived biochars can reach the CO_2 adsorption capacities in the range of 14–220 $mg\ g^{-1}$, depending on the features of the precursors, synthesis methods, experimental parameters, and further surface modifications. It was concluded that CO_2 adsorption capacities were strongly related to microporosity on the biochars surface, i.e., the higher amounts of micropores were formed the more CO_2 adsorption capacity was found. The recovery of biowastes to CO_2 adsorbents may have a promising future because the cost of the CO_2 capture process should be remarkably decreased. As such, large amounts of wastes originally taken to landfills can be avoided. Because of the surpassing results in the waste-derived biochars, these low-cost adsorbents can possibly be used in the practical applications for CO_2 capture.

References

Alabadi, A., Razzaque, S., Yang, Y.W., Chen, S., Tan, B., 2015. Chem. Eng. J. 281, 606–612.
Alonso, A., Moral-Vico, J., Markeb, A.A., Busquets-Fite, M., Komilis, D., Puntes, V., et al., 2017. Sci. Total Environ. 595, 51–62.
Arenillas, A., Rubiera, F., Parra, J.B., Ania, C.O., Pis, J.J., 2005. Appl. Surf. Sci. 252, 619–624.
Bandosz, T.J., Ania, C.O., 2006. Surface Chemistry of Activated Carbons and Its Characterization. Elsevier, Amsterdam, pp. 159–230.
Boonpoke, A., Chiarakorn, S., Laosiripojana, N., Towprayoon, S., Chidthaisong, A., 2011. J. Sust. Energ. Environ. 2, 77–81.
Boonpoke, A., Chiarakorn, S., Laosiripojana, N., Towprayoon, S., Chidthaisong, A., 2012. Korean J. Chem. Eng. 29, 89–94.
Chen, J., Yang, J., Hu, G., Hu, X., Li, Z., Shen, S., et al., 2016. Chem. Eng. 4, 1439–1445.
Cheng, F., Liang, J., Zhao, J., Tao, Z., Chen, J., 2008. Chem. Mater. 20, 1889–1895.
Coromina, H.M., Walsh, D.A., Mokaya, R., 2016. J. Mater. Chem. A 4, 280–289.
Creamer, A.E., Gao, B., Zhang, M., 2014. Chem. Eng. J. 249 (2014), 174–179.
D'Alessandro, D.M., Smit, B., Long, J.R., 2010. Angew. Chem. Int. Ed. 49, 6058–6082.
Figueroa, D., Fout, T., Plasynski, S., Mcllvried, H., Srivastava, R.D., 2008. Int. J. Greenh. Gas Control 2, 9–20.
Foo, K.Y., Hameed, B.H., 2013. Chem. Eng. J. 187, 53–62.
García, S., Gil, M.V., Martín, C.F., Pis, J.J., Rubiera, F., Pevida, C., 2011. Chem. Eng. J. 171, 549–556.
Hao, G.H., Li, W.C., Lu, A.H., 2011. J. Mater. Chem. 21, 6447–6451.
Haszeldine, R.S., 2009. Science 325, 1647–1652.
Heidari, A., Younesi, H., Rashidi, A., Ghoreyshi, A., 2014a. Chem. Eng. J. 254, 503–513.
Heidari, A., Younesi, H., Rashidi, A., Ghoreyshi, A., 2014b. J. Taiwan Inst. Chem. Eng. 45, 579–588.
Huang, Y.F., Chiueh, P.T., Shih, C.H., Lo, S.L., Sun, L., Zhong, Y., et al., 2015. Energy 84, 75–82.
Kinney, T., Masiello, C., Dugan, B., Hockaday, W., Dean, M., Zygourakis, K., et al., 2012. Biomass Bioenergy 41, 34–43.
Kwon, S., Fan, M., Da Costa, H.F.M., Russell, A.G., Berchtold, K.A., Dubey, M.K., 2011. Coal Gasification and Its Applications. Elsevier Inc, Oxford, UK.
Lehmann, J., Joseph, S., 2012. Biochar for Environmental Management: Science and Technology. Routledge, Abingdon.
Liu, S.H., Huang, Y.Y., 2018. J. Clean Prod. 162, 1376–1387.
Maroto-Valer, M.M., Tang, Z., Zhang, Y., 2005. Fuel Process Technol. 86, 1487–1502.
Maroto-Valer, M.M., Lu, Z., Zhan, Y., Tang, Z., 2008. Waste Manage. 28, 2320–2328.
McCarl, B.A., Peacocke, C., Chrisman, R., Kung, C.-C., Sands, R.D., 2009. Environ. Manag. Sci. Technol. 341–358.
Metz, B., Davidson, O., de Coninck, H., Loos, M., Meyer, L., 2005. Intergovernmental Panel on Climate Change (IPCC) on Carbon Capture and Storage. Cambridge University Press, Cambridge, UK.
Nasri, N.S., Hamza, U.D., Ismail, S.N., Ahmed, M.M., Mohsin, R., 2014. J. Clean Prod. 71, 148–157.
Olivares-Marín, M., Maroto-Valer, M., 2011. Fuel Process Technol. 92, 322–329.
Olivares-Marín, M., Garcia, S., Pevida, C., Wong, M.S., Maroto-Valer, M., 2011. J. Environ. Manage. 92, 2810–2817.
Oschatz, M., Antonietti, M., 2018. Energ Environ. Sci. 11, 57–70.
Patnukao, P., Pavasant, P., 2008. Bioresour. Technol. 99, 8540–8543.
Pevida, C., Drage, T.C., Snape, C.E., 2008a. Carbon 46, 1464–1474.
Pevida, C., Arias, B., Fermoso, J., Rubiera, F., Pis, J.J., 2008b. Appl. Surf. Sci. 254, 7165–7172.
Plaza, M.G., Pevida, C., Arias, B., Fermoso, J., Casal, M.D., Martín, C.F., et al., 2009. Fuel 88, 2442–2447.
Plaza, M.G., Gonzalez, A.S., Pis, J.J., Rubiera, F., Pevida, C., 2014. Appl. Energy 114, 551–562.
Plaza, M.G., Pevida, C., Martín, C.F., Fermoso, J., Pis, J.J., Rubiera, F., 2010. Sep. Purif. Technol. 71, 102–106.

Plaza, M.G., García, S., Rubiera, F., Pis, J.J., Pevida, C., 2011. Sep. Purif. Technol. 80, 96–104.
Rashidi, N.A., Yusup, S., Hameed, B.H., 2013. Energy 61, 440–446.
Rezaei, F., Webley, P., 2010. Sep. Purif. Technol. 70, 243–256.
Sayari, A., Belmabkhout, Y., Serna-Guerrero, R., 2011. Chem. Eng. J. 171, 760–774.
Sevilla, M., Fuertes, A.B., 2011. Energy Environ. Sci. 4, 1765–1771.
Sjostrom, S., Krutka, H., Starns, T., Campbell, T., 2011. Energy Procedia 4, 1584–1592.
Son, S.J., Choi, J.S., Choo, K.Y., Song, S.D., Vijayalakshmi, S., Kim, T.H., 2005. Korean J. Chem. Eng. 22, 291–297.
Thote, J.A., Iyer, K.S., Chatti, R., Labhsetwar, N.K., Biniwale, R.B., Rayalu, S.S., 2010. Carbon 48, 396–402.
Timur, S., Kantarli, I.C., Onenc, S., Yanik, J., 2010. J. Anal. Appl. Pyrolysis 89, 129–136.
Wang, J.F., Qian, W.Z., He, Y.F., Xiong, Y.B., Song, P.F., Wang, R.M., 2017. Waste Manage. 65, 11–21.
Wang, Q., Luo, J., Zhong, Z., Borgna, A., 2011. Energ. Environ. Sci. 4, 42–55.
Wilcox, J., 2012. Carbon Capture. Springer, New York.
Yang, M.L., Guo, L.P., Hu, G.S., Hu, X., Xu, L.Q., Chen, J., et al., 2015. Environ. Sci. Technol. 49, 7063–7070.
Zhang, Z., Maroto-Valer, M.M., Tang, Z., 2004. ACS National Meeting Book of Abstracts. 227. FUEL–133.
Zhang, C.M., Song, W., Sun, G.H., Xie, L.J., Wang, J.L., Li, K.X., et al., 2013. Energy Fuels 27, 4818–4823.
Zhang, X., Zhang, S., Yang, H., Feng, Y., Chen, Y., Wang, X., et al., 2014. Chem. Eng. J. 257, 20–27.
Zhu, X.L., Wang, P.Y., Peng, C., Yang, J., Yan, X.B., 2014. Chin. Chem. Lett. 25, 929–932.
Zhu, B.J., Shang, C.X., Guo, Z.X., 2016. ACS Sustain. Chem. Eng. 4, 1050–1057.

CHAPTER

17

The Use of Biochar-Coated Lime Plaster Pellets for Indoor Carbon Dioxide Sequestration

Harn Wei Kua and Su Yun Gladys Choo

Department of Building, National University of Singapore, Singapore, Singapore

17.1 INTRODUCTION

For the past decade or so, direct air capture (DAC) of atmospheric carbon dioxide (CO_2) has been widely regarded as a novel technological approach to control the CO_2 concentration in the atmosphere directly. Identified as an alternative method to conventional methods of carbon sequesteration at the points of emission, DAC usually involves the use of an aqueous solvent (mostly, sodium hydroxide) to dissolve ambient CO_2 and produce sodium carbonate in the process. However, most of the ongoing discussions on the deployment of DAC technologies have been linked to mitigating large-scale effects of climate change. Comparatively, fewer discussions have been initiated on the possibility of using DAC technique to sequester indoor CO_2, for addressing the negative impacts of elevated levels of indoor CO_2 concentration.

Indoor CO_2 levels are controlled by a few key factors, such as increased number of occupants in the buildings, the ventilation rate of the buildings, and the exterior CO_2 concentration. In Singapore, the acceptable limit for CO_2, to achieve acceptable indoor air quality (IAQ), is 1000 ppm (Institute of Environmental Epidemiology, 1996). Studies have found that high air re-circulation rate can worsen the IAQ, especially if only 10% of fresh air is introduced into the indoor space (Sekhar et al., 2007); part of this effect can be attributed to indoor CO_2 concentration. Other studies have shown that indoor CO_2 concentration ranging from 500 to 800 ppm can reduce IAQ and cause discomfort. High indoor CO_2 concentrations can lead to discomfort, comprising headache, dizziness, sweating,

Biochar from Biomass and Waste
DOI: https://doi.org/10.1016/B978-0-12-811729-3.00017-0

increasing breathing rate, shortness of breath, drowsiness, fast heart rate, and even loss of consciousness (Bull, 2010).

One possible solution to this IAQ-related problem is controlling indoor CO_2 concentration using DAC approach. However, one of the biggest technical challenges faced by the most effective DAC technologies is the need to regenerate the solvents used for absorbing the CO_2. This has led to ongoing studies that explore the use of porous materials to effect physical sorption as a mean of directly removing CO_2 from the ambient air.

The CO_2 sorption nature of biochar has been extensively studied. The most important results of these studies point the way to how the sorption capability of biochar can be enhanced and optimized. Ghani et al. (2013) identified the proportion of hemicellulose, cellulose, and lignin in the biomass as the key factors determining the amount of stabilized carbon in the biochar formed. In general, the higher the carbon content in the biomass, the better is the carbon adsorption ability. Other than the carbon content, the structure of biochar, especially its total surface area, determines its CO_2-adsorption capability. In the case of biochar produced from pyrolysis, the pyrolysis temperature, pyrolysis rate, and pressure under which the pyrolysis occurs are the most important factors determining char morphology (Newalkar et al., 2014). Further studies on how the CO_2 sorption nature of biochar can influence its other properties as a building materials can be found in recent works by Gupta et al. (2018a,b, 2018c,d, 2017a,b,c,d), Gupta and Kua (2017), and Gupta and Kua (2016).

The main aim of this work is to build on these existing works by studying the potential of using biochar-coated pellets for adsorbing CO_2 in an enclosed indoor environment, in which the CO_2 concentration exceeds the limit of 1000 ppm. These pellets are deployed in the cavity of non-structural walls that can be made from wooden or gypsum panels.

The two research objectives are to compare the effectiveness of CO_2 sorption by biochar-coated pellets that are made of lime plaster and to estimate how mixing biochar with pellet materials can affect the original CO_2 sorptivity of biochar (herein called the "embedment enhancement" of CO_2 sorption caused by the lime plaster).

17.2 MATERIALS AND METHODS

17.2.1 Estimating the Embedment Enhancement of Sorptivity

Fig. 17.1 shows the surfaces of two pellets, with and without a biochar particle embedded in it. The part of the biochar particle that is embedded in the pellet (in Fig. 17.1B) is

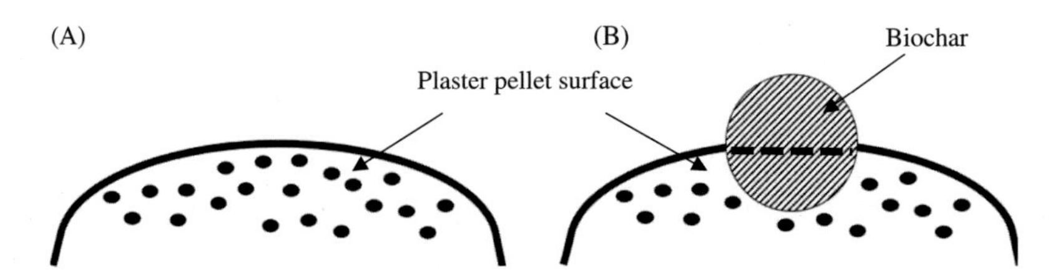

FIGURE 17.1 Representations of the surfaces of pellets (A) with and (B) without a biochar particle. The part of biochar particle underneath the dotted line is embedded in the pellet in (B).

expected to have different CO_2 sorptivity from the part that is exposed. The total volume and mass of plaster in both (A) and (B) are the same, and biochar particles are sprinkled on to the surface of the pellets.

The total moles of CO_2 (in mmol) that can be sequestered by the biochar-coated pellet (Fig. 17.1B) can be expressed as

$$\begin{aligned} m_{CO_2\_\text{coated pellet}} &= \left(\rho_{\text{plaster}} V_{\text{plaster in pores}}\right) S^* \\ &+ (m_{\text{total plaster}} - \rho_{\text{plaster}}\, V_{\text{plaster in pores}}) S_{\text{plaster}} + m_{CO_2\_\text{biochar\_exposed}} \\ &+ m_{CO_2\_\text{biochar\_embedded}} \\ &= \left(\rho_{\text{plaster}} V_{\text{plaster in pores}}\right) S^* + m_{CO_2\_\text{pellet}} - \left(\rho_{\text{plaster}}\, V_{\text{medium in pores}}\right) S_{\text{plaster}} \\ &+ m_{CO_2\_\text{biochar\_exposed}} + m_{CO_2\_\text{biochar\_embedded}} \end{aligned} \tag{17.1}$$

where $m_{CO_2\_\text{pellet}}$, $m_{CO_2\_\text{biochar\_exposed}}$ and $m_{CO_2\_\text{biochar\_embedded}}$ are the amount (in mmol) of CO_2 sequestered by the uncoated pellet (in Fig. 17.1A), exposed part of the biochar, and embedded part of the biochar, respectively. ρ_{plaster} and S_{plaster} are the density and CO_2 sorptivity of the lime plaster. $V_{\text{plaster in pores}}$ represents the volume of lime plaster that is absorbed into the pores of the part of the biochar immersed in the plaster, and S^* is the net *combined* CO_2 sorptivity of biochar with pores that are filled with plaster (shown in Fig. 17.1B as the part of the biochar underneath the dotted line). $m_{CO_2\_\text{biochar\_embedded}}$ can be redefined as

$$m_{CO_2\_\text{biochar\_embedded}} = \rho_{\text{biochar}} V_{\text{biochar\_embedded}} S^* \tag{17.2}$$

where $V_{\text{biochar-embedded}}$ is the volume of the biochar that was embedded in the pellet.

Substituting Eq. (17.2) into (17.1), Δm can be defined from (17.1) as

$$\begin{aligned} \Delta m &= m_{CO_2\_\text{coated pellet}} - m_{CO_2\ \text{pellet}} \\ &= \left(\rho_{\text{plaster}} V_{\text{plaster in pores}} + \rho_{\text{biochar}} V_{\text{biochar\_embedded}}\right) S^* \\ &\quad - \left(\rho_{\text{plaster}} V_{\text{plaster in pores}}\right) S_{\text{plaster}} + m_{CO_2\_\text{biochar\_exposed}} \end{aligned} \tag{17.3}$$

The total amount of CO_2 that can be sequestered by pure biochar can be found separately as

$$m_{CO_2\_\text{biochar}} = m_{\text{biochar}} \times S_{\text{biochar}} \tag{17.4}$$

where $m_{CO_2\_\text{biochar}}$ is the total mole of CO_2 sequestered by the total amount of biochar (m_{biochar}) used in the experiment (but were not coated on to pellets), and S_{biochar} is the CO_2 sorptivity of fully exposed biochar. $m_{CO_2\text{biochar}\ \text{exposed}}$ can be defined as

$$m_{CO_2\_\text{biochar-exposed}} = (1-f).m_{\text{biochar}} S_{\text{biochar}} \tag{17.5}$$

where f is the fraction of the total mass or volume of the biochar that is immersed in the pellet medium. By substituting Eq. (17.5) into (17.3), and calculating the difference between Eqs. (17.3) and (17.4), one gets an expression for the embedment enhancement

$$\begin{aligned} m_{ee} &= (m_{biochar} \times S_{biochar}) - \Delta m \\ &= f.m_{biochar}S_{biochar} - \left(\rho_{plaster}V_{plaster\ in\ pores} + \rho_{biochar}V_{biochar\_embedded}\right)S^* \\ &\quad + \left(\rho_{plaster}V_{plaster\ in\ pores}\right)S_{plaster} \end{aligned} \tag{17.6}$$

The porosity of biochar ($\varphi_{biochar}$) has been defined by De Souza Casta and Sandberg (2004) as

$$\varphi_{biochar} = 1 - \frac{\rho_{biochar}}{\rho_{Carbon}} \tag{17.7}$$

where the density of carbon, ρ_{carbon}, is 1957 kg m^{-3}. Porosity can also be defined in terms of the ratio of the volume of space in the biochar to the total volume of the biochar particles,

$$\varphi_{biochar} = \frac{V_{space}}{V_{biochar}} \tag{17.8}$$

Equating (17.7) and (17.8), and assuming that whenever part of a biochar particle is immersed in the pellet, the empty space (or, pore) in the particle will be fully filled up by the plaster solution—that is, $V_{plaster\ in\ pores} = f.V_{space}$—then $V_{plaster\ in\ pores}$ can be expressed as

$$V_{plaster\ in\ pores} = f.V_{biochar}\left(1 - \frac{\rho_{biochar}}{\rho_{carbon}}\right) \tag{17.9}$$

Substituting Eq. (17.9) into (17.6), the embedment enhancement can be redefined as

$$\begin{aligned} m_{ee} &= (m_{biochar} \times S_{biochar}) - \Delta m \\ &= f.m_{biochar}S_{biochar} \\ &\quad - f.\left(\rho_{plaster}.V_{biochar}\left(1 - \frac{\rho_{biochar}}{\rho_{carbon}}\right) + m_{biochar}\right)S^* \\ &\quad + \left(f.\rho_{plaster}.V_{biochar}\left(1 - \frac{\rho_{biochar}}{\rho_{carbon}}\right)\right)S_{plaster} \\ &= f.m_{biochar}S_{biochar} - f.\left(\frac{\rho_{plaster}m_{biochar}}{\rho_{biochar}}\left(1 - \frac{\rho_{biochar}}{\rho_{carbon}}\right) + m_{biochar}\right)S^* \\ &\quad + f.\left(\frac{\rho_{plaster}m_{biochar}}{\rho_{biochar}}\left(1 - \frac{\rho_{biochar}}{\rho_{Carbon}}\right)\right)S_{plaster} \end{aligned} \tag{17.10}$$

The f fraction is unknown and is dependent on how the biochar powder is applied to the pellet surface. In this study, it is treated as an independent variable; with Eq. (17.10), for a given value of f, S^* can be derived. In other words, when the mass/volume fraction of the biochar that is embedded in the pellet is known, the net *combined* CO_2 sorptivity of embedded part of the biochar can be estimated. This will also allow us to estimate how embedment of biochar changes its original CO_2 sorptivity. Eq. (17.10) shows that this estimation depends on the density and CO_2 sorptivity of plaster.

17.2.2 Sample Preparation and Study

The biochar samples and lime plaster pellets were made separately. Sawdust obtained from mixed wood sources was used as the feedstock for producing the biochar. Sawdust is a widely available waste from the local carpentry industry. The produced sawdust was gound down to a powder, after it was cooled in a controlled indoor environment.

The particle size distribution (PSD) of biochar was determined by polarized intensity differential scattering using laser diffraction particle size analyzer. Elemental composition of the biochar was determined by energy dispersive X-ray spectroscopy (EDX). Surface morphology of the biochar was analyzed using the Brunauer–Emmett–Teller (BET) methodology. Scanning electron microscopy (SEM) images were used to supplement the findings from these methods.

Plaster samples were prepared using a water–plaster ratio of 1:3. At this ratio, the plaster mixture could be easily rolled into pellets by hand. The shapes of all pellets were kept as uniform as possible—each having a diameter of approximately 15 mm. While they were still moist, the pellets were rolled in a container that was filled with the ground biochar powder. It was ensured by visual inspection that the entire surface of these pellets were covered with biochar powder. Freshly coated pellets were then air-dried for 24 h to ensure maximum adherence between the coating and the pellets.

Gypsum boards were used to prepare a sample of a cavity wall—two panels of gymsum boards that were drilled with holes was held 30-mm apart and held in place using metal clips and wire gauze. Each of these holes measured 10 mm in diameter and they were kept roughly 2 cm apart, center to center. The purpose of having these holes on the gypsum boards was to facilitate contact of the pellets with the ambient CO_2.

Once the biochar-coated pellets had cooled down after 24 h, they were heated in a furnace under the same conditions at which the biochar was produced. This was done so that gases that were adsorbed during the cooling and drying process were desorbed from the biochar before the start of the experiment.

17.2.3 Carbon Dioxide Sorption Measurement

The glass tank method was used to measure the amount of CO_2 adsorbed by the pellets. The advantage of this method is that it allows the sorption of the entire cavity wall sample to be measured. The gypsum cavity wall was positioned inside a glass tank with a width of 30 mm (and total volume of 0.062 m^3). Telaire 7001 hand held CO_2 sensors, which operate based on absorption infrared technology, were placed in the tank as well. The 30-cm wide cavity in the gypsum wall sample was filled with the desorped biochar-coated pellets. The tanks were then sealed by compressing upper and lower acrylic sheets that sandwiched the glass tank.

Pure CO_2 was then introduced into the glass tank through a thin plastic tube that was passed through a small 1-cm hole drilled into the upper acrylic sheet. The tube was disconnected when the CO_2 concentration in tank (internal concentration) reached around 1100 ± 100 ppm. Using a stopwatch, the total time taken for the in-tank concentration to reach the indoor concentration of 400 ppm was measured. Once this was reached—which was considered as one full cycle—more CO_2 was introduced into the tank again until the in-tank concentration reached around 1100 ± 100 ppm again; and the above process was repeated. This sequence of CO_2 dosage and measurement were carried for a total duration of 8–9 h; that is, if a particular full cycle ended between 8 and 9 h after the start of the experiment, it was considered the final cycle and the experiment was halted after that. To ensure methodological rigor, all experiments were conducted in triplicate.

Two kinds of control measurements—CO_2 leakage from a sealed tank without any samples, and CO_2 sorption of uncoated plaster pellets—were also taken. During control measurements of the uncoated pellets, the pellets also underwent identical desorption treatment as those biochar coated (as stated above). Taking these measurements allows the CO_2 sorption by the biochar coating to be derived, which also provides two important pieces of information: (1) the percentage contribution of the biochar to the overall CO_2 sorption of the pellets, and (2) the CO_2 sorption of the biochar particles due to mixing with the plaster.

17.2.4 Approximation of Amount of Carbon Dioxide Sequestered by Biochar Coating

For all the measurements taken, to convert CO_2 concentration (in ppm) to mass of CO_2 (m_{CO_2} in kg), the ideal gas equation is applied. That is

$$PV_{CO_2} = \frac{m_{CO_2}}{M_{CO_2}} RT \tag{17.11}$$

where P is taken as the atmospheric pressure in this case (101,325 Pa), V_{CO_2} is the net volume of CO_2 sequestered by the biochar-coated pellets from the glass tank, m_{CO_2} is the net mass of CO_2 sequestered (in kg), M_{CO_2} is the molar mass of CO_2, (R/M_{CO_2}) is the gas constant of CO_2 (189 J kg^{-1} K^{-1}), and T is the absolute temperature (which is the room temperature of 298 K). Applying the concept of partial pressure, V_{CO_2} can be expressed in terms of the volume of air in the glass tank, V_{air}:

$$V_{CO_2} = \frac{n_{CO_2}}{n_{air}} V_{air} \tag{17.12}$$

in which the quality (n_{CO_2}/n_{air}) is measured in unit of ppm. Substituting Eq. (17.12) into (17.11), and noting the volume of air in the tank as 0.062 m^3, m_{CO_2} can be estimated as

$$m_{CO_2} = \frac{(\text{concentration reduction due to pellets} - \text{concentration reduction due to leakage from tank}) \times 10^{-6} \times 101{,}325 \times 0.062}{189 \times 298} \tag{17.13}$$

Finally, m_{CO_2} is converted into molar unit (in mmol) by multiplying it by a factor of $(1000\ (g/kg)/44(g))$.

17.3 EXPERIMENTAL RESULTS

17.3.1 Physical Properties and Morphology of Biochar

Table 17.1 shows the elemental composition of the produced biochar determined by energy dispersive EDX. It has a distinctively high percentage of carbon, which is a signature of char made from woody biomass. Fig. 17.2 shows the PSD of biochar. It can be observed from PSD (Fig. 17.2) that about 45% of the biochar particles are finer than 10 μm. D_{10} and D_{50} are about 4.0 and 10.7 μm, respectively.

The SEM images in Fig. 17.3 shows that biochar prepared from sawdust has an elongated shape, with distinct ridges and honeycomb clusters of pores; these are the typical features of tropical softwood. Well defined pores (shown in the dotted circle in Fig. 17.3B) are expected to enhance the amount of CO_2 sorption.

BET analysis results show that the biochar contains a wide range of meso- and micropores, with diameters spanning from 1.67 to 255.71 nm, to yield a total surface area of $196.92 \pm 5.95\ m^2\ g^{-1}$ (Table 17.2). These results fall within the range of values obtained by

TABLE 17.1 Elemental Composition of Biochar

Carbon	Oxygen	Calcium	Magnesium	Potassium	Silica	Aluminum
87.13	7.21	0.65	0.51	0.42	0.45	1.35

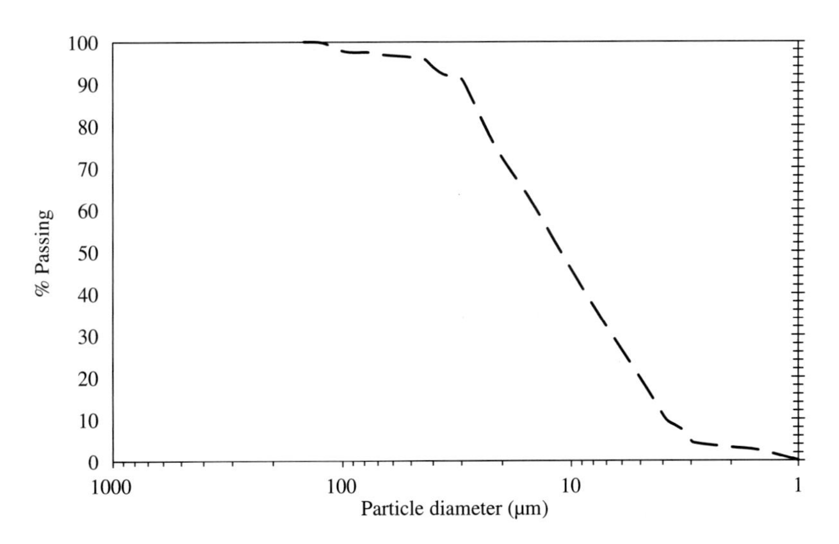

FIGURE 17.2 Particle size distribution of biochar (Gupta and Kua, 2018).

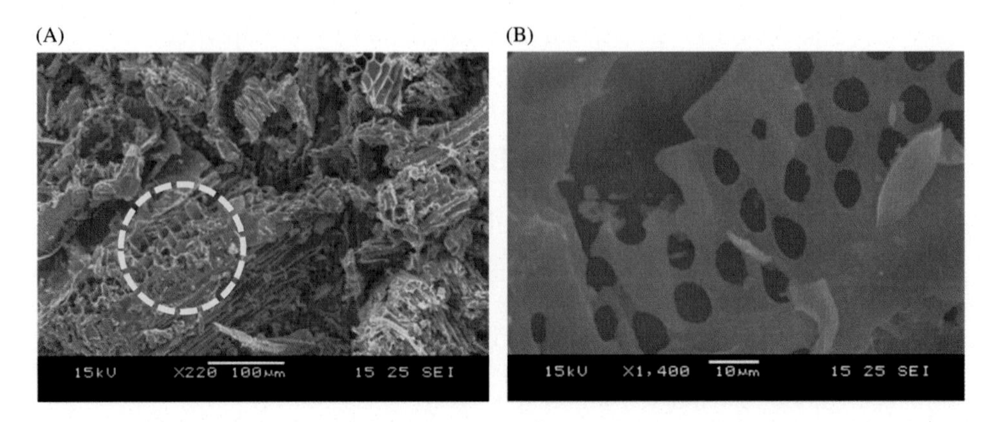

FIGURE 17.3 Scanning electron microscope images of the (A) surface of the biochar and (B) pores formed.

TABLE 17.2 Surface Properties and Pore-Related Morphology of Biochar

BET Total Surface Area ($m^2 g^{-1}$)	Level of Uncertainty ($m^2 g^{-1}$)	Micropore Area ($m^2 g^{-1}$)	External Surface Area ($m^2 g^{-1}$)	Average Particle Size (Å)	Sorption Average Pore Width (Å)	Micropore Volume ($cm^3 g^{-1}$)
196.92	± 5.95	142.89	54.03	304.70	23.85	0.08

TABLE 17.3 Masses of Biochar Used in the Coating on Both Types of Pellets

Types of Pellets	Experimental Trial	Total Mass of Biochar Used in the Coating on the Pellets (g)
Plaster	1	6.27
	2	6.19
	3	6.31

Ghani et al. (2013) from a variety of softwood—10–200 m^2 g—which was obtained from a pyrolysis temperature between 450 and 850°C.

17.3.2 Carbon Dioxide Sorption of Pellets and Biochar Coating

The masses of the biochar present in the coatings of the plaster pellets are shown in Table 17.3; that is, on approximately 1.6 kg of pellets used to fill the cavity wall, biochar accounted for only about 0.38% of the mass. The total amount of CO_2 sequestered by the biochar-coated pellets and the biochar coating are shown in Table 17.4. A few key observations must be highlighted.

Table 17.4 shows that the total number of rounds of sorption by biochar-coated plaster pellets was similar to those by uncoated plaster pellets. Furthermore, over the 8–9 h

TABLE 17.4 Comparison on the Carbon Dioxide Sorptivity of Biochar-Coated Plaster Pellets and Uncoated Plaster Pellets

Trial 1				Trial 2				Trial 3			
Plaster Pellet + Biochar		Plaster Pellet		Plaster Pellet + Biochar		Plaster Pellet		Plaster Pellet + Biochar		Plaster Pellet	
Time Taken (min)	CO_2 Absorbed (ppm)	Time Taken (min)	CO_2 Absorbed (ppm)	Time Taken (min)	CO_2 Absorbed (ppm)	Time Taken (min)	CO_2 Absorbed (ppm)	Time Taken (min)	CO_2 Absorbed (ppm)	Time Taken (min)	CO_2 Absorbed (ppm)
15	731	15	719	12	728	13	729	14	730	8	730
20	735	25	727	14	736	17	737	18	729	15	738
22	725	27	731	19	720	18	739	20	735	14	725
30	722	29	724	27	735	20	725	25	729	19	726
40	722	32	730	37	734	28	736	38	731	20	729
38	714	37	726	40	737	32	710	41	735	31	715
36	724	38	720	38	726	33	730	43	730	33	716
38	717	38	723	37	719	33	725	39	728	35	739
40	729	40	719	41	731	38	722	37	732	36	728
27	728	40	729	42	720	37	722	38	734	34	729
30	728	43	727	38	727	40	735	40	729	41	736
40	740	47	723	43	729	42	736	42	731	41	719
38	719	60	717	41	733	44	729	39	726	45	714
34	718	48	733	37	731	44	718	36	728	49	710
36	737			37	728	54	712	41	734	51	725
Average CO_2 removed (ppm) in the time period	Biochar-coated plaster pellet: 10,928 ppm										
	Uncoated plaster pellet: 10,888.67 ppm										
	Maximum difference in average: 139.33 ppm										

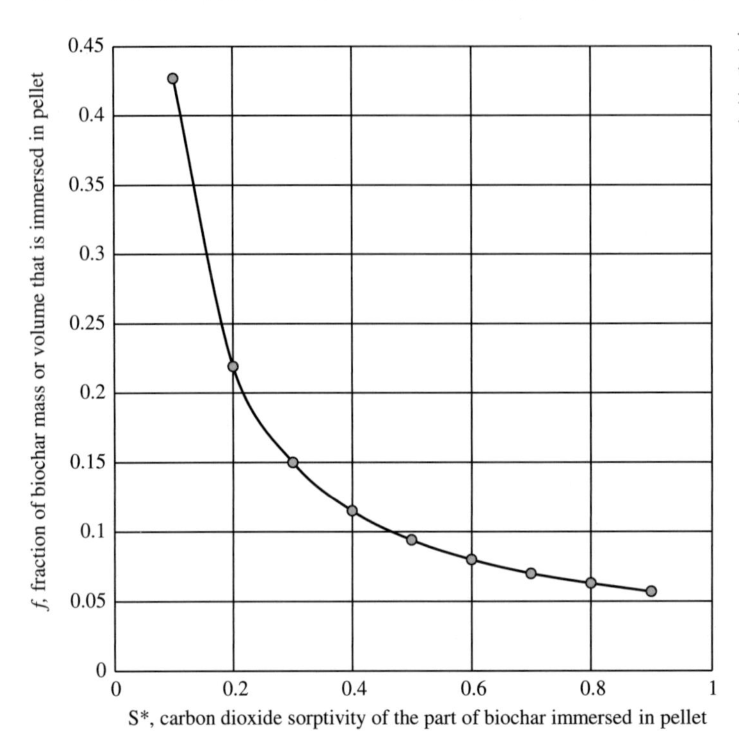

FIGURE 17.4 The variation of the fraction of biochar that is immersed, *f*, with the sorptivity of the immersed part of biochar, *S**.

period, biochar-coated plaster pellets removed a maximum average of 139.33 ppm of CO_2 more than uncoated pellets. This implies that although biochar contributes to only 0.38% of the total mass of the pellet, it contributes about 1.3% of the total sorption of the coated pellets.

Using Eq. (17.3) for conversion, this concentration is equivalent to about 0.06 mmol $(CO_2)\ g^{-1}$ of biochar sequestered. The sorptivity of the biochar ($S_{biochar}$) used in this study was measured separately to be 0.01 mmol $(CO_2)\ g^{-1}$ of biochar. Using Eq. (17.3), the average embedment enhancement caused by the plaster pellets was found for values of f ranging from 0.1 to 0.9 (Fig. 17.4).

For example, if the calculated CO_2 sorption of the biochar coating is due to biochar particles with 90% of their volume or mass immersed in the pellets (that is, $f = 0.9$), the sorptivity of the immersed parts of the biochar is 0.057 mmol g^{-1}, or 5.71 times higher than the sorptivity of pure biochar. However, if the value of f is lower (that is, less of the biochar mass or volume was immersed), the value of S^* has to be higher to account for the calculated sorptivity of the biochar coating.

17.4 DISCUSSIONS

The CO_2 sorption of plaster is primarily driven by the carbonation process. The relative humidity of the air trapped in the sealed glass tanks (in which the CO_2 sorption of the pellets were measured) was 60%–70%. Other than sequestering CO_2, the heated (and desorbed) plaster pellets absorbed water from the trapped air as well. The dissolution of gaseous CO_2 can be expressed as

$$CO_2(g) \leftrightarrow CO_2(aq) \tag{17.14}$$

Concurrently, dissolution of calcium hydroxide present in the plaster paste facilitates its reaction with the dissolved CO_2, which can be described as

$$Ca(OH)_2(crys) \leftrightarrow Ca^{2+}(aq) + 2OH^-(aq) \tag{17.15}$$

$$CO_2(aq) + Ca^{2+}(aq) + 2OH^-(aq) \rightarrow CaCO_3(crys) + H_2O \tag{17.16}$$

Although the actual values of f were not measured in this study, as shown in Fig. 17.4, immersing biochar in pellets increases the sorptivity of the biochar. In fact, if the entire biochar particle is immersed (that is, $f = 1$), the sorptivity of the immersed biochar becomes 0.052 mmol g^{-1}—5.25 times higher than the sorptivity of pure biochar.

CO_2 is present in the surrounding air and it diffuses into the interconnected pore system of the biochar through those pores that reside on the surface of the biochar particles. A possible cause for the increase in sorptivity when the pore of biochar is filled with plaster solution—thus bringing about embedment enhancement—is that when the plaster solution is absorbed onto the inner surface of the pores, the total surface area over which the plaster solution can come into contact with surrounding CO_2 is increased (Fig. 17.5). The CO_2 sorptivity of plaster (0.017 mmol g^{-1}, which is affected by carbonation process) is slightly higher than that of pure biochar (0.010 mmol g^{-1}); therefore, when the plaster solution comes into contact with the CO_2 in the pore system of the biochar, there is high likelihood that the CO_2 molecules will be transferred from the pore walls into the solution. In other words, the pores and pore system of the biochar may serve as a "conduit" or "highway" that transfers CO_2 from the surrounding to the plaster solution within the pore system (illustrated in Fig. 17.5).

The amount of plaster pellets and biochar needed for a life-sized wall can be deduced from this study. The cavity wall sample used in our experiments has an average area of 1300 cm^2 and it contains 1.6 kg of pellets. In Singapore, a typical interior wall measuring

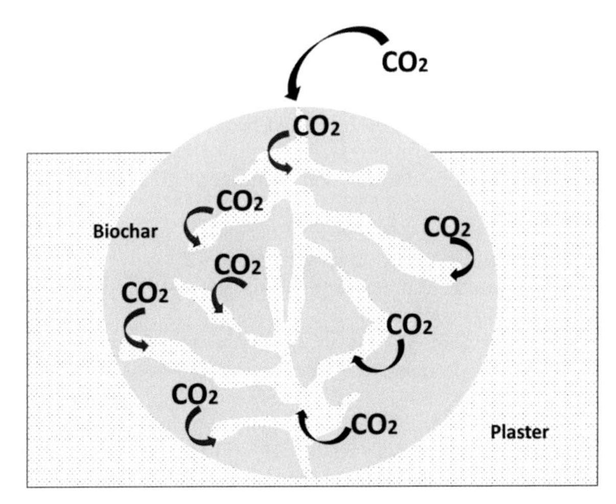

FIGURE 17.5 A likely explanation for the phenomenon of embedment enhancement of carbon dioxide (CO_2) sorptivity. Biochar pore system acting as "conduits" or "highway" to transfer CO_2 molecules from the surrounding air into the biochar, and then from the biochar into the plaster solution in the pore system (shown as curved arrows in the biochar).

2.5 m-by-2.5 m (6.25 m^2 or 62,500 cm^2) will contain around 300 g of biochar and 76.9 kg of pellets. If placed in an indoor environment where the interior CO_2 concentration is about 1000 ppm, this wall has the potential to remove 58 g of CO_2 in just 8 h.

If these pellets can be removed and replaced periodically, so that the CO_2 sorption by this cavity can be sustained throughout the service lifespan of the building, there is a potential of removing 63.5 kg of CO_2 in a year by only one wall. Even though this magnitude of CO_2 removal is substantial, the deadload placed on the building by the mass of the wall and pellets is large.

Therefore, future research efforts should focus on redesigning the pellets and selection of pellet materials to decrease the overall mass and increase the CO_2 sorption respectively. Selection of alternate pellet material is crucial, because the manufacturing of lime plaster produces substantial amounts of greenhouse gases.

Nonetheless, the results from this present study show a potential large-scale application for biochar for DAC, thus further promoting the benefits of biochar as a material of choice for climate change mitigation and adaptation in the built environment.

References

Bull, S., 2010. Carbon Dioxide General Information, https://www.gov.uk/government/uploads/system/uploads/attachment_data/file/341404/hpa_Carbon_Dioxide_General_Information_v1.pdf (accessed 04.01.16).

De Souza Casta, F., Sandberg, D., 2004. Mathematical model of a smoldering log. Combust. Flame 139, 227–238.

Ghani, W.A.W.A.K., Mohd, A., da Silva, G., Bachmann, R.T., Yun, H., Yap, T., et al., 2013. Biochar production from waste rubber-wood-sawdust and its potential use in C sequestration: chemical and physical characterization. Ind. Crops Prod. 44, 18–24.

Gupta, S., Kua, H.W., 2016. Encapsulation technology and techniques in self-healing concrete. J. Mater. Civil Eng. 28, 12.

Gupta, S., Kua, H.W., 2017. Factors determining the potential of biochar as a carbon capturing and sequestering construction material: critical review. J. Mater. Civil Eng. 29 (9), 04017086.

Gupta, S., Kua, H.W., 2018. Effect of water entrainment by pre-soaked biochar particles on strength and permeability of cement mortar. Construct. Build. Mater. 159, 107–125.

Gupta, S., Kua, H.W., Low, C.Y., 2018d. Use of biochar as carbon sequestering additive in cement mortar. Cement Concr. Compos. 87, 110–129.

Gupta, S., Kua, H.W., Tan, S.Y.C., 2017a. Use of biochar-coated polypropylene fibers for carbon sequestration and physical improvement of mortar. Cement Concr. Compos. 83, 171–187.

Gupta, S., Kua, H.W., Pang, S.D., 2017b. Combination of polypropylene fibre and superabsorbent polymer to improve physical properties of cement mortar. Mag. Concr. Res. 70, 1–15.

Gupta, S., Pang, S.D., Kua, H.W., 2017c. Autonomous healing in concrete by bio-based healing agents–a review. Constr. Build. Mater. 146, 419–428.

Gupta, S., Kua, H.W., Koh, H.J., 2018a. Application of biochar from food and wood waste as green admixture for cement mortar. Sci. Total Environ. 619, 419–435.

Gupta, S., Kua, H.W., Pang, S.D., 2018b. Healing cement mortar by immobilization of bacteria in biochar: an integrated approach of self-healing and carbon sequestration. Cement Concr. Compos. 86, 238–254.

Gupta, S., Kua, H.W., Dai Pang, S., 2018c. Biochar-mortar composite: Manufacturing, evaluation of physical properties and economic viability. Construct. Build. Mater. 167, 874–889.

Institute of Environmental Epidemiology, 1996. Guidelines for good indoor air quality in office premises, in guidelines for good indoor air quality in office premises, https://www.bca.gov.sg/Greenmark/others/NEA_Office_IAQ_Guidelines.pdf (accessed 23.02.16).

Newalkar, G., Iisa, K., D'Amico, A.D., Sievers, C., Agrawal, P., 2014. Effect of temperature, pressure, and residence time on pyrolysis of pine in an entrained flow reactor. Energy Fuels 28 (8), 5144–5157.

Sekhar, S.C., Bin, Y., Tham, K.W., Cheong, K.W.D., 2007. IAQ and energy performance of the newly developed single coil twin fan air-conditioning and air distribution system—results of a field trial. Proceedings of Clima 2007.

Further Reading

Busch, A., Alles, S., Gensterblum, Y., Prinz, D., Dewhurst, D.N., Raven, M.D., et al., 2008. Carbon dioxide storage potential of shales. Int. J. Greenhouse Gas Control 2 (3), 297–308.

Romanov, V.N., 2013. Evidence of irreversible CO_2 intercalation in montmorillonite. Int. J. Greenhouse Gas Control 14, 220–226.

Romanov, V., Ackman, T., Soong, Y., Kleinman, R., 2009. CO_2 storage in shallow underground and surface coal mines: challenges and opportunities. Environ. Sci. Technol. 43, 561–564.

Romanov, V., Soong, Y., Carney, C., Rush, G., Nielsen, B., O'Connor, W., 2015. Mineralization of carbon dioxide: literature review. ChemBioEng 2 (4), 231–256.

Slabaugh, W., 1971. Surface chemistry of thermally decomposed organo-montmorillonite complexes. Clays Clay Miner. 19 (3), 201–204.

CHAPTER

18

Novel Application of Biochar in Stormwater Harvesting

Daniel C.W. Tsang, Iris K.M. Yu and Xinni Xiong

Department of Civil and Environmental Engineering, The Hong Kong Polytechnic University, Hung Hom, Kowloon, Hong Kong, P.R. China

18.1 INTRODUCTION

Stormwater is a complex matrix with myriad chemical and biological constituents, depending upon the watershed characteristics, catchment area, surrounding hydrogeology, etc. The challenge to stormwater harvesting is adequately to remove biological and chemical contaminants (pathogens, metals, and organic pollutants), besides basic water quality parameters (suspended matter, organic matter and nutrients). The chemical characteristics of stormwater depend on the nature of surfaces (roads or roofs, pavement/building materials, etc.) with which it comes into contact as well as natural processes and anthropogenic activities in the catchments (agricultural/industrial/residential uses, traffic intensity, etc.) (Eriksson et al., 2007; Zgheib et al., 2012).

Bioretention cell (also known as rain garden) is a passive treatment for capturing stormwater at the source (BCC, 2006; RI DEM, 2010), where surface runoff gradually percolates down through filter layer (i.e., bioretention soil), transition layer, and drainage layer. The treated stormwater is conveyed to a storage tank subsequently. Bioretention cell has demonstrated significant metal retention and thus a decrease in outflow concentrations of 80%–97% (Muthanna et al., 2007; Sun and Davis, 2007; Li and Davis, 2008). The soil filter layer can provide physical-chemical (filtration and adsorption) and biological (biodegradation and phytoextraction) treatment. The process is chemical-free and self-sufficient with little demand for maintenance and energy. However, the performance of bioretention cell has been found variable. Furthermore, first flush effect and transient wetting and drying between rainfall events may hinder the contaminant attenuation and re-mobilize the adsorbed contaminants, thus increasing the uncertainties about the performance and

Biochar from Biomass and Waste
DOI: https://doi.org/10.1016/B978-0-12-811729-3.00018-2

reliability of decentralized, on-site bioretention cell. Therefore, advanced design of bioretention cell using novel and green materials is imperative for more efficient removal of contaminants.

In light of the successful application of biochar in soil remediation, biochar shows a high-potential in serving as a remediation agent for the stormwater matrix. Biochar that can be produced from various types of biomass (e.g., forestry/agricultural waste) is an economically appealing and environmentally benign option. Recent literature has demonstrated the use of biochar in removing a wide range of metals/metalloids and organic pollutants in aqueous media. Adsorption kinetics, adsorption isotherms, and removal mechanisms have been examined via batch and column studies. With its tunable surface functionality, the use of biochar showcases opportunities in the engineering of cost-effective and environmentally compatible stormwater harvesting systems.

This chapter highlights the current developments and understandings of biochar-based remediation system for stormwater contaminated by metals/metalloids and organic compounds, with a focus on the significance of interactions among complex matrices and varying flow conditions in a real-life situation. Considerations in design of biochar-based bioretention system are also illustrated, using Hong Kong as an example.

18.2 STORMWATER HARVESTING

18.2.1 Stormwater Quality

The hydrologic cycle for a natural system is characterized by a water mass balance equation: Precipitation = Runoff + Infiltration + Evapotranspiration + ΔStorage. As there is much lower infiltration in urban areas due to impermeable structures (e.g., pavement roads, sidewalks, parking lots), the volume and contamination of surface runoff have been increasing along with urban development (Aryal et al., 2010). In addition to urbanization, the quality of surface runoff (in terms of concentrations of chemical and biological pollutants) also depends on land use and human activities (Zgheib et al., 2012). For instance, the concentrations of total coliform (TC), fecal coliform, and *Escherichia coli* (*E. coli*) were found to vary with respect to different land use and stormwater managements in the order of: combined sewer overflow > agricultural land use > separate sewer overflow > forestry land use (Kim et al., 2005). Moreover, diffused pollution is another significant source of contamination to surface water, which results in a high level of organic pollutants and nutrients in forestry and agricultural runoff as well as unacceptably high coliform content in river water (Kim et al., 2007; Kim and Yoon, 2011). In Hong Kong, surface runoff is often polluted because of the densely populated land use. For example, the performance indicators of stormwater runoff (e.g., *E. coli*, suspended solids, turbidity, and color) in Happy Valley, Hong Kong, are far beyond the recommended water quality standards for non-potable reuse (Table 18.1). There is a pressing need to develop a simple, cost-effective, and green technology as an alternative to the current practices in stormwater management, and prove its environmental and field compatibility in the local context.

TABLE 18.1 Stormwater Quality for the Water Harvesting Scheme in Happy Valley, Hong Kong (HK DSD, 2014)

Parameter	Unit	Sample 1	Sample 2	Sample 3	Sample 4	Sample 5
E. coli	cfu/100 mL	60,000	27,000	35,000	78,000	37,000
Total residual chlorine	mg L^{-1}	<0.2	<0.2	<0.2	<0.2	<0.2
Dissolved oxygen	mg L^{-1}	8.4	6.4	6.1	5.7	6.0
Suspended solids	mg L^{-1}	180	37	32	38	32
Color	Hazen unit	75.0	37.5	25.0	25.0	25.0
Turbidity	NTU	66	10	6	6	8
pH	pH unit	7.3	7.1	7.1	7.1	6.9
Threshold odor number	TON	<1	1	1	1	1
Biochemical oxygen demand	mg L^{-1}	8	12	9	11	10
Ammonia as N	mg L^{-1} as N	0.36	0.16	0.53	0.56	0.50
Total surfactants	mg L^{-1}	<1.0	<1.0	<1.0	<1.0	<1.0

18.2.2 Blue-Green Infrastructure for Water Resilience

In view of the importance of drainage planning and revitalization of water bodies, blue-green infrastructure is regarded as a key to offer effective measures to incorporate the surface water management into a greenspace to improve sustainability and resilience of drainage systems. Previous studies have demonstrated that blue-green infrastructures can provide various benefits to the urban cities in environmental and stormwater management aspects, for instance, improvement on air quality by green roofs (Li et al., 2010), reduction in peak flow during high intensity rainfall events by green channel cover (Palanisamy and Chui, 2015), and enhancement in stormwater infiltration and groundwater recharge (Chui and Trinh, 2016).

Bioretention systems (Fig. 18.1), one of the blue-green infrastructures, have been adopted for conveyance and treatment for stormwater in the United States, United Kingdom, Australia, etc. This is a passive treatment technology for capturing stormwater at the source. A portion of the stormwater returns to the atmosphere by evapotranspiration, while the remainder can be harvested, stored, and ultimately used for non-potable purposes.

There are four major configurations for bioretention systems, including bioretention swales (i.e., bioswales), bioretention basins (i.e., biofiltration basins), biopods, and bioretention street trees (Water by Design, 2014). Bioswale consists of two main components, i.e., swale component for conveying and pre-treating stormwater runoff to remove coarse to medium-sized sediments, and filter media for removing finer particulates and contaminants followed by collecting infiltrated runoff via perforated sub-drain for discharging to receiving waters or conveying to storage tank. Although biofiltration basin has a similar drainage profile to bioswale, the primary purpose of biofiltration is to treat and harvest

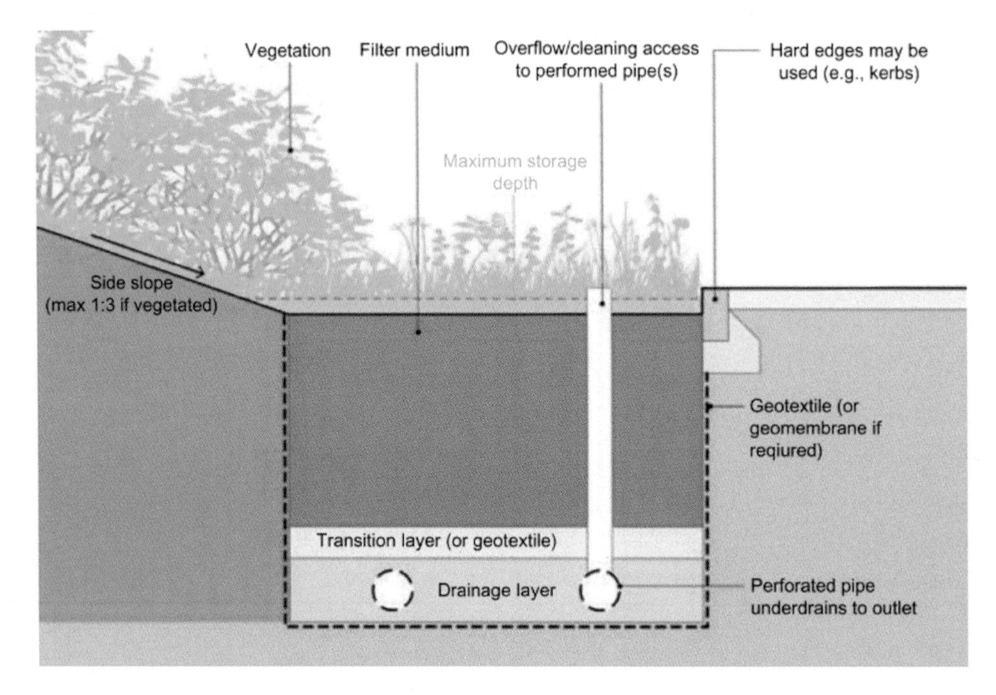

FIGURE 18.1 Components of a bioretention system (CIRIA, 2015).

surface runoff. There are three layers in the system, including a filter layer (i.e., bioretention soil), transition layer, and drainage layer. After flowing through these layers, the treated stormwater is ready for reuse at the storage tank. Excess water during heavy rainstorms overflows from the detention space to the overflow pit.

With a view to ensuring safe utilization of harvested stormwater, jurisdictions such as Hong Kong have established Technical Specifications on Grey Water Reuse and Rainwater Harvesting (HK WSD, 2015) that suggested a list of water quality standards for non-potable reuse of stormwater (Table 18.2). As one of fecal indicator bacteria, *E. coli* indicates the likelihood of occurrence of viruses, protozoans, and pathogenic bacteria, and it is linked to public waterborne illnesses due to exposure to water contaminated with urban runoff during swimming and other forms of recreation. The concentration of *E. coli* in stormwater runoff varies from 10^1 to 10^6 CFU mL^{-1}, as shown in recent studies in Hong Kong and around the world (Lundy et al., 2011; Kim et al., 2010; HK DSD, 2014). Moreover, other priority pollutants (Table 18.3), constituting a potential risk to receiving waters or possible non-potable reuse, should also be considered (Zgheib et al., 2012).

In fact, the typical pollutant removal efficiencies in typical bioretention systems are not always satisfactory, particularly for the nutrient removal (e.g., 60%–80% for total phosphorus and 20%–50% for total nitrogen) (CIRIA, 2015). Phosphorus is primarily removed by adsorption onto bioretention soils and precipitation (Roy-Poirier et al., 2010a,b), while nitrogen is removed via plant uptake, nitrification, denitrification, immobilization, adsorption, storage in soil organic matter (Collins et al., 2010). A poor net nitrogen removal is attributed to nitrate (NO_3-N) leaching (Davis et al., 2001; Cho et al., 2009; Hatt et al., 2009,

TABLE 18.2 Water Quality Standards for Treated Grey Water and Rainwater Effluent (HK WSD, 2015)

Parameter	Unit	Recommended Water Quality Standards
E. coli	cfu/100 mL	Non detectable
Total residual chlorine	mg L^{-1}	≥ 1 exiting treatment system
		≥ 0.2 at user end
Dissolved oxygen in reclaimed water	mg L^{-1}	≥ 2
Total suspended solids (TSS)	mg L^{-1}	≤ 5
Color	Hazen unit	≤ 20
Turbidity	NTU	≤ 5
pH		6–9
Threshold odor number (TON)		≤ 100
5-Day biochemical oxygen demand (BOD_5)	mg L^{-1}	≤ 10
Ammoniacal nitrogen	mg L^{-1} as N	≤ 1
Synthetic detergents	mg L^{-1}	≤ 5

TABLE 18.3 Priority Pollutants and Their Concentrations Found in Stormwater

Pollutants		Concentrations	WHO Guidelines for Drinking-water Quality (WHO, 2011)
Metals (mg L^{-1})	Zn	0.1–4.4	
	Pb	0.1–2.4	0.01
	Cu	0.08–0.9	2
	Cd	0.01–0.8	0.003
	Ni	0.01–0.5	0.07
	Cr	0.01–0.1	0.05
Herbicides (μg L^{-1})	2,4-D (2,4-dichlorophenoxyacetic acid)	11.5–67	30
	Isoproturon	0.05–0.2	9
	MCPA (4-(2-methyl-4-chlorophenoxy) acetic acid)	0.065–2.2	2

Fletcher, T.D., Duncan, H., Poelsma, P., Lloyd, S., 2004. Stormwater flow and quality, and the effectiveness of non-proprietary stormwater treatment measures—a review and gap analysis. Technical Report 04/8, December 2004, Cooperative Research Centre for Catchment Hydrology; Lau, S.L., Han, Y., Kang, J.H., Kayhanian, M., Stenstrom, M.K., 2009. Characteristics of highway stormwater runoff in Los Angeles: metals and polycyclic aromatic hydrocarbons. Water Environ. Res. 81, 308–318 (Lau et al., 2009); Gan, H., Zhuo, M., Li, D., Zhou, Y., 2008. Quality characterization and impact assessment of highway runoff in urban and rural area of Guangzhou, China. Environ. Monit. Assess. 140, 147–159 (Gan et al., 2008); Lundy, L., Ellis, J.B., Revitt, D.M., 2011. Risk prioritisation of stormwater pollutant sources. Water Res. 46, 6589–6600; MDA, 2011. Summary Report: Urban Storm Water Runoff Monitoring for Pesticides and Nitrate in Helena and Billings, 2011. Montana: Montana Department of Agriculture (MDA, 2011); Zgheib, S., Moilleron, R., Chebbo, G., 2012. Priority pollutants in urban stormwater: Part 1 – Case of separate storm sewers. Water Res. 46(20), 6683–6692; LeFevre, G., Paus, K., Natarajan, P., Gulliver, J., Novak, P., Hozalski, R., 2015. Review of dissolved pollutants in urban storm water and their removal and fate in bioretention cells. J. Environ. Eng. 141, 04014050 (LeFevre et al., 2015).

2007; Li and Davis, 2014), despite a high ammonia (NH_3-N) removal via adsorption and ion exchange shown in laboratory (Davis et al., 2001; Cho et al., 2009) and field studies (Dietz and Clausen, 2005; Hunt et al., 2007; Li and Davis, 2014). Nitrification may enhance NH_3-N removal but reduce NO_3-N removal under aerobic conditions between wetting and drying cycles (Hatt et al., 2007; Hsieh et al., 2007; Cho et al., 2009). The above findings indicate a necessity of improving the traditional bioretention system (e.g., bioswale and biofiltration basin) for effective implementation.

18.3 BIOCHAR AS A NOVEL STORMWATER REMEDIATION AGENT

18.3.1 Emerging Application

In view of the well-recognized limitations of typical bioretention system, filter media could be modified by substituting a part of bioretention soil by biochar (Fig. 18.2). Biochar is an effective absorbent for phosphate (Jung et al., 2015) with medium to high surface area that may be comparable to activated carbon (AC), while its breakeven price is only around USD246 per ton (i.e., one-sixth of AC) (Ahmad et al., 2012). Moreover, oxygen-containing functional groups (e.g., carboxylate and hydroxyl groups) on the biochar surfaces can enable a strong binding affinity toward metals/metalloids (e.g., copper, cadmium, and lead) (Tong et al., 2011; Uchimiya et al., 2011). Biochar has also been proven effective for removing organic pollutants (e.g., pesticides, pharmaceutical and personal care products, dyes, humic acid, perfluorooctane sulfonate) via a wide range of mechanisms, such as pore filling, hydrophobic partitioning, aromatic−π and cation−π interaction, electrostatic interaction, and hydrogen bonding (Inyang and Dickenson, 2015; Vithanage

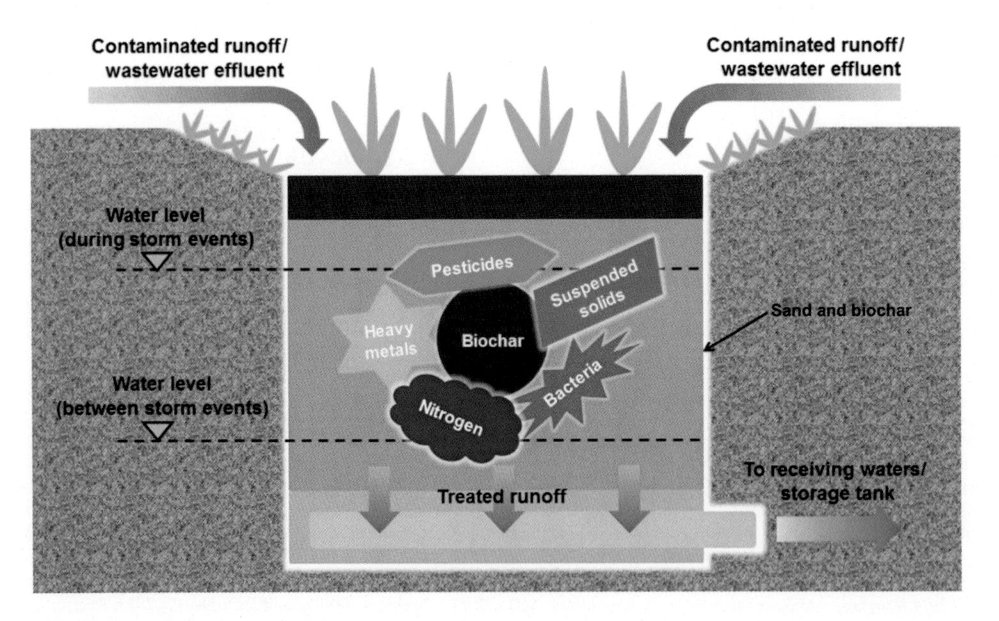

FIGURE 18.2 Biochar as an absorbent in filter media of bioswale.

et al., 2016). Besides, biochar can provide carbon substrates as electron donor for denitrification, which stimulates both nitrification and denitrification while reducing N_2O (greenhouse gas) emission (Wang et al., 2013; Xu et al., 2014). A comprehensive summary of biochar properties and performance can be found in recent publications (Zhang et al., 2015a; Ahmad et al., 2017; Beiyuan et al., 2017; Fang et al., 2016; Rajapaksha et al., 2016a, b). More importantly, more than 99% bacteria removal (e.g., *E. coli*) from synthetic stormwater can be achieved by locally available wood biochars with and without chemical modifications in bioretention columns as shown in the latest studies (Lau et al., 2017).

18.3.2 Production Methods

Biochar is usually produced from biomass under thermochemical processes, including pyrolysis, gasification, and hydrothermal carbonization, among which pyrolysis that requires limited oxygen supply and temperature of 300–800°C appeared as the most common and sustainable practice. Slow pyrolysis (conventional pyrolysis) yields biochar as the major product in company with syngas (CO, CH_4, and H_2) and bio-oil with high energy density, which demands long vapor residence time (>1 h) and slow heating rate (5–7°C min^{-1}) (Liu et al., 2015). Although fast pyrolysis produces biochar within short residence time (>10 s) at high heating rate (>200°C min^{-1}), the yield of biochar is lower due to the preferential formation of bio-oil (Qian et al., 2015). Similarly, gasification that operates at >700°C in O_2 or steam as the oxidants gives syngas as the main product with biochar as the solid co-product. Comparatively, hydrothermal carbonization is a less energy-intensive process, which requires a temperature ranging from 180 to 250°C and yields hydrochar as an analog of biochar (Libra et al., 2011). During the hydrothermal process, only a small amount of gas is generated (1%–5%) (Libra et al., 2011). The details of these thermal processes have been articulated in the existing reviews on biochar production and use (Libra et al., 2011; Liu et al., 2015; Qian et al., 2015).

18.3.3 Physical and Chemical Properties

Surface area and porosity are the two important physical properties as they determine the accessibility of biochar surface to contaminants. The pore structure of the feedstock retains in biochar during pyrolysis. It was reported that softwood-derived biochar tended to have more vesicles and pores and higher surface area compared with the hardwood-derived biochar (Mukome et al., 2013). This is because the former had less dense composition that were more susceptible to thermal decomposition. As pyrolysis temperature increases, the micropore volume and surface area of biochar enlarge upon the release of volatiles and formation of internal pores as well as channel structures, but pyrolysis over 700°C with a low heating rate would lead to micropore blockage or cracking and reduction of surface area (Ahmad et al., 2012, 2014). Post-activation using acid/alkaline and oxidants (e.g., steam and CO_2/NH_3) exposes more pores to increase the surface area (Xiong et al., 2013; Chia et al., 2015; Rajapaksha et al., 2016a,b).

As for chemical properties, high cation exchange capacity (CEC), i.e., the parameter representing biochar's negative surface charge, favors the adsorption of cations such as metals/metalloids. Comparatively, biochars with high anion exchange capacity (AEC) show strong affinity for anions, such as phosphate. Low pyrolysis temperature (e.g., 250–350°C) may lead to high CEC as a result of the considerable volatile organic matter remaining on the biochar (Mukherjee et al., 2011; Suliman et al., 2017). These properties are also governed by the biochar feedstock. For example, Mukherjee et al. (2011) concluded that grass biochar generally had higher CEC than oak and pine wood biochar. The ion exchange capacity is sensitive to medium pH (particularly for biochars produced at low pyrolysis temperature), which will depend on the stormwater composition. Furthermore, upon oxidation of the biochar surface during storage in air (Wang et al., 2017), the content of acidic and carboxyl functional groups increases, whereas the amount of basic and lactones groups decreases, leading to an increasing CEC. Hydrophobicity of biochar is another important chemical property that enhances the pollutant adhesion/deposition on the surface. It increases with pyrolysis temperature as a result of increased carbonization that raises the fixed carbon content and reduces the O content (Ahmad et al., 2012). Note that high pyrolysis temperature also results in high ash content due to the concentration effect induced by carbonization, and sometimes leads to low CEC, depending on the feedstock (Ahmad et al., 2012; Suliman et al., 2017). Hydrophobicity was also determined by the surface functional groups originated from the feedstock. For instance, non-wood feedstock (e.g., grass and manure) yields biochar with less aromatic but more aliphatic groups compared with wood-derived biochar (Mukome et al., 2013). In addition, the surface oxygen functional groups can be grafted by acid/alkaline as well as oxidants (Rajapaksha et al., 2016a,b). Therefore, biochar feedstock and pyrolysis conditions play an important role in determining the physicochemical properties, which are critical for the efficiency of contaminant removal.

18.3.4 Removal Mechanisms

Filtration is one of the major mechanisms of contaminant removal. Biochar filter physically removes the large solid particles in wastewater by size exclusion, depending on the physical properties, such as pore size, filter thickness, etc.

Removal can be also achieved by *sorption* in forms of various interactions between biochar surface and contaminants. For example, *hydrogen bonding* takes place between the surface polar functional groups and organic compounds containing electronegative elements (Sun et al., 2012). With the pH increasing, the hydrogen bond donors tend to change from –COOH and –OH to –OH alone in regard of sorption of ClO_4^- (Fang et al., 2013). *Electrostatic interactions* occur between positively charged organic pollutants and negatively charged biochar surface (Ahmad et al., 2014; Mukherjee et al., 2011). The *π-electron donor–accepter (EDA) interaction* arises between polar compounds (e.g., nitroaromatics) and the basal plane of the graphene sheets of biochar (Zhu et al., 2005). Biochar produced by pyrolysis <500°C containing electron-withdrawing entities (e.g., alkyl, carbohydrate, carboxyl) tend to act as electron-acceptors, while biochar with electron-donating species produced at high temperature (e.g., phenolic hydroxyls, poly-condensed aromatic rings, or electron-rich graphene sheets) are electron donors (Sun et al., 2012). However, the oxygen containing functional groups played a relatively minor role (Zhu et al., 2005). Besides, metal(loid)s can be adsorbed via *ion exchange* on the biochar with carboxylic groups and sulfonic groups as well as *precipitation* and *complexation*, in which metal(loid)s react with the anions (e.g., carbonate, phosphate, hydroxide) released from biochar (Gwenzi et al., 2017). *Reduction* of metal species also occurs concurrently (Dong et al., 2011).

Bacteria *attachment* is the adhesion of microorganism to the biochar surface. Biochar with smaller size favors the bacterial removal efficiency (Mohanty and Boehm, 2014). Such biofilm facilitates *biodegradation* of some contaminants with the enzyme-containing extracellular matrix released by microbes. For instance, biotic biochar achieved 50%–59% higher removal of naphthenic acids from wastewater than the sterile biochar, suggesting the significance of the synergistic biofilm–biochar system (Frankel et al., 2016). Interestingly, it has been reported that *E. coli* in cultivable soil can be inactivated by biochar produced at short residence time and low temperature, probably because of particular volatile matters (Gurtler et al., 2014). Similar *microbial inactivation* may apply in the aqueous environment, which warrants future investigations.

18.4 REMOVAL EFFICIENCIES IN BIOCHAR-BASED SYSTEMS

18.4.1 Metals/Metalloids

Metals/metalloids in stormwater result from intensive automobile use, weathering of building materials, and atmospheric deposition. Bioretention cell (i.e., rain garden) is a passive treatment for capturing stormwater at the source (BCC, 2006; RI DEM, 2010), which comprises filter layer (i.e., bioretention soil), transition layer, and drainage layer. Although bioretention cell has demonstrated a significant decrease of metal concentrations in the effluents (80%–97%; Muthanna et al., 2007; Sun and Davis, 2007; Li and Davis, 2008), the performance of bioretention cell has been reported to vary. For example, low and varying removal efficiency has been reported: 77% for Zn, 54% for Cu, and 31% for Pb, respectively (Fletcher et al., 2004; Hunt et al., 2008). It should be noted that the top layer accounts for over 70% of metal accumulation in the filter media (Muthanna et al., 2007; Blecken et al., 2009), probably because the metals were largely bound with runoff particles that were removed by surface filtration. The accumulation of metal-bound particulate matter may lead to buildup problems that limit the lifetime of a bioretention cell (Roy-Poirier et al., 2010a,b). The particulate-bound metals (As, Cd, Cr, Cu, Pb, and Zn) became labile in dissolved form when exposed to acidic rainfall, thus accumulation of metal-enriched runoff particles in a bioretention cell could negate the effectiveness over time or even present a significant source of metal leaching (Roy-Poirier et al., 2010a,b). Moreover, first flush effect and transient wetting and drying between rainfall events were reported as significant factors in controlling the reliability of the bioretention cell. In particular, biofilters were reported to show low performance in case of a 3-week or longer antecedent drying before a rainfall (Blecken et al., 2009).

In view of the low durability of bioretention cell, there is a need for developing an alternative novel retention bed to recover stormwater. Biochar shows a high potential in serving as the sustainable and powerful absorbent, in view of their promising adsorption performance towards metals/metalloids in aqueous solutions as demonstrated by the recent column studies (Table 18.4). However, there were insufficient studies on the

TABLE 18.4 Removal of Metals/Metalloids by Biochar

Biochar Type	Contaminants (Initial Concentration)	Removal	References
Peanut hull hydrochar	Pb^{2+} (50 ppm)	~1[a]	Xue et al. (2012)
H_2O_2 activated peanut hull hydrochar		~0.5[a]	
H_2O_2 activated peanut hull hydrochar	$Ni(NO_3)_2$, $Ca(NO_3)_2$, $Cu(NO_3)_2$, $Pb(NO_3)_2$ (0.25 mmol L^{-1} each)	~1[a] (Cd^{2+}), ~0.9[a] (Cu^{2+}), ~1[a] (Ni^{2+}), ~0.7[a] (Pb^{2+})	
Wood pellet biochar from gasification (520°C)	$CdSO_4$ (20–30 mg L^{-1})	18% (Cd), 19% (Cr), 65% (Cu), 75% (Pb), 17% (Ni), 24% (Zn)	Reddy et al. (2014)
	K_2CrSO_4 (1–5 mg L^{-1})		
	$CuSO_4 \cdot 5H_2O$ (1–5 mg L^{-1})		
	$PbCl_2$ (0.5–5 mg L^{-1})		
	$NiCl_2$ (100–120 mg L^{-1})		
	$ZnSO_4 \cdot 7H_2O$ (50–60 mg L^{-1})		
	Naphthalene (10–700 μg L^{-1})		
	Phenanthrene (10–100 μg L^{-1})		
	Benzopyrene (90–100 μg L^{-1})		
	TSS (145–150 μg L^{-1})		
	E. coli (3500–8200 MPN/100 mL)		
Hickory wood biochar modified with NaOH	Pb^{2+}, Cd^{2+}, Cu^{2+}, Zn^{2+}, Ni^{2+} (100 mg L^{-1} each)	~35 min[b] (Pb^{2+}), ~25 min[b] (Cu^{2+}), ~7 min[b] (Zn^{2+}), ~7 min[b] (Cd^{2+}), ~7 min[b] (Ni^{2+})	Ding et al. (2016)
Chicken bone biochar	Cd, Cu, Zn (50 mg L^{-1} each)	14 day[b] (Cu), 8 day[b] (Cd), 7 day[b] (Zn)	Park et al. (2015)
Buffalo weed biochar- alginate beads	Cd^{2+} (20 mg L^{-1})	~160 h (0.25 mL min^{-1})[c]	Roh et al. (2015)
		~20 h (0.50 mL min^{-1})[c]	
		<10 h (1.0 mL min^{-1})[c]	
	(10 mg L^{-1})	~160 h (0.25 mL min^{-1})[c]	

Teak leaves biochar	Ni^{2+}	(50 mg L^{-1})	Bed height	1 cm	250.7 min (1 mL min^{-1})[c]	Vilvanathan and Shanthakumar (2017)
					141.6 min (2.5 mL min^{-1})[c]	
					123.3 min (5 mL min^{-1})[c]	
				2 cm	242.9 min (1 mL min^{-1})[c]	
				3 cm	188.2 min (1 mL min^{-1})[c]	
		(25 mg L^{-1})		1 cm	295.7 min (1 mL min^{-1})[c]	
		(75 mg L^{-1})		1 cm	219.5 min (1 mL min^{-1})[c]	
	Co^{2+}	(50 mg L^{-1})		1 cm	266.7 min (1 mL min^{-1})[c]	
					135.8 min (2.5 mL min^{-1})[c]	
					121.8 min (5 mL min^{-1})[c]	
				2 cm	254.8 min (1 mL min^{-1})[c]	
				3 cm	148 min (1 mL min^{-1})[c]	
		(25 mg L^{-1})		1 cm	304.2 min (1 mL min^{-1})[c]	
		(75 mg L^{-1})		1 cm	191.6 min (1 mL min^{-1})[c]	
Water hyacinth biochar	As (565.9 mg kg^{-1}), Cd (24.2 mg kg^{-1}), Pb (317.1 mg kg^{-1})[d]				−58% Cd, +18.7% Pb, ~0% As[e]	Yin et al. (2016)
Hardwood biochar	As (96 mg kg^{-1}), Cd (119 mg kg^{-1}), Zn (249 mg kg^{-1})[f]				≤8 μg L^{-1} As (negligible effect)[g]	Beesley and Marmiroli (2011)
					<2 μg L^{-1} Cd (300-fold reduction)[g]	
					<15 μg L^{-1} Zn (45-fold reduction)[g]	
Hickory chip Fe-impregnated biochar	TiO_2 nanoparticle (10 ppm)				36%	Inyang et al. (2013)
	Ag nanoparticle (10 ppm)				60.9%	

[a]*Effluent concentration/influent concentration at 400 bed volume.*
[b]*Breakthrough time, at which effluent concentration = 0.05 × influent concentration.*
[c]*The time required for 50% adsorbate breakthrough at flow rate indicated in brackets.*
[d]*Leaching column test on biochar-amended soil with acidic solution.*
[e]*Leaching compared to no-biochar system.*
[f]*Leaching column test on biochar-amended soil with deionized water.*
[g]*Concentration in eluate, comparison to leaching of soil without biochar amendment in brackets.*

recovery of the complicated stormwater using biochar. Reddy et al. (2014) prepared synthetic stormwater containing total suspended solids, nutrients, metals/metalloids, and organic compounds for column experiments. The results indicated that biochar produced from wood pellet gasification at 520°C achieved wide-ranging removal efficiencies for different metals, e.g., from 17% (Ni) to 75% (Pb). Similarly, hydrochar that was produced from hydrothermal carbonization of peanut hull followed by H_2O_2 treatment proved metal removal ability in an order of $Pb^{2+} > Cu^{2+} > Cd^{2+} > Ni^{2+}$ in column experiments (Xue et al., 2012). Other studies also demonstrated the preferential adsorption of Pb^{2+} and Cu^{2+} on biochar in mixed metal systems (Ding et al., 2015). The greater adsorption tendency for Pb and Cu was possibly accounted by their higher hydrolysis constant associated with a lower degree of solvation, which favored the metal adsorption on the solid face (Ding et al., 2015). Thus, there may be competition for binding sites in a system with multiple metals. For instance, the adsorption capacity for Zn in single-metal conditions dropped most significantly in ternary-metal conditions (from 178 to 92 mg g^{-1}), compared to Cu (from 210 to 156 mg g^{-1}) and Cd (from 192 to 123 mg g^{-1}) (Park et al., 2015). Therefore, it is necessary to identify the variety of metals/metalloids in stormwater collected from specific locations to tailor the biochar-based remediation system.

In addition to the metal combination, the column design and inlet metal ion concentration also influence the biochar adsorption performance. It was shown that higher bed height lengthened the life span of adsorbent bed, while increasing flow rate and inlet metal ion concentration would accelerate the exhaustion (Vilvanathan and Shanthakumar, 2017). The flow rate also determines the adsorption efficiency. For example, buffalo weed biochar-alginate beads displayed higher maximum adsorption capacity for Cd^{2+} at a lower flow rate of 0.25 mL min^{-1} (13.4 mg g^{-1}) compared to 0.5 mL min^{-1} (9.7 mg g^{-1}) and 1.0 mL min^{-1} (3.5 mg g^{-1}) (Roh et al., 2015). Besides, the size of biochar particles in the column emerges as a significant parameter. Biochar beads wrapped by alginate avoided swelling and crumbling problem and regeneration difficulty, which were often associated with biochar powder (Roh et al., 2015).

To enhance the removal efficiency, surface modification via chemical treatments can be carried out to enrich the functionality of biochar. Peanut hull hydrochar treated by H_2O_2 exhibited Pb sorption capacity over 20 times the untreated hydrochar (22.82 vs 0.88 mg g^{-1}) (Xue et al., 2012). Despite the similar mineral compositions, the modified hydrochar using 10% H_2O_2 solution (2 h, room temperature) presented higher oxygen content (22.3% vs 16.4%) on the surface and less carbon content (48% vs 56%) than untreated hydrochar. This was ascribed to the oxidation of carbonized surfaces by H_2O_2 that increased the oxygen-containing functional groups, especially the carboxyl groups, which played a key role in Pb sorption through the surface exchange reaction. Similarly, post modification by NaOH on hickory wood biochar also showed much higher (2.6–5.8 times) adsorption capacities for Pb, Cu, Cd, Zn, and Ni, probably because of the significantly larger surface area (873 vs 256 m^3 g^{-1}) and hence improved cation-exchange capacity (Ding et al., 2015). In addition, impregnation of nanomaterials on biochar as strong sorbents can enhance metal removal. Tang et al. (2015) achieved the remarkable mercury removal efficiency of 98.1% using the nano-graphene-incorporated biochar in batch experiments, which was attributed to the enhanced surface complexation between mercury and the oxygen containing functional groups and C = C groups.

Although the current column studies demonstrated successful removal of metals/metalloids from aqueous solutions using biochar, its performance in stormwater applications remains to be confirmed, where the co-existing contaminants including pathogens and organic pollutants may compete for the sorption sites on biochar. More opportunities toward surface modification should be also explored to tailor the surface properties for stormwater harvesting, which is more challenging compared with metal sorption in simple aqueous solutions.

18.4.2 Organic Contaminants

Organic contaminants are also present in stormwater runoff, including pesticides, herbicides, and fungicides, which have been widely used in agricultural activities. Since sand-based infiltration infrastructure could not retain and degrade most of the organic contaminants in the rainfall runoff, more effective sorbents should be explored to enhance the removal efficiency. Sand caps amended with AC significantly reduced the concentration of polycyclic aromatic hydrocarbons in column experiments by 2–3 orders of magnitude compared to the control (Gidley et al., 2012). This suggested the great potential of the more cost-effective biochar for removing organic contaminants from stormwater beyond soil applications. Table 18.5 summarizes the column studies on immobilization of organic contaminants on biochar in aqueous mediums, which elucidates the design of biochar-based retention cell or filter for stormwater harvesting.

An integrated recirculating wetland, as a surrogate of bioretention cell, partly constructed by Fe-impregnated biochar accomplished over 99% removal efficiencies toward four pesticides, including chlorpyrifos, endosulfan, fenvalerate, and diuron (Tang et al., 2016). Similarly, constructed wetland simulated by columns packed with corncob biochar or wood biochar achieved removal of 59%–72% COD and 75%–83% BOD_5 (Kizito et al., 2017). The porous biochar with large surface area not only held high adsorption capacity for organic contaminants, but also provided suitable microbial cultivation habitat for biodegradation processes. There were higher microbial diversity in the biochar-packed constructed wetlands compared with that packed with gravels, according to the 16S rDNA pyrosequencing that revealed bacteria from the *Clostridiaceae* and *Xanthomamonodaceae* families were responsible for the degradation of organics (Kizito et al., 2017). Creating cycles of sequential adsorption (flooded phase) and microbial oxidation (drained phase) further enhanced the organic removal efficiencies (8%–14% higher removal for both COD and BOD_5) because such practice enabled sufficient oxygen supply for metabolizing the adsorbed contaminants, regenerating microbial active sites for further adsorption and degradation.

Biochar-packed columns were also promising for long-term anaerobic biodegradation of, for example, pentabromodiphenyl ether (BDE-99) as an emerging trace toxic organic compound (Yan et al., 2017). The high sorption capacity (79.04%–86.04%) of biochar (derived from the strong affinity for hydrophobic poly-BDEs) lowered the toxicity to archaea (microbes) and enhanced the biodegradability of organic matters, improving the archaeal biodiversity in the biochar-amended silty clay column. As the recharge time and column depth increased, the dominating removal mechanism evolved from adsorption to biodegradation, as the result of restrained biochar sorption capacity and increased stability of the microbial community.

TABLE 18.5 Removal of Organic Contaminants by Biochar

	Contaminants			
Biochar Type	**Name**	**Initial Concentration ($mg\ L^{-1}$)**	**Removal**	**Reference**
Buffalo weed biochar-alginate beads	2,4,6-Trinitrotoluene (TNT)	10	~300 h[a]	Roh et al. (2015)
		20	~100 h[a]	
	1,3,5-Trinitro-1,3,5-triazacyclohexane (RDX)	10	~180 h[a]	
		20	<10 h[a]	
Magnetic activated sawdust hydrochar	Tetracycline	100	280 bed volume[b]	Chen et al. (2017)
Fe-impregnated biochar	Chlorpyrifos	0.0789	99%	Tang et al. (2016)
	Endosulfan	0.0781	99%	
	Fenvalerate	0.0592	100%	
	Diuron	0.0662	99%	
Pine wood biochar from gasification (MCG-biochar)	Atrazine	0.02	~160 pore volume[c]	Ulrich et al. (2015)
	Tris(3-chloro-2-propyl) phosphate	0.02	~160 pore volume[c]	
	Benzotriazole	0.02	~80 pore volume[c]	
Pine wood biochar from slow pyrolysis (BN-biochar)	Atrazine	0.02	<10 pore volume[c]	
	Tris(3-chloro-2-propyl) phosphate	0.02	<10 pore volume[c]	
	Benzotriazole	0.02	<10 pore volume[c]	
Biochar-amended silty clay	Pentabromodiphenyl ether (BDE-99)	0.025	77.2%–100%[d]	Yan et al. (2017)
Rich hull biochar	Fomesafen	5 $mg\ kg^{-1}$ soil[e]	76% (retention)	Khorram et al. (2015)
Alder wood biochar	Azimsulfuron	1 $kg\ ha^{-1}$ soil[e]	20%[f] (retention)	Trigo et al. (2016)
Apple wood biochar			18%[f] (retention)	
Forest wood biochar	Naphthalene	0.4 $mg\ kg^{-1}$ soil[e]	−99.3%[g]	Zand and Grathwohl (2016)
	Fluorine	13.24 $mg\ kg^{-1}$ soil[e]	−91.5%[g]	
	Pyrene	8.27 $mg\ kg^{-1}$ soil[e]	+26.6%[g]	
	Benzo(b)fluoranthene	0.54 $mg\ kg^{-1}$ soil[e]	+900%[g]	
	Indeno(1,2,3-cd)pyrene	0.06 $mg\ kg^{-1}$ soil[e]	Negligible[g]	

(Continued)

TABLE 18.5 (Continued)

	Contaminants			
Biochar Type	**Name**	**Initial Concentration ($mg\ L^{-1}$)**	**Removal**	**Reference**
Soybean stover biochar (BC; pyrolyzed at 300°C or 700°C)	Trichloroethylene (TCE)	100	27 h[c] (BC700) 1.1 h[c] (BC300)	Zhang et al. (2015b)
Wood chip biochar	Atrazine	2.2 kg ha^{-1} soil[e]	−52% [g] (laboratory) −53% (field; concentration in groundwater)	Delwiche et al. (2014)
Wood-based biochar (*Skogens kol*)	Chlorpyrifos	1 kg ha^{-1} soil[e]	~85% (retention)	Cederlund et al. (2017)
	Glyphosate	2.25 kg ha^{-1} soil[e]	~100% (retention)	
	Diuron	8 kg ha^{-1} soil[e]	~100% (retention)	
	4-Chloro-2-methylphenoxyacetic acid (MCPA)	2.25 kg ha^{-1} soil[e]	~70 (retention)	
Hickory chip Fe-impregnated biochar	Carbon nanotubes	10 ppm	22%	Inyang et al. (2013)
Wheat straw-derived biochar	Sulfamethoxazole, sulfamethazine, sulfadiazine	10[e]	+12–20% (retention compared to untreated soil)	Liu et al. (2017)
Olive-mill waste biochar	R-metalaxyl	2 kg ha^{-1} soil[e]	99% (retention)	Gamiz et al. (2016)
	S-metalaxyl		89% (retention)	
Softwood biochar	Phenanthrene	0.1[e]	74%–84% (retention; 100 mL flush)	Trinh et al. (2017)
	Isoproturon	0.1[e]	23%–56% (retention; 200 mL flush)	
	Sulfamethazine	0.1[e]	20%–45% (retention; 100 mL flush)	

[a]*The time required for 50% adsorbate breakthrough.*
[b]*Effective bed volume was 8 cm^3.*
[c]*Breakthrough time, at which effluent concentration = 0.05 × influent concentration.*
[d]*Calculated based on the initial and final concentration of contaminant reported in the literature.*
[e]*Leaching column test on biochar-amended soil.*
[f]*Calculated by subtracting 100% by the total amount of azimsulfuron leached.*
[g]*Change in accumulated amount of leaching compared to unamended soil.*

The retention of organic contaminants depends on the physicochemical properties of biochar, which are highly related to the biochar production process. It was found that higher dissolved organic carbon (DOC) content and lower mesopore volume would result in slower sorption kinetics for organic contaminants, as the DOC may impede the intraparticle mass transfer via pore blockage (Ulrich et al., 2015). The same study also indicated that 0.2 wt% gasification biochar with the higher surface area of 318 $m^2 g^{-1}$ and pore volume of 0.31 $cm^3 g^{-1}$ (usually produced at >1100°C) retained toxic organic contaminants more effectively than 1.0 wt% pyrolysis biochar with the surface area of 108 $m^2 g^{-1}$ and pore volume of 0.07 $cm^3 g^{-1}$ (usually produced at >600°C), underscoring the significance of these parameters in determining the adsorption capacity.

The characteristics of biochar can be also enhanced by post-synthesis modifications, such as alkaline activation (KOH) followed by magnetization, which increased the surface area of sawdust hydrochar from 1.7 to 1710 $m^2 g^{-1}$ and pore volume from 0.003 to 0.97 $cm^3 g^{-1}$ (Chen et al., 2017). As the result of the favorable physical structures in addition to the enhanced chemical interactions, the adsorption capacity for tetracycline reached 423.7 $mg\ g^{-1}$ regardless of the varying pH of 5–9. Besides, Fe-impregnated biochar also effectively reduced the pesticides in contaminated water via adsorption and activation of microbes for biodegradation (Tang et al., 2016). It was prepared by soaking the biochar with $Fe(NO_3)_3 \cdot 9H_2O$ solution for 12 h under strong agitation. Another magnetic biochar derived from chemical co-precipitation of Fe^{3+}/Fe^{2+} on biomass before pyrolysis at 400°C achieved removal efficiency at 99.6% and 87.1% for naphthalene (NAPH) and p-nitrotoluene (p-NT), respectively (Chen et al., 2011). Nevertheless, for removal of the multiple contaminants in stormwater, more research efforts should be placed on the compatibility of chemical modifications and nanomaterial impregnations to maximize the biochar effectiveness in recovering the authentic or synthetic stormwater. These findings elucidate the viability of biochar application for organic pollutant removal from stormwater. However, the impacts and dynamics of the co-existing contaminants in the complex stormwater system warrant further investigations.

18.5 SURFACE MODIFICATION FOR IMPROVING REMOVAL

Biochar surface modification either on the chemical or physical scale could effectively functionalize surface properties with larger surface area or functional groups that enhance sorption capacity in aqueous solutions, such as oxidation for more acidic oxygen functional groups (Huff and Lee, 2016), and alkalization for larger surface area and adsorption capacity for tetracycline (Liu et al., 2012). Since different contaminants were removed via different mechanisms, selection of modification methods depends on specific target contaminants.

There is limited information on the biochar modifications aiming for stormwater harvesting, which may require engineered biochar with an outstanding adsorption capacity for removing a complex matrix of contaminants. However, insights may be derived from previous studies on biochar for soil application or simple aqueous system. Impregnation of nanomaterials or metal oxides on biochar as strong sorbents may potentially facilitate stormwater harvesting. For example, incorporating nano-graphene by carbonizing biomass in graphene suspensions at 600°C in N_2 achieved the mercury removal efficiency of 98.1%

in water (Tang et al., 2015). Another magnetic biochar derived from chemical co-precipitation of Fe^{3+}/Fe^{2+} with biomass before pyrolysis at 400°C achieved a removal efficiency of 99.6% and 87.1% for NAPH and p-NT, respectively (Chen et al., 2011). The Fe-modified biochar may also facilitate bacteria removal, in view that the iron-oxide coated sand achieved a significantly higher efficiency of *E. coli* removal than the conventional bioretention media (99% vs 82%), due to the higher positive surface charge and greater roughness (Zhang et al., 2010). On the other hand, electro-modified biochar carrying nano-sized aluminum crystals presented phosphate adsorption capacity of 31.28 mg-P g^{-1} (Jung et al., 2015).

Hence, biochar impregnated with different nanomaterials display high adsorption capacity for metals/metalloids, organic contaminants, bacteria, as well as nutrients, respectively. Nevertheless, for removal of the co-existing contaminants in stormwater, more research efforts should be placed on the practicability of co-impregnation of multiple nanomaterials on biochar, and the effectiveness of the resultant biochar in recovering the authentic or synthetic stormwater.

Enhancement of surface area and porosity by means of chemical activation should continue as the trend of study as these are the important biochar parameters in controlling the removal efficiency. It was found that H_2SO_4-treatment enhanced the surface area, increasing *E. coli* retention from 96.6% to 98.7% (Lau et al., 2017). In comparison, H_3PO_4- and KOH-modification showed insignificant enhancement in terms of removal, while amino-modification introduced abundant oxygen containing groups on biochar, leading to increased surface hydrophilicity that hindered the adsorption of the hydrophobic *E. coli* (92.1% removal). Therefore, the associated alteration of the surface chemistry should be taken into consideration when we determine the porosity modification approach. In addition, despite the successful KOH treatment upfront, the surface area and pore volume dropped by 22% and 26%, respectively, after magnetization with $FeCl_3$ as the last modification step (Chen et al., 2017). Hence, the compatibility of chemical modifications and nanomaterial impregnations should be evaluated to maximize the removal of multiple contaminants (e.g., metals/metalloids, organic contaminants, and nutrients) in stormwater harvesting system. The long-term stability of the modified biochar against various conditions (e.g., stormwater matrix, weather variations) should be further studied.

18.6 ROLES OF BIOCHAR AGING

The aging of biochar is avoidable in the real-life application. Biochar aging would cause variations on its physical and chemical properties, such as surface area reduction and pore destruction (Baltrėnas et al., 2015), thus deteriorating its performance in contaminant removal. AEC would decrease by 54% in average with biochar aging, which served as a parameter for reduction of nutrient leaching in soil environment or removal of anionic contaminants in aqueous environment (Lawrinenko et al., 2016). The AEC reduction was relatively less significant for biochar produced at higher temperature as a result of resistant oxonium groups bridged with arene carbon. However, carboxyl groups would increase as biochar aged, with the average carbon ratios of O = C–O increasing from 3.0% to 7.5% after 1 year and 8.9% after 2 years, which might result in a higher CEC in soil (Singh et al., 2014).

When artificial aging was simulated using H_2O_2, biochar displayed higher sorption capacity for Cd but less for Cu (Frišták et al., 2015). It was illustrated that Cd ions from aqueous solution was bound to the surface exchange active sites on biochar, while Cu ions were coordinated with the organic matter fraction and sulphides. Therefore, aging process improved Cd sorption with the newly created oxygen-based functional groups and enhanced biochar porosity, but restrained Cu sorption due to the removal of residual organic matter on the surface. Similarly, favorable effect of aging on Pb(II) and As(III) removal from soil has also been reported (Wang et al., 2017). For soil application, the CEC of biochar increases over time with the development of carbonyl, carboxylate, ether, and hydroxyl groups, reducing the possibility of metal ion leaching (Lawrinenko, 2014). Such benefit of aging should be assessed in the stormwater system.

Similar change in adsorption preference was also observed for organic pollutants, for example, increased sorption of di-alkyl phthalate (Ghaffar et al., 2015) and decreased capacity for fomesafen (Khorram et al., 2016) as the consequence of aging. The changing preference may pertain to the long-term stability of the modified biochar against various conditions (e.g., growth of microbes, weathering), which deserve more investigations. Nevertheless, the biochar aging effect varies among different contaminants and environments. It is thus necessary to expend our understanding on the change of biochar physicochemical properties as well as adsorption capacity for each contaminant under the context of a dynamic complex system.

The surface and pore structure of biochar favor the sorption of microbial community in forms of flocculation, adsorption, covalent bonding, cell cross-linking, encapsulation in polymer-gel, and entrapment in a matrix (Lehmann et al., 2011). Over aging, biofilm would form within pore spaces in the biologically aged filters, producing extracellular polymeric substances simultaneously, which would alter the biochar filter's properties, such as hydrophobicity and electrokinetic properties. It has been proved in laboratory-scale that formation of biofilm may impair bacterial removal, decreasing *E. coli* log removal from 1.53 ± 0.11 to 1.01 ± 0.14 when there was a biofilm in the filter comprising sand as well as biochar (Afrooz and Boehm, 2016). It might be interpreted that biofilm weakens hydrophobic interactions between biochar surfaces and bacterial cells by reducing the hydrophobicity of bare biochar surfaces, and deposition of biofilm on biochar particles reduces available sorption sites for bacterial attachment (Afrooz and Boehm, 2016). Nevertheless, as an advantage of biofilm, the total nitrate removal capacity could be enhanced from $-6.8 \pm 16.6\%$ to $26.8 \pm 7.2\%$, due to biological uptake or microbial conversion processes, such as denitrification and anaerobic ammonium oxidation (Afrooz and Boehm, 2017). However, the present studies are only limited to laboratory scale or using certain bacteria to represent biofilms that are complex in real life. Further research is needed on effects of biochar aging on field-scale with the consideration of different environmental factors.

18.7 CASE STUDY—HONG KONG

Due to a sub-tropical climate, Hong Kong has rainy and humid summers that contribute to an average annual rainfall of 2400 mm (HKO, 2016a). Therefore, the bioswale should be able to at least cater to minor storm events (i.e., 2–5-year return periods) and a rain intensity for Amber rainstorm (i.e., larger than 30 mm h^{-1}) (HKO, 2016b). Vegetation is another critical consideration for designing a suitable bioswale in blue-green cities (e.g., future New Development Areas) because plants are essential for pollutant removal during stormwater conveyance in bioretention systems or even capable of achieving a satisfactory wastewater treatment efficiency in the constructed wetland systems (Wu et al., 2008; Chow et al., 2015). There are some criteria to be considered to ensure a sustainable and cost-effective design of soft landscape, for example, low maintenance requirements, high permeability of the filter media with free draining nature (relatively low water holding capacity), resilience in short periods of inundation punctuated by long dry periods, provision of planting medium to support long-term healthy growth of selected plants, evergreen foliage to provide year-round greenery, ornamental flowers/assorted leaf textures/colorful foliage for aesthetic concerns.

It is imperative to establish a technical guideline to offer recommendations on planning, design, operation, and maintenance of bioswale, including:

- Design considerations (e.g., ratio of bioretention area to catchment area, flow rate);
- Engineered bioretention filter media;
- Expected improvement in water quality (SS, DO, turbidity, color, *E. coli*, etc., as summarized in Table 18.2);
- Expected improvement in removal of priority pollutants (such as metals/metalloids and herbicides);
- Maintenance requirement for bioswale;
- Service life of bioswale;
- Economic analysis and life cycle assessment.

18.7.1 Components of Bioretention System

In order to maintain good drainage conditions, the filter media should achieve a saturated hydraulic conductivity of 180 mm h^{-1}, which is within the recommended range of 100–300 mm h^{-1} as suggested in international guidelines (PUB, 2011; Water by Design, 2014; CIRIA, 2015). The thickness of filter media should be at least 400–700 mm as the minimum depths for plants or trees. The saturated hydraulic conductivity of filter media can be determined in laboratory following BS 1377-5:1990 and an in situ test BS EN ISO 22282-5:2012. The silt and clay content should attain the range of 2%–6% (w/w) (Water by Design, 2014) and organic carbon content should range between 3% and 10% (w/w) (PUB, 2011). Therefore, the tailored filter media will have maximum 10% of biochar and/or compost content with a well-graded particle size distribution (Table 18.6) as well as $D_{50} = 0.45$ mm.

Biochar as the filter layer can be treated, e.g., via acid-modification to increase the surface area, pore size distribution, and functional group density. The physio-chemical properties of biochar can be tailored to increase the removal efficiencies of various contaminants in stormwater. Characterization of biochar is necessary, for example:

- Elemental composition for C, H, N, S, and O (elemental analyzer)
- Surface area and pore size distribution (BET and BJH analysis)

TABLE 18.6 Example of Well-Graded Particle Size Distribution

Sieve Size (mm)	%Passing
6	100
2.0	90–100
0.6	40–70
0.2	5–20
0.063	<5

- Microscopic imaging and elemental mapping (SEM-EDS)
- Surface chemistry analysis (Boehm's titration, XPS and FTIR)

Vegetation is essential to the overall treatment efficiency of a bioretention system because of its role in pre-treatment during conveyance of surface runoff. Bioretention plants should be commercially available in local gardening shops. Chinese silvergrass (*Miscanthus sinensis*), purple heart (*Setcreasea purpurea*), purple loosestrife (*Lythrum salicaria*), and slender brake (*Pteris ensiformis*) are possible options for Hong Kong and many Asian cities.

18.7.2 Considerations for Design

Design flow rate, the flow pattern, and duration should be calculated to match with the official precipitation record (e.g., Hong Kong Observatory's record). In Hong Kong, for example, the lowest, median, and highest mean monthly rainfall (Table 18.7) should be selected for designing flow regime (i.e., duration and flow pattern) based on the precipitation data from the relevant weather stations located close to the target areas (e.g., Anderson Road Quarry New Development Area, Hung Hui Kiu New Development Area, and Tung Chung New Town Extension in Fig. 18.3). The total design flow of real and synthetic stormwater is suggested to be equivalent to 3 years of mean total annual rainfall in the selected areas (e.g., Fig. 18.4).

18.7.3 Guidelines for Sizing

Bioswale and biofiltration basin can be designed and sized in accordance with international design guidelines, i.e., SuDS Manual in the UK (CIRIA, 2015), Bioretention Technical Design Guidelines in Australia (Water by Design, 2014), and ABC Waters Design Guidelines in Singapore (PUB, 2011). Some design parameters that govern the size of bioswale and biofiltration basin are listed in Table 18.8 (PUB, 2011; Water by Design, 2014; CIRIA, 2015).

TABLE 18.7 Mean Monthly Rainfall at Yuen Long Station, Hong Kong International Airport and Tseung Kwan O Stations Between 1996 and 2016 (HKO, 2017)

	Mean Rainfall (mm)		
Month	**Yuen Long Station**	**Hong Kong International Airport Station**	**Tseung Kwan O Station**
01	24.1	43.4	35.7
02	29.9	32.5	40.1
03	55.0	67.3	82.2
04	132.3	152.2	162.3
05	213.3	255.9	316.3
06	276.6	399.3	488.1
07	181.2	280.9	346.1
08	200.2	339.9	396.0
09	183.2	207.7	291.0
10	36.4	53.6	86.9
11	28.0	42.1	47.2
12	33.2	38.9	27.5
Min	24.1	32.5	27.5
Max	276.6	399.3	488.1
Median	93.7	109.8	124.6

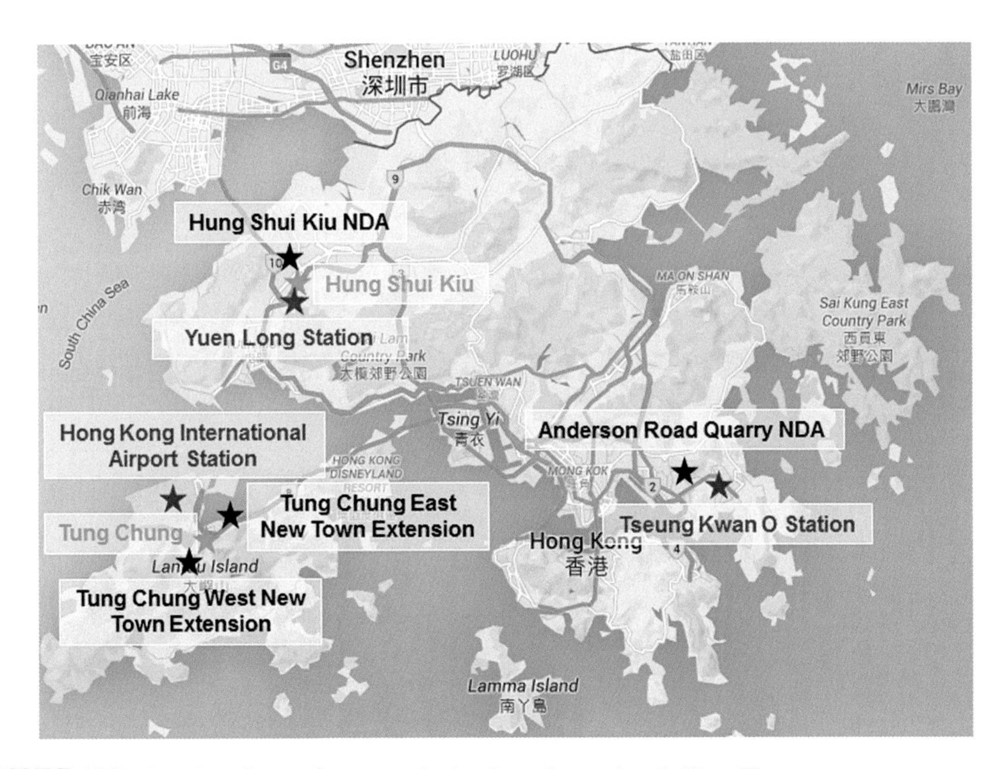

FIGURE 18.3 Location of example areas and related weather stations in Hong Kong.

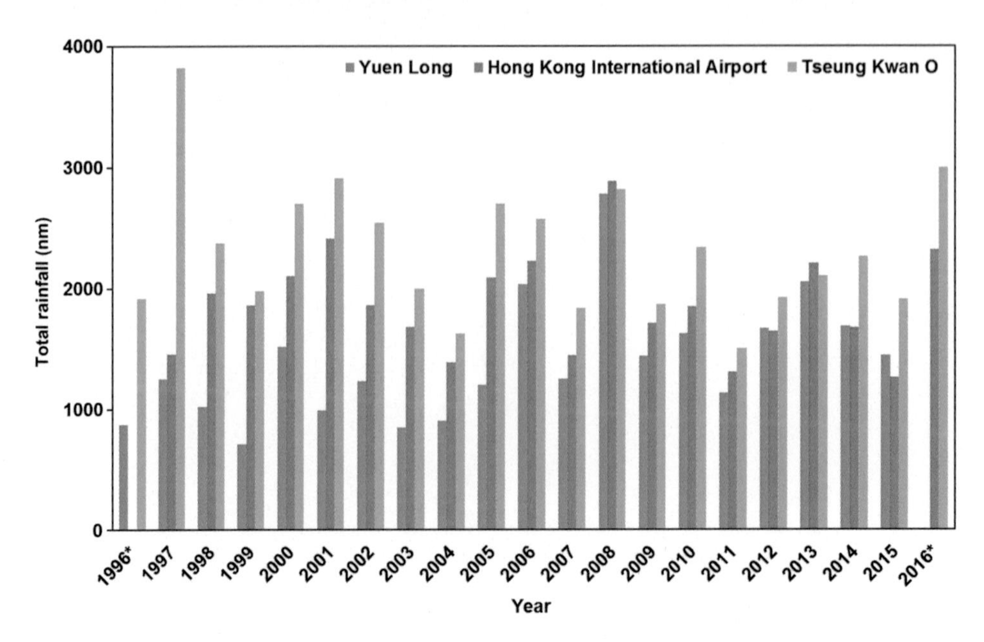

FIGURE 18.4 Total annual rainfall at Yuen Long Station, Hong Kong International Airport, and Tseung Kwan O Station between 1996 and 2016 (HKO, 2017).

TABLE 18.8 Key Parameters for Bioswale Design

Design Parameters	Values	References
Filter media width	600–2000 mm	PUB (2011), Water by Design (2014), CIRIA (2015)
Filter media length	<40 m	Water by Design (2014), CIRIA (2015)
Slope of swale (i.e., side slope) for bioswale	1:10–1:4	PUB (2011)
	<1:4	Water by Design (2014), CIRIA (2015)
Longitudinal slope for bioswale	1:200–1:50	Water by Design (2014)
	≥ 1:25	PUB (2011)
	<1:40	CIRIA (2015)
Maximum velocity to prevent scouring in minor events (2–10 year ARI)	0.5 m s^{-1}	PUB (2011); CIRIA (2015)
	1.0 m s^{-1}	Water by Design (2014)
Maximum velocity to prevent scouring in major events (50–100 year ARI)	1.0 m s^{-1}	Water by Design (2014)
	1.5 m s^{-1}	CIRIA (2015)
	2.0 m s^{-1}	PUB (2011)
Manning's *n* (with flow depth lower than vegetation height)	0.15–0.3	PUB (2011)
	0.35	CIRIA (2015)
Manning's *n* (with flow depth greater than vegetation height)	0.03–0.05	PUB (2011)
Bioretention area to catchment area ratio	<3% (catchment < 10 ha)	Water by Design (2014)
	>2% (catchment < 50 ha)	PUB (2011)
	2%–4%	CIRIA (2015)

18.8 CONCLUSIONS

In order to safeguard the water security and resilience for sustainable development, there is an imminent need to develop blue-green infrastructures to cater to the environmentally friendly drainage networks worldwide. A bioretention system for stormwater harvesting (e.g., bioswale and biofiltration basin) is a simple and passive measure for surface runoff treatment and flow attenuation. In view of the unstable performance of typical bioretention systems, an engineered filter media mixed with biochar is advocated to improve the harvested stormwater quality to the authority standards. The physical and chemical properties of biochar, including surface area, porosity, elemental composition, pH value, functional groups, etc., are significant in determining its ability to remove contaminants. To enhance sorption capacity, post activation (e.g., oxidation or gas purging) as well as pre-/post-synthesis modifications (e.g., nanoparticles formation or metal impregnation) can be applied for better functionality. Nevertheless, more investigations are needed to verify the biochar performance in the complex matrix of stormwater containing various bacteria, metals/metalloids, and herbicides, as well as the impacts of real-life occasions such as first flush effect and transient wetting and drying. As for the design of a bioretention system, flow attenuation, design, and maintenance requirements (such as saturated hydraulic conductivity and particle size distribution of the engineered filter media) should be evaluated vigorously to ensure the field compatibility (i.e., subtropical climate) and sustainability of the system.

Acknowledgment

The authors appreciate the financial support from the Hong Kong Environment and Conservation Fund (ECF Project 87/2017) and the Hong Kong Research Grants Council (PolyU 538613).

References

Afrooz, A.N., Boehm, A.B., 2016. Escherichia coli removal in biochar-modified biofilters: effects of biofilm. PloS one 11, e0167489.

Afrooz, A.N., Boehm, A.B., 2017. Effects of submerged zone, media aging, and antecedent dry period on the performance of biochar-amended biofilters in removing fecal indicators and nutrients from natural stormwater. Ecol. Eng. 102, 320–330.

Ahmad, M., Lee, S.S., Dou, X., Mohan, D., Sung, J.K., Yang, J.E., et al., 2012. Effects of pyrolysis temperature on soybean stover- and peanut shell-derived biochar properties and TCE adsorption in water. Bioresour. Technol. 118, 536–544.

Ahmad, M., Rajapaksha, A.U., Lim, J.E., Zhang, M., Bolan, N., Mohan, D., et al., 2014. Biochar as a sorbent for contaminant management in soil and water: a review. Chemosphere 99, 19–33.

Ahmad, M., Lee, S.S., Al-Wabel, M., Tsang, D.C.W., Ok, Y.S., 2017. Biochar-induced changes in soil properties affected immobilization/mobilization of metals/metalloids in contaminated soils. J. Soil Sed. 17, 717–730.

Aryal, R., Vigneswaran, S., Kandasamy, J., Naidu, R., 2010. Urban stormwater quality and treatment. Korean J. Chem. Eng. 27, 1343–1359.

Baltrėnas, P., Baltrėnaitė, E., Spudulis, E., 2015. Biochar from pine and birch morphology and pore structure change by treatment in biofilter. Water Air Soil Pollut. 226, 69.

BCC, 2006. Water Sensitive Urban Design – Technical Design Guidelines for South East Queensland, Brisbane City Council, Australia, June 2006.

Beesley, L., Marmiroli, M., 2011. The immobilisation and retention of soluble arsenic, cadmium and zinc by biochar. Environ. Pollut. 159, 474–480.

Beiyuan, J., Tsang, D.C.W., Yip, A.C.K., Zhang, W., Ok, Y.S., Li, X.D., 2017. Risk mitigation by waste-based permeable reactive barriers for groundwater pollution control at e-waste recycling sites. Environ. Geochem. Health 39, 75–88.

Blecken, G.T., Zinger, Y., Deletic, A., Fletcher, T.D., Viklander, M., 2009. Influence of intermittent wetting and drying conditions on heavy metal removal by stormwater biofilters. Water Res. 43, 4590–4598.

Cederlund, H., Börjesson, E., Stenström, J., 2017. Effects of a wood-based biochar on the leaching of pesticides chlorpyrifos, diuron, glyphosate and MCPA. J. Environ. Manage. 191, 28–34.

Chen, B., Chen, Z., Lv, S., 2011. A novel magnetic biochar efficiently sorbs organic pollutants and phosphate. Bioresour. Technol. 102, 716–723.

Chen, S.Q., Chen, Y.L., Jiang, H., 2017. Slow pyrolysis magnetization of hydrochar for effective and highly stable removal of tetracycline from aqueous solution. Ind. Eng. Chem. Res. 56, 3059–3066.

Chia, C.H., Downie, A., Munroe, P., 2015. Characteristics of Biochar: Physical and Structural Properties. Biochar for Environmental Management: Science and Technology. Earthscan Books Ltd, London, pp. 89–109.

Cho, K.W., Song, K.G., Cho, J.W., Kim, T.G., Ahn, K.H., 2009. Removal of nitrogen by a layered soil infiltration system during intermittent storm events. Chemosphere 76, 690–696.

Chow, K.L., Man, Y.B., Tam, N.F.Y., Liang, Y., Wong, M.H., 2015. Removal of decabromodiphenyl ether (BDE-209) using a combined system involving TiO_2 photocatalysis and wetland plants. J. Hazard. Mater. 322, 263–269.

Chui, T.F.M., Trinh, D.H., 2016. Modelling infiltration enhancement in a tropical urban catchment for improved stormwater management. Hydrol. Process 30, 4405–4419.

CIRIA, 2015. The SuDS Manual. CIRIA, London.

Collins, K.A., Lawrence, T.J., Stander, E.K., Jontos, R.J., Kaushal, S.S., Newcomer, T.A., et al., 2010. Opportunities and challenges for managing nitrogen in urban stormwater: a review and synthesis. Ecol. Eng. 36, 1507–1519.

Davis, A.P., Shokouhian, M., Sharma, H., Minami, C., 2001. Laboratory study of biological retention for urban stormwater management. Water Environ. Res. 73, 5–14.

Delwiche, K.B., Lehmann, J., Walter, M.T., 2014. Atrazine leaching from biochar-amended soils. Chemosphere 95, 346–352.

Dietz, M.E., Clausen, J.C., 2005. A field evaluation of rain garden flow and pollutant treatment. Water Air Soil Pollut. 167, 123–138.

Ding, Z., Hu, X., Wan, Y., Wang, S., Gao, B., 2016. Removal of lead, copper, cadmium, zinc, and nickel from aqueous solutions by alkali-modified biochar: Batch and column tests. J. Ind. Eng. Chem. 33, 239–245.

Ding, Z., Hu, X., Wan, Y., Wang, S., Gao, B., 2016. Removal of lead, copper, cadmium, zinc, and nickel from aqueous solutions by alkali-modified biochar: batch and column tests. Ind. Eng. Chem. Res. 33, 239–245.

Dong, X., Ma, L.Q., Li, Y., 2011. Characteristics and mechanisms of hexavalent chromium removal by biochar from sugar beet tailing. J. Hazard. Mater. 190, 909–915.

Eriksson, E., Baun, A., Scholes, L., Ledin, A., Ahlman, S., Revitt, M., et al., 2007. Selected stormwater priority pollutants – a European perspective. Sci. Total Environ. 383, 41–51.

Fang, Q., Chen, B., Lin, Y., Guan, Y., 2013. Aromatic and hydrophobic surfaces of wood-derived biochar enhance perchlorate adsorption via hydrogen bonding to oxygen-containing organic groups. Environ. Sci. Technol. 48, 279–288.

Fang, S., Tsang, D.C.W., Zhou, F., Zhang, W., Qiu, R., 2016. Stabilization of cationic and anionic metal species in contaminated soils using sludge-derived biochar. Chemosphere 149, 263–271.

Fletcher, T.D., Duncan, H., Poelsma, P., Lloyd, S., 2004. Stormwater flow and quality, and the effectiveness of non-proprietary stormwater treatment measures – a review and gap analysis. Technical Report 04/8, December 2004, Cooperative Research Centre for Catchment Hydrology.

Frankel, M.L., Bhuiyan, T.I., Veksha, A., Demeter, M.A., Layzell, D.B., Helleur, R.J., et al., 2016. Removal and biodegradation of naphthenic acids by biochar and attached environmental biofilms in the presence of co-contaminating metals. Bioresour. Technol. 216, 352–361.

Frišták, V., Friesl-Hanl, W., Wawra, A., Pipíška, M., Soja, G., 2015. Effect of biochar artificial ageing on Cd and Cu sorption characteristics. J. Geochem. Explor. 159, 178–184.

Gamiz, B., Pignatello, J.J., Cox, L., Hermosin, M.C., Celis, R., 2016. Environmental fate of the fungicide metalaxyl in soil amended with composted olive-mill waste and its biochar: an enantioselective study. Sci. Total Environ. 541, 776–783.

Gan, H., Zhuo, M., Li, D., Zhou, Y., 2008. Quality characterization and impact assessment of highway runoff in urban and rural area of Guangzhou, China. Environ. Monit. Assess. 140, 147–159.

Ghaffar, A., Ghosh, S., Li, F., Dong, X., Zhang, D., Wu, M., et al., 2015. Effect of biochar aging on surface characteristics and adsorption behavior of dialkyl phthalates. Environ. Pollut. 206, 502–509.

Gidley, P.T., Kwon, S., Yakirevich, A., Magar, V.S., Ghosh, U., 2012. Advection dominated transport of polycyclic aromatic hydrocarbons in amended sediment caps. Environ. Sci. Technol. 46, 5032–5039.

Gurtler, J.B., Boateng, A.A., Han, Y., Douds Jr, D.D., 2014. Inactivation of *E. coli* O157: H7 in cultivable soil by fast and slow pyrolysis-generated biochar. Foodborne Pathog. Dis. 11, 215–223.

Gwenzi, W., Chaukura, N., Noubactep, C., Mukome, F.N., 2017. Biochar-based water treatment systems as a potential low-cost and sustainable technology for clean water provision. J. Environ. Manage. 197, 732–749.

Hatt, B.E., Fletcher, D., Deletic, A., 2007. Hydraulic and pollutant removal performance of stormwater filters under variable wetting and drying regimes. Water Sci. Technol. 56, 11–19.

Hatt, B.E., Fletcher, T.D., Deletic, A., 2009. Hydraulic and pollutant removal performance of stormwater biofiltration systems at the field scale. J. Hydrol. 365, 310–321.

HK DSD, 2014. An integrated approach in flood prevention and water reclamation in a densely populated metropolitan: will the innovation work? In: Paper Presented at the DSD International Conference 2014, HK Drainage Services Department, Hong Kong.

HK WSD, 2015. Technical Specifications on Grey Water Reuse and Rainwater Harvesting. Water Supplies Department, Hong Kong.

HKO, 2016a. Monthly Meteorological Normals for Hong Kong. Retrieved from http://www.hko.gov.hk/cis/normal/1981_2010/normals_e.htm#table2.

HKO, 2016b. Rainstorm Warning System. Retrieved from http://www.hko.gov.hk/wservice/warning/rainstor.htm.

HKO, 2017. Climatological Information. Retrieved from http://www.hko.gov.hk/cis/climat_e.htm.

Hsieh, C.H., Davis, A.P., Needelman, B.A., 2007. Nitrogen removal from urban stormwater runoff through layered bioretention columns. Water Environ. Res. 79, 2404–2411.

Huff, M.D., Lee, J.W., 2016. Biochar-surface oxygenation with hydrogen peroxide. J. Environ. Manage. 165, 17–21.

Hunt, W.F., Smith, J.T., Jadlocki, S.J., Hathaway, J.M., Eubanks, P.R., 2008. Pollutant removal and peak flow mitigation by a bioretention cell in Urban Charlotte, N.C. J. Environ. Eng. ASCE 134, 403–408.

Inyang, M., Dickenson, E., 2015. The potential role of biochar in the removal of organic and microbial contaminants from potable and reuse water: a review. Chemosphere 134, 232–240.

Inyang, M., Gao, B., Wu, L., Yao, Y., Zhang, M., Liu, L., 2013. Filtration of engineered nanoparticles in carbon-based fixed bed columns. Chem. Eng. J. 220, 221–227.

Jung, K.W., Hwang, M.J., Jeong, T.U., Ahn, K.H., 2015. A novel approach for preparation of modified-biochar derived from marine macroalgae: dual purpose electro-modification for improvement of surface area and metal impregnation. Bioresour. Technol. 191, 342–345.

Khorram, M.S., Wang, Y., Jin, X.X., Fang, H., Yu, Y.L., 2015. Reduced mobility of fomesafen through enhanced adsorption in biochar-amended soil. Environ. Toxicol. Chem. 34, 1258–1266.

Khorram, M.S., Zheng, Y., Lin, D., Zhang, Q., Fang, H., Yu, Y., 2016. DZEissipation of fomesafen in biochar-amended soil and its availability to corn (Zea mays L.) and earthworm (Eisenia fetida). J. Soils Sediments 16, 2439–2448.

Kim, G., Yoon, J., 2011. Development and application of total coliform load duration curve for the Geum River, Korea. KSCE J. Civ. Eng. 15, 239–244.

Kim, G., Choi, E., Lee, D., 2005. Diffuse and point pollution impacts on the pathogen indicator organism level in the Geum River, Korea. Sci. Total Environ. 350, 94–105.

Kim, G., Chung, S., Lee, C., 2007. Water quality of runoff from agricultural-forestry watersheds in the Geum River Basin, Korea. Environ. Monit. Assess. 134, 441–452.

Kim, G., Jong, H., Lee, J., Kong, D., 2010. Fecal indicator concentrations of surface runoff in rural watersheds, Korea. Desalin. Water Treat. 19, 26–31.

Kizito, S., Lv, T., Wu, S., Ajmal, Z., Luo, H., Dong, R., 2017. Treatment of anaerobic digested effluent in biochar-packed vertical flow constructed wetland columns: role of media and tidal operation. Sci. Total Environ. 592, 197–205.

Lau, S.L., Han, Y., Kang, J.H., Kayhanian, M., Stenstrom, M.K., 2009. Characteristics of highway stormwater runoff in Los Angeles: metals and polycyclic aromatic hydrocarbons. Water Environ. Res. 81, 308–318.

Lau, A.Y.T., Tsang, D.C.W., Graham, N.J.D., Ok, Y.S., Yang, X., Li, X.D., 2017. Surface-modified biochar in a bioretention system for *Escherichia coli* removal from stormwater. Chemosphere 169, 89–98.

Lawrinenko, M., 2014. Anion Exchange Capacity of Biochar. Iowa State University, Iowa.

Lawrinenko, M., Laird, D.A., Johnson, R.L., Jing, D., 2016. Accelerated aging of biochars: impact on anion exchange capacity. Carbon 103, 217–227.

LeFevre, G., Paus, K., Natarajan, P., Gulliver, J., Novak, P., Hozalski, R., 2015. Review of dissolved pollutants in urban storm water and their removal and fate in bioretention cells. J. Environ. Eng. 141, 04014050.

Lehmann, J., Rillig, M.C., Thies, J., Masiello, C.A., Hockaday, W.C., Crowley, D., 2011. Biochar effects on soil biota—a review. Soil Biol. Biochem. 43, 1812–1836.

Li, H., Davis, A.P., 2008. Heavy metal capture and accumulation in bioretention media. Environ. Sci. Technol. 42, 5247–5253.

Li, L., Davis, A.P., 2014. Urban stormwater runoff nitrogen composition and fate in bioretention systems. Environ. Sci. Technol. 48, 3403–3410.

Li, J., Wai, O.W.H., Li, Y.S., Zhan, J., Ho, Y.A., Li, J., et al., 2010. Effect of green roof on ambient CO_2 concentration. Build. Environ. 45, 2644–2651.

Libra, J.A., Ro, K.S., Kammann, C., Funke, A., Berge, N.D., Neubauer, Y., et al., 2011. Hydrothermal carbonization of biomass residuals: a comparative review of the chemistry, processes and applications of wet and dry pyrolysis. Biofuels 2, 71–106.

Liu, P., Liu, W.J., Jiang, H., Chen, J.J., Li, W.W., Yu, H.Q., 2012. Modification of biochar derived from fast pyrolysis of biomass and its application in removal of tetracycline from aqueous solution. Bioresour. Technol. 121, 235–240.

Liu, W.J., Jiang, H., Yu, H.Q., 2015. Development of biochar-based functional materials: toward a sustainable platform carbon material. Chem. Rev. 115, 12251–12285.

Liu, Z., Han, Y., Jing, M., Chen, J., 2017. Sorption and transport of sulfonamides in soils amended with wheat straw-derived biochar: effects of water pH, coexistence copper ion, and dissolved organic matter. J. Soils Sediments 17, 771–779.

Lundy, L., Ellis, J.B., Revitt, D.M., 2011. Risk prioritisation of stormwater pollutant sources. Water Res. 46, 6589–6600.

MDA, 2011. Summary Report: Urban Storm Water Runoff Monitoring for Pesticides and Nitrate in Helena and Billings, 2011. Montana Department of Agriculture, Montana.

Mohanty, S.K., Boehm, A.B., 2014. Escherichia coli removal in biochar-augmented biofilter: Effect of infiltration rate, initial bacterial concentration, biochar particle size, and presence of compost. Environ. Sci. Technol. 48, 11535–11542.

Mukherjee, A., Zimmerman, A.R., Harris, W., 2011. Surface chemistry variations among a series of laboratory-produced biochars. Geoderma 163, 247–255.

Mukome, F.N., Zhang, X., Silva, L.C., Six, J., Parikh, S.J., 2013. Use of chemical and physical characteristics to investigate trends in biochar feedstocks. J. Agric. Food Chem. 61, 2196.

Muthanna, T.M., Viklander, M., Blecken, G., Thorolfsson, S.T., 2007. Snowmelt pollutant removal in bioretention areas. Water Res. 41, 4061–4072.

Palanisamy, B., Chui, T.F.M., 2015. Rehabilitation of concrete canals in urban catchments using low impact development techniques. J. Hydrol. 523, 309–319.

Park, J.H., Cho, J.S., Ok, Y.S., Kim, S.H., Kang, S.W., Choi, I.W., et al., 2015. Competitive adsorption and selectivity sequence of heavy metals by chicken bone-derived biochar: batch and column experiment. J. Environ. Sci. Health A 50, 1194–1204.

PUB, 2011. ABC Waters Design Guidelines, second ed. Public Utilities Board, Singapore.

Qian, K., Kumar, A., Zhang, H., Bellmer, D., Huhnke, R., 2015. Recent advances in utilization of biochar. Renew. Sust. Energ. Rev. 42, 1055–1064.

Rajapaksha, A.U., Chen, S.S., Tsang, D.C.W., Zhang, M., Vithanage, M., Mandal, S., et al., 2016a. Review on engineered/designer biochar for contaminant removal from soil and water: potential and implication of biochar modification. Chemosphere 148, 276–291.

Rajapaksha, A.U., Vithanage, M., Lee, S.S., Seo, D.C., Tsang, D.C.W., Ok, Y.S., 2016b. Steam activation of biochars facilitates kinetics and pH-resilience of sulfamethazine sorption. J. Soil Sed. 16, 889–895.

Reddy, K.R., Xie, T., Dastgheibi, S., 2014. Evaluation of biochar as a potential filter media for the removal of mixed contaminants from urban storm water runoff. J. Environ. Eng. 140, 04014043.

RI DEM, 2010. Rhode Island Stormwater Design and Installation Standards Manual. Rhode Island Department of Environmental Management, USA, December 2010.

Roh, H., Yu, M.R., Yakkala, K., Koduru, J.R., Yang, J.K., Chang, Y.Y., 2015. Removal studies of Cd(II) and explosive compounds using buffalo weed biochar-alginate beads. J. Ind. Eng. Chem. 26, 226–233.

Roy-Poirier, A., Champagne, P., Filion, Y., 2010a. Bioretention processes for phosphorus pollution control. Environ. Rev. 18, 159–173.

Roy-Poirier, A., Champagne, P., Filion, Y., 2010b. Review of bioretention system research and design: past, present, and future. J. Environ. Eng. ASCE 136, 878–889.

Singh, B., Fang, Y., Cowie, B.C., Thomsen, L., 2014. NEXAFS and XPS characterisation of carbon functional groups of fresh and aged biochars. Org. Geochem. 77, 1–10.

Suliman, W., Harsh, J.B., Fortuna, A.-M., Garcia-Pérez, M., Abu-Lail, N.I., 2017. Quantitative effects of biochar oxidation and pyrolysis temperature on the transport of pathogenic and nonpathogenic *Escherichia coli* in biochar-amended sand columns. Environ. Sci. Technol. 51, 5071–5081.

Sun, K., Jin, J., Keiluweit, M., Kleber, M., Wang, Z., Pan, Z., et al., 2012. Polar and aliphatic domains regulate sorption of phthalic acid esters (PAEs) to biochars. Bioresour. Technol. 118, 120–127.

Sun, Z., Davis, A.P., 2007. Heavy metal fates in laboratory bioretention systems. Chemosphere 66, 1601–1609.

Tang, J., Lv, H., Gong, Y., Huang, Y., 2015. Preparation and characterization of a novel graphene/biochar composite for aqueous phenanthrene and mercury removal. Bioresour. Technol. 196, 355–363.

Tang, X., Yang, Y., Tao, R., Chen, P., Dai, Y., Jin, C., et al., 2016. Fate of mixed pesticides in an integrated recirculating constructed wetland (IRCW). Sci. Total Environ. 571, 935–942.

Tong, S.J., Li, J.Y., Yuan, J.H., Xu, R.K., 2011. Adsorption of Cu(II) by biochars generated from three crop straws. Chem. Eng. J. 172, 828–834.

Trigo, C., Cox, L., Spokas, K., 2016. Influence of pyrolysis temperature and hardwood species on resulting biochar properties and their effect on azimsulfuron sorption as compared to other sorbents. Sci. Total Environ. 566, 1454–1464.

Trinh, B.S., Werner, D., Reid, B.J., 2017. Application of a full-scale wood gasification biochar as a soil improver to reduce organic pollutant leaching risks. J. Chem. Technol. Biotechnol. 92, 1928–1937.

Uchimiya, M., Chang, S.C., Klasson, K.T., 2011. Screening biochars for heavy metal retention in soil: role of oxygen functional groups. J. Hazard. Mater. 190, 432–441.

Ulrich, B.A., Im, E.A., Werner, D., Higgins, C.P., 2015. Biochar and activated carbon for enhanced trace organic contaminant retention in stormwater infiltration systems. Environ. Sci. Technol. 49, 6222–6230.

Vilvanathan, S., Shanthakumar, S., 2017. Column adsorption studies on nickel and cobalt removal from aqueous solution using native and biochar form of *Tectona grandis*. Environ. Prog. Sustain. Energy 36, 1030–1038.

Vithanage, M., Mayakaduwa, S.S., Herath, I., Ok, Y.S., Mohan, D., 2016. Kinetics, thermodynamics and mechanistic studies of carbofuran removal using biochars from tea waste and rice husks. Chemosphere 150, 781–789.

Wang, C., Lu, H., Dong, D., Deng, H., Strong, P.J., Wang, H., et al., 2013. Insight into the effects of biochar on manure composting: evidence supporting the relationship between N_2O emission and denitrifying community. Environ. Sci. Technol. 47, 7341–7349.

Wang, H., Feng, M., Zhou, F., Huang, X., Tsang, D.C.W., Zhang, W., 2017. Effects of atmospheric ageing under different temperatures on surface properties of sludge-derived biochar and metal/metalloid stabilization. Chemosphere 184, 176–184.

Water by Design, 2014. Bioretention Technical Design Guidelines (Version1.1). Healthy Waterways Ltd, Brisbane.

WHO, 2011. Guidelines for drinking-water quality. WHO chronicle, 38, 104–8.

Wu, Y., Chung, A., Tam, N.F.Y., Pi, N., Wong, M.H., 2008. Constructed mangrove wetland as secondary treatment system for municipal wastewater. Ecol. Eng. 34, 137–146.

Xiong, Z., Shihong, Z., Haiping, Y., Tao, S., Yingquan, C., Hanping, C., 2013. Influence of NH_3/CO_2 modification on the characteristic of biochar and the CO_2 capture. Bioenergy Res. 6, 1147–1153.

Xu, H., Wang, X., Li, H., Yao, H., Su, J., Zhu, Y., 2014. Biochar impacts soil microbial community composition and nitrogen cycling in an acidic soil planted with rape. Environ. Sci. Technol. 48, 9391–9399.

Xue, Y., Gao, B., Yao, Y., Inyang, M., Zhang, M., Zimmerman, A.R., et al., 2012. Hydrogen peroxide modification enhances the ability of biochar (hydrochar) produced from hydrothermal carbonization of peanut hull to remove aqueous heavy metals: batch and column tests. Chem. Eng. J. 200, 673–680.

Yan, Y., Ma, M., Liu, X., Ma, W., Li, M., Yan, L., 2017. Effect of biochar on anaerobic degradation of pentabromodiphenyl ether (BDE-99) by archaea during natural groundwater recharge with treated municipal wastewater. Int. Biodeterior. Biodegrad. 124, 119–127.

Yin, D., Wang, X., Chen, C., Peng, B., Tan, C., Li, H., 2016. Varying effect of biochar on Cd, Pb and As mobility in a multi-metal contaminated paddy soil. Chemosphere 152, 196–206.

Zand, A.D., Grathwohl, P., 2016. Enhanced immobilization of polycyclic aromatic hydrocarbons in contaminated soil using forest wood-derived biochar and activated carbon under saturated conditions, and the importance of biochar particle size. Pol. J. Environ. Stud. 25, 427–441.

Zgheib, S., Moilleron, R., Chebbo, G., 2012. Priority pollutants in urban stormwater: part 1 – case of separate storm sewers. Water Res. 46, 6683–6692.

Zhang, L., Seagren, E.A., Davis, A.P., Karns, J.S., 2010. The capture and destruction of *Escherichia coli* from simulated urban runoff using conventional bioretention media and iron oxide-coated sand. Water Environ. Res. 82, 701–714.

Zhang, W., Zheng, J., Zheng, P., Tsang, D.C.W., Qiu, R., 2015a. Sludge-derived biochar for As(III) immobilization: effects of solution chemistry on sorption behavior. J. Environ. Qual. 44, 1119–1126.

Zhang, M., Ahmad, M., Al-Wabel, M.I., Vithanage, M., Rajapaksha, A.U., Kim, H.S., et al., 2015b. Adsorptive removal of trichloroethylene in water by crop residue biochars pyrolyzed at contrasting temperatures: continuous fixed-bed experiments. J. Chem. 2015. Available from: https://doi.org/10.1155/2015/647072.

Zhu, D., Kwon, S., Pignatello, J.J., 2005. Adsorption of single-ring organic compounds to wood charcoals prepared under different thermochemical conditions. Environ. Sci. Technol. 39, 3990–3998.

Further Reading

Carstea, E.M., Bridgeman, J., Baker, A., Reynolds, D.M., 2016. Fluorescence spectroscopy for wastewater monitoring: a review. Water Res. 95, 205–219.

Hale, S., Hanley, K., Lehmann, J., Zimmerman, A., Cornelissen, G., 2011. Effects of chemical, biological, and physical aging as well as soil addition on the sorption of pyrene to activated carbon and biochar. Environ. Sci. Technol. 45, 10445–10453.

HK DSD, 2013. Stormwater Drainage Manual (with Eurocodes incorporated) – Planning, Design and Management (Fourth Edition, May 2013). Drainage Services Department, Hong Kong.

HK DSD, 2015. Practice Note No. 1/2015 – Guidelines on Environmental and Ecological Considerations for River Channel Design. Drainage Services Department, Hong Kong.

Xiong, X., Yu, I.K.M., Cao, L., Tsang, D.C.W., Zhang, S., Ok, Y.S., 2017. A review of biochar-based catalysts for chemical synthesis, biofuel production, and pollution control. Bioresour. Technol. 2017, 254–270.

CHAPTER

19

Potential Toxic Compounds in Biochar: Knowledge Gaps Between Biochar Research and Safety

Hao Zheng[1], *Bingjie Liu*[1], *Guocheng Liu*[2], *Zhaohui Cai*[1] and *Chenchen Zhang*[1]

[1]**Institute of Coastal Environmental Pollution Control, Key Laboratory of Marine Environment and Ecology, Ministry of Education, College of Environmental Science and Engineering, Ocean University of China, Qingdao, P.R. China** [2]**College of Resource and Environment, Qingdao Agricultural University, Qingdao, P.R. China**

19.1 INTRODUCTION

With the growing risks of human-induced climate change, soil degradation, food crisis, environmental pollution, and energy crisis, an environmentally sustainable and greener revolution is urgently needed (Tilman et al., 2001). This greener revolution should bring sustainable benefits, such as safety food production, soil conservation, resource utilization of agricultural residues, and/or mitigation of greenhouse gas emission. Biochar, as a soil amendment embedded into a regional carbon cycle, seems to provide an opportunity for developing such a greener revolution because of its multiple benefits (Glaser et al., 2009; Lehmann, 2007; Lehmann and Joseph, 2015).

Biochar, a porous carbon-rich solid material, is produced by thermochemical conversion of organic materials in the absence of or presence of limited air at 350 − 700°C (Lehmann and Joseph, 2015). Biochar product is intentionally produced for abating climate change by sequestrating carbon into soils and mitigating greenhouse gas (e.g., CO_2, CH_4, and N_2O) emission, improving soil properties, and increasing crop growth and productivity (Liu et al., 2015). Thus, biochar is now increasingly receiving attention as an "environmentally-friendly"

Biochar from Biomass and Waste
DOI: https://doi.org/10.1016/B978-0-12-811729-3.00019-4

approach for scientists and policy makers in many fields, such as agriculture, environment, energy, and climate. This is clearly proved by the exponential growth of the numbers of published literatures on biochar (Fig. 19.1). However, biochar is not a silver bullet (Zheng et al., 2016), and incorporation of biochar into soils does not always obtain positive results (Rogovska et al., 2012; Spokas et al., 2011a). For example, Spokas et al. (2011a) reported that approximately 20% of the reviewed studies ($n = 46$) showed negative yield or growth impacts that resulted from biochar incorporation. Furthermore, Sigmund et al. (2017) reported that biochar particles even have a cytotoxic effect on the fibroblast cells (NIH 3T3), likely to relate to the particulate nature and size distribution of the biochar. However, although the mechanisms underlying such negative effects are yet lacking, the inherent contaminants within biochar may partly contribute to these adverse effects following application to soil (Smith et al., 2013a; Wang et al., 2017b). Moreover, these negative findings associated with the contaminants in biochar promoted the development of biochar guidelines and/or standards to ensure its safety application (EBC, 2012; IBI, 2012).

The properties and qualities of biochars are mainly controlled by feedstock types and pyrolysis conditions (Zheng et al., 2013a), which have been extensively reviewed in much literature (Gul et al., 2015; Lian and Xing, 2017; Tripathi et al., 2016) and previous chapters. Theoretically, all the biomass resources, including wood waste, aquatic plants (e.g., giant reed, algae), sawdust, agricultural straw, and animal waste, can be used as parent material to produce biochar (Kuppusamy et al., 2016; Özçimen and Ersoy-Meriçboyu, 2010). Several technologies have been used to produce biochars, and their yields and properties vary with the employed technologies (Table 19.1). A threshold for pyrolysis temperature of 300–450°C was suggested for charring biochar to amend acidic soils due to the higher yield, pH, cation exchange capacity (CEC), and nutrients (Wang et al., 2017c; Zheng et al., 2013a), while biochar produced at higher temperatures (≥500°C) is recommended for soil contaminant remediation because of its greater capacity to adsorb pollutants, such as antibiotics (Zheng et al., 2013b) and heavy metals (Wang et al., 2015a). However, unwanted, several phytotoxic and potentially carcinogenic aromatic compounds like polycyclic aromatic hydrocarbons (PAHs) and dioxins can be formed during charring process, although modern pyrolyzers are designed to capture or eliminate these contaminants (Zheng et al., 2016). Additionally, some biochar may contain high concentrations of heavy metals

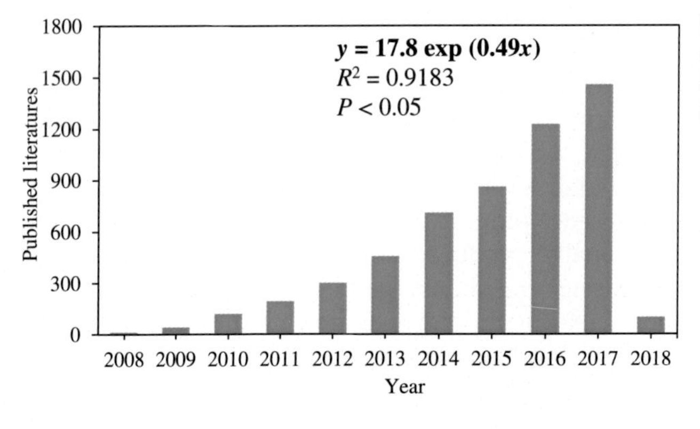

FIGURE 19.1 Exponential increment of the number of biochar related papers. Source: *The data was gathered from* Web of Science, *15.2.2018.*

TABLE 19.1 Comparison of the Technologies Used to Produce Biochar[a]

Technology Type	Heating Temperature (°C)	Heating Rate (°C s^{-1})	Feed Stock Particle Size (mm)	Residence Time (s)	Biochar Yield (%)	Bio-Oil Yield (%)	Gas Yield (%)
Slow pyrolysis	350–700	1–100	Variable[b]	450–550	15–40	20–55	20–60
Fast pyrolysis	450–550	>1000	<1	0.5–10	10–30	50–70	5–15
Flash pyrolysis	300–800	>1000	<0.2	<0.5	30–40		60–70
Gasification	800–1800	Variable	<2	10–20	0–10		90–100
Hydrothermal carbonization	150–400	<1	<1	Variable	5–40	20–40	2–10

[a]*The data in the date were cited from Manyà (2012), Spokas et al. (2011a), Tripathi et al. (2016).*
[b]*The particle size for the feedstock used in slow pyrolysis largely varies with the equipment, from meter or centimeter to millimeter.*

originated from the feedstock (e.g., sludge, sediment, and furfural residue) that contained high content of heavy metals (Liu et al., 2017a). Unfortunately, these toxic compounds in biochar may have adverse impacts on soil quality and biota functions, and plant growth (Rogovska et al., 2012; Wang et al., 2017b). Because of the aforementioned multiple benefits, biochar is expected to be increasingly used in a broad range of situations, and, therefore, it is urgently needed is evaluation of the toxicity potential of biochar before its large-scale application as a soil conditioner, or as an additive for livestock bedding and feeding.

In this chapter, we aim to provide a comprehensive overview on the research advances in the field of inherent pollutants formed during biochar production or charring. Specifically, the main objectives are to (1) understand the formation of organic and inorganic pollutants that produced in biochar production, (2) compare the total and bioavailable content of these contaminants in the reported biochars, (3) analyze the effects of feedstock and pyrolysis conditions on generation and content of these pollutants, (4) summarize the environmental risk and standard of biochar associated with these contaminants, and (5) propose the possible ways to mitigate these contaminants of biochar. Finally, armed with the complication knowledge of biochar's pollutants in these studies, the future development trends and research directions in this field are further addressed.

19.2 ORGANIC POLLUTANTS IN BIOCHAR

To date, the studies have showed that pyrolysis of biomass may form a wide variety of organic chemicals, including volatile organic compounds (VOC), PAHs, and dioxins (Ghidotti et al., 2017a; Hale et al., 2012; Spokas et al., 2011b). These compounds in biochar might be a potential drawback due to their toxicity to soil microorganisms and plants (Wang et al., 2017b). Recently, researchers identified biochar's potential organic pollutants by qualitative and quantitative methods (Hale et al., 2012; Spokas et al., 2011b). However, if biochar application has negative effects on agricultural systems or ecotoxicity risk resulting from biochar's inherent contaminants, as discussed in the following sections, then its economic benefits and environmental safety are questionable. Therefore, a comprehensive understanding of the formation, total, and bioavailable content of organic pollutants within biochar such as VOCs, PAHs, and dioxins would be very important for safety application of biochar into soil.

19.2.1 Volatile Organic Compounds

VOCs are defined as organic compounds that have boiling points of $\leq 250°C$ and vapor pressures ≤ 10.3 Pa at 293.15 K and 101.325 kPa (Kamal et al., 2016). Because of their volatility and toxicity, VOCs have been regarded as contributors of haze, photochemical smog, ozone depletion, and global warming, especially in China (Zhang et al., 2018), and thus environment and human health are seriously damaged by VOCs. Therefore, increasing efforts have been made to develop efficient VOC abatement techniques, such as catalysis oxidation, condensation, biological degradation, and adsorption to minimize VOC emission (Zhang et al., 2017).

VOCs are originated from both natural emissions, such as volcano eruption and forest fire, and anthropogenic emissions, such as exploitation and usage of fossil fuels, combustion, and refinement of biomass (e.g., pyrolysis, gasification, and liquidation) (Zhang et al., 2018). Pyrolysis or gasification is usually used to carbonize plant- or animal waste-based biomass to obtain the desired biochar products, and the VOCs may retain or sorb on these biochars (Ghidotti et al., 2017a; Spokas et al., 2011b). So far, several studies have confirmed that VOCs released from biochar into soils possess stimulatory or inhibitory effects on plant productivity (Spokas et al., 2010) and microbial communities and processes (Wang et al., 2013). Obviously, VOCs within biochar should not be ignored during their production, storage, and utilization. However, compared with heavy metals and PAHs in biochar, only a limited number of studies have examined the profiles, content, and toxic effects of VOCs in biochar (Cole et al., 2012; Ghidotti et al., 2017a; Smith et al., 2013b; Spokas et al., 2011b). There are still lots of uncertainties in the formation, toxic effect, and fate of VOCs within biochar, which largely limit the safety application of biochars in soil. However, to emphasize the progress and further to address the environmental implications of the VOCs in biochars, the information on inherent VOCs in biochar is discussed.

19.2.1.1 Formation of Volatile Organic Compounds in Biochar

VOCs in biochars are generally derived from re-condensation of pyrolysis vapors and liquids, the pyrolytic products named syngas and bio-oil (also is called wood vinegar or pyroligneous acid, respectively) (Buss et al., 2015). Lignocellulosic biomass, such as wood chip, sawdust, peanut shell, giant reed, and straw, are the most naturally abundant feedstock for biochar production (Thines et al., 2017; Wang et al., 2015b; Zheng et al., 2017, 2013a). The lignocellulosic biomass is mainly composed of cellulose, hemicelluloses, lignin, and inorganic minerals, which varies in different biomass (Liu et al., 2017b; Mohan et al., 2006). These chemical components largely affect yields and characteristics of the decomposition products such as biochar or bio-oil. Several excellent reviews have described the classes of mechanisms that had been previously proposed for biomass pyrolysis to produce biochar, bio-oil, or syngas (Chen et al., 2017; Huber et al., 2006; Liu et al., 2017b; Mohan et al., 2006). Liu et al. (2017b) comprehensively performed an overview of the transformation, distribution, and release of the main chemical elements (C, H, O, N, P, Cl, S, and metals) during biomass pyrolysis process, as well as evolution of biochar, bio-oil, gas, and tar. These compiled studies highlighted the chemical mechanisms for biomass pyrolysis based on the three components of cellulose, hemicelluloses, and lignin, which are conducive to understand organic contaminant formation in biochar.

Generally, thermal degradation of biomass components follows the order of hemicelluloses > cellulose > lignin (Babu, 2008). Briefly, hemicelluloses first break down at temperatures of 200–260°C, producing more volatiles, less tars and chars than cellulose, and cellulose follows at the temperatures of 240–350°C to produce anhydrocellulose and levoglucosan, whereas lignin is the last component to decomposition at temperatures of 280–500°C and with a maximum rate being observed at 350–450°C, which yields phenols via cleavage of ether and carbon linkages (Babu, 2008; Huber et al., 2006; Liu et al., 2017b; Mohan et al., 2006). For instance, much acetic acid liberated from wood biomass during pyrolysis is attributed to deacetylation of the hemicellulose (Mohan et al., 2006). Moreover, biomass with higher content of lignin produces higher yield of biochars than pure cellulose or biomass contained lower content lignin (Lee et al., 2013). More details of biomass degradation were well reviewed by Mohan et al. (2006) and Liu et al. (2017b).

In the biomass pyrolysis process, two-stage models are proposed to describe the pyrolysis mechanisms of wood and other cellulosic materials (Babu, 2008). Pyrolysis of biomass was considered to be a two-stage reaction, in which the products of the first stage break up Reactions (19.1) and (19.2), further in the presence of each other to occur Reactions (19.3) to produce secondary pyrolysis products.

Parallel reactions:

$$\text{Virgin biomass} \rightarrow (\text{volatiles} + \text{gases})_1 \tag{19.1}$$

$$\text{Virgin biomass} \rightarrow (\text{char})_1 \tag{19.2}$$

Secondary interactions:

$$(\text{Volatiles} + \text{gases})_1 + (\text{char})_1 \rightarrow (\text{volatiles} + \text{gases})_2 + (\text{char})_2 \tag{19.3}$$

Obviously, VOCs release and react with chars, gases, and liquid fraction during pyrolysis. Therefore, a huge variety of VOCs, such as low molecular weight organic acids, alcohols, ketones, ethene, and phenols, are re-condensed and trapped in biochar pores (Spokas et al., 2011b). These VOCs in biochar normally are associated with the pyrolysis liquid fraction (Buss et al., 2015; Ghidotti et al., 2017b). For example, Ghidotti et al. (2017b) extracted water-soluble organic compounds (WSOCs) from corn stalk biochar and bio-oil produced at 350 – 650°C, and comprehensively characterized them with spectroscopic, chromatographic, and mass spectrometry techniques, in order to establish the linkage between WSOC patterns and biochar bulk properties in relation to the bio-oil composition. Their findings confirmed that even at the pilot plant scale these aromatic units are formed by interaction between the pyrolysis vapors and biochars and survived into biochar pores, thus determining the suitability of biochar for environmental applications.

19.2.1.2 Content of Volatile Organic Compounds in Biochars

Only a limited number of studies have qualitatively or quantitatively examined VOCs in biochar (Buss and Mašek, 2014; Ghidotti et al., 2017b; Spokas et al., 2010). We searched for literature that reported VOC concentrations and profiles in biochar, regardless of what kind of pyrolysis technologies used, and identified roughly 12 papers listed in Table 19.2. The feedstock and pyrolysis parameters such as pyrolysis technology, heating temperature, and equipment was also contained in Table 19.2, as well as the methods used for extraction and detection of VOCs in biochars. Moreover, it should be noted that among these studies, most of them were conducted under controlled laboratory conditions to qualitatively determine the effect of various factors on VOC formation, less attention has paid to quantitative analysis.

To the best of our knowledge, Spokas et al. (2010) were the first to document the formation of VOCs from biochar, and they hypothesized that the ethylene in biochars could be an additional potential mechanism for the soil and plant responses observed from biochar amendments. To test this hypothesis, the ethylene production potential of 12 biochars was examined through sealed aerobic laboratory incubations, and they found that the biochars produced ethylene ranging from 1.0 to 54.4 ng g^{-1} char d^{-1}. Subsequently, Spokas et al. (2011b) qualitatively identified more than 140 VOCs sorbed on 77 biochars encompassing a variety of parent feedstocks and manufacturing processes by headspace thermal desorption coupled to capillary gas chromatographic–mass spectrometry. However, the quantitative analysis of VOCs was not conducted in this study, thus VOC contents in these biochars were not clear. The total content of VOCs in biochar was first reported by Becker et al. (2013), who compared the relative content and composition of VOCs in the hydrochars derived from wheat straw, biogas digestate, and four woody materials by hydrothermal carbonization at 190–270°C. They reported that the total amount of VOCs in the char samples produced at 270°C ranged between 2000 and 16000 μg g^{-1} (0.2–1.6 wt %), and 50–9000 μg g^{-1} of benzenes and 300–1800 μg g^{-1} of phenols were observed. Based on this quantitative determination of VOCs, the authors proposed that the fresh hydrochar should be optimized for the application as soil amendment or for water purification. Unfortunately, compared to the qualitative analysis, fewer studies were conducted on quantitative analysis of VOCs sorbed on biochars. Consequently, Dutta et al. (2017) reviewed VOCs formation in biochars, in which they only listed the identified VOCs in biochar and their retention times by the headspace thermal desorption–GC/MS method, without any concentration ranges. Overall, however, the total VOC content in 152 biochars varied widely, ranging between 0.34 and 16000 μg g^{-1}. The total VOC content also changed with pyrolysis type, temperature, and feedstock, which will be discussed in the following section. Additionally, the types of VOCs observed in biochar largely vary with the feedstock and pyrolysis parameters. For example, biochar produced from hydrothermal treatment at higher temperature generally contained more types of VOCs than those from lower temperature (Becker et al., 2013).

TABLE 19.2 Reported Total and Available Concentrations of VOCs in Biochar

Biochar Sample	Technology	Heating Temperature (°C)	Equipment	VOC Extraction and Detect Method	Type	Total Concentration ($\mu g\ g^{-1}$)	Available Concentration ($\mu g\ g^{-1}$)	Reference
Twelve biochars from wood, corn cobs, peanut hulls, etc.	Slow pyrolysis Fast pyrolysis	350–500		Incubation and gas chromatography–mass spectrometer	1 (ethylene)			Spokas et al. (2010)
Seventy-seven different biochars from macadamia shells, oak hardwood sawdust corn stover, etc.	Fast pyrolysis slow pyrolysis gasification hydrothermal carbonization Microwave assisted pyrolysis	250–800	Gasifier	Headspace thermal desorption and gas chromatographic–mass spectrometry	>140			Spokas et al. (2011b)
Switch grass biochar	Fast pyrolysis	450	Fluidized-bed reactor	Toluene extraction	Not detected			Becker et al. (2013)
	Gasification	824		Laser desorption ionization and mass spectrometry				
	Slow pyrolysis	500						
Pinewood	Fast pyrolysis	450	Auger reactor	Aqueous extraction electrospray ionization coupled to Fourier transform ion cyclotronresonance mass spectrometry	Three fractions based on their molecular sizes and electric charges			Ghidotti et al. (2017a)
Chicken litter	Fast pyrolysis	424	Superheated steam pyrolysis unit					
Peanut shell		411						
Digestate	Hydrothermal carbonization	190–270	125-mL unstirred parr pressure vessels	Headspace gas chromatography	25–78	2000–16,000		Becker et al. (2013)
Wheat straw	Hydrothermal carbonization	190–270			29–87			
Pine wood	Hydrothermal carbonization	190–270			51–90			
Poplar wood	Hydrothermal carbonization	190–270			11–92			

(*Continued*)

TABLE 19.2 (Continued)

Biochar Sample	Technology	Heating Temperature (°C)	Equipment	VOC Extraction and Detect Method	Type	Total Concentration ($\mu g\ g^{-1}$)	Available Concentration ($\mu g\ g^{-1}$)	Reference
Masanduba wood	Hydrothermal carbonization	150–270			8–79			
Garapa wood	Hydrothermal carbonization	150–270			8–71			
Softwood pellets	Slow pyrolysis	550	Rotary kiln	Water extraction	Not detected			Buss and Mašek (2014)
				Mettler-Toledo thermo-gravimetric analysis	Pot experiment			
Softwood pellets	Slow pyrolysis	550	Not detected	Carbon disulphide extraction	33	Below detection limit (20 $\mu g\ g^{-1}$)–1166		Buss et al. (2015)
				Semi-quantitative analysis				
Mushroom medium	Slow pyrolysis	450	Traditional biochar kiln	Aqueous extraction gas chromatographic–mass spectrometry	20	Not detected	5200	Ghidotti et al. (2017a)
Corn stalk					41		7700	
Rice straw					25		2100	
Softwood pellets	Slow pyrolysis	550	Rotary kiln	Water extraction	8	0.9–13.7		Buss and Mašek (2014)
				MiniRAE lite VOC analyzer				
Cellulose	Slow pyrolysis	300–500	Parr batch reactor	2D gas chromatography mass spectrometry and electrospray ionization Fourier transform ion cyclotron resonance mass spectrometry	51		28.58–1251	Ghidotti et al. (2017a)
Lignin								
Pine								
Cold spot pine								
Pelletized corn stalks	Slow pyrolysis	350–650		Aqueous extraction chromatographic–mass spectrometry	88		35–3000	Spokas et al. (2010)

19.2.1.3 Bioavailable Content of Volatile Organic Compounds in Biochar

Sustainable application of biochar to soil requires careful consideration of bioavailable fractions of VOCs, which may directly determine the adverse effects of biochar in soils (Smith et al., 2016). Similar to nutrients such as N, P, and K (Zheng et al., 2013a), total VOCs in biochar may not necessarily reflect the actual availability of these chemical compounds to soil microbes and plants. For example, Smith et al. (2013a) reported that water-extractable substances of pinewood-derived biochar showed a significant inhibitory effect on both blue-green alga (cyanobacteria *Synechococcus*) and eukaryotic green alga (*Desmodesmus*), likely due to a 500-delta (or smaller) organic chemical species that carries at least one carboxyl group. Rombolà et al. (2015) proved that toxic compounds such as volatile fatty acids in the biochar from poultry litter responsible for the toxicity to seed germination are biodegradable. However, only a few studies have paid attention to the water-extractable or bioavailable VOCs in biochar (Table 19.2). Smith et al. (2013b) characterized molecular level compositions of water-extractable substances of pinewood-derived biochar using electrospray ionization (ESI) coupled to Fourier transform ion cyclotron (ESI-FTICR-MS). Finally, they concluded that growth inhibition of aquatic photosynthetic microorganisms most likely resulted from the unique carbohydrate ligneous components and sulfur containing condensed ligneous components. However, the content of these chemicals was not quantified. Buss et al. (2015) compared the VOC content between high-VOC and low-VOC biochars, which were contaminated by a high and low dose of re-condensed pyrolysis liquids and gases, respectively. As addressed by them, this study quantitatively analyzed the composition and concentration of organics sorbed to biochars, and the WSOCs with the highest concentrations in the two high-VOC biochars were acetic, formic, butyric and propionic acids, all with concentrations over 100 $\mu g\ g^{-1}$. Similarly, WSOCs of biochars from pelletized corn stalks and their corresponding bio-oil were comprehensively characterized to investigate the evolution of VOCs in biochars, but the composition and the content was not determined (Ghidotti et al., 2017b). Overall, the reported water-soluble VOC content in the studied biochars ranged from 2100 to 7700 $\mu g\ g^{-1}$. Therefore, more studies are urgently needed to quantify the available biochar-derived VOCs, because of the concentration of VOCs in biochar could be a good indicator for potential phytotoxic effects (Buss and Mašek, 2016; Buss et al., 2015).

19.2.1.4 Factors Affected Volatile Organic Compound Profiles and Contents in Biochars

As well reviewed, characteristics of biochars and bio-oil are highly dependent on individual pyrolysis parameters, such as feedstock, heating temperature, duration time, and pyrolysis type (Babu, 2008; Huber et al., 2006; Liu et al., 2017b; Mohan et al., 2006; Pandey and Kim, 2011). As a result, it is reasonable to hypothesize that these parameters also have influence on VOC contents and types in biochars. However, because of the rather fragmentary results from limited studies related to VOCs within biochar (Table 19.2), it is difficult to draw a concise picture.

Although a variety of biomass materials were used to produce biochar, no clear effect of the feedstock type on VOC composition and content was observed (Table 19.2). This was confirmed by Spokas et al. (2011b), who found over 140 individual chemical compounds thermally desorbed from 77 different biochars and no clear feedstock dependencies to the sorbed VOC composition, but the contents of these VOCs was not measured. Smith et al. (2016) reported that the toxic WSOC of pinewood-derived biochar was likely due to mono-, di-, and trisubstituted phenolic compounds derived from lignin, while toxic WSOC of cellulose-derived biochar was acidic and bio-oil like in nature. Only a study reported by Rombolà et al. (2015) found that significantly higher concentrations of volatile fatty acids in poultry litter biochars (4–9 mg g^{-1}) than those of corn stock biochars (1–4 mg g^{-1}). Additionally, Spokas et al. (2011b) speculated that elemental composition of the feedstock appeared to influence some of the VOC species, which need further investigation.

Pyrolysis conditions largely affected VOC formation and content in biochars. Spokas et al. (2011b) were first to qualitatively evaluate effects of pyrolysis technology and pyrolysis temperatures on the profiles of the inherent VOCs in different biochars. They concluded several general findings on VOC profiles associated to these pyrolysis conditions: (1) hydrothermal carbonization and fast pyrolysis typically produce the greatest number of VOCs in biochar, while gasification result in low to non-detectable levels of VOCs; (2) biochar prepared at lower temperatures (≤350°C) contains the VOCs consisting of short carbon chain aldehydes, furans, and ketones, whereas biochar prepared at higher temperature (400–900°C) typically retains aromatic compounds and longer carbon chain hydrocarbons; (3) post-production processing may reduce the amount of sorbed VOCs on biochars. However, these finding were obtained from qualitative analysis of VOCs on biochars, thus more quantitative analysis is urgently needed to confirm them. Notably, higher content of VOCs in biochar prepared from hydrothermal carbonization, which is widely attributed to contact with the liquid phase in hydrothermal reactor (Becker et al., 2013). Generally, VOCs follow a decreasing trend as a function of production temperature. Wang et al. (2013) measured water-soluble and total phenolic compounds in six bicohars produced from a wetland plant (giant reed) using slow pyrolysis. They noted that the water-soluble and total phenolic compounds significantly decreased from 2.01 to 0.016 mg g^{-1} and 7.01 to 0.019 mg g^{-1}, respectively, with increasing temperature from 200 to 600°C. These findings confirmed the higher pyrolysis temperature, the lower content of VOCs. A general decrease trend in the relative amount of the aldehyde, furan, and ketone with increasing production temperature was also observed (Spokas et al., 2011b). With increasing pyrolytic temperature, the carbonization degree of biochar increase, which can be estimated by the hydrogen to carbon (H/C) ratios (Zheng et al., 2013a). Consequently, significantly negative correlations were observed between the quantity of all VOC classes and the H/C ratios of biochars, reported by Ghidotti et al. (2017a), who thus proposed that the risk of VOCs emissions can be reduced by producing biochar with a high carbonization degree.

19.2.1.5 Negative Impact and Standard of Biochar Associated With Volatile Organic Compounds

The amount of VOCs within biochar is very low (Table 19.2), but its potential negative effects on organisms such as plants, soil microbe, and alga, have proven to be significant (Rogovska et al., 2012; Smith et al., 2013a; Wang et al., 2017b). For example, seed germination significantly decreased by exposure of poultry litter biochars, tentatively ascribed to

WSOC in biochar (Rombolà et al., 2015). Spokas et al. (2010) declared that ethylene in biochar could be a contributing factor to the observed plant and soil microbial effects. Smith et al. (2013a) stated that water extract of pinewood-derived biochar showed a significant inhibitory effect on aquatic photosynthetic microorganism growth in a dose-dependent manner, probably due to some type of 500-dalton (or smaller) organic chemicals containing at least one carboxyl group. Among these studies, germination test, and a relatively inexpensive and fast method, has been highly proposed for screening biochar for potential toxic effects (Bouqbis et al., 2017; Rogovska et al., 2012; Rombolà et al., 2015).

VOCs in biochar also affect cycling of soil nutrient, such as N and P, due to their participation in abiotic and biotic reactions known to influence soil quality. Wang et al. (2013) first reported that peanut shell derived biochar weakened nitrification process, due to the reduced abundance and diversity of ammonia-oxidizing bacteria (AOB) resulted from toxicity of inherent phenolic compounds in biochar to AOB. Sun et al. (2015) found that phenols comprised a fraction of the VOCs in biochar that potentially could be toxic to some microbes and inhibit their growth in the short time, and thus the VOCs shaped the structure of soil microbial communities. Similar result was also reported by Farrell et al. (2013). Additionally, such negative effects induced by fresh biochar might be short-lived, which could disappear for the aged biochars (Dutta et al., 2017). Still, there is no clear conclusion on the underlying mechanisms responsible for these adverse effects of biochar contained VOCs, which deserve further investigation in the future.

It has been widely accepted that VOCs are very important for the determination of biochar quality. However, although several biochar quality guidelines have been proposed by the International Biochar Initiative (IBI, 2012) and the European Biochar Certificate (EBC, 2012), no quantitative information or threshold of VOCs was included. For the safety and sustainable application of biochar, we strongly propose that VOCs should be included among the criteria for assessment of biochar quality, which needs more attention on this topic.

19.2.2 Polycyclic Aromatic Hydrocarbons

PAHs, a group of semivolatile compounds with two or more aromatic rings fused in linear (e.g., anthracene), angular (e.g., phenanthrene), cluster (e.g., triphenylene), and cyclic (e.g., coronene) arrays, are persistent organic pollutants that carry great environmental risk and health concerns such as cancer, mutations, and malformations (Ma and Harrad, 2015; Rubio-Clemente et al., 2014). In order to reduce emission of these compounds into the environment, 16 PAHs have been classified in the priority list by the United States Environmental Protection Agency (USEPA) and other governments. PAHs are released into the environment from anthropogenic and natural sources. For natural source, forest fires, and grass land fires produce more PAHs than those from vegetation synthesis, microbial synthesis, and volcanic activity (Lamichhane et al., 2016). Anthropogenic activities include vehicle exhaust, agricultural production, straw burning, and combustion of fossil fuels produce a significant amount of PAHs into the environment (Lamichhane et al., 2016; Ma and Harrad, 2015). Among these sources, burning biomass waste is proved as the major source of PAH emission (Samburova et al., 2016). This is one of the reasons that straw burning is not recommended, and the Chinese government made strict policies to forbid this agricultural activity. To reduce pollutants (e.g., PAHs and fly ash) and enhance high-value utilization of biomass, pyrolysis is considered as a relatively environmentally benign approach (Liu et al., 2017b; Zheng et al., 2013a). However, inevitably, certain amounts of PAHs still can be detected in biomass pyrolysis products such as gas, bio-oil, tar, and biochar (Liu et al., 2017b; Mohan et al., 2006; Wang et al., 2013). Therefore, recently, more attention has shifted to PAH concentration in biochar because of the potential application of biochar to soil for soil improvement or remediation, and carbon sequestration. For example, Hale et al. (2012) observed that the total PAH concentrations ranged from 0.07 to 45 $\mu g\ g^{-1}$ in a suite of over 50 biochars produced at 250–900°C using various methods and biomass. Lin et al. (2018) reported that the content of ten nitro-PAHs was 0.058 $\mu g\ g^{-1}$ for the raw sludge, but ranged between 0.141 and 0.744 $\mu g\ g^{-1}$ for the produced chars, confirming that pyrolysis enriched PAHs in biochar. Because of the toxic, mutagenic, and carcinogenic natures of PAHs and adverse impacts of biochar, significant interest in detecting formation, concentrations, and compositions of PAHs within biochar has been increasingly raised (Freddo et al., 2012; Hale et al., 2012; Hilber et al., 2017; Keiluweit et al., 2012). Consequently, more studies have been published on biochar associated with PAHs than those of VOCs. For more information, Table 19.3 summarizes the studies reported PAH formation and concentration in biochars that prepared under various biomass and pyrolysis conditions.

19.2.2.1 Polycyclic Aromatic Hydrocarbon Formation in Biochars

Although a large amount of studies reported PAH formation in biomass pyrolysis, the picture for PAHs in biochar is very complex. Lots of reaction mechanisms have been proposed (Bucheli et al., 2015; Liu et al., 2017b; Lu and Mulholland, 2004; Shukla and Koshi, 2012). Liu et al. (2017b) and Bucheli et al. (2015) well reviewed formation the process of PAHs during biomass pyrolysis. As addressed by them, the widely accepted mechanism is hydrogen abstraction acetylene addition, in which gaseous C_2Hx radicals or intermediates such as ethyne and ethene that generated from cracking of cellulose, hemicelluloses, and lignin in biomass undergo a series of bimolecular reactions to form larger (poly-)aromatic ring structures. That is why the content of PAHs sorbed on fresh biochar increase with increasing pyrolysis temperature and residence time. Additionally, Keiluweit et al. (2012) presented that PAH formed by two main pathways based on pyrolytic temperature. At temperature $<500°C$, PAHs produce from unimolecular cyclization, dehydrogenation, dealkylation, and aromatization of ligneous and cellulosic components in feedstock. The native compounds, such as H_2O, CO_2, CH_4, and H_2S, are eliminated and the aromatized structures are retained, which then experienced direct nuclear condensation with further cyclization. At temperatures $>500°C$, a free radical pathway followed by pyrosynthesis into larger aromatic structures generates PAHs.

19.2.2.2 Total Concentrations of Polycyclic Aromatic Hydrocarbon in Biochar

Recently, a few studies have examined the content of PAHs in a large amount of biochars derived from various feedstock and pyrolysis conditions under laboratory and industry scale (Table 19.3). Compared with those of VOCs, these results have depicted a

TABLE 19.3 Reported Total and Available Concentrations of PAHs in Biochars

Feedstock	Technolgy	Heating Temperature (°C)	PAH Extraction Method	Total PAH Concentration ($\mu g\ g^{-1}$)	Available PAH Concentration ($\mu g\ g^{-1}$)	Reference
Polyethylene	Gasification	450–1050	Dichloromethane extraction	0.598–16.63		Buss and Mašek (2014)
Varnish wastes						
Olive oil						
Solid waste						
Waste lube oils						
Paper waste						
Sewage sludges						
Wheat straw	Slow pyrolysis	400–525		33.7		Buss et al. (2015)
Poplar wood						
Spruce wood						
Redwood	Slow pyrolysis	300–600	Pressurized liquid extraction	0.08–8.7		Freddo et al. (2012)
Rice straw						
Maize						
Bamboo						
Hardwood	Slow pyrolysis	300–450	Dimethylsulfoxide extraction	10	<2	Fagernäs et al. (2012)
Vine wood	Slow pyrolysis	350–750	Toluene, methanol, dichloromethane, acetone, ethanol, propanol, hexane, heptane extraction	9.1–355		Hilber et al. (2012)
Elephant grass						
Coniferous wood						
Coniferous an deciduous						
Digested dairy manure	Slow pyrolysis	250–900	Toluene extraction	0.07–45	0.17–10.0	Hale et al. (2012)

(Continued)

TABLE 19.3 (Continued)

Feedstock	Technolgy	Heating Temperature (°C)	PAH Extraction Method	Total PAH Concentration (μg g^{-1})	Available PAH Concentration (μg g^{-1})	Reference
Food waste	Fast pyrolysis					
Paper mill waste	Gasification					
Corn stover						
Wheat straw						
Rubberwood sawdust						
Lodgepole pine						
Pine wood						
Switch grass						
Laurel oak						
Loblolly pine						
Eastern gamma grass						
Hardwood						
Heartland pine						
Empty fruit bunches						
Coconut shell						
Tall fescue straw	Slow pyrolysis	100–700	Toluene-methanol extraction	0.05–30.2		Keiluweit et al. (2012)
Ponderosa pine wood						
Distiller grains	Slow pyrolysis	350–400	Acetone/ cyclohexane extraction	1.2–19		Fabbri et al. (2013)
Elephant grass	Slow pyrolysis	350–650	Accelerated solvent extractor	1.124–28.339		Buss et al. (2015)
Coconut shell						

Wicker						
Wheat straw						
Fraxinus excelsior	Slow pyrolysis	300–600	Hexane, acetone and triethylamine extraction	9.56–64.65		Quilliam et al. (2013)
Rice husk						
Pulp sludge	Slow pyrolysis	450–550	Sodium sulfate and hexane extraction	0.4–236		Khan et al. (2015)
Elephant grass	Slow pyrolysis	350–650	Accelerated solvent extractor	3.5–39.9		Kołtowski and Oleszczuk (2015)
Willow						
Wheat straw						
Straw pellets	Slow pyrolysis	350–750	Toluene extraction	1.2–100		Buss et al. (2016)
Softwood pellets						
Willow chips						
Miscanthus chips						
Demolition wood						
Arundo donax						
Sewage sludge						
Willow	Slow pyrolysis	500–700	Toluene extraction	0.6–1.5		Madej et al. (2016)
Miscanthus						
Wheat straw						
Sida hermaphrodita						
Pine wood	Slow pyrolysis	250–700	Dichloromethane extraction	0.19–0.86		Lyu et al. (2016)
Sewage sludge	Slow pyrolysis	500–700	Acetone and heptane extraction	0.6–1.1	81–126 ng L^{-1}	Zielinska and Oleszczuk (2016)

(*Continued*)

TABLE 19.3 (Continued)

Feedstock	Technolgy	Heating Temperature (°C)	PAH Extraction Method	Total PAH Concentration ($\mu g\ g^{-1}$)	Available PAH Concentration ($\mu g\ g^{-1}$)	Reference
Pinewood	Slow pyrolysis	400–750	Toluene extraction	0.4–1987	$12-85 \pm 11$ ng L^{-1}	Hale et al. (2012)
Spruce wood						
Beech wood						
Sugar beet						
Elephant grass						
Wheat husks						
Paper sludge						
Sewage sludge						
Rice husk	Slow pyrolysis	400–800		1.0–11.3	1.0–1113 $\mu g\ mL^{-1}$	Dunnigan et al. (2017)
Sludge	Microwave heating pyrolysis	400–800	Acetone and dichloromethane extraction	23–65		Lin et al. (2018)

more comprehensive picture on the levels and availability of PAHs in biochar. To our best knowledge, the earliest report on PAHs in biochar was conducted by Singh et al. (2010). They tried to measure total PAHs of 11 biochars derived from five feedstocks, including wood, leaves, paper sludge, poultry litter, and cow manure, but the concentrations were below the limit of detection. Fagernäs et al. (2012) reported that the total content of PAHs dominated with naphthalene and methylnaphthalenes in a birch derived biochar was 10 $\mu g\ g^{-1}$, but benzo[*a*]pyrene and benzo[*a*]anthracene were not observed or their concentrations were below the detection limit ($<0.1\ \mu g\ g^{-1}$). Hale et al. (2012) were the first deeply to quantify total PAHs of over 50 biochars produced at 250–900°C using various methods and biomass, with the total concentrations ranged from 0.07 to 45 $\mu g\ g^{-1}$. A similar level of total PAHs with 0.078–2.125 μg^{-1} in biochars produced from giant reed at 200–600 was also observed by Wang et al. (2013). After these reports, more and more studies were conducted to quantify PAH concentration and profiles in biochar. Oleszczuk et al. (2013) reported that the level of the sum of PAHs varied from 1.124 to 28.339 $\mu g\ g^{-1}$. Similarly, Fabbri et al. (2013) observed that the total PAH level of four biochars ranged between 1.2 and 19 $\mu g\ g^{-1}$, and Khan et al. (2015) found that the total PAH content of nine pulp sludge derived biochars varied from 0.4 to 236 $\mu g\ g^{-1}$. The biochar sample generated at 450°C for 60 min was found to contain the highest concentration of PAHs (236 $\mu g\ g^{-1}$) in the study reported by Khan et al. (2015). Recently, the highest concentration of PAH sum of 1987 $\mu g\ g^{-1}$ was noted by Hilber et al. (2017). Overall, the sum of PAH content in these biochars varied widely over six orders of magnitude, from 0.05 to 1987 $\mu g\ g^{-1}$ (Table 19.3). Moreover, Dutta et al. (2017) reviewed several studies and found that naphthalene was the most abundant PAH in biochar, followed by phenanthrene and fluorene.

19.2.2.3 Bioavailable Content of Polycyclic Aromatic Hydrocarbons in Biochar

As we know, bioavailable fractions of pollutants in soil or water are vital to consider because they are actually exposed to target organisms. From Table 19.3, we can see that these studies mainly focused on total concentrations of PAHs formed in biochar, and the bioavailable fractions of these compounds have received less attention. However, a clear increasing trend of the studies with respect to bioavailable content of PAHs are observed (Table 19.3). Obviously, the earlier reports about PAHs in biochar only focused on the total concentrations (Fagernäs et al., 2012; Keiluweit et al., 2012), but the newer studies began to focus on both total and bioavailable content of PAHs within biochar because of development of analytical technique (Dunnigan et al., 2017; Hilber et al., 2017). For example, Zielinska and Oleszczuk (2016) evaluated the effect of pyrolysis temperature on freely dissolved PAH contents in sewage sludge derived biochars and found that their concentrations is very low, with the range of 81–126 $ng\ L^{-1}$. Similar lower contents of bioavailable PAHs relative to the total PAHs were observed in other studies (Hale et al., 2012). Obviously, the bioavailable content of PAHs in biochars is much lower than those of the sum, but the concentrations are difficult to compare due to the different units in the different studies, such as 0.17–10.0 $\mu g\ g^{-1}$ (Hale et al., 2012) vs 12–81 $ng\ L^{-1}$ (Hilber et al., 2017). The much lower content of PAHs in biochar can be explained by two aspects. On one hand, PAHs strongly bond to biochar via $\pi-\pi$ interactions among the planar, aromatic PAHs, and condensed, aromatic sheets of biochar, which have been extensively proved in studies of PAH sorption by biochars (Lian and Xing, 2017). On the other hand, a portion of PAHs are occluded and completely locked up within the biochar micropores during their formation, rendering them completely unavailable (Hale et al., 2012). Overall, the limited data on bioavailability of PAHs in biochar is not very sufficient to draw definitive conclusions. Further studies are still needed to determine the bioavailable fractions of PAHs in biochars derived from various feedstock and pyrolysis conditions.

19.2.2.4 Influence of Feedstock and Pyrolysis Parameters on Polycyclic Aromatic Hydrocarbon Content in Biochar

Clearly, the concentration and profiles of PAHs retained in biochar during pyrolysis highly vary with biomass feedstock and pyrolytic conditions such as heating temperature, production method, and reactor. Almost all of the studies tried their best to establish the linkage between the formation and concentration of PAHs in biochar and the feedstock types and pyrolysis parameters. For example, to address the important knowledge gap of PAHs in biochar, Hale et al. (2012) measured total and bioavailable PAHs in 59 biochars that produced from 23 different feedstocks through slow pyrolysis using various laboratory and in situ equipment and via fast pyrolysis and gasification. They found that increasing pyrolysis time and temperature decreased PAH concentrations in biochars, and total PAH concentrations in the fast pyrolysis and gasification biochar were higher than those of the slow pyrolysis (Table 19.3). However, although several of these studies investigated the relationship between physicochemical properties and PAHs content or composition of biochars, no clear pattern of how strongly PAHs were bound to different biochars was found based on the biochars' physicochemical properties (Hale et al., 2012).

The type of feedstock is one of the key factors that controls biochar physicochemical characteristics. For the concentration and composition of PAHs present in the fresh biochar, they varied on the feedstock types, but the results from these studies were not always clear and consistent. Buss et al. (2016) reported that straw-derived biochar contained 5.8 times higher PAH concentrations than softwood-derived biochar. Hale et al. (2012) noted that total PAH concentrations in corn stover-derived biochars (3.27 $\mu g\ g^{-1}$) was higher than those in pine wood-derived biochars (0.07 $\mu g\ g^{-1}$). However, Keiluweit et al. (2012) found that the yield of total PAHs is similar for grass and wood biochars across the temperature range of 100–700°C, but significant differences was observed in the PAH composition. At the same pyrolysis temperature of 300°C, the most abundant PAHs in grass biochar are pyrene and chrysene, while 1,7-dimethylphenanthrene and retene dominated the PAHs in wood biochar. Oleszczuk et al. (2013) showed that the total PAH content clearly depended on the kind of material tested. It was reported that the highest concentration of the total PAHs was measured in miscanthus-derived biochar (28.339 $\mu g\ g^{-1}$), whereas the concentration in the coconut shell and wicker derived biochars with 1.124 and 1.224 $\mu g\ g^{-1}$ did not differ significantly from each other, respectively. Moreover, they also found that the composition of PAHs varied with the feedstock type, and 3-ring compounds (e.g., phenanthrene) accounted for 64.6%–82.6% of total PAHs content. Therefore, more studies are needed to explore the effect of feedstock type on PAH content and composition.

Pyrolysis temperature largely affected PAH concentrations in biochar. In general, temperatures of pyrolysis between 350 and 500°C produced maximum concentrations of PAHs in biochars (Table 19.3). For instance, Hale et al. (2012) observed that the PAH sum in biochar produced from pine wood at 900°C was significantly lower than those of other temperatures. Wang et al. (2013) noted the similar temperature-dependent trend in their biochars produced from giant reed. With increasing pyrolytic temperature, total PAH concentrations of the giant reed biochars significantly increased from 0.119 $\mu g\ g^{-1}$ at 200°C to 2.126 $\mu g\ g^{-1}$ at 350°C, then decreased to 1.041 $\mu g\ g^{-1}$ at 500°C and 0.166 $\mu g\ g^{-1}$ at 600°C. Consistently, Keiluweit et al. (2012) reported that the maximum for wood and grass-derived biochar occurs at 400 and 500°C, respectively. On the contrary, Buss et al. (2016) found that the heating temperature had no significant effect on PAH content in biochars produced at 350–650°C, regardless of feedstock type. An apparent increasing trend of PAHs along with the increasing pyrolysis temperature was reported by Dunnigan et al. (2017), which was explained as a complex pyrosynthetic formation route, the rate of which increases at elevated pyrolysis temperatures. Clear trends with regard to temperature affecting the composition of PAHs were not observed, probably due to the limited studies. For example, reported by Hale et al. (2012), the most abundant PAHs extracted from grass biochar are pyrene and chrysene at 300°C, phenanthrene and pyrene at 400°C, and phenanthrene and benzofluoranthrenes at 500°C. Buss et al. (2016) stated that the predominant PAH in their biochars were 3-ring PAHs, primarily phenanthrene. The temperature dependent variations of PAH content or composition can be explained by the two process that PAHs formed in biochar as discussed above.

Apart from pyrolytic temperature, other parameters such as duration time, pyrolysis type, equipment, and sweep gas also determine PAH concentration and composition in biochar (Table 19.3). Several studies have demonstrated that longer residence time possessed lower amounts of sorbed PAHs (Hale et al., 2012; Lyu et al., 2016). Fabbri et al. (2013) and Hale et al. (2012) proved that the biochars produced by slow pyrolysis from woody biomass possessed the lowest level of sorbed PAHs than those by fast pyrolysis and gasification from grass. This was because that PAHs generated by slow pyrolysis (generally with longer duration time, Table 19.1), may escape to the gaseous phase, whereas PAHs that are produced by gasification and fast pyrolysis may condense on the biochar material itself (Hale et al., 2012; Hilber et al., 2017). Additionally, biochars with the lower PAH concentrations were those produced in controlled reactors, not traditional kiln, drum, and stove setups in field conditions (Freddo et al., 2012; Hale et al., 2012; Hilber et al., 2012). Moreover, biochar is usually produced under nitrogen atmosphere, thus increasing gas flow significantly decreased PAH concentrations in biochar, because the formed PAHs can be swept out from the rectors (Buss et al., 2016). Additionally, CO_2 was also used as a sweep gas to control the PAH content in biochars. For example, Cho et al. (2015) found that the generation of condensable hydrocarbons was significantly reduced under presence of CO_2 at 500°C due to its blow and the unknown reaction between CO_2 and condensable hydrocarbons. Overall, it was concluded that besides feedstock type, pyrolysis conditions that addressed above jointly determined the PAH concentration in biochar, and more efforts are warranted further to explore the optimum conditions and biomass feedstock to produce biochar with little PAHs.

19.2.2.5 Negative Impact of Biochar Associated With Polycyclic Aromatic Hydrocarbons

Although low total levels and much lower levels of bioavailable PAHs have generally been detected in biochar (Table 19.3), a few studies have been greatly paid on negative impact of PAHs released from biochar due to their carcinogenic, teratogenic, and mutagenic characters. A few organisms, such as plants, alga, protozoa, and earthworm were used to evaluate the toxicological effects of biochar associated with PAHs (Bouqbis et al., 2017; Rombolà et al., 2015; Smith et al., 2013a; Wang et al., 2017b). For example, Rogovska et al. (2012) found that PAHs in aqueous extracts of biochar prepared by high-temperature gasification are at least partly responsible for the reduction in seedling growth. To our best knowledge, Oleszczuk et al. (2013) were the first to perform toxicological estimation of different biochars on different organisms including plant (*Lepidium sativum*), bacteria (*Vibrio fischeri* and 11 different strains from Microbial Assay for Risk Assessment (MARA)), alga (*Selenastrum capricornutum*), protozoa (*Tetrahymena thermophila*), and crustaceans (*Daphnia magna*). They noted that the most sensitive organism was *D. magna*, and a significant correlation was observed between the content of PAHs and toxicity in the case of *D. magna*, confirming the toxicological impact resulted from the inherent PAHs in biochars. Systematically, to assess the possible effects of biochar water extracted solution types on different organisms, Wang et al. (2017b) conducted biotoxicity tests of three fast pyrolytic biochar extract solutions (rice husk, saw dust, and *Acorus calamus*) on a microorganism (*Pseudomonas aeruginosa*), a plant (*Triticum* spp.), and an animal (*Caenorhabditis elegans*). Little toxic effect on all the tested organisms was observed for the biochars derived from rice husk and sawdust, whereas the biochar derived from *A. calamus* shows significant toxicity on all the tested organisms, probably due to that certain small aromatic molecules. Recently, the other approaches, such as Microtox test (Lyu et al., 2016) and cytotoxicity test (Sigmund et al., 2017), were also used to characterize toxic effects of biochar, which are greatly helpful to understand the underlying mechanisms responsible for the negative effects induced by biochar and its inherent PAHs.

Similar to VOCs discussed above, PAHs can have bactericidal properties that would adversely affect structure and function of soil microbial community (Zhu et al., 2017). For instance, Wang et al. (2013) first demonstrated that the remaining PAHs in low-temperature biochars (300–400°C) played a major role in reducing N_2O emission via inhibiting denitrification. Oleszczuk et al. (2013) used a MARA, a multi-species assay which allows measurement of toxic effects of chemicals and environmental samples, to test the toxicological effect of biochar to bacteria, including *V. fischeri* and 11 different strains from MARA. The results showed that the biochars had a toxic character toward more than a half of the test organisms, due to the relatively high level of PAHs (1.124–28.339 $\mu g\ g^{-1}$) retained in the biochars. However, although lots of studies attributed the negative impact of biochar on soil microorganism to the sorbed contaminants such as PAHs and VOCs (Lehmann et al., 2011; Zhu et al., 2017), limited evidence is available on this topic. It is evident that further research is required to understand fully the adverse effect of biochar sorbed PAHs and VOCs on the structure and function of the soil microbial community, as well as the underlying mechanisms. These efforts will be greatly helpful to produce high quality biochar and to use the biochar in safety.

19.2.2.6 Guideline of Biochar Associated With Polycyclic Aromatic Hydrocarbons

Acquisition of knowledge with relation to biochar related with PAHs will result in development of suitable standards that will allow the safety of biochar production and utilization. To date, several quality guidelines of biochar have been proposed by several biochar academic association, such as the International Biochar Initiative (IBI, 2012), European Biochar Certificate (EBC, 2012), and UK Biochar Center (UBC). However, only the permissible level of total concentrations of PAHs and heavy metals were included. For example, IBI uses threshold values of 6–300 mg kg^{-1} of 16 USEPA PAHs for biochars. The EBC set up the limit as 12 mg kg^{-1} of 16 USEPA PAHs for basic grade biochar, and up to 4 mg kg^{-1} for premium grade biochar. These threshold values are beneficial to select the high quality biochar as soil amendment. According to the above guideline, Khan et al. (2015) concluded that their biochars generated at 500–550°C for 30 or 120 min meet the IBI criterion. Obviously, most of the total concentrations of PAHs in biochars fall within these limit threshold values, with the exception of several biochars generated from gasification (Table 19.3). However, the PAHs and heavy metal in the biochars do not fully reflect the overall real ecotoxicity of all the inherent contaminants. Thus, more efforts are still needed to improve these biochar standards based on comprehensively exploring the bioavailability of all the contaminants in biochar and their combined toxicity contaminants.

19.2.3 Dioxins in Biochar

Dioxins, a family of chlorinated compounds, including polychlorinated dibenzo dioxins (PCDD) and polychlorinated dibenzo furans (PCDF) that share chemical structures and characteristics, are mostly generated on solid surfaces during pyrolysis when temperature is 200 – 400°C and the pyrolysis time is seconds (Stanmore, 2004). Dioxins are highly toxic and ubiquitous compounds, and originally, they were thought to be totally of anthropogenic origin (Kulkarni et al., 2008). To date, very little data has been published on dioxin concentrations in biochar. Apart from determining PAHs concentration, Hale et al. (2012) also detected dioxins retained in the biochars, and found that the total dioxin concentrations were very low (up to 92 pg g^{-1}) and the bioavailable concentrations were below the analytical limit of detection. Moreover, the food waste biochars contained higher levels of dioxins than the other biochars, possibly due to the higher content of chlorine in the food wastes. Lyu et al. (2016) reported that the concentrations of PCDD/DF varied among biochars produced at various temperatures of pyrolysis, and the highest concentrations of PCDD/DF of 612 pg g^{-1} were observed in a biochar produced at 300°C. However, Khan et al. (2015) considered that dioxins should be analyzed in their biochars that contained the highest concentration of PAHs (236 μg g^{-1}), but no compounds were detected within the minimum detection limit (0.1 pg g^{-1}), which might be attributed to the clean nature of the feedstock without chlorine and heavy metals. Additionally, the maximum allowed thresholds for dioxins within biochars have been established by IBI, EBC, and UBC, and they were 17, 20, and <20 ng kg^{-1}, respectively (EBC, 2012; IBI, 2012). Still, the data for dioxins in biochar is limited to get any general conclusion, and further studies are needed in future.

19.3 HEAVY METALS IN BIOCHAR

19.3.1 Total Content of Heavy Metals in Biochar

Biochar products typically consist of carbon and mineral components, and the minerals include different species of heavy metals, such as Pb, Cd, Cu, As, etc., which are generally derived from the feedstock. Heavy metals in biomass are mainly accumulated and concentrated in biochar during charring. Hossain et al. (2011) reported that the concentrations of toxic metals (Pb, Cd, Ni, Cr, As, and Se) in the biochars from wastewater sludge varied with the pyrolytic temperature, and these toxic metals were found to be enriched in the biochars. The main concerns of heavy metals in biochar primarily involve Pb, Cd, Cu, Mn, Zn, Ni, Cr, and As. The total contents of these heavy metals in biochars produced from different feedstock at different temperatures are summarized as drawn in Fig. 19.2. There are great differences in the total contents of heavy metals among the biochars, which are closely related to the inherent minerals in sewage sludge, animal manure, production waste, and green waste. Green waste biochars (e.g., crop straw, wood dust, and grass) contain rather lower contents of the above-mentioned heavy metals compared to the biochars from sewage sludge, animal manure, and production waste, implying their lower potential risk as soil amendments. The total contents of Pb, Cd, Cu, Mn, Zn, Ni, Cr, and As in the sewage sludge-derived biochars were in the range of 44–506, 2.6–10, 148–2361, 403–1543, 542–3368, 48–924, 55–1378, 3–51 mg kg^{-1}, respectively (Fig. 19.2). The concentrations of Cu, Ni, Cr, Cd, and Zn in some of biochars derived from sewage sludge and animal manure, are excessively great. These biochar samples should be forbidden to fertilize vegetables and food crops (CJ/T309-2009, China), thus possibly limiting land application of biochars. Notably, As contents in a large portion of biochars from sewage sludge and manure exceed its harmless concentration (300 mg kg^{-1}) for agricultural use (GB8172, China). But, rather limited samples can be summarized and referenced. Therefore, a thorough risk analysis of heavy metals in biochar products is critical before their application as soil amendments or remediation.

19.3.2 Speciation of Heavy Metals in Biochar

The bioavailability/eco-toxicity of heavy metals in environment is well known to predominantly depend on the chemical speciation of heavy metals. Tessier and BCR sequential extraction are widely used to determine the chemical speciation of heavy metals in soils and sediments (Rauret et al., 1999; Tessier et al., 1979). The corresponding chemical speciation of heavy metals is presented in Table 19.4. Presently, these two methods were also employed for examining the direct and potential effect fractions of heavy metals in biochar. Numerous works demonstrated that the biochars displayed lower content of direct effect fraction of heavy metals (F1 + F2 + F3 fractions in Tessier extraction; F1 + F2 fractions in BCR extraction) compared with the feedstock, and the direct effect fraction was transformed into relatively stable fractions (Devi and Saroha, 2014; Leng et al., 2014; Yuan et al., 2011). Devi and Saroha (2014) found that the direct effect fraction in the biochar prepared through slow pyrolysis gradually decreased with increasing temperature. By contrast, the liquefaction biochar produced at relatively low temperatures (<400°C) did not follow this temperature-dependent tendency (Chen et al., 2014; Leng et al., 2014).

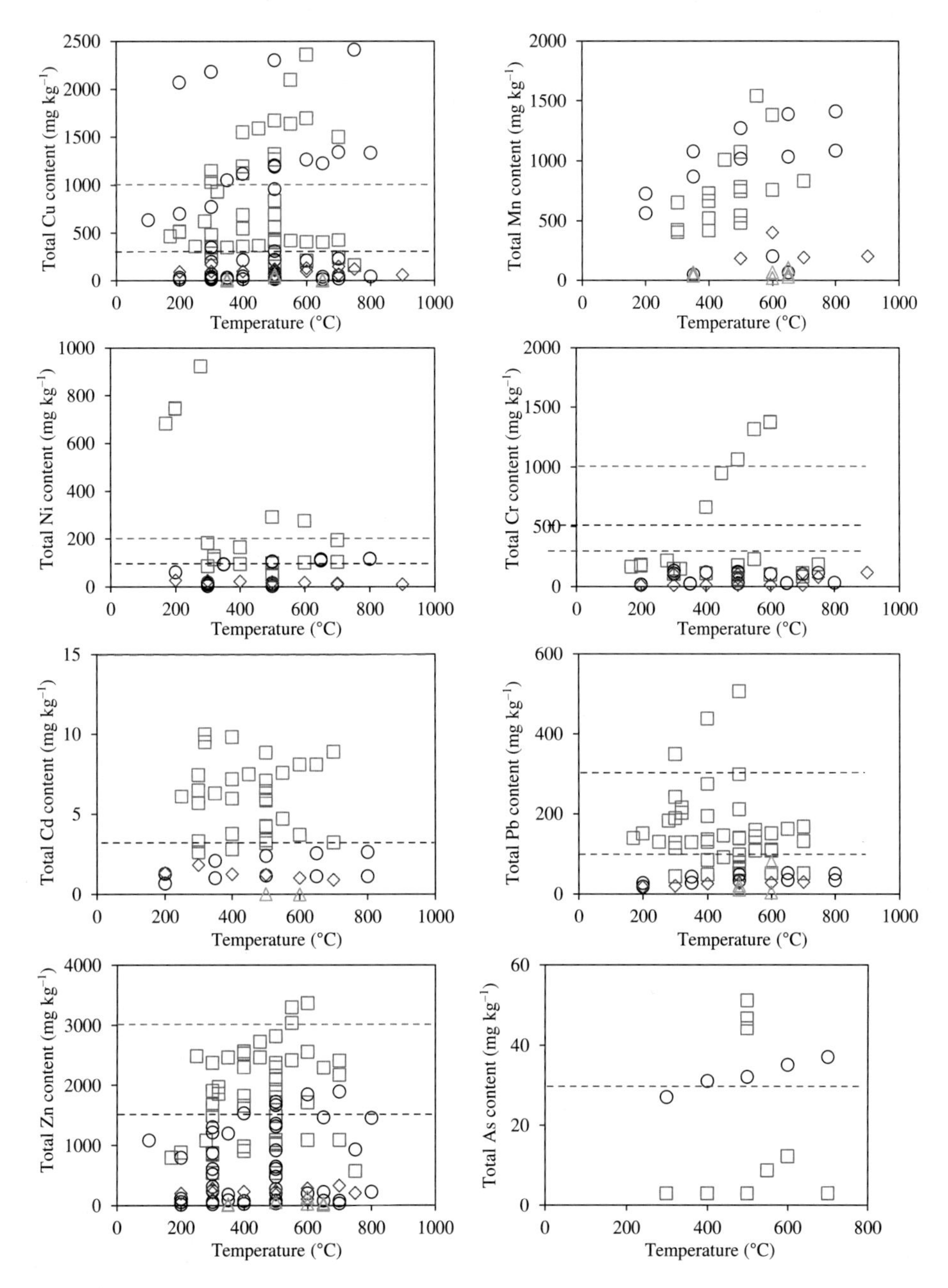

FIGURE 19.2 Total content of heavy metals in the reported biochars from: (□) sewage sludge; (○) manure; (◇) production waste; (△) green waste. A control standard for agricultural use (CJ/T309-2009 and GB8172, China): black dotted line, sludge is permitted to fertilize vegetables, food crops, oil crops, fruit trees, feed crops, and fiber crops; red dotted line, sludge is permitted to fertilize oil crops, fruit trees, feed crops, and fiber crops, and is forbidden to fertilize vegetables and food crops; blue dotted line, harmless concentration for organic fertilizer. Sources of dataset: *Ahmad et al. (2018); Cely et al. (2015); Chen et al. (2008); Chen et al. (2014); Devi and Saroha (2014); He et al. (2010); Higashikawa et al. (2016); Hossain et al. (2010); Hossain et al. (2011); Inyang et al. (2012); Jin et al. (2014); Jin et al. (2016); Jin et al. (2017); Leng et al. (2014); Leng et al. (2018); Lin et al. (2013); Lu et al. (2013); Lu et al. (2015); Meng et al. (2017); Peng et al. (2017); Shao et al. (2015); Shi et al. (2013); Yuan et al. (2011); Yuan et al. (2015); Zeng et al. (2018); Zhao et al. (2013).*

TABLE 19.4 Relationships Between Chemical Speciation (Tessier and BCR Extraction) and Bioavailability/Eco-Toxicity of Heavy Metals

Tessier Extraction (Tessier et al., 1979)	BCR Extraction (Rauret et al., 1999)	Bioavailability Eco-Toxicity
Exchangeable fraction (F1)	Exchangeable and acid soluble fraction (F1′)	Direct effect
Carbonate fraction (F2)		
Fe/Mn oxide fraction (F3)	Reducible fraction (F2′)	
Organic matter bound fraction (F4)	Oxidizable fraction (F3′)	Potential effect
Residue fraction (F5)	Residue fraction (F4′)	No effect

Whether the pyrolysis biochars or the liquefaction biochars, no specific pattern can be ruled between their direct effect fraction of all the toxic metals in this chapter and the pyrolytic temperature (Fig. 19.2).

The chemical speciation of heavy metals in biochar may be likely governed by the molecular speciation of inherent heavy metals in feedstock and the species transformation during the biomass charring. However, for the present, the reports regarding the molecular species of heavy metals in biochars are relatively limited. Cu-citrate and Cu-glutathione were dominant species in animal manure and, for the pyrolyzed products, the content of these two species of Cu was reduced, while CuO, CuS, and Cu_2S could be detected using synchrotron-based X-ray spectroscopy (Lin et al., 2013, 2017). Also, the fraction of Zn bound to organic matter (i.e., Zn-citrate and Zn-acetate) in the carbonaceous materials decreased relative to the feedstock (26.7%–69.2% vs 40.2%–76.7%), and ZnS increased by 6.6%–25%. The transformation of Cu- and Zn-organic matter fraction into their mineral fraction is mainly due to the formation of crystalline and aromatic carbon phase from carbonizing the amorphous organic carbon. Liu et al. (2017a) used X-ray diffractometer to determine the mineral crystals in the furfural production residue-derived biochars, such as $Al(H_2PO_4)_3$ in relatively low-temperature biochars (300–400°C), $AlPO_4$ in relatively high-temperature biochars (500–600°C), and $Pb_2(SO_4)O$ and $Pb_2P_2O_7$ in the biochars (300–600°C). Furthermore, Zeng et al. (2018) reported presence of numerous species of heavy minerals in the manure-produced biochars, including Cr_2O_3, $ZnMn_2O_4$, $Cu_3(PO_4)(OH)_3$, $Ca_{7.29}Pb_{2.21}(PO_4)_3(OH)$, and $C_{10}H_{12}Cr_2N_2O_7$. In addition, part of mineral fraction may be encapsulated in the carbon matrixes and porous structures of biochars, and even the minerals react with biomass carbon to form inorganic–organic composites (Lin et al., 2013; Liu et al., 2017a; Zhao et al., 2016). These fractions of heavy metals in biochar can be considered to be relatively resistant to solubilization and bio-utilization. Overall, the conversion of molecular speciation of heavy metals during the different carbonization procedures and conditions of feedstock into biochar and the corresponding mechanisms need to be further investigated. These are crucial and useful for understanding and predicting the potential fate and bioavailability of heavy metals in biochar in the environment.

19.3.3 Environmental Risk of Biochar Associated With Heavy Metals

Examining contamination degree and assessing environmental risk of heavy metals in biochar are essential before their application as soil amendments. Little study paid attention on heavy metals' environmental risk of biochars from green waste (e.g., crop, wood, and grass residues), mainly due to low content of heavy metals in these biochars (Fig. 19.2). The potential risk of heavy metals was assessed for the sewage sludge and manure biochars in most of previous works. Risk assessment code (RAC) (Huang and Yuan, 2016) is a speciation index that assesses ecological risk of single-metal basing on BCR and Tessier extractions (Table 19.5). As shown in Fig. 19.3, for a given metal, the sewage sludge biochars had different RAC values, implying different risks in the environment. Most of our gathered data show that the biochars from sewage sludge have the risk range from low to very high. But, there were several studies that reported no risk of Pb, Cd, Cu, and Cr in the biochars. Compared with the pyrolysis biochars from sewage sludge, the liquefaction biochars had relatively lower RAC values of heavy metals.

The index of geo-accumulation (I_{geo}), potential ecological risk factor (ER) and risk index (RI), are often employed to indicate the potential risk of all total content of heavy metals (Table 19.5). Zhai et al. (2014) reported different I_{geo} values of <0 for Cd, Cu, and Zn (uncontaminated) and 4.33–4.56 for Pb (heavily to extremely contaminated); different ER values of 0–4.82 (low risk) for Cd, Cu, and Zn, and 150–176 for Pb (considerable to high risk); and different ER values of 155–177 for all the tested heavy metals (moderate risk) in the liquefaction biochars from sewage sludge at 300–350°C. Dissimilarly, Cd was at heavily contaminated and very high risk level with I_{geo} and ER of 3.83–4.03 and 638–733, while Pb was at uncontaminated and low risk level with I_{geo} and ER below 0 in the sewage sludge biochars prepared by liquefaction at 300–350°C (Chen et al., 2014). For the pyrolysis biochars, it is commonly considered that the contamination degree and potential risk of heavy metals can be decreased with increasing temperature, as indicated with lower I_{geo}, ER and RI indexes for lower-temperature biochar (Devi and Saroha, 2014; Jin et al., 2016; Yuan et al., 2015). The biochars pyrolyzed from different sewage sludge exhibited various potential ecological risk. Low potential ecological risk (ER, 1.98–17.0) of Pb, Cu, Zn, Ni, and Cr in the sewage sludge biochars (200–700°C) was reported by Devi and Saroha (2014), while Cd showed high risk (ER, 371–455) in the low-temperature biochars (300–400°C) and low or moderate risk (ER, 48.6–107) in the high-temperature biochars (500–700°C). In the study reported by Jin et al. (2016), Pb, Ni, and Cr in the sewage sludge biochars (400–600°C) also possessed low potential ecological risk (ER, 0.81–40); the potential ecological risk of Cu was high for the biochar (400 °C, ER, 276), considerable for the biochars (450–500°C, ER, 129–153), and low for the biochars (550–600°C, ER, 59–76); and Zn was at low risk level with the ER range of 51–59. These differences in the contamination degree and potential risk of heavy metals in the biochars from sewage sludge may be closely associated with the original species of heavy metals in the feedstock and the transformation of heavy metals' species during pyrolysis process.

In addition, many studies performed toxicity characteristic leaching procedure (TCLP) and diethylenetriaminepentaacetic acid (DTPA) extraction to investigate the leaching characteristics and estimate the bioavailability and possible risk of heavy metals in biochar. For instance, the TCLP concentration of Cr in the biochars pyrolyzed from paper mill effluent treatment plant sludge at 200–500°C was in the range of 2.39–2.75 mg L^{-1}, and that of Cu

TABLE 19.5 Indices for the Ecological Risk Assessment

I_{goe}	Contamination Degree	ER	Risk Degree	RI	Risk Degree	RAC	Risk Degree
≤0	Uncontaminated	<40	Low	<150	Low	≤1	No
0–1	Uncontaminated to moderately uncontaminated	40–80	Moderate	150–300	Moderate	1–10	Low
1–2	Moderately contaminated	80–160	Considerable	300–600	Considerable	10–30	Middle
2–3	Moderately to heavily contaminated	160–320	High	≥600	High	30–50	High
3–4	Heavily contaminated	≥320	Very high			>50	Very high
4–5	Heavily to extremely contaminated						
>5	Extremely contaminated						
Reference	Müller (1969)	Hakanson (1980)				Perin et al. (1985)	

ranged from 3.72 to 16.7 mg L^{-1}, greater than the permissible limits by Indian standards for industrial and sewage effluents discharge (2.0 mg L^{-1} for Cr and 3.0 mg L^{-1} for Cu), as well as Zn for the biochars at 200 and 300°C; but, other heavy metals showed insignificant risk to surface water, including Pb, Cd, and Ni (Devi and Saroha, 2014). For the liquefaction biochars from sewage sludge at 200 and 280°C, the concentrations of Pb, Cd, Cu, Zn, and Cr in the TCLP tests were lower than the permissible upper limits (USEPA, SW-846, the test methods for evaluating solid waste, physical/chemical methods), except Ni in the biochar at 200°C. Lu et al. (2013) reported that charring sewage sludge into biochars (300–500°C) could significantly mitigate the environmental risk of heavy metals, supported by the decreased percentage of DTPA-extractable metal content to the total content of Pb, Cd, Zn, Cu, and Mn from 16%, 54%, 82%, 43%, and 34% in sewage sludge to 0%–4%, 0%–1.2%, 0%–9%, 0%–3%, and 0%–4% in the biochars, respectively. Similarly, Zeng et al. (2018) also reported that the bioavailability and leaching toxicity of heavy metals (Pb, Cd, Zn, Cu, Ni, and Cr) in swine and goat manures were reduced after they were carbonized into biochars (200–800°C), according to the decreased data of DTPA and TCLP concentrations. However, in the studies, the TCLP concentrations of Pb, Zn, and Ni from these manure-derived biochars exceeded the threshold values (USEPA, 1992). Overall, the specific and exact pattern for the environmental risk of heavy metals in biochar cannot be drawn based on the current reports, thus more investigations are still needed.

19.4 PERSISTENT FREE RADICALS IN BIOCHAR

Apart from the conventional pollutants, such as PAHs and heavy metals addressed above, some emerging contaminants, such as nanoparticles (Xu et al., 2017) and persistent

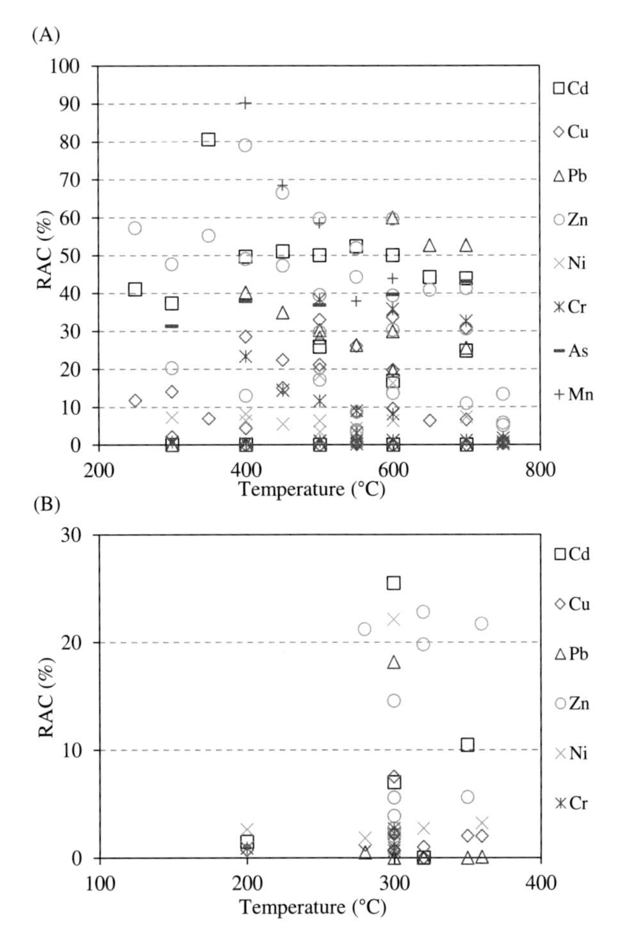

FIGURE 19.3 Risk assessment code (RAC) of heavy metals in the reported biochars from sewage sludge and manure produced by pyrolysis (A) and liquefaction (B). RAC (%) $= 100 \times \frac{F1 + F2 + F3}{F1 + F2 + F3 + F4 + F5}$ or $\frac{F1' + F2'}{F1' + F2' + F3' + F4'}$. The risk index of heavy metals in the biochars can be classified with RAC as no risk, low risk, medium risk, high risk, and very high risk, when the RAC value ranges <1, 1–10, 11–30, 31–50, and >50, respectively. Sources of dataset: *Chen et al. (2014); He et al. (2010); Jin et al. (2016); Jin et al. (2017); Leng et al. (2014); Leng et al. (2018); Lin et al. (2017); Meng et al. (2017); Shao et al. (2015); Shi et al. (2013); Yuan et al. (2011); Yuan et al. (2015); Zhai et al. (2014); Zeng et al. (2018).*

free radicals (PFR) (Fang et al., 2014; Liao et al., 2014; Yang et al., 2016) have recently attracted a great deal of research attention. PFRs, such as semiquinones, cyclopentadienyls, and phenoxyls, with the half-lives of hours and days, are generated and stabilized during the pyrolysis of biochar (Fang et al., 2015; Liao et al., 2014). These PFRs can induce toxicity to microbes and plants. For example, Liao et al. (2014) first reported generation of PFRs in biochars. They found that all the biochars derived from three common types of biomass, namely corn stalks, rice, and wheat straws, two model biopolymer materials, including cellulose and lignin, contained significant amount of PFRs. Even after 1 month, the PFRs only decreased less than 10%. Furthermore, they proved that these radicals, with the relatively long lifetime in biochar particles, could induce reactive oxygen species in water, and thus have significant inhibitions on plant seed germination. Similarly, Fang et al. (2015) examined the effects of metals such as Fe^{3+} and Cu^{2+} and phenolic compounds (hydroquinone, catechol, and phenol) loaded on biomass on the formation of PFRs in biochar. They reported that the metal and phenolic compound treatments control on both the contents and types of PFRs in biochar, thus it was proposed that manipulating the amount of metals and phenolic compounds in biomass may be an efficient method to regulate PFRs in biochar. Additionally, PFRs in biochar may have a positive contribution to organic contaminant degradation (Fang et al., 2014; Ouyang et al., 2017; Yang et al., 2016). For instance, Yang et al. (2016) deeply proved that the obvious degradation of *p*-nitrophenol in the presence of biochar was attributed to PFRs contained in biochar. Fang et al. (2014) observed that biochar could efficiently degrade 2-chlorobiphenyl in the presence of H_2O_2 via the enhanced generation of •OH, mainly resulting from PFR in biochars. These studies showed that the degradation of organic pollutants by PFR in biochar can be a common process, which is needed to consider in evaluating their fate and risk. However, few available studies limited drawing a general conclusion of PFR in biochar. Bicohar produced from different feedstock using different technology may show different PFR characteristics, and they may have different effects on soil organisms and organic contaminants. Therefore, the PFR characteristics of biochar and their associated risks must be evaluated individually, and more efforts are still needed.

19.5 POSSIBLE WAYS TO MITIGATE THE CONTAMINATION OF BIOCHAR

In order to avoid or mitigate possible negative effects on soil ecosystems, it is essential to manufacture biochars with low concentrations of contaminants, such as PAHs and heavy metals. However, most of the published studies have focused on formation, transformation, and concentration of these contaminants (Hale et al., 2012; Keiluweit et al., 2012; Zielinska and Oleszczuk, 2016), but few of them cared about how to suppress or avoid their formation. From the previous studies, biochar could potentially contain two types of contaminants, i.e., organic contaminant such as VOCs and PAHs, and heavy metals. On one hand, for heavy metals in biochar, they were generally originated from the feedstock that contained heavy metals, which are left in biochar during pyrolysis. For dioxins in biochar, they are frequently formed in the chlorine-rich feedstock (e.g., food waste). Therefore, careful selection of clean feedstock is very necessary to avoid or minimize these contaminants. On the other hand, pyrolysis conditions (e.g., temperature and residence time) largely affect formation and concentration of contaminants in biochar, thus the good feedstock should carefully match with conversion technology that has appropriate operating conditions, such as specifically temperature range and residence time. For example, based on data collected on 46 biochars contained PAHs, Buss et al. (2016) simply suggested that feed-stock selection and careful matching with conversion technology is necessary to ensure production of clean biochar. Similarly, Buss et al. (2016) suggested that biochar produced with clean feedstock at 500–550°C for 30 or 120 min meet the IBI criterion of PAHs. In this respect, industrial reactor, or controlled equipment is highly recommended to employ, rather than the traditional kiln.

In general, organic pollutants are retained on biochar during its production from re-condensation of pyrolysis vapors and liquids (Buss et al., 2015). Thus, one likely strategy to reduce PAHs and VOCs in biochar is to divert and collect gases and liquids separately, as proposed by Bucheli et al. (2015), which is a big challenge for designing the modern and scaled reactors for biochar production. Madej et al. (2016) proved that the complete removal of gas-phase pyrosynthesized PAHs resulted in biochars with low PAH concentrations. Apart from the reactor, inert gases such as N_2 and CO_2 are used to sweep out gases from the pyrolysis system, which also is a good strategy to minimize PAHs or VOCs in biochar (Buss et al., 2016). However, not all the carrier gas may reduce these contaminants. For example, presence of O_2 may enhance the PAH yields (Kwon and Castaldi, 2012). After biochar production, post-treatment strategies, such as drying and composting, are used to reduce the contaminants in biochar. Kołtowski and Oleszczuk (2015) dried biochar samples and found that increase of drying temperature from 100 to 300°C caused the total reduction of all PAHs content in biochars. Composting of biochar with organic materials, such as dairy manure, agricultural straw, and biogas residue, is recommended as a practicable method for producing biochar-based amendments to overcome these inherent deficiencies associated with contaminants, as well as to reduce the needs for synthetic fertilizer (Wang et al., 2017a; Xiao et al., 2017). However, available data about contents of PAHs or VOCs in the composted biochar is limited. Moreover, the post-treatment will increase cost of biochar manufacture, which is not conducive for biochar application. Therefore, apart from selecting clean feedstock, it is very necessary to minimize the contaminants during biochar production.

19.6 CONCLUSION AND RECOMMENDATION

Biochar is promoted not only as a technology that can improve soil quality and productivity and sequestrate carbon, but also a management tool that can achieve the multiple goals, including mitigation of global climate change, high-value utilization of waste resources, environmental remediation, and food safety. However, not all biochars can obtain the good outcomes, because of the complicated interaction between biochar and soil and the inherent contaminants in biochar. Therefore, the contaminants generated in biochar should be paid more attention. From the published literature, it is clearly shown that the conventional pollutants such as VOCs, PAHs, dioxins, and heavy metals, and the emerging contaminants, such as nanoparticles and PFRs, can be formed during biochar production. The total and bioavailable contents of these conventional pollutants in biochar are dependent on feedstock and pyrolysis conditions, but few general conclusions can be obtained due to the limited studies, especially for the bioavailable contents of these contaminants. For the emerging contaminants in biochar, such as PFRs, it seems that they have negative effects on plant or soil microorganism, but positive effects on organic pollutant degradation. Several studies have proved that the inherent contaminants in biochar may caused adverse influence on organisms, such as plant, soil microbe, and alga, and even one study reported showing direct observation of the cytotoxicity of biochar particles. These studies highlighted the environmental risk that resulted from biochar.

Consequently, several quality guidelines of biochar and possible strategies to mitigate biochar contamination are highly proposed to minimize the risk of biochar exposure during its production and application.

In summary, the content and composition of contaminants in fresh biochar must be thoroughly considered before biochar application, together with the interaction between biochar and soil biota. Currently, there are insufficient data in the literature to draw conclusions concerning contaminants in biochar, and thus further studies need to be fill these gaps. First, the content and composition of contaminants should be thoroughly characterized for more biochars derived from various feedstock and pyrolysis conditions. This will help to elucidate the link between thermal conversion of biomass and resulted biochar that contained contaminants. Meanwhile, it is also very beneficial for testing the content and composition of contaminants in the biochar-based functional materials, such as biochar-based nano-composite, magnetic biochar, composted biochar. Moreover, besides the total content of the contaminants in biochar, the bioavailable fractions must be well measured. Second, it is urgent to develop standard methods for extracting and measuring these contaminants. Various methods that were used to extract and detect VOCs and PAHs in biochar resulted in difficulty in comparison of the available data in the published literature. Third, biological tests, which were frequently treated as an alternative to chemical analyses, should be mandatorily performed before biochar application. The biological tests with more species of organisms may permit knowing the possible toxic effect on organisms induced by the contaminants that are contained in biochar. Additionally, several simple and fast biological tests, such as germination test and earthworm avoidance test, should be proposed to check quickly the quality and safety of biochar in practical applications. Moreover, long-term risk assessments of biochar contained contaminants such as PAHs and VOCs in soils are needed. Based on the above-mentioned aspects, last but not the least, the guideline or standard for biochar related to the inherent contaminant should be developed, which should involve all the contaminants reported in biochar.

References

Ahmad, Z., Gao, B., Mosa, A., Yu, H., Yin, X., Bashir, A., et al., 2018. Removal of Cu(II), Cd(II) and Pb(II) ions from aqueous solutions by biochars derived from potassium-rich biomass. J. Clean. Prod. 180, 437–449.

Babu, B.V., 2008. Biomass pyrolysis: A state-of-the-art review. Biofuels Bioprod. Biorefining 2 (5), 393–414.

Becker, R., Dorgerloh, U., Helmis, M., Mumme, J., Diakite, M., Nehls, I., 2013. Hydrothermally carbonized plant materials: Patterns of volatile organic compounds detected by gas chromatography. Bioresour. Technol. 130, 621–628.

Bouqbis, L., Daoud, S., Koyro, H.W., Kammann, C.I., Ainlhout, F.Z., Harrouni, M.C., 2017. Phytotoxic effects of argan shell biochar on salad and barley germination. Agric. Nat. Resour. 51 (4), 247–252.

Bucheli, T.D., Hilber, I., Schmidt, H.-P., 2015. Polycyclic aromatic hydrocarbons and polychlorinated aromatic compounds in biochar. In: Lehmann, J., Joseph, S. (Eds.), Biochar for Environmental Management: Science, Technology and Implementation. Earthscan Publications Ltd, London, UK, pp. 595–624.

Buss, W., Mašek, O., 2014. Mobile organic compounds in biochar – A potential source of contamination – Phytotoxic effects on cress seed (*Lepidium sativum*) germination. J. Environ. Manage. 137 (0), 111–119.

Buss, W., Mašek, O., 2016. High-VOC biochar—effectiveness of post-treatment measures and potential health risks related to handling and storage. Environ. Sci. Pollut. Res. 23 (19), 19580–19589.

Buss, W., Mašek, O., Graham, M., Wüst, D., 2015. Inherent organic compounds in biochar–Their content, composition and potential toxic effects. J. Environ. Manage. 156, 150–157.

Buss, W., Graham, M.C., MacKinnon, G., Mašek, O., 2016. Strategies for producing biochars with minimum PAH contamination. J. Anal. Appl. Pyrol. 119, 24–30.

Cely, P., Gascó, G., Paz-Ferreiro, J., Méndez, A., 2015. Agronomic properties of biochars from different manure wastes. J. Anal. Appl. Pyrol. 111, 173–182.

Chen, Y., Liu, G., Wang, L., Kang, Y., Yang, J., 2008. Occurrence and fate of some trace elements during pyrolysis of Yima coal, China. Energ. Fuel 22 (6), 3877–3882.

Chen, H., Zhai, Y., Xu, B., Xiang, B., Zhu, L., Qiu, L., et al., 2014. Fate and risk assessment of heavy metals in residue from co-liquefaction of *Camellia oleifera* cake and sewage sludge in supercritical ethanol. Bioresour. Technol. 167, 578–581.

Chen, Y., Zhang, X., Chen, W., Yang, H., Chen, H., 2017. The structure evolution of biochar from biomass pyrolysis and its correlation with gas pollutant adsorption performance. Bioresour. Technol. 246, 101–109.

Cho, D.W., Cho, S.H., Song, H., Kwon, E.E., 2015. Carbon dioxide assisted sustainability enhancement of pyrolysis of waste biomass: a case study with spent coffee ground. Bioresour. Technol. 189, 1–6.

Cole, D.P., Smith, E.A., Lee, Y.J., 2012. High-resolution mass spectrometric characterization of molecules on biochar from pyrolysis and gasification of switchgrass. Energy Fuels 26 (6), 3803–3809.

Devi, P., Saroha, A.K., 2014. Risk analysis of pyrolyzed biochar made from paper mill effluent treatment plant sludge for bioavailability and eco-toxicity of heavy metals. Bioresour. Technol. 162, 308–315.

Dunnigan, L., Morton, B.J., van Eyk, P.J., Ashman, P.J., Zhang, X., Hall, P.A., et al., 2017. Polycyclic aromatic hydrocarbons on particulate matter emitted during the co-generation of bioenergy and biochar from rice husk. Bioresour. Technol. 244 (Part 1), 1015–1023.

Dutta, T., Kwon, E., Bhattacharya, S.S., Jeon, B.H., Deep, A., Uchimiya, M., et al., 2017. Polycyclic aromatic hydrocarbons and volatile organic compounds in biochar and biochar-amended soil: a review. GCB Bioenergy 9, 990–1004.

EBC, 2012. European Biochar Certificate – Guidelines for a Sustainable Production of Biochar. European Biochar Foundation (EBC), Arbaz, Switzerland. Available from: https://doi.org/10.13140/RG.2.1.4658.7043.

Fabbri, D., Rombolà, A.G., Torri, C., Spokas, K.A., 2013. Determination of polycyclic aromatic hydrocarbons in biochar and biochar amended soil. J. Anal. Appl. Pyrol. 103 (0), 60–67.

Fagernäs, L., Kuoppala, E., Simell, P., 2012. Polycyclic aromatic hydrocarbons in birch wood slow pyrolysis products. Energy Fuels 26 (11), 6960–6970.

Fang, G., Gao, J., Liu, C., Dionysiou, D.D., Wang, Y., Zhou, D., 2014. Key role of persistent free radicals in hydrogen peroxide activation by biochar: implications to organic contaminant degradation. Environ. Sci. Technol. 48 (3), 1902–1910.

Fang, G., Liu, C., Gao, J., Dionysiou, D.D., Zhou, D., 2015. Manipulation of persistent free radicals in biochar to activate persulfate for contaminant degradation. Environ. Sci. Technol. 49 (9), 5645–5653.

Farrell, M., Kuhn, T.K., Macdonald, L.M., Maddern, T.M., Murphy, D.V., Hall, P.A., et al., 2013. Microbial utilisation of biochar-derived carbon. Sci. Total Environ. 465, 288–297.

Freddo, A., Cai, C., Reid, B.J., 2012. Environmental contextualisation of potential toxic elements and polycyclic aromatic hydrocarbons in biochar. Environ. Pollut. 171, 18–24.

Ghidotti, M., Fabbri, D., Hornung, A., 2017a. Profiles of volatile organic compounds in biochar: insights into process conditions and quality assessment. ACS Sustain. Chem. Eng. 5 (1), 510–517.

Ghidotti, M., Fabbri, D., Mašek, O., Mackay, C.L., Montalti, M., Hornung, A., 2017b. Source and biological response of biochar organic compounds released into water; relationships with bio-oil composition and carbonization degree. Environ. Sci. Technol. 51 (11), 6580–6589.

Glaser, B., Parr, M., Braun, C., Kopolo, G., 2009. Biochar is carbon negative. Nat. Geosci. 2 (1), 2.

Gul, S., Whalen, J.K., Thomas, B.W., Sachdeva, V., Deng, H., 2015. Physico-chemical properties and microbial responses in biochar-amended soils: mechanisms and future directions. Agric. Ecosyst. Environ. 206, 46–59.

Hakanson, L., 1980. Ecological risk index for aquatic pollution control: a sedimentological approach. Water Res. 14, 975–1001.

Hale, S.E., Lehmann, J., Rutherford, D., Zimmerman, A., Bachmann, R.T., Shitumbanuma, V., et al., 2012. Quantifying the total and bioavailable PAHs and dioxins in biochars. Environ. Sci. Technol. 46 (5), 2830–2838.

He, Y., Zhai, Y., Li, C., Yang, F., Chen, L., Fan, X., et al., 2010. The fate of Cu, Zn, Pb and Cd during the pyrolysis of sewage sludge at different temperatures. Environ. Technol. 31 (5), 567–574.

Higashikawa, F.S., Conz, R.F., Colzato, M., Cerri, C.E.P., Alleoni, L.R.F., 2016. Effects of feedstock type and slow pyrolysis temperature in the production of biochars on the removal of cadmium and nickel from water. J. Clean. Prod. 137, 965–972.

Hilber, I., Blum, F., Leifeld, J., Schmidt, H.-P., Bucheli, T.D., 2012. Quantitative determination of PAHs in biochar: a prerequisite to ensure its quality and safe application. J. Agric. Food Chem. 60 (12), 3042–3050.

Hilber, I., Mayer, P., Gouliarmou, V., Hale, S.E., Cornelissen, G., Schmidt, H.-P., et al., 2017. Bioavailability and bioaccessibility of polycyclic aromatic hydrocarbons from (post-pyrolytically treated) biochars. Chemosphere 174, 700–707.

Hossain, M.K., Strezov, V., Chan, K.Y., Ziolkowski, A., Nelson, P.F., 2011. Influence of pyrolysis temperature on production and nutrient properties of wastewater sludge biochar. J. Environ. Manage. 92 (1), 223–228.

Huang, H., Yuan, X., 2016. The migration and transformation behaviors of heavy metals during the hydrothermal treatment of sewage sludge. Bioresour. Technol. 200, 991–998.

Huber, G.W., Iborra, S., Corma, A., 2006. Synthesis of transportation fuels from biomass: chemistry, catalysts, and engineering. Chem. Rev. 106 (9), 4044–4098.

IBI, 2012. Guidelines for Specifications of Biochars for Use in Soils. <https://www.biochar-international.org>.

Inyang, M., Gao, B., Yao, Y., Xue, Y., Zimmerman, A.R., Pullammanappallil, P., et al., 2012. Removal of heavy metals from aqueous solution by biochars derived from anaerobically digested biomass. Bioresour. Technol. 110, 50–56.

Jin, H., Arazo, R.O., Gao, J., Capareda, S., Chang, Z., 2014. Leaching of heavy metals from fast pyrolysis residues produced from different particle sizes of sewage sludge. J. Anal. Appl. Pyrol. 109, 168–175.

Jin, J., Li, Y., Zhang, J., Wu, S., Cao, Y., Liang, P., et al., 2016. Influence of pyrolysis temperature on properties and environmental safety of heavy metals in biochars derived from municipal sewage sludge. J. Hazard. Mater. 320, 417–426.

Jin, J., Wang, M., Cao, Y., Wu, S., Liang, P., Li, Y., et al., 2017. Cumulative effects of bamboo sawdust addition on pyrolysis of sewage sludge: Biochar properties and environmental risk from metals. Bioresour. Technol. 228, 218–226.

Kamal, M.S., Razzak, S.A., Hossain, M.M., 2016. Catalytic oxidation of volatile organic compounds (VOCs): a review. Atmos. Environ. 140, 117–134.

Keiluweit, M., Kleber, M., Sparrow, M.A., Simoneit, B.R.T., Prahl, F.G., 2012. Solvent-extractable polycyclic aromatic hydrocarbons in biochar: influence of pyrolysis temperature and feedstock. Environ. Sci. Technol. 46 (17), 9333–9341.

Khan, A., Mirza, M., Fahlman, B., Rybchuk, R., Yang, J., Harfield, D., et al., 2015. Mapping thermomechanical pulp sludge (TMPS) biochar characteristics for greenhouse produce safety. J. Agric. Food Chem. 63 (5), 1648–1657.

Kołtowski, M., Oleszczuk, P., 2015. Toxicity of biochars after polycyclic aromatic hydrocarbons removal by thermal treatment. Ecol. Eng. 75, 79–85.

Kulkarni, P.S., Crespo, J.G., Afonso, C.A.M., 2008. Dioxins sources and current remediation technologies: a review. Environ. Int. 34 (1), 139–153.

Kuppusamy, S., Thavamani, P., Megharaj, M., Venkateswarlu, K., Naidu, R., 2016. Agronomic and remedial benefits and risks of applying biochar to soil: current knowledge and future research directions. Environ. Int. 87, 1–12.

Kwon, E.E., Castaldi, M.J., 2012. Mechanistic understanding of polycyclic aromatic hydrocarbons (PAHs) from the thermal degradation of tires under various oxygen concentration atmospheres. Environ. Sci. Technol. 46 (23), 12921–12926.

Lamichhane, S., Bal Krishna, K.C., Sarukkalige, R., 2016. Polycyclic aromatic hydrocarbons (PAHs) removal by sorption: a review. Chemosphere 148, 336–353.

Lee, Y., Park, J., Ryu, C., Gang, K.S., Yang, W., Park, Y.-K., et al., 2013. Comparison of biochar properties from biomass residues produced by slow pyrolysis at 500°C. Bioresour. Technol. 148, 196–201.

Lehmann, J., 2007. A handful of carbon. Nature 447 (5), 143–144.

Lehmann, J., Joseph, S., 2015. Biochar for Environmental Management: Science, Technology and Implementation, 2nd ed Earthscan Publications Ltd., London, UK.

Lehmann, J., Rillig, M.C., Thies, J., Masiello, C.A., Hockaday, W.C., Crowley, D., 2011. Biochar effects on soil biota – a review. Soil Biol. Biochem. 43 (9), 1812–1836.

Leng, L., Yuan, X., Huang, H., Jiang, H., Chen, X., Zeng, G., 2014. The migration and transformation behavior of heavy metals during the liquefaction process of sewage sludge. Bioresour. Technol. 167, 144–150.

Leng, L., Leng, S., Chen, J., Yuan, X., Li, J., Li, K., et al., 2018. The migration and transformation behavior of heavy metals during co-liquefaction of municipal sewage sludge and lignocellulosic biomass. Bioresour. Technol. 259, 156–163.

Lian, F., Xing, B., 2017. Black carbon (biochar) in water/soil environments: molecular structure, sorption, stability, and potential risk. Environ. Sci. Technol. 51 (23), 13517–13532.

Liao, S., Pan, B., Li, H., Zhang, D., Xing, B., 2014. Detecting free radicals in biochars and determining their ability to inhibit the germination and growth of corn, wheat and rice seedlings. Environ. Sci. Technol. 48 (15), 8581–8587.

Lin, Q., Liang, L., Wang, L.H., Ni, Q.L., Yang, K., Zhang, J., et al., 2013. Roles of pyrolysis on availability, species and distribution of Cu and Zn in the swine manure: chemical extractions and high-energy synchrotron analyses. Chemosphere 93 (9), 2094–2100.

Lin, Q., Xu, X., Wang, L.H., Chen, Q., Fang, J., Shen, X.D., et al., 2017. The speciation, leachability and bioaccessibility of Cu and Zn in animal manure-derived biochar: effect of feedstock and pyrolysis temperature. Front. Environ. Sci. Eng. 11 (3), 5.

Lin, K., Lai, N., Zeng, J., Chiang, H., 2018. Residue characteristics of sludge from a chemical industrial plant by microwave heating pyrolysis. Environ. Sci. Pollut. Res. 25 (7), 6487–6496.

Liu, W., Jiang, H., Yu, H., 2015. Development of biochar-based functional materials: toward a sustainable platform carbon material. Chem. Rev. 115 (22), 12251–12285.

Liu, G., Chen, L., Jiang, Z., Zheng, H., Dai, Y., Luo, X., et al., 2017a. Aging impacts of low molecular weight organic acids (LMWOAs) on furfural production residue-derived biochars: porosity, functional properties, and inorganic minerals. Sci. Total Environ. 607–608, 1428–1436.

Liu, W., Li, W., Jiang, H., Yu, H., 2017b. Fates of chemical elements in biomass during its pyrolysis. Chem. Rev. 117 (9), 6367–6398.

Lu, M., Mulholland, J.A., 2004. PAH growth from the pyrolysis of CPD, indene and naphthalene mixture. Chemosphere 55 (4), 605–610.

Lu, H., Zhang, W., Wang, S., Zhuang, L., Yang, Y., Qiu, R., 2013. Characterization of sewage sludge-derived biochars from different feedstocks and pyrolysis temperatures. J. Analyt. Appl. Pyrol. 102, 137–143.

Lu, T., Yuan, H., Wang, Y., Huang, H., Chen, Y., 2015. Characteristic of heavy metals in biochar derived from sewage sludge. J. Mater. Cycles Waste 18 (4), 725–733.

Lyu, H., He, Y., Tang, J., Hecker, M., Liu, Q., Jones, P.D., et al., 2016. Effect of pyrolysis temperature on potential toxicity of biochar if applied to the environment. Environ. Pollut. 218, 1–7.

Müller, G., 1969. Index of geoaccumulation in sediments of the Rhine River. GeoJournal 108 (2), 108–118.

Ma, Y., Harrad, S., 2015. Spatiotemporal analysis and human exposure assessment on polycyclic aromatic hydrocarbons in indoor air, settled house dust, and diet: a review. Environ. Int. 84, 7–16.

Madej, J., Hilber, I., Bucheli, T.D., Oleszczuk, P., 2016. Biochars with low polycyclic aromatic hydrocarbon concentrations achievable by pyrolysis under high carrier gas flows irrespective of oxygen content or feedstock. J. Analyt. Appl. Pyrol. 122, 365–369.

Manyà, J.J., 2012. Pyrolysis for biochar purposes: a review to establish current knowledge gaps and research needs. Environ. Sci. Technol. 46 (15), 7939–7954.

Meng, J., Wang, L., Zhong, L., Liu, X., Brookes, P.C., Xu, J., et al., 2017. Contrasting effects of composting and pyrolysis on bioavailability and speciation of Cu and Zn in pig manure. Chemosphere 180, 93–99.

Mohan, D., Pittman, C.U., Steele, P.H., 2006. Pyrolysis of wood/biomass for bio-oil: a critical review. Energy Fuels 20 (3), 848–889.

Oleszczuk, P., Jośko, I., Kuśmierz, M., 2013. Biochar properties regarding to contaminants content and ecotoxicological assessment. J. Hazard. Mater. 260, 375–382.

Ouyang, D., Yan, J., Qian, L., Chen, Y., Han, L., Su, A., et al., 2017. Degradation of 1,4-dioxane by biochar supported nano magnetite particles activating persulfate. Chemosphere 184, 609–617.

Özçimen, D., Ersoy-Meriçboyu, A., 2010. Characterization of biochar and bio-oil samples obtained from carbonization of various biomass materials. Renew. Energy 35 (6), 1319–1324.

Pandey, M.P., Kim, C.S., 2011. Lignin depolymerization and conversion: a review of thermochemical methods. Chem. Eng. Technol. 34 (1), 29–41.

Peng, N., Li, Y., Liu, T., Lang, Q., Gai, C., Liu, Z., 2017. Polycyclic aromatic hydrocarbons and toxic heavy metals in municipal solid waste and corresponding hydrochars. Energ. Fuel 31 (2), 1665–1671.

Perin, G., Craboledda, L., Cirillo, M., Dotta, L., Zanette, M.L., Orio, A.A., 1985. Heavy metal speciation in the sediments of Northern Adriatic Sea: a new approach for environmental toxicity determination. Heavy Metal in the Environment. CEP Consultant, Edinburgh, pp. 454–456.

Quilliam, R.S., Rangecroft, S., Emmett, B.A., Deluca, T.H., Jones, D.L., 2013. Is biochar a source or sink for polycyclic aromatic hydrocarbon (PAH) compounds in agricultural soils? GCB Bioenergy 5, 96–103.

Rauret, G., Lopez-Sanchez, J.F., Sahuquillo, A., Rubio, R., Davidson, C., Ure, A., et al., 1999. Improvement of the BCR three step sequential extraction procedure prior to the certification of new sediment and soil reference materials. J. Environ. Monit. 1 (1), 57–61.

Rogovska, N., Laird, D., Cruse, R.M., Trabue, S., Heaton, E., 2012. Germination tests for assessing biochar quality. J. Environ. Qual. 41 (4), 1014–1022.

Rombolà, A.G., Marisi, G., Torri, C., Fabbri, D., Buscaroli, A., Ghidotti, M., et al., 2015. Relationships between chemical characteristics and phytotoxicity of biochar from poultry litter pyrolysis. J. Agric. Food Chem. 63 (30), 6660–6667.

Rubio-Clemente, A., Torres-Palma, R.A., Peñuela, G.A., 2014. Removal of polycyclic aromatic hydrocarbons in aqueous environment by chemical treatments: a review. Sci. Total Environ. 478, 201–225.

Shao, J., Yuan, X., Leng, L., Huang, H., Jiang, L., Wang, H., et al., 2015. The comparison of the migration and transformation behavior of heavy metals during pyrolysis and liquefaction of municipal sewage sludge, paper mill sludge, and slaughterhouse sludge. Bioresour. Technol. 198, 16–22.

Samburova, V., Connolly, J., Gyawali, M., Yatavelli, R.L.N., Watts, A.C., Chakrabarty, R.K., et al., 2016. Polycyclic aromatic hydrocarbons in biomass-burning emissions and their contribution to light absorption and aerosol toxicity. Sci. Total Environ. 568, 391–401.

Shi, W., Liu, C., Shu, Y., Feng, C., Lei, Z., Zhang, Z., 2013. Synergistic effect of rice husk addition on hydrothermal treatment of sewage sludge: fate and environmental risk of heavy metals. Bioresour. Technol. 149, 496–502.

Shukla, B., Koshi, M., 2012. A novel route for PAH growth in HACA based mechanisms. Combust. Flame 159 (12), 3589–3596.

Sigmund, G., Huber, D., Bucheli, T.D., Baumann, M., Borth, N., Guebitz, G.M., et al., 2017. Cytotoxicity of biochar: a workplace safety concern? Environ. Sci. Technol. Lett. 4 (9), 362–366.

Singh, B., Singh, B.P., Cowie, A.L., 2010. Characterisation and evaluation of biochars for their application as a soil amendment. Soil Res. 48 (7), 516–525.

Smith, C.R., Buzan, E.M., Lee, J.W., 2013a. Potential impact of biochar water-extractable substances on environmental sustainability. ACS Sustain. Chem. Eng. 1 (1), 118–126.

Smith, C.R., Sleighter, R.L., Hatcher, P.G., Lee, J.W., 2013b. Molecular characterization of inhibiting biochar water-extractable substances using electrospray ionization Fourier transform ion cyclotron resonance mass spectrometry. Environ. Sci. Technol. 47 (23), 13294–13302.

Smith, C.R., Hatcher, P.G., Kumar, S., Lee, J.W., 2016. Investigation into the sources of biochar water-soluble organic compounds and their potential toxicity on aquatic microorganisms. ACS Sustain. Chem. Eng. 4 (5), 2550–2558.

Spokas, K., Baker, J., Reicosky, D., 2010. Ethylene: potential key for biochar amendment impacts. Plant Soil 333 (1), 443–452.

Spokas, K.A., Cantrell, K.B., Novak, J.M., Archer, D.W., Ippolito, J.A., Collins, H.P., et al., 2011a. Biochar: a synthesis of its agronomic impact beyond carbon sequestration. J. Environ. Qual. 41, 973–989.

Spokas, K.A., Novak, J.M., Stewart, C.E., Cantrell, K.B., Uchimiya, M., DuSaire, M.G., et al., 2011b. Qualitative analysis of volatile organic compounds on biochar. Chemosphere 85 (5), 869–882.

Stanmore, B.R., 2004. The formation of dioxins in combustion systems. Combust. Flame 136 (3), 398–427.

Sun, D., Meng, J., Liang, H., Yang, E., Huang, Y., Chen, W., et al., 2015. Effect of volatile organic compounds absorbed to fresh biochar on survival of *Bacillus mucilaginosus* and structure of soil microbial communities. J. Soils Sediment. 15 (2), 271–281.

Tessier, A., Campbell, P.G.C., Bisson, M., 1979. Sequential extraction procedure for the speciation of particulate trace metals. Anal. Chem. 51 (7), 844–851.

Thines, K.R., Abdullah, E.C., Mubarak, N.M., Ruthiraan, M., 2017. Synthesis of magnetic biochar from agricultural waste biomass to enhancing route for waste water and polymer application: a review. Renew. Sustain. Energy Rev. 67, 257–276.

Tilman, D., Fargione, J., Wolff, B., D'Antonio, C., Dobson, A., Howarth, R., et al., 2001. Forecasting agriculturally driven global environmental change. Science 292 (5515), 281–284.

Tripathi, M., Sahu, J.N., Ganesan, P., 2016. Effect of process parameters on production of biochar from biomass waste through pyrolysis: a review. Renew. Sustain. Energy Rev. 55, 467–481.

USEPA, 1992. U.S. EPA, Offices of Water and Wastewater and compliance (Ed.) Guidelines for water reuse. U.S. EPA, Washington.

Wang, Z., Zheng, H., Luo, Y., Deng, X., Herbert, S., Xing, B., 2013. Characterization and influence of biochars on nitrous oxide emission from agricultural soil. Environ. Pollut. 174, 289–296.

Wang, Z., Liu, G., Zheng, H., Li, F., Ngo, H.H., Guo, W., et al., 2015a. Investigating the mechanisms of biochar's removal of lead from solution. Bioresour. Technol. 177, 308–317.

Wang, Z., Zong, H., Zheng, H., Liu, G., Chen, L., Xing, B., 2015b. Reduced nitrification and abundance of ammonia-oxidizing bacteria in acidic soil amended with biochar. Chemosphere 138, 576–583.

Wang, H., Zheng, H., Jiang, Z., Dai, Y., Liu, G., Chen, L., et al., 2017a. Efficacies of biochar and biochar-based amendment on vegetable yield and nitrogen utilization in four consecutive planting seasons. Sci. Total Environ. 593–594, 124–133.

Wang, Y., Jing, X., Li, L., Liu, W., Tong, Z., Jiang, H., 2017b. Biotoxicity evaluations of three typical biochars using a simulated system of fast pyrolytic biochar extracts on organisms of three kingdoms. ACS Sustain. Chem. Eng. 5 (1), 481–488.

Wang, Z., Chen, L., Sun, F., Luo, X., Wang, H., Liu, G., et al., 2017c. Effects of adding biochar on the properties and nitrogen bioavailability of an acidic soil. Eur. J. Soil Sci. 68 (4), 559–572.

Xiao, R., Awasthi, M.K., Li, R., Park, J., Pensky, S.M., Wang, Q., et al., 2017. Recent developments in biochar utilization as an additive in organic solid waste composting: a review. Bioresour. Technol. 246, 203–213.

Xu, F., Wei, C., Zeng, Q., Li, X., Alvarez, P.J.J., Li, Q., et al., 2017. Aggregation behavior of dissolved black carbon: implications for vertical mass flux and fractionation in aquatic systems. Environ. Sci. Technol. 51 (23), 13723–13732.

Yang, J., Pan, B., Li, H., Liao, S., Zhang, D., Wu, M., et al., 2016. Degradation of p-nitrophenol on biochars: role of persistent free radicals. Environ. Sci. Technol. 50 (2), 694–700.

Yuan, X., Huang, H., Zeng, G., Li, H., Wang, J., Zhou, C., et al., 2011. Total concentrations and chemical speciation of heavy metals in liquefaction residues of sewage sludge. Bioresour. Technol. 102 (5), 4104–4110.

Yuan, X., Leng, L., Huang, H., Chen, X., Wang, H., Xiao, Z., et al., 2015. Speciation and environmental risk assessment of heavy metal in bio-oil from liquefaction/pyrolysis of sewage sludge. Chemosphere 120, 645–652.

Zhao, L., Cao, X., Masek, O., Zimmerman, A., 2013. Heterogeneity of biochar properties as a function of feedstock sources and production temperatures. J. Hazard. Mater. 256–257, 1–9.

Zeng, X., Xiao, Z., Zhang, G., Wang, A., Li, Z., Liu, Y., et al., 2018. Speciation and bioavailability of heavy metals in pyrolytic biochar of swine and goat manures. J. Analyt. Appl. Pyrol. 132, 82–93.

Zhai, Y., Chen, H., Xu, B., Xiang, B., Chen, Z., Li, C., et al., 2014. Influence of sewage sludge-based activated carbon and temperature on the liquefaction of sewage sludge: yield and composition of bio-oil, immobilization and risk assessment of heavy metals. Bioresour. Technol. 159, 72–79.

Zhang, X., Gao, B., Creamer, A.E., Cao, C., Li, Y., 2017. Adsorption of VOCs onto engineered carbon materials: a review. J. Hazard. Mater. 338, 102–123.

Zhang, S., You, J., Kennes, C., Cheng, Z., Ye, J., Chen, D., et al., 2018. Current advances of VOCs degradation by bioelectrochemical systems: a review. Chem. Eng. J. 334, 2625–2637.

Zhao, L., Cao, X., Zheng, W., Scott, J.W., Sharma, B.K., Chen, X., 2016. Copyrolysis of biomass with phosphate fertilizers to improve biochar carbon retention, slow nutrient release, and stabilize heavy metals in soil. ACS Sustain. Chem. Eng. 4 (3), 1630–1636.

Zheng, W., Holm, N., Spokas, K.A., 2016. Research and application of biochar in North America. In: Guo, M., He, Z., Uchimiya, S.M. (Eds.), Agricultural and Environmental Applications of Biochar: Advances and Barriers. Soil Science Society of America, Inc, Madison, WI, pp. 475–494.

Zheng, H., Sun, C., Hou, X., Wu, M., Yao, Y., Li, F., 2017. Pyrolysis of *Arundo donax* L. to produce pyrolytic vinegar and its effect on the growth of dinoflagellate Karenia brevis. Bioresour. Technol. 247, 273–281.

Zheng, H., Wang, Z., Deng, X., Zhao, J., Luo, Y., Novak, J., et al., 2013a. Characteristics and nutrient values of biochars produced from giant reed at different temperatures. Bioresour. Technol. 130, 463–471.

Zheng, H., Wang, Z., Zhao, J., Herbert, S., Xing, B., 2013b. Sorption of antibiotic sulfamethoxazole varies with biochars produced at different temperatures. Environ. Pollut. 181, 60–67.

Zhu, X., Chen, B., Zhu, L., Xing, B., 2017. Effects and mechanisms of biochar–microbe interactions in soil improvement and pollution remediation: a review. Environ. Pollut. 227, 98–115.

Zielinska, A., Oleszczuk, P., 2016. Effect of pyrolysis temperatures on freely dissolved polycyclic aromatic hydrocarbon (PAH) concentrations in sewage sludge-derived biochars. Chemosphere 153, 68–74.

CHAPTER

20

On the Carbon Abatement Potential and Economic Viability of Biochar Production Systems: Cost-Benefit and Life Cycle Assessment

Siming You[1] *and Xiaonan Wang*[2]

[1]Division of Systems, Power & Energy, School of Engineering, University of Glasgow, Glasgow, United Kingdom [2]Department of Chemical and Biomolecular Engineering, National University of Singapore, Singapore

20.1 INTRODUCTION

Human beings are constantly challenged by increasing demands for raw material, energy, and food, in the face of rapid population expansion and economic growth. The U. S. Energy Information Administration (EIA) projected that the world energy consumption would increase by 28% between 2015 and 2040 (Conti et al., 2016). Fossil fuels are expected to remain the dominant source of world energy consumption over the next few decades, but it was predicted that oil, coal, and gas reserves would be depleted in 35, 107, and 37 years, respectively (Shafiee and Topal, 2009). Meanwhile, there are widespread concerns over the climate change associated with the high carbon footprint of fossil fuel-based energy production. The diminution and supply uncertainty of traditional fossil fuels, together with increasing energy demand and climate change concerns, have urged a worldwide exploration of alternative energy sources.

Biochar from Biomass and Waste
DOI: https://doi.org/10.1016/B978-0-12-811729-3.00020-0

Energy production from biomass, that is, bioenergy is regarded as one of the promising candidates to displace fossil fuels in the long run for both energy supply and climate change mitigation. Biomass resources are formed from or derived from living species like plants and animals and include virgin and waste biomass (Basu, 2010). A typical process involved by virgin biomass is photosynthesis, through which plants use sunlight to convert carbon dioxide and water to carbohydrate and oxygen. Unlike fossil fuel, biomass does not take millions of years to develop and thus is renewable. The thermochemical processing of biomass releases the carbon dioxide absorbed which is subsequently absorbed by biomass again. This processing alone does not add to the Earth's carbon dioxide inventory, and biomass is commonly considered carbon neutral under the assumption of rapid biomass re-growth (Searchinger, 2010). As a result, bioenergy is treated as an important measure of global climate change mitigation policies. Bioenergy makes up the largest portion ($\sim$77%) of the world's renewable energy and has been extensively employed in developed, developing, and underdeveloped countries (Atewamba and Boimah, 2017). The International Energy Agency (IEA) predicted that the demand for bioenergy would increase three- to tenfold by 2050 (Graves et al., 2016).

In recent years, significant research interests in biomass are triggered by increased findings of the environmental and energy applications of biomass-derived product: biochar (You et al., 2017a). Biochar is the carbon-rich solid co-product of thermochemical reactions of carbonaceous materials under the conditions with limited or without oxygen. Biochar refers to the same carbon-rich product as charcoal, while the latter is conventionally used when the solid product is used as a fuel. Biochar mainly consists of recalcitrant carbon, labile carbon, and ash. The recalcitrant carbon made of stable aromatic rings has a high resistance to decomposition and was reported to stay stable in soil for hundreds or even thousands of years (Haberstroh et al., 2006; Lehmann et al., 2008). Since the carbon in biochar is originated from the atmosphere via photosynthesis, the stability suggests that biochar could be used to sequester carbon for climate change mitigation. In addition, biochar has an indirect impact on carbon dioxide reduction by improving soil quality and crop productivity and reducing soil greenhouse gas emissions.

The practical implementation of a biochar production system is ultimately subject to the environmental and economic feasibility of the whole system, which entails a carbon footprint assessment from a life cycle perspective and a systematic cost-benefit analysis (CBA). For example, the whole process of biomass utilization is not necessarily carbon-neutral or carbon-negative, from a life cycle perspective, when transport and harvesting processes of biomass are also taken into consideration. The income from waste biomass tipping fee and biochar income does not ensure the whole biochar system is economically viable. Furthermore, biochar is being increasingly used for other energy and environmental applications such as tar removal, adsorbent, electrochemical materials, catalysts, etc. (You et al., 2017a), and bears a less obvious carbon footprint impact due to the lack of relevant data. This urges significant efforts to map the environmental and economic feasibility of biochar production systems under different application scenarios. Economics and environmental sustainability are essential indexes to evaluate the biochar production systems. Considering the critical impact of biochar on carbon abatement, we exclusively consider equivalent carbon dioxide (CO_2e) emission as the environmental impact of life cycle assessment (LCA) in this chapter.

20.2 METHODS OF ENVIRONMENTAL AND ECONOMIC ANALYSIS

20.2.1 Life Cycle Assessment

LCA is a widely exploited framework that helps decision makers to understand the environmental performance associated with specific products, processes, or activities. Instead of only accounting for the technical performance and economic costs, more and more governments and plant operators have now realized and attempted to address the environmental sustainability of process systems at the same time. This has stimulated the development of methods for assessing the availability of and impacts from the consumption of energy and raw materials in manufacturing and use. LCA attempts to quantify the overall impacts associated with energy and materials and to provide information on trade-offs among different impacts in the life cycle.

20.2.1.1 Standardization

Life cycle analysis can be dated back to the 1960s when worldwide concerns on the rapid depletion of limited raw materials and energy resources sparked interest in finding ways to understand and forecast the supply and utilization of energy and resources in the future. During the 1970s, the energy crisis caused by the oil shortage prompted a critical review of the energy-intensive nature of process industries. This motivated the need for a system-oriented tool, such as LCA, to track materials and energy flows in industrial systems. An important milestone for LCA to be generally accepted by a wide range of stakeholders and the international community was the development of international standards for LCA methodology. Starting in 1997, the International Standards Organization (ISO) 14000 series was progressively released to address various aspects of environmental management (International Organization for Standardization, 1997, 1998). The ISO 14000 series standards included a mandate to conduct an LCA, and provided guidance on procedures and methods. However, to eliminate inconsistencies among this series of ISO standards and to improve readability, two new international standards for LCA were published in 2006 (International Organization for Standardization, 2006). ISO 14040 defines the Principles and Framework without requirements elaborated, while ISO 14044 discusses all requirements and guidelines. The ISO 14040:2006 Environmental Management documentation further replaces the previous standards by adding several definitions, clarifications, and applications of LCA, as well as describing the technical content in detail.

The proliferation of LCA tools along with the market explosion for LCA software became problematic with different studies being based on different datasets and methods. To standardize LCA methods through a consensus approach, in 2002, the United Nations Environment Program (UNEP), in an international partnership with the Society of Environmental Toxicology and Chemistry (SETAC), launched the Life Cycle Initiative. This cross-country cooperation brought new vigor into life cycle-based methods for evaluating materials, products, and systems. Further, the LCA Global Coordinating Group (GCG) was founded in 2011 within SETAC as a communication center among the regional advisory committees for all members in the SETAC LCA Community, including Europe, North America, and the other three SETAC Geographic Units (Latin America, Asia-Pacific, and Africa) (Klöpffer, 2014). The United States Environmental Protection Agency (USEPA) published "Life-cycle assessment: principles and practice" (Curran, 2006) to summarize the general LCA methodologies as well as life cycle impact assessment (LCIA) tools. Extensive case studies have also been conducted to educate and promote LCA in a sustainability context (Andrae, 2009; Geibig and Socolof, 2005).

20.2.1.2 *Procedure*

LCA consists of four basic steps, that is, goal and scope definition, inventory analysis, impact assessment, and interpretation and improvement. As an iterative methodology, all the stages of LCA need to be performed sequentially. The results from previous steps essentially impact the next phase as a comprehensive and consistent analysis. With the primary goal of choosing the best product or process with the least effect on the environment and human health, all attributes within one LCA study should be considered to realize an overall reduction of energy and resource consumption and corresponding emissions (Sonnemann and Vigon, 2011).

20.2.1.2.1 DEFINE GOAL AND SCOPE

The goal and scope definition step in LCA first requires the decision on system boundaries by setting criteria that specify which unit processes are included or excluded from a product system. This is sometimes not easy in practice because subjectivity in terms of boundary selection cannot be avoided (Suh et al., 2004). A simplified approach is to draw an artificial system boundary that only incorporates a few upstream and downstream processes that accounts for the majority of the resources consumed and products achieved (Hendrickson et al., 1998).

Another essential step in scope definition is to choose the functional unit that defines the boundary of the analysis. Taking the building sector as an example, it is more reasonable to compare among construction products at the level of a whole building instead of simply on the basis of a per weight unit in order to understand how material selection can impact the overall building design in terms of ancillary materials, component weight and thickness, building footprint, and so forth, to meet certain design criteria (Ortiz et al., 2009).

20.2.1.2.2 LIFE CYCLE INVENTORY

Life cycle inventory (LCI) is the quantification of energy and raw material requirements for the entire life cycle of a product, process, or activity, as well as emissions (e.g., atmosphere, water, solid wastes, and other releases) from the same object of study (International Organization for Standardization, 2006). ISO standards require that unit processes should be interconnected to allow calculations over the complete system, based on flowcharts and system boundaries, which is accomplished by normalizing all system input and output flows to the functional unit. The data on both elementary flows from and to the environment and product flows from and to other unit processes are collected and quantitatively normalized. An evaluation step is also suggested for LCI quality assessment.

20.2.1.2.3 LIFE CYCLE IMPACT ASSESSMENT

LCIA follows LCI as an evaluation process with respect to the components identified in the inventory analysis. First, related impact categories are selected. The LCI results from the previous phase are then imported to match the materials or energy flow with specific impact categories; this is the classification step. The following characterization step requires science-based methods to convert the inventory within each category into impact indicators. Impact indicators are typically calculated by weighting the quantity of inventory with a characterization factor, through the simplest but creditable linear summation by assumption. An optional normalization step can also be conducted to make the results comparable among all impact categories. The obtained indicators can be sorted by scale or location (e.g., local, regional, and global). The weighting step again is optional and aimed at identifying the most important potential impacts. Finally, LCIA results can be evaluated to get a better understanding of the reliability and uncertainty.

20.2.1.3 Challenges and Roadmap

One major dilemma is that all analyses through LCA stages should be based on scientific criteria while value-based judgments are also necessary, especially in weighing the relative significance among all impact categories. As in the first step of LCIA, related impact categories need to be selected. If the LCA practitioners believe that a certain category is not relevant in the given study, the corresponding effects will be underestimated (Reap et al., 2008). A safer way is to incorporate as many categories as possible at the beginning stage and screen later when the evaluation is completed, even though this may cause more effort.

Also, the practice of LCA inevitably involves considerable uncertainties. For example, the quality of an LCA relies heavily on the LCI data it uses. With careful analysis of the underlying data and production processes considering possible correlations, the uncertainties might be reduced and the relative desirability of design alternatives can be assessed with greater confidence. In LCIA, it is a common strategy to adopt deterministic models to approximate life cycle impacts. However, these require various assumptions and weighting criteria that introduce bias and subjectivity when defining the model structure or assigning values to each parameter in the model. The deterministic modeling approach also fails to capture the variability and uncertainty inherent in LCA. Decision makers need to understand these limitations and uncertainties in LCA outcomes when comparing several different product systems especially when diversified LCIA results appear (Hauschild et al., 2008; Pennington et al., 2004).

Another noticeable issue is the transparency of an LCA study. Although potential environmental tradeoffs can be identified from LCA results, the trend of converting the impact results to a single score increases the difficulties in analyzing life-cycle impacts (Curran, 2006). It is a consensus that the analyses should be as transparent as possible. But an aggregation step is also meaningful to demonstrate comparable results. To make a balance among the creditability, interpretability, and transparency, the level of aggregation should be carefully determined.

Last, despite the current availability of electronic databases and software packages, LCA can still be data and labor intensive. When a decision needs to be made within a corporation, limited time, budget, data, and analytic approaches may create obstacles to conducting a comprehensive LCA. Streamlined LCA that does a screening first to eliminate the less important components or processes can be used to tackle some issues caused by analytical burdens to some extent (Graedel and Graedel, 1998). The well-developed database and software nowadays also make the LCA practice more efficient than in the past. However, this creates another problem with the consistency among analyses results, which cannot be neglected. While efforts have been made toward harmonization of the diversified LCA tools, there remains numerous data gaps and inconsistent interpretation.

To ameliorate the issues with LCA, which are not limited to those described above, a wide range of research programs are being pursued. More and more publications, workshops, and international cooperation have emerged to increase the understanding and applicability of LCA to make full use of this versatile tool. The most recent trend has also adopted machine learning methods to fit the data gap in LCA inventory construction and impact evaluation.

Numerous life-cycle impact assessment methods or LCA software tools have also been developed further to convert inventory data to life-cycle impact. EPA's Tool for the Reduction and Assessment of Chemical and other environmental Impacts (TRACI) is an important one of these tools, characterizing impact categories at the midpoint level to support consistency in environmental decision making (Bare et al., 2012). The environmental impact categories, such as global warming, ozone depletion, acidification, eutrophication, photochemical smog, human health cancer, human health non-cancer, human health criteria, eco-toxicity, fossil fuel use, land use, and water use are considered to form a comprehensive analysis of the potential impacts associated with the raw material usage and chemical releases resulting from the production processes. As a result, a preliminary comparison of two or more product options can be made. Another characterization tool that is commonly adopted to assess the releases of pollutants of a product or process is the USEtox model developed under the UNEP-SETAC Life Cycle Initiative (Rosenbaum et al., 2008). USEtox is designed to describe the fate, exposure, and effects of chemicals by calculating characterization factors for carcinogenic impacts and non-carcinogenic impacts for chemical emissions to all media. The toxicity impact score is calculated by multiplying the mass of a substance emitted in a given compartment with the corresponding toxicity characterization factor. This transparent matrix format allows identification of the relative importance of potential impacts and the main exposure pathways in the overall score, and therefore makes it a versatile model that can characterize toxicity at midpoint and/or endpoint levels.

Apart from the previously mentioned tools developed by the government agencies, numerous academic or commercial software packages are also available in the market, such as GaBi and SimaPro. Among these, it is worth noting the development of open source software, which provides source codes free of charge, and can be modified by anyone. This recent innovation in LCA software bears the mission of making the life-cycle approaches visually attractive and popular in standard programming languages for both sophisticated and simple models.

20.2.2 Cost-Benefit Analysis

A CBA is a systematic way to evaluate the desirability of a project by considering the wide impact on relevant groups over all time periods (Boardman et al., 2017). CBA of novel processes or retrofitting existing systems is paramount and pervasive using both lab testing results and scaling-up data. It clearly shows the monetized costs and benefits associated with the studied processes, and the values are continuously updated through the lifecycle, especially compared with the other state-of-art technologies or current practice in accordance to the prevailing industry. The novel pathways of biochar production and utilization are able to produce high value-added products in a much less carbon-intensive avenue, thereby allowing cost recovery and climate change mitigation. It is meaningful to quantify these costs and benefits associated with the focused process and product.

20.2.2.1 Common Practice

There are two major types of CBA, that is, ex-ante and ex-post CBA, conducted before the start and after the implementation of projects, respectively (Sassone and Schaffer, 1978). Both direct and indirect costs and benefits are qualified in multiple scenarios. To conduct CBA, a qualitative evaluation is first conducted to identify all categories of cost and benefits, based on the inventory of materials and energy. Their relative importance is also denoted to judge the key contributing factors. The time value of money needs to be quantified through certain criteria, such as the net present value (NPV) or benefit–cost ratio (BCR), and internal rate of return (IRR). Generally, NPV is adopted as the proper decision criterion to be used in CBA, which is calculated to measure the monetized benefits and costs of the project (Söderqvist et al., 2015).

$$\mathrm{NPV} = \sum_{t=0}^{T} \frac{1}{(1+r_t)^t}(B_t - C_t) \tag{20.1}$$

Benefit (B) and cost (C) items of the systems to be evaluated normally include typical categories such as materials and energy consumptions of each step, waste emissions, etc. The results of BCR and IRR are generally consistent with that of NPV (You et al., 2016).

Moreover, scenario analysis and advanced computational methodologies can provide additional insights and further improvements on the processes and technologies. Besides the basic economic and environmental analysis, a series of scenarios with respect to the completion or investment levels can be analyzed. A multi-objective optimization can also be conducted using stochastic multi-criteria decision analysis to obtain the optimal whole-system design to support sustainable management of the energy systems (Wang et al., 2018). We illustrated the typical steps for CBA and a template calculating costs and benefits in Fig. 20.1.

20.2.2.2 Quantification Techniques and Discussion

Direct costs include capital investment and operational expenditure, alternatively operating and maintenance (O&M) costs (generally comprising machinery costs, staff training, and salary). These are straightforward in the corporate accounting system and can be used directly in CBA, while the remainder of indirect costs bear more uncertainties and difficulties to be quantified, which will be focused in this subsection.

For the cost pre-estimation or post-evaluation, the procedures of total cost assessment can be followed on the basis of LCI analysis results (Curkovic and Sroufe, 2007). It considers five types of costs, which are direct costs (including both capital and O&M costs), indirect costs (costs not allocated to the product or process), contingent costs (fines and

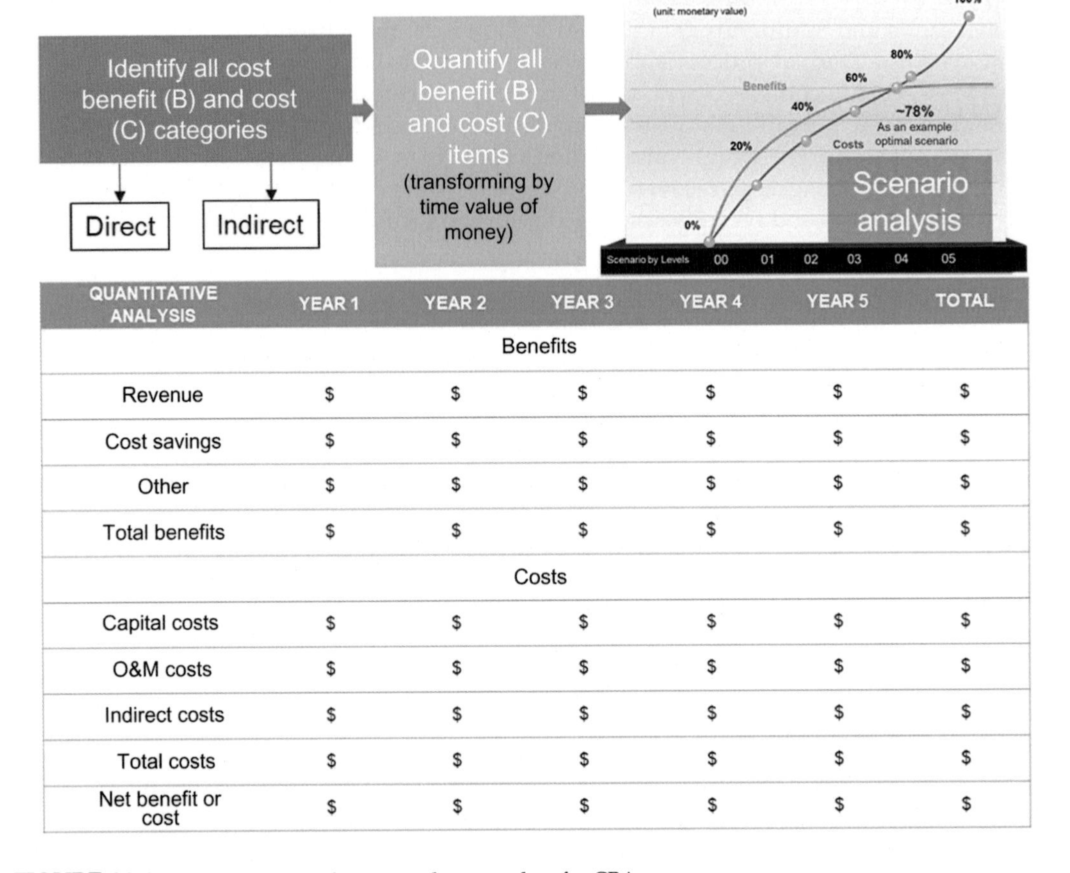

QUANTITATIVE ANALYSIS	YEAR 1	YEAR 2	YEAR 3	YEAR 4	YEAR 5	TOTAL
			Benefits			
Revenue	$	$	$	$	$	$
Cost savings	$	$	$	$	$	$
Other	$	$	$	$	$	$
Total benefits	$	$	$	$	$	$
			Costs			
Capital costs	$	$	$	$	$	$
O&M costs	$	$	$	$	$	$
Indirect costs	$	$	$	$	$	$
Total costs	$	$	$	$	$	$
Net benefit or cost	$	$	$	$	$	$

FIGURE 20.1 Typical steps and an exemplary template for CBA.

penalties), intangible costs, and external costs, among which the external costs such as the social costs associated with air, water, and land pollutions are most difficult to quantify. Both qualitative and quantitative analyses are comprised following the procedure of classification, characterization, and valuation of impacts to ecosystems, human health, natural resources and other essential categories. Another systematic method of environmental cost accounting (ECA), or simplified as environmental accounting (EA), can also be adopted in CBA as a special accounting system that addresses both private costs and societal costs. It is a term for impacts on society and the environment not reflected in a firm's normal accounting systems. ECA allocates costs incurred in order to comply with regulatory standards and reduce or eliminate releases of hazardous substances as well as environmental impacts, and costs associated with the failure of addressing these issues. Environmental accounting information is sometimes treated as an extra cost of doing business in the decision-making context, which has a unique role in choosing alternatives as an important scenario to consider (Odum, 1996). "Environmental liabilities," which includes compliance remediation, compensation fines, and penalties, punitive damages, and natural resource damages, is another term referring to the potentials for fines, penalties, and even more serious results for violations of environmental regulations and laws (Cormier and Magnan, 1997). Many tools or techniques have been developed for application in all scales and sectors, which become especially valuable when decision-making is in concern (Bieber et al., 2018; Triantafyllidis et al., 2018; Wang et al., 2014).

20.2.3 Scenario Setting

The environmental impacts and economic feasibility are often evaluated across different scenarios for comparison. For example, Clare et al. (2015) compared the economic and carbon abatement potential of straw pyrolysis for biochar and electrical energy production with two design scenarios (straw briquetting and combustion for heat generation and straw gasification for electrical energy production) and two baseline scenarios (straw reincorporation into the soil and straw burning on the field). Harsono et al. (2013) compared the energy balances, GHG emissions, and economics of biochar production from the slow pyrolysis of empty fruit brunches in a palm oil mill with that of a baseline case where empty fruit brunches were applied to the trees in the palm oil plantation. Homagain et al. (2016) compared the economics of four scenarios of biochar-based bioenergy production and soil application in terms of high or low feedstock availabilities and with or without soil application. The scenario setting also echoes the reference scenario selection during LCA that may significantly affect the resulting carbon abatement potential. Shabangu et al. (2014) compared the economic feasibility of pine to biochar and methanol based on three technological scenarios,that is, slow pyrolysis at 300 and 450°C, and gasification at 800°C, with the same processing of the volatiles into syngas and the conversion of the syngas into methanol. Baseline scenarios could be selected based on the existing biomass utilization practice together with multiple design scenarios for system optimization and selection. The LCA and CBA results could vary significantly across different scenarios.

20.3 THERMOCHEMICAL PROCESSES

The application, performance, and economic values of biochar are dependent on its physiochemical properties, which are closely related to the thermochemical conversion conditions (i.e., heating rate, temperature, and reactor configurations). In the absence of oxygen, the thermochemical process is referred to as pyrolysis whose co-products include syngas (mainly hydrogen and carbon monoxide), bio-oil (tarry non-condensable volatile matters), and biochar. Depending on the reaction conditions (temperature, residence time, and heating rate), pyrolysis could be further categorized into, e.g., slow and fast pyrolysis. Slow pyrolysis features low temperatures (300–600°C), long residence times (5–30 min), and relatively high biochar production. Compared with slow pyrolysis, fast pyrolysis involves a higher heating rate and shorter residence time and has been widely employed for bio-oil production (Pootakham and Kumar, 2010). Limited oxygen is involved in the gasification process, whose co-products include predominantly syngas, and a small amount of biochar and bio-oils (tar). The reaction temperature of gasification is generally

TABLE 20.1 The Processing Conditions and Co-Products Production of Different Thermochemical Processes

	Processing Conditions			Co-Products (%)			
Processes	**Temperature**	**Heating rate**	**Residence Time**	**Syngas**	**Bio-Oil**	**Biochar**	**References**
Slow pyrolysis	300–600°C	2–15°C min^{-1}	5–30 min	35	30	35	Sharma et al. (2015)
Fast pyrolysis	400–600°C	10–200°C s^{-1}	1 s	13	75	12	Bridgwater (2012), Sharma et al. (2015)
Gasification	550–1600°C	–	10–20 s	85	5	10	Arena (2012b), Bridgwater (2012), Sharma et al. (2015)

in the range of 550–1600°C, which is higher than that of pyrolysis, leading to a less biochar production. The processing conditions and typical mass percentages of the co-products of different processes are listed in Table 20.1.

20.3.1 Co-Products

Bio-oil and syngas are the common co-products of biochar from the thermochemical processes. The bio-oil could be used directly as fuel oils for electricity generation, blended with diesel fuels or as valuable chemicals due to its high content of phenol and phenol derivatives (Islam et al., 2010; Quan et al., 2010). The heating value of bio-oil could be up to 40%–50% of hydrocarbon fuels (Jahirul et al., 2012). This suggests a great commercial potential for bio-oil. The bio-oil produced from the pyrolysis of Cryptomeria residue was reported to have a 19.6% higher value than low-sulfur fuel oil with an equivalent energy value (Ning et al., 2013). The bio-oil from the fast pyrolysis of woody biomass had a price around \$19–25 GJ^{-1} on an energy basis, which is comparable to heating oil in Europe (Oasmaa et al., 2015). The price of pyrolysis oil was also reported to be in the range between \$125 and 210 ton^{-1} (Piemonte et al., 2016). Bio-oil derived transportation fuels were projected to have a minimum price of \$1.74–3.09 $gallon^{-1}$ (Brown et al., 2011). However, Wongkhorsub and Chindaprasert (2013) found that the pyrolysis oil could hardly replace diesel in terms of engine performance and energy output unless its price was lower than 85% of diesel oil. On another study (Field et al., 2013), the bio-oil produced from a slow pyrolysis system was considered to displace heavy fuel oil use on an energy-equivalent basis and dominated the system economic return at a 21% yield and a higher heating value of 34.1 MJ kg^{-1}.

Syngas could be used for power generation, upgraded into value-added biofuel and chemicals (e.g., ethanol, acetate, formate, and butanol) via either catalytic conversion or anaerobic fermentation, and used for producing hydrogen via the water gas shift reaction ($CO(g) + H_2O(g) \leftrightarrow CO_2(g) + H_2(g)$) (Brunetti et al., 2015; Martin et al., 2016; Xiong et al., 2017). Combustion of syngas for heat and electricity generation is one of the most simple and popular ways of syngas utilization. The syngas from biomass pyrolysis was found to have a heating value between 12 and 14 MJ Nm^{-3} (He et al., 2010; Zhang et al., 2015). The heating value of syngas from air, oxygen, and steam gasification ranges from 4 to 7, 10 to 15, and 15 to 20 MJ Nm^{-3}, respectively (Arena, 2012a). The heat could be used to support the process of pyrolysis which is endothermic, dry feedstock, or output to heat storage for downstream usage (Harsono et al., 2013; Swanson et al., 2010). Field et al. (2013) considered that 0.21 kg of pyrolysis gas was needed to supply the energy for the pyrolysis of 1 kg of dry biomass. The auto-thermal nature of gasification suggests that no energy is needed to maintain gasification except that external fuel is used to initiate the process and waste heat is recovered to dry biomass. Part of the electricity could be used for the demand of biochar production systems, while the remainder is either for end-user consumption or fed into the national electricity grid (You et al., 2017b). For a slow pyrolysis system, roughly 10% of the generated electricity was required to support the process with an additional 10%–15% lost in the process (Gaunt and Lehmann, 2008; Roberts et al., 2009). A slow pyrolysis-based system with an annual biochar production of 960 ton in Malaysia was reported to have an electricity demand of 1.08 kWh day^{-1} as a result of operating computer panels, lamps, and other equipment (Harsono et al., 2013).

20.3.2 Biochar

Although slow pyrolysis has a relatively large biochar productivity, much of the biochar available is in practice produced as a co-product from fast and gasification systems that are currently aimed to maximize energy production (Field et al., 2013). The physiochemical properties (e.g., specific surface area, total porous volume, pH, and organic contents) and yield of biochar vary significantly depending on the operating parameters, which offers an opportunity for designing bespoke biochar for specific applications but suggests a variability in biochar value. Regarding the composition of biomass, high lignin and low moisture contents are favorable conditions for pyrolysis biochar production (Tripathi et al., 2016). Temperature is the major parameter affecting biochar production for both gasification and pyrolysis. The most suitable temperature range for pyrolysis biochar production was between 450 and 600°C (Tripathi et al., 2016). The biochar quality is low when the temperature of pyrolysis was lower than 350°C (Ahmed et al., 2016). For slow pyrolysis, increasing peak temperature facilitated the escape of volatile matters from feedstock and favored the formation of biochar with a higher aromatic character, fixed carbon content, and porosity which were potentially related to the carbon stability and nutrient retention function of biochar upon soil application (Manyà, 2012).

However, a higher reaction temperature generally leads to a decrease in biochar yields for both gasification and pyrolysis. The relatively high temperature during gasification leads to a lower biochar yield (<200 g kg^{-1}) than pyrolysis, but significantly more syngas production (You et al., 2017a). The yield of safflower seed biochar from gasification decreased by 3%–8% as the temperature increased from 400 to 600°C (Angın et al., 2013). To achieve both high yields of bio-oil and biochar with the enhanced physiochemical properties, catalysts are required, which adds to the cost of the biochar production process (Manyà, 2012). Note that the gasification process using O_2, steam or CO_2 as the gasifying reagents could pose an "activation" effect on the biochar produced and increase the specific surface area and total pore volume, which means an improvement in the soil-water retention and soil aeration ability of biochar, but a reduction in the carbon retention and sequestration capacity (Manyà, 2012; You et al., 2017a).

The stable carbon content in biochar is typically ranged from 50% to 85% and directly associated with its carbon sequestration capacity (Downie et al., 2009). Existing studies considered 90% gasification biochar and 80% pyrolysis biochar were stable over the 100-year time-scale upon soil application (Crombie et al., 2013; Cross and Sohi, 2013). As mentioned earlier, the higher stability of gasification biochar is resulting from the higher reaction temperature of gasification compared to pyrolysis (You et al., 2017a). However, there is a potential concern over the contamination of gasification biochar by polycyclic aromatic hydrocarbon (PAH) that are commonly formed under higher temperatures and may affect the soil applicability of gasification biochar. Biochar was found effectively to reduce the PAH concentration present in soils by 50% (Beesley et al., 2010), but relevant evaluation is still rare for its impact on PAH originated from biochar itself. In addition to the operating conditions, the carbon stability of biochar is also depended on crop types, soil conditions, and biochar's exposure to weathering (Sohi, 2013). For example, the agronomic value of biochar was likely to be higher when it was applied to poorer quality, acidic, or highly degraded soils (Kimetu et al., 2008; Liu et al., 2013). The carbon sequestration of biochar could be expressed by a carbon stability factor (Field et al., 2013):

$$CO_{2,\text{sequest}} = 3.66(1 - e^{(t_{1/2}\ln(0.5))/\text{TH}}) \tag{20.2}$$

where $t_{1/2}$ is the half-life of biochar in soil and TH is the analytical time horizon (e.g., 100 years).

Due to its unique physiochemical properties (e.g., high porosity, high surface area, and inorganic content), biochar could also be used to reduce soil N_2O emissions (Ameloot et al., 2016; Zheng et al., 2012), increase soil organic carbon (SOC) contents (Liu et al., 2016; Zimmerman et al., 2011), facilitate fertilizer application efficiency (Van Zwieten et al., 2010), mitigate soil contamination (Bian et al., 2013; Houben et al., 2013), and enhance crop productivity (Genesio et al., 2015; Schmidt et al., 2015) upon its application as a soil amendment. Woody and herbaceous biochar were shown to have a soil N_2O reduction potential of 27% for a 1%–2% (by soil weight) biochar application rate (Cayuela et al., 2014). Due to the high pH of biochar, it could pose a liming effect on the soil (especially acidic soil) that could in return ameliorate Al toxicity and P deficiency (Qiao-Hong et al., 2014). The liming effect could be expressed by calcium carbonate equivalence (CCE) representing the acid-neutralizing capacity of biochar with respect to pure $CaCO_3$. CCE could be estimated based on the base elements (Ca, Mg, K, and Na) and ash contents of biochar (Field et al., 2013) as

$$\text{CCE}(\%) = 5.378 + 1.582B - 0.2136A \tag{20.3}$$

where B is the percentage of base elements and A is the total ash content. The liming effect of biochar suggests that biochar could be used to displace the agricultural limestone application on a CCE-equivalent basis or reduce fertilizer application to maintain a given crop yield. The latter will involve an indirect carbon reduction in terms of fertilizer application-associated N_2O emission and pH-mediated N-to-N_2O emission (Zheng et al., 2012). The impact of biochar on the mineralization of SOC is often called a priming effect. Both positive and negative effects were found and no definite conclusion has been reached yet (Fang et al., 2015), while the magnitude of this impact was also reported to be small (Field et al., 2013).

20.4 CARBON ABATEMENT—LIFE CYCLE ASSESSMENT

For the LCA of biochar production systems with biochar used for soil amendment, both direct and indirect emission or abatement components need to be accounted for. Direct components include biomass sourcing operations (e.g., land preparation, fertilizer, herbicides, and pesticides applications, plantation, and harvesting), biomass transport, biomass pre-treatment (e.g., drying, chopping, and grinding), thermochemical processing of biomass, energy storage and transmission, biochar transport, and biochar applications, while indirect emissions abatement components include land-use changes, reduced N_2O emissions, avoided fertilizer application, increased fertilizer use efficiency, and enhanced crop productivity. The LCA of slow pyrolysis-based biochar production systems, including both direct and indirect components, showed that the systems could achieve a carbon abatement potential of around 1.0 ton CO_2e per ton of feedstock consumed (Hammond et al., 2011; Woolf et al., 2010).

20.4.1 Emissions

Biomass selection is the first step in the design of biochar production systems. There could be a huge difference between the life-cycle carbon footprints of a biochar production system when different biomass types are considered. For example, the slow pyrolysis systems based on stover and yard waste had a net carbon abatement potential of -864 and -885 kg CO_2e ton^{-1} dry feedstock, respectively, while the system based on switchgrass was a carbon emitter of $+36$ kg CO_2e ton^{-1} dry feedstock (Roberts et al., 2009). Corn fodder was a more economical biomass than forest residue and switchgrass for biochar production from slow pyrolysis (Dutta and Raghavan, 2014; Roberts et al., 2009). The production and properties of co-products could also potentially vary significantly for different types of biomass. For example, the biochar production from animal manure could pose a great impact to crop productivity than that from wood, while it was less stable in the soil than those from wood (Jeffery et al., 2015). It is important to consider the suitability, availability, quantity, existing applications, impacts to the food supply, and costs of biomass. Biomass whose plantation requires vast extra agricultural lands could pose a threat to food crop plantation or existing eco-systems (forest) and thus adverse environmental and social impacts (Popp et al., 2014).

Without including the potential effect of the land-use change, the slow pyrolysis processing of biomass was found to contribute 51.2% and 47.3% of the total GHG emissions for forest residue- and corn fodder-based slow pyrolysis systems (Dutta and Raghavan, 2014). A similar result was obtained for a palm oil empty fruit brunches-based slow pyrolysis system where the slow pyrolysis process accounted for around 60% of the GHG emissions due to the intensive consumption of diesel fuel (Harsono et al., 2013). Other studies showed that the emissions related to land-use changes (cropland diversion from annual crops to perennial energy crops and land conversion to cropland to offset the crops lost due to the plantation of energy crops) and biomass production served as the dominant GHG emission contributors (Gaunt and Lehmann, 2008; Roberts et al., 2009).

Transport generally involves the transport between biomass field and biochar production site, and between the site and biochar application site. The carbon emissions associated with the consumption of diesel fuel during transport should include both tailpipe CO_2 emissions and embodied emissions resulting from the extraction, refining, and distribution of diesel fuel (Field et al., 2013). Existing studies showed that the transport emission did not constitute a significant contributor to the overall carbon footprint of biochar production systems (Dutta and Raghavan, 2014).

20.4.2 Abatement

The direct carbon sequestration generally accounts for the largest contributor to the overall carbon abatement potential of a biochar production system. For a slow pyrolysis-based biochar production system, a carbon abatement of 0.76 ton CO_2e per ton feedstock processed was achieved under a half-life of 240 years and 29% biochar mass yield (Field et al., 2013). Roberts et al. (2009) showed the stable carbon could contribute 54%–66% of the net carbon emission reductions across a variety of biomass types (early and late stover, switchgrass, and yard waste). The carbon retention in biochar could contribute to around half of the carbon emission reductions of a slow pyrolysis system (Gaunt and Lehmann, 2008).

In existing studies of LCA, biochar was also considered as a fuel to displace fossil coal on an energy-equivalent basis (Field et al., 2013). However, the carbon abatement potential of biochar upon the soil application was two to five times greater than that upon the energy application (Gaunt and Lehmann, 2008). The carbon emissions per unit electricity generation for a biochar production system could be around 2–10 lower than that for fossil fuel-based electricity generation (Gaunt and Lehmann, 2008). The soil application of biochar resulted in 29% more carbon emission reduction than its application as a fuel (Roberts et al., 2009). From a global standpoint, biochar had the potential to achieve an annual net emission reduction up to 1.8 Pg CO_2e which accounted for around 12% of current anthropogenic CO_2e emissions, and a total net emissions of 130 Pg CO_2e over the course of a century, without endangering food security, habitat, or soil conservation (Woolf et al., 2010).

Clare et al. (2015) considered that biochar could be used to replace 25% NPK fertilizer for the case of China in their LCA. They assumed that nitrogen fertilizer contributed to a standard NPK (16:16:16) mix and excluded the emissions from potassium (K) and phosphorus (P) production in the synthetic due to their one-order-of-magnitude smaller contributions. Hence, the carbon intensity of fertilizer production industry will be an important input parameter to account for the impact of biochar to reduce fertilizer use.

Generally, much higher levels of uncertainty are involved with the estimation of indirect emissions when relevant processes are affected by human behaviors, such as fertilizer application methods and tillage methods and soil types. From a life-cycle perspective, the carbon abatement potential of a biochar production system was reported to be similar among the cases where biochar was used as soil amendment and those where biochar was used for energy production via combustion, if the indirect emission components were not taken into consideration (Jeffery et al., 2015). The impacts of some of these indirect emissions on the overall system carbon footprint are limited. For example, the biochar-induced N_2O emission reduction only contributes 1%–4% of the carbon abatement potential of a pyrolysis-based biochar production system (Clare et al., 2015; Roberts et al., 2009). The impacts of some other indirect emissions could not be neglected. Clare et al. (2015) showed that each ton of biochar that replaced chemical fertilizer could alone abate 1.33 ton CO_2e corresponding to 0.39 ton CO_2e for each ton of dry straw feedstock, which could amount to 30% of the overall carbon abatement potential. Hence, factoring the indirect emission components will generally make systems with the soil application of biochar superior to comparison systems with the energy application of biochar in terms of carbon abatement.

System expansion is used to allocate the carbon abatement impact of co-products during the LCA. The allocation has also been done based on the economic value of the co-products generated by a system. For example, Harsono et al. (2013) allocated the energy demand, the GHG emissions, and the costs of the agricultural production processes with the economic values of the products from palm oil production,that is, crude palm oil, kernel oil, and empty fruit brunches.

20.5 ECONOMIC FEASIBILITY—COST-BENEFIT ANALYSIS

20.5.1 Costs

Capital costs include those related to system construction, land procurement, and interest payments. Capital costs could make up around 30% of total biochar production costs from forestry residues, depending on system scale (Shackley et al., 2011). To account for the depreciation of tangible assets, such as plant, equipment, and machines, a straight-line method that assumes a constant depreciation could be used (Harsono et al., 2013). Compared to conventional power plants, biomass power plants generally have a higher capital cost of around \$1575 kW^{-1} because of more auxiliary equipment (e.g., chopper, packer, and feeding system) and larger land areas involved for the case of China (Zhang et al., 2014). Note that the capital investment may be subject to loans that could be up to 70%–80% of the total investment with a repayment period of 10 years (Zhang et al., 2014).

The O&M cost generally comprises staff training and salary, biomass cost, equipment maintenance cost, and biochar application cost. For a slow pyrolysis-based biochar production system with a capacity of 1 MWh, the cost of pyrolysis (36%) is the most expensive stage, followed by storage and processing (14%) including pelletization, feedstock collection (12%), and transport cost (9%) (Homagain et al., 2016). The high costs involved in the pyrolysis process and feedstock production were regarded as the major barriers to the economical implementation of pyrolysis-based biochar systems (Roberts et al., 2009). The costs of biomass could dominate the O&M cost, especially for the types of biomass that have multiple application and marketing values. This is the reason why it is normally

TABLE 20.2 Biomass Costs

Biomass	Price ($ ton^{-1})[a]	Country or Region	Year	References
Rice straw	39	Philippines	2016	Nguyen et al. (2016)
Rice straw	39	India	2017	Lal et al. (2017)
Rice straw	40	China	2007–09	Zhang et al. (2014)
Wheat straw	45	China	2007–09	Zhang et al. (2014)
Wheat straw	65	UK	2013	Littlewood et al. (2013)
Wheat straw	47	India	2013	Jat et al. (2014)
Wheat straw	50	Denmark	2014	Janssen et al. (2014)
Cryptomeria	34	Taiwan	2012	Ning et al. (2013)
Corn stover briquette	139	Denmark	2017	Redl et al. (2017)
Palm oil empty fruit brunches	15.8	Malaysia	2013	Harsono et al. (2013)

[a]*In case that the raw data is not in U.S. dollars, it is converted to U.S. dollars based on the exchange rate in January 2018.*

suggested that a biochar production system based on the waste biomass with tipping subsidies is economically preferable. A list of the costs of biomass is given in Table 20.2. The carbon emissions associated with biomass sourcing, transport, and pre-treatment were relatively small, while they may account for a significant part of the overall cost in a slow pyrolysis-based system (Dutta and Raghavan, 2014). The latter is especially an economic concern if the biomass feedstock is being used for other marketing applications (e.g., spent grains as a valuable animal feed), potentially incurring a significant opportunity cost (Field et al., 2013). Biomass that has some existing applications (e.g., animal feed and composts) and thus a high opportunity cost should not be suitable for biochar production. The biomass price could vary significantly, depending on the time, region, supply, and demand, and transportation costs. The transport cost is highly dependent on the distance, but generally has a minor effect on the overall economics of biochar production systems (Dutta and Raghavan, 2014). Biochar application costs involve the costs of implementation, fuel, equipment, and labor. Depending on a long (300 km) or short (100 km) biomass transport distance, a cost of $156,739 years^{-1} or $133,228 years^{-1} was incurred for the soil application of biochar (Homagain et al., 2016).

20.5.2 Benefits

Without a definite pricing system for biochar and carbon abatement, heat and electricity, or syngas and bio-oil derived products selling, serve as the major benefit for a biochar production system. The values of the heat and electricity generated in biochar production systems are generally determined according to their market values, or in many cases based on relevant financial subsidies such as feed-in-tariff. Currently, financial subsidies still play a critical role in the profitability of biochar production systems. Some countries also provide biomass-burning avoidance subsidies to reduce related air pollution hassles and encourage the recycling and energy utilization of waste biomass. For example, bioelectricity from agricultural and waste biomass is financially sponsored by a feed-in-tariff of $0.12 kWh^{-1} in China in addition to grant subsidies and loan discount (Zhang et al., 2014). Businesses will also receive a $28 ton^{-1} subsidy upon using straw for livestock rearing, paper production or bioenergy generation in some parts of China where straw-burning was causing significant air quality deterioration (Clare et al., 2015). The subsidies could make a gasification-based system profitable because significantly more electricity was produced by gasification, even though carbon mitigation credit and biochar selling were not included. Without the bioelectricity and straw-burning avoidance subsidies, the rice straw-based pyrolysis biochar system in China had an NPV of $ − 20.3 million over a course of 20 years, while avoidance and bioelectricity subsidies increased the NPV to $ − 1.84 million. Pyrolysis-based biochar and methanol production was found to be not profitable without valuing the biochar as a product, and the projected break-even biochar prices were about $220 ton^{-1} and $280 ton^{-1} for the pyrolysis at 300 and 450°C, respectively (Shabangu et al., 2014).

Without the existence of carbon mitigation credibility and biochar marketing value, a system with a higher energy efficiency is generally economically more viable but could potentially have a dip in the overall carbon abatement potential due to less biochar production, and vice versa. Most of the existing gasification systems attempt to increase their profitability via improving their syngas and power generation efficiency, which inevitably leads to the decrease in biochar production. This is especially true for decentralized, small-scale biochar production systems where the benefit from the harvesting of the small fraction co-products is negligible and could hardly justify an extra investment for sophisticated separation, filtration, and clean-up units. As a technology producing more biochar, slow pyrolysis generally leads to a system with higher carbon abatement potential but lower profitability, and for gasification, biochar for the soil application would not be more valuable than biochar for energy production if the carbon price is lower than $49 ton^{-1} CO_2e under the conditions of high conversion temperatures and soils with low buffering capacity (Field et al., 2013). Fast pyrolysis systems tend to sequester less carbon than slow pyrolysis systems, but the fast pyrolysis systems are more desirable because of the higher value of the ultimate energy products (e.g., transportation fuels) from the fast pyrolysis systems than the products (biochar and syngas) from the slow pyrolysis systems (Brown et al., 2011). This suggested a common trade-off between the economic feasibility (biofuel production) and carbon saving potential (biochar production) of biochar production systems (Fowles, 2007; Jeffery et al., 2015). To achieve parity of economics and carbon abatement, the carbon should be priced according to (Field et al., 2013)

$$P_c = (R_{\mathrm{HIGH}} - R_{\mathrm{LOW}})/(\mathrm{GHG}_{\mathrm{LOW}} - \mathrm{GHG}_{\mathrm{HIGH}}) \qquad (20.3)$$

R denotes financial returns ($ ton^{-1} biomass) while GHG denotes carbon abatement potential of difference arrangements. HIGH and LOW denote the arrangements with high and low returns, respectively.

As carbon trading markets become mature and the value of biochar significantly increases on its higher value applications, the profitability of the system should be optimized at a point where the syngas- or energy-based profit is not necessarily maximized

TABLE 20.3 Biochar Prices

Price ($ ton^{-1})	Country or Region	Year	Reference
259	China	2014	Clare et al. (2015)
120	Taiwan	2010–12	Ning et al. (2013)
90	Philippines	2016	Ahmed et al. (2016)
8850	UK	2016	Ahmed et al. (2016)
80	India	2016	Ahmed et al. (2016)
13,480	USA	2016	Ahmed et al. (2016)

(You et al., 2017a). That is, an optimal condition exists regarding the generations of bioenergy and biochar to ensure both maximum profitability and carbon abatement potential, with the existence of carbon mitigation credibility and biochar selling benefit. Note that financial subsidies are closely correlated with the overall economic and environmental performance of biochar production systems. The implementation and improvement of biochar production systems in many cases are highly contingent upon subsidies, while the development of financial supports is mostly aimed at the systems with substantiated and superior economic and environmental performance (Clare et al., 2015).

The reported biochar price varied significantly across different countries (Table 20.3) and there is still a great uncertainty and volatility about this value. A list of biochar retail prices is given in Table 20.3. The break-even price of biochar for a profitable system was reported to be \$128 ton^{-1} (with bioenergy subsidies), \$206 ton^{-1} (without bioenergy subsidies) for a pyrolysis biochar system, which suggested that there was a demand further to increase the agronomic value of biochar (Clare et al., 2015). For a palm oil empty fruit brunches-based slow pyrolysis system, a biochar price higher than \$533 ton^{-1} was needed to ensure the economic viability of the system (Harsono et al., 2013). A carbon price of \$51–71 ton^{-1} CO_2e was also needed for the break-even of the pyrolysis biochar system. CAD \$60 ton^{-1} of CO_2e is needed for the profitability of a slow pyrolysis-based system where the distance of feedstock transport was less than 200 km, bio-oil and syngas were used for electricity generation, and biochar was used as a soil amendment (Homagain et al., 2016).

There hasn't been a consistent framework for accounting for the value of biochar. Clare et al. (2015) considered biochar's agronomic value as the value of the yield improvement in one growing year without including the costs of biochar spreading and transportation. Based on a 10% yield increase for a 3 ton ha^{-1} application rate, the agronomic value was \$110 ton^{-1} compared to \$33 ton^{-1} for a 10 ton ha^{-1} application rate. Without accounting for carbon mitigation credit, the NPV of a pyrolysis biochar system decreased from $\$-1.84$ million to $\$-10.1$ million when the agronomic value of biochar decreased from \$110 to \$33 ton^{-1} (Clare et al., 2015). A 40% increase in the crop yield was also reported, which led to a biochar value of \$5740 ton^{-1} and gave pyrolysis biochar systems an unparalleled economic advantage against gasification biochar systems (Clare et al., 2015). However, such a high crop productivity enhances needs to be validated further by more field trials. Additionally, the value of biochar was also assigned based on the liming effect and the opportunistic cost of soil products, such as aglime and fertilizer displaced by biochar, which led to a biochar value of \$0.53 and \$1.48 ton^{-1} under a displacement rate of 61 and 3.9 kg ton^{-1} biochar, respectively (Field et al., 2013). Roberts et al. (2009) valued biochar based on the P and K content, improved fertilizer use efficiency, and carbon emission reduction. A complete accounting of the value of biochar should also consider other mechanisms, such as increased water and nutrient retention and availability, increased cation exchange capacity, organic matter addition, and increased microbial (e.g., mycorrhizal fungi) activity, which are subject to the inadequacy of experimental data. Anyway, the value of biochar should reflect the price that customers are willing to pay.

Waste has been considered as a good source for biochar production in view of its low cost, high availability, and pressure from the waste pile-up. The tipping fee received upon waste disposal may constitute an important income for biochar production systems and one of the major drives for the financial viability of biochar production systems (Ng et al., 2017; Roberts et al., 2009; You et al., 2016). For example, Roberts et al. (2009) found that yard waste had the highest profitability potential (\$69 ton^{-1} dry feedstock at \$80 ton^{-1} CO_2e) for a slow pyrolysis system among corn stover, yard waste, and switchgrass, and suggested that biochar could hardly be profitable except for a distributed system using waste biomass (Roberts et al., 2009). Using waste as the feedstock for biochar production could also bring the benefit of lowering the cost of raw material collection (Homagain et al., 2016). However, biochar production from waste could be potentially contaminated by various pollutants, such as PAHs, heavy metals, perfluorooctanoic acid/perfluorooctanesulfonic acid, etc. This is especially a concern for waste that may contain significant pollutants itself, such as demolition waste, sludge, and industrial waste. Some of these pollutants (e.g., PAHs) could not be fully decomposed and will even be enriched by the thermochemical processes. For example, most of the heavy metals in original waste, such as sewage sludge and biosolids, would be retained in gasification and pyrolysis biochar with higher concentrations in smaller biochar particles due to vaporization and re-condensation of metal salts (Chan and Xu, 2009; Nzihou and Stanmore, 2013). Hence, additional leaching or clean-up process may be needed prior to the soil application of waste-derived biochar if the pollutants in biochar are bioavailable, which incurs additional capital and O&M costs to the system.

20.6 UNCERTAINTIES AND DIRECTIONS

20.6.1 Uncertainties

For both LCA and CBA, due to the lack of data and essential variability of data (e.g., high price volatility of co-products, fertilizers, and fuels), the uncertainty of input data is a common problem and generally a sensitivity and/or uncertainty analysis is needed.

Clare et al. (2015) conducted an uncertainty analysis to determine the median values and 95% confidence intervals of ton CO_2e emitted per ton feedstock based on a Monte Carlo method and tested the impact of each parameter on economic and carbon abatement results (sensitivity analysis) by varying parameters by $\pm 20\%$, while keeping all other parameters constant. They found that the prices of co-products and capital cost had major impacts on the economics of biochar production systems, while direct emissions from the gasification and pyrolysis of straw, and offset emissions from avoided fossil fuel use, had the greatest impact on the carbon abatement potential of relevant systems. You et al. (2017b, 2016) conducted a sensitivity analysis using the design-of-experiments method and found that it is more favorable to reduce the construction and O&M costs, and increase the electricity efficiency to improve the economics of gasification-based systems for various types of biomass (oil palm residue, food waste, and sewage sludge).

Sensitivity analysis was also conducted based on the method of standardized perturbation (1%) regarding key input parameters, which showed that the physical properties (carbon content and heating value) of biochar were primary influential factors followed by labor and energy prices, liming effect parameters, carbon stability in soil, and baseline soil N_2O emission rates (Field et al., 2013). For a palm oil empty fruit brunches-based slow pyrolysis system, the economic viability is most sensitive to the cost of empty fruit brunches, price of diesel fuel, and price of biochar (Harsono et al., 2013). A 5% increase in the diesel fuel and 15% increase in the empty fruit brunches price would make the system economically unviable (Harsono et al., 2013). For a fast pyrolysis system that converted bio-oil into transportation fuels, its profitability was most sensitive to bio-oil yield, followed by biomass cost, fuel yield, and capital cost, while the selling prices of bio-oil and syngas had a minor effect on its profitability (Brown et al., 2011).

20.6.2 Future Life Cycle Assessment and Cost-Benefit Analysis

The LCA and CBA results are sensitive to the properties of biochar that are further subject to the variability of operating conditions. Hence, ideal LCA and CBA should accurately formulate the function between the economic indicators and environmental impacts, and the operating conditions involved with the production of biochar. However, various simplifications were inevitably adopted in real LCA and CBA due to the availability of limited data. Field et al. (2013) modeled the properties, recalcitrance, and agronomic responses of biochar as a function of reaction temperature for LCA and CBA. Based on existing database from bench-top scale experiments, functions between the yields and heating rate of the co-products and reaction temperature were developed. Significant uncertainty exists underlying the functions because of the composite nature of the data and potential error of scaling-up. However, this simplified modeling was one of the few attempts to conduct LCA and CBA in terms of the variation of operating conditions, which should be the future direction for the development of system optimization platform in terms of economic and environmental sustainability.

References

Ahmed, M.B., Zhou, J.L., Ngo, H.H., Guo, W., 2016. Insight into biochar properties and its cost analysis. Biomass Bioenergy 84, 76–86.

Ameloot, N., Maenhout, P., De Neve, S., Sleutel, S., 2016. Biochar-induced N_2O emission reductions after field incorporation in a loam soil. Geoderma 267, 10–16.

Andrae, A.S., 2009. Global Life Cycle Impact Assessments of Material Shifts: The Example of a Lead-Free Electronics Industry. Springer Science & Business Media, Berlin, Germany.

Angın, D., Altintig, E., Köse, T.E., 2013. Influence of process parameters on the surface and chemical properties of activated carbon obtained from biochar by chemical activation. Bioresour. Technol. 148, 542–549.

Arena, U., 2012a. Process and technological aspects of municipal solid waste gasification: a review. Waste Manag. 32 (4), 625–639.

Arena, U., 2012b. Process and technological aspects of municipal solid waste gasification: a review. Waste Manag. 32 (4), 625–639.

Atewamba, C., Boimah, M., 2017. Policy forum: potential options for greening the Concessionary Forestry Business Model in rural Africa. Forest Policy Econ. 85, 46–51.

Bare, J., Young, D., Qam, S., Hopton, M., Chief, S., 2012. Tool for the Reduction and Assessment of Chemical and Other Environmental Impacts (TRACI). US Environmental Protection Agency, Washington, DC.

Basu, P., 2010. Biomass Gasification and Pyrolysis: Practical Design and Theory. Academic Press, Cambridge, MA.

Beesley, L., Moreno-Jiménez, E., Gomez-Eyles, J.L., 2010. Effects of biochar and greenwaste compost amendments on mobility, bioavailability and toxicity of inorganic and organic contaminants in a multi-element polluted soil. Environ. Pollut. 158 (6), 2282–2287.

Bian, R., Chen, D., Liu, X., Cui, L., Li, L., Pan, G., et al., 2013. Biochar soil amendment as a solution to prevent Cd-tainted rice from China: results from a cross-site field experiment. Ecol. Eng. 58, 378–383.

Bieber, N., Ker, J.H., Wang, X., Triantafyllidis, C., van Dam, K.H., Koppelaar, R.H., et al., 2018. Sustainable planning of the energy-water-food nexus using decision making tools. Energy Policy 113, 584–607.

Boardman, A.E., Greenberg, D.H., Vining, A.R., Weimer, D.L., 2017. Cost-Benefit Analysis: Concepts and Practice. Cambridge University Press, Cambridge, United Kingdom.

Bridgwater, A.V., 2012. Review of fast pyrolysis of biomass and product upgrading. Biomass Bioenergy 38, 68–94.

Brown, T.R., Wright, M.M., Brown, R.C., 2011. Estimating profitability of two biochar production scenarios: slow pyrolysis vs fast pyrolysis. Biofuels Bioprod. Biorefining 5 (1), 54–68.

Brunetti, A., Caravella, A., Fernandez, E., Tanaka, D.P., Gallucci, F., Drioli, E., et al., 2015. Syngas upgrading in a membrane reactor with thin Pd-alloy supported membrane. Int. J. Hydrogen Energy 40 (34), 10883–10893.

Cayuela, M., Van Zwieten, L., Singh, B., Jeffery, S., Roig, A., Sánchez-Monedero, M., 2014. Biochar's role in mitigating soil nitrous oxide emissions: a review and meta-analysis. Agric. Ecosyst. Environ. 191, 5–16.

Chan, K.Y., Xu, Z., 2009. Biochar: nutrient properties and their enhancement. Biochar Environ. Manage. Sci. Technol. 1, 67–84.

Clare, A., Shackley, S., Joseph, S., Hammond, J., Pan, G., Bloom, A., 2015. Competing uses for China's straw: the economic and carbon abatement potential of biochar. Gcb Bioenergy 7 (6), 1272–1282.

Conti, J., Holtberg, P., Diefenderfer, J., LaRose, A., Turnure, J.T., Westfall, L., 2016. International energy outlook 2016 with projections to 2040. USDOE Energy Information Administration (EIA). Office of Energy Analysis, Washington, DC (United States).

Cormier, D., Magnan, M., 1997. Investors' assessment of implicit environmental liabilities: an empirical investigation. J. Account. Public Policy 16 (2), 215–241.

Crombie, K., Mašek, O., Sohi, S.P., Brownsort, P., Cross, A., 2013. The effect of pyrolysis conditions on biochar stability as determined by three methods. Gcb Bioenergy 5 (2), 122–131.

Cross, A., Sohi, S.P., 2013. A method for screening the relative long-term stability of biochar. Gcb Bioenergy 5 (2), 215–220.

Curkovic, S., Sroufe, R., 2007. Total quality environmental management and total cost assessment: an exploratory study. Int. J. Prod. Econ. 105 (2), 560–579.

Curran, M.A., 2006. Life-Cycle Assessment: Principles and Practice. National Risk Management Research Laboratory, Office of Research and Development, US Environmental Protection Agency, Cincinnati, OH.

Downie, A., Crosky, A., Munroe, P., 2009. Physical properties of biochar. In: Lehmann, J., Joseph, S. (Eds.), Biochar for enviromental management: Science and Technology. Earthscan, London, UK, pp. 13–32.

Dutta, B., Raghavan, V., 2014. A life cycle assessment of environmental and economic balance of biochar systems in Quebec. Int. J. Energy Environ. Eng. 5 (2–3), 106.

Fang, Y., Singh, B., Singh, B.P., 2015. Effect of temperature on biochar priming effects and its stability in soils. Soil Biol. Biochem. 80, 136–145.

Field, J.L., Keske, C.M., Birch, G.L., DeFoort, M.W., Cotrufo, M.F., 2013. Distributed biochar and bioenergy coproduction: a regionally specific case study of environmental benefits and economic impacts. Gcb Bioenergy 5 (2), 177–191.

Fowles, M., 2007. Black carbon sequestration as an alternative to bioenergy. Biomass Bioenergy 31 (6), 426–432.

Gaunt, J.L., Lehmann, J., 2008. Energy balance and emissions associated with biochar sequestration and pyrolysis bioenergy production. Environ. Sci. Technol. 42 (11), 4152–4158.

Geibig, J.R., Socolof, M.L., 2005. Solders in Electronics: A Life-Cycle Assessment. National Service Center for Environmental Publications U.S. Environmental Protection Agency, Cincinnati, OH, grant# X-82931801.

Genesio, L., Miglietta, F., Baronti, S., Vaccari, F.P., 2015. Biochar increases vineyard productivity without affecting grape quality: results from a four years field experiment in Tuscany. Agric. Ecosyst. Environ. 201, 20–25.

Graedel, T.E., Graedel, T.E., 1998. Streamlined Life-Cycle Assessment. Prentice Hall, Upper Saddle River, NJ.

Graves, R.A., Pearson, S.M., Turner, M.G., 2016. Landscape patterns of bioenergy in a changing climate: implications for crop allocation and land-use competition. Ecol. Appl. 26 (2), 515–529.

Haberstroh, P.R., Brandes, J.A., Gélinas, Y., Dickens, A.F., Wirick, S., Cody, G., 2006. Chemical composition of the graphitic black carbon fraction in riverine and marine sediments at sub-micron scales using carbon X-ray spectromicroscopy. Geochim. Cosmochim. Acta 70 (6), 1483–1494.

Hammond, J., Shackley, S., Sohi, S., Brownsort, P., 2011. Prospective life cycle carbon abatement for pyrolysis biochar systems in the UK. Energy Policy 39 (5), 2646–2655.

Harsono, S.S., Grundman, P., Lau, L.H., Hansen, A., Salleh, M.A.M., Meyer-Aurich, A., et al., 2013. Energy balances, greenhouse gas emissions and economics of biochar production from palm oil empty fruit bunches. Resour. Conserv. Recycl. 77, 108–115.

Hauschild, M.Z., Huijbregts, M., Jolliet, O., MacLeod, M., Margni, M., van de Meent, D., et al., 2008. Building a model based on scientific consensus for life cycle impact assessment of chemicals: the search for harmony and parsimony. Environ. Sci. Technol. 42 (19), 7032–7037.

He, M., Xiao, B., Liu, S., Hu, Z., Guo, X., Luo, S., et al., 2010. Syngas production from pyrolysis of municipal solid waste (MSW) with dolomite as downstream catalysts. J. Anal. Appl. Pyrol. 87 (2), 181–187.

Hendrickson, C., Horvath, A., Joshi, S., Lave, L., 1998. Peer reviewed: economic input–output models for environmental life-cycle assessment. Environ. Sci. Technol. 32 (7), 184A–191A.

Homagain, K., Shahi, C., Luckai, N., Sharma, M., 2016. Life cycle cost and economic assessment of biochar-based bioenergy production and biochar land application in Northwestern Ontario, Canada. Forest Ecosyst. 3 (1), 21.

Houben, D., Evrard, L., Sonnet, P., 2013. Mobility, bioavailability and pH-dependent leaching of cadmium, zinc and lead in a contaminated soil amended with biochar. Chemosphere 92 (11), 1450–1457.

International Organization for Standardization, 1997. Environmental Management: Life Cycle Assessment: Principles and Framework. ISO, Geneva, Switzerland.

International Organization for Standardization, 1998. ISO 14041. Environmental Management – Life Cycle Assessment – Goal and Scope Definition and Inventory Analysis. ISO, Geneva, Switzerland.

International Organization for Standardization, 2006. ISO 14044: Environmental Management, Life Cycle Assessment, Requirements and Guidelines. ISO, Geneva, Switzerland.

Islam, M.R., Parveen, M., Haniu, H., Sarker, M.I., 2010. Innovation in pyrolysis technology for management of scrap tire: a solution of energy and environment. Int. J. Environ. Sci. Dev. 1 (1), 89.

Jahirul, M.I., Rasul, M.G., Chowdhury, A.A., Ashwath, N., 2012. Biofuels production through biomass pyrolysis—a technological review. Energies 5 (12), 4952–5001.

Janssen, M., Tillman, A.-M., Cannella, D., Jørgensen, H., 2014. Influence of high gravity process conditions on the environmental impact of ethanol production from wheat straw. Bioresour. Technol. 173, 148–158.

Jat, R.K., Sapkota, T.B., Singh, R.G., Jat, M., Kumar, M., Gupta, R.K., 2014. Seven years of conservation agriculture in a rice–wheat rotation of Eastern Gangetic Plains of South Asia: yield trends and economic profitability. Field Crops Res. 164, 199–210.

Jeffery, S., Bezemer, T.M., Cornelissen, G., Kuyper, T.W., Lehmann, J., Mommer, L., et al., 2015. The way forward in biochar research: targeting trade-offs between the potential wins. Gcb Bioenergy 7 (1), 1–13.

Kimetu, J.M., Lehmann, J., Ngoze, S.O., Mugendi, D.N., Kinyangi, J.M., Riha, S., et al., 2008. Reversibility of soil productivity decline with organic matter of differing quality along a degradation gradient. Ecosystems 11 (5), 726.

Klöpffer, W., 2014. Background and Future Prospects in Life Cycle Assessment. Springer Science & Business Media, Berlin, Germany.

Lal, B., Gautam, P., Panda, B., Raja, R., Singh, T., Tripathi, R., et al., 2017. Crop and varietal diversification of rainfed rice based cropping systems for higher productivity and profitability in Eastern India. PLoS One 12 (4), e0175709.

Lehmann, J., Skjemstad, J., Sohi, S., Carter, J., Barson, M., Falloon, P., et al., 2008. Australian climate–carbon cycle feedback reduced by soil black carbon. Nat. Geosci. 1 (12), 832.

Littlewood, J., Murphy, R.J., Wang, L., 2013. Importance of policy support and feedstock prices on economic feasibility of bioethanol production from wheat straw in the UK. Renew. Sustain. Energy Rev. 17, 291–300.

Liu, X., Zhang, A., Ji, C., Joseph, S., Bian, R., Li, L., et al., 2013. Biochar's effect on crop productivity and the dependence on experimental conditions—a meta-analysis of literature data. Plant Soil 373 (1–2), 583–594.

Liu, S., Zhang, Y., Zong, Y., Hu, Z., Wu, S., Zhou, J., et al., 2016. Response of soil carbon dioxide fluxes, soil organic carbon and microbial biomass carbon to biochar amendment: a meta-analysis. Gcb Bioenergy 8 (2), 392–406.

Manyà, J.J., 2012. Pyrolysis for biochar purposes: a review to establish current knowledge gaps and research needs. Environ. Sci. Technol. 46 (15), 7939–7954.

Martin, M.E., Richter, H., Saha, S., Angenent, L.T., 2016. Traits of selected Clostridium strains for syngas fermentation to ethanol. Biotechnol. Bioeng. 113 (3), 531–539.

Ng, W.C., You, S., Ling, R., Gin, K.Y.-H., Dai, Y., Wang, C.-H., 2017. Co-gasification of woody biomass and chicken manure: syngas production, biochar reutilization, and cost-benefit analysis. Energy 139, 732–742.

Nguyen, V., Topno, S., Balingbing, C., Nguyen, V., Röder, M., Quilty, J., et al., 2016. Generating a positive energy balance from using rice straw for anaerobic digestion. Energy Rep. 2, 117–122.

Ning, S.-K., Hung, M.-C., Chang, Y.-H., Wan, H.-P., Lee, H.-T., Shih, R.-F., 2013. Benefit assessment of cost, energy, and environment for biomass pyrolysis oil. J. Cleaner Prod. 59, 141–149.

Nzihou, A., Stanmore, B., 2013. The fate of heavy metals during combustion and gasification of contaminated biomass—a brief review. J. Hazard. Mater. 256, 56–66.

Oasmaa, A., van de Beld, B., Saari, P., Elliott, D.C., Solantausta, Y., 2015. Norms, standards, and legislation for fast pyrolysis bio-oils from lignocellulosic biomass. Energy Fuels 29 (4), 2471–2484.

Odum, H.T., 1996. Environmental Accounting: Emergy and Environmental Decision Making. Wiley, Hoboken, NJ.

Ortiz, O., Castells, F., Sonnemann, G., 2009. Sustainability in the construction industry: a review of recent developments based on LCA. Construct. Build. Mater. 23 (1), 28–39.

Pennington, D., Potting, J., Finnveden, G., Lindeijer, E., Jolliet, O., Rydberg, T., et al., 2004. Life cycle assessment Part 2: current impact assessment practice. Environ. Int. 30 (5), 721–739.

Piemonte, V., Capocelli, M., Orticello, G., Di Paola, L., 2016. Bio-oil production and upgrading: New challenges for membrane applications. Membrane Technologies for Biorefining. Elsevier, Amsterdam, pp. 263–287.

Pootakham, T., Kumar, A., 2010. A comparison of pipeline versus truck transport of bio-oil. Bioresour. Technol. 101 (1), 414–421.

Popp, J., Lakner, Z., Harangi-Rakos, M., Fari, M., 2014. The effect of bioenergy expansion: food, energy, and environment. Renew. Sustain. Energy Rev. 32, 559–578.

Qiao-Hong, Z., Xin-Hua, P., Huang, T.-Q., Zu-Bin, X., Holden, N., 2014. Effect of biochar addition on maize growth and nitrogen use efficiency in acidic red soils. Pedosphere 24 (6), 699–708.

Quan, C., Li, A., Gao, N., 2010. Characterization of products recycling from PCB waste pyrolysis. J. Anal. Appl. Pyrol. 89 (1), 102–106.

Reap, J., Roman, F., Duncan, S., Bras, B., 2008. A survey of unresolved problems in life cycle assessment. Int. J. Life Cycle Assess. 13 (5), 374.

Redl, S., Sukumara, S., Ploeger, T., Wu, L., Jensen, T.Ø., Nielsen, A.T., et al., 2017. Thermodynamics and economic feasibility of acetone production from syngas using the thermophilic production host *Moorella thermoacetica*. Biotechnol. Biofuels 10 (1), 150.

Roberts, K.G., Gloy, B.A., Joseph, S., Scott, N.R., Lehmann, J., 2009. Life cycle assessment of biochar systems: estimating the energetic, economic, and climate change potential. Environ. Sci. Technol. 44 (2), 827–833.

Rosenbaum, R.K., Bachmann, T.M., Gold, L.S., Huijbregts, M.A., Jolliet, O., Juraske, R., et al., 2008. USEtox – the UNEP-SETAC toxicity model: recommended characterisation factors for human toxicity and freshwater ecotoxicity in life cycle impact assessment. Int. J. Life Cycle Assess. 13 (7), 532.

Sassone, P.G., Schaffer, W.A., 1978. Cost-Benefit Analysis: A Handbook. Academic Press, New York.

Schmidt, H.P., Pandit, B.H., Martinsen, V., Cornelissen, G., Conte, P., Kammann, C.I., 2015. Fourfold increase in pumpkin yield in response to low-dosage root zone application of urine-enhanced biochar to a fertile tropical soil. Agriculture 5 (3), 723–741.

Searchinger, T.D., 2010. Biofuels and the need for additional carbon. Environ. Res. Lett. 5 (2), 024007.

Shabangu, S., Woolf, D., Fisher, E.M., Angenent, L.T., Lehmann, J., 2014. Techno-economic assessment of biomass slow pyrolysis into different biochar and methanol concepts. Fuel 117, 742–748.

Shackley, S., Hammond, J., Gaunt, J., Ibarrola, R., 2011. The feasibility and costs of biochar deployment in the UK. Carbon Manag. 2 (3), 335–356.

Shafiee, S., Topal, E., 2009. When will fossil fuel reserves be diminished? Energy Policy 37 (1), 181–189.

Sharma, A., Pareek, V., Zhang, D., 2015. Biomass pyrolysis – a review of modelling, process parameters and catalytic studies. Renew. Sustain. Energy Rev. 50, 1081–1096.

Söderqvist, T., Brinkhoff, P., Norberg, T., Rosén, L., Back, P.-E., Norrman, J., 2015. Cost-benefit analysis as a part of sustainability assessment of remediation alternatives for contaminated land. J. Environ. Manage. 157, 267–278.

Sohi, S.P., 2013. Pyrolysis bioenergy with biochar production – greater carbon abatement and benefits to soil. Gcb Bioenergy 5, 2.

Sonnemann, G., Vigon, B., 2011. Global guidance principles for Life Cycle Assessment (LCA) databases: a basis for greener processes and products. United Nations Environment Programme, Nairobi, Kenya.

Suh, S., Lenzen, M., Treloar, G.J., Hondo, H., Horvath, A., Huppes, G., et al., 2004. System boundary selection in life-cycle inventories using hybrid approaches. Environ. Sci. Technol. 38 (3), 657–664.

Swanson, R.M., Platon, A., Satrio, J.A., Brown, R.C., 2010. Techno-economic analysis of biomass-to-liquids production based on gasification. Fuel 89, S11–S19.

Triantafyllidis, C.P., Koppelaar, R.H., Wang, X., van Dam, K.H., Shah, N., 2018. An integrated optimisation platform for sustainable resource and infrastructure planning. Environ. Modell. Softw. 101, 146–168.

Tripathi, M., Sahu, J.N., Ganesan, P., 2016. Effect of process parameters on production of biochar from biomass waste through pyrolysis: a review. Renew. Sustain. Energy Rev. 55, 467–481.

Van Zwieten, L., Kimber, S., Downie, A., Morris, S., Petty, S., Rust, J., et al., 2010. A glasshouse study on the interaction of low mineral ash biochar with nitrogen in a sandy soil. Soil Res. 48 (7), 569–576.

Wang, X., Tong, C., Palazoglu, A., El-Farra, N.H., 2014. Energy management for the chlor-alkali process with hybrid renewable energy generation using receding horizon optimization. In: Decision and Control (CDC), 2014 IEEE 53rd Annual Conference on IEEE, pp. 4838–4843.

Wang, X., Guo, M., Koppelaar, R.H., van Dam, K.H., Triantafyllidis, C.P., Shah, N., 2018. A nexus approach for sustainable urban energy-water-waste systems planning and operation. Environ. Sci. Technol. 52 (5), 3257–3266.

Wongkhorsub, C., Chindaprasert, N., 2013. A comparison of the use of pyrolysis oils in diesel engine. Energy Power Eng. 5 (04), 350.

Woolf, D., Amonette, J.E., Street-Perrott, F.A., Lehmann, J., Joseph, S., 2010. Sustainable biochar to mitigate global climate change. Nat. Commun. 1, 56.

Xiong, X., Iris, K., Cao, L., Tsang, D.C., Zhang, S., Ok, Y.S., 2017. A review of biochar-based catalysts for chemical synthesis, biofuel production, and pollution control. Bioresour. Technol. 246, 254–270.

You, S., Wang, W., Dai, Y., Tong, Y.W., Wang, C.-H., 2016. Comparison of the co-gasification of sewage sludge and food wastes and cost-benefit analysis of gasification-and incineration-based waste treatment schemes. Bioresour. Technol. 218, 595–605.

You, S., Ok, Y.S., Chen, S.S., Tsang, D.C., Kwon, E.E., Lee, J., et al., 2017a. A critical review on sustainable biochar system through gasification: energy and environmental applications. Bioresour. Technol. 246, 242–253.

You, S., Tong, H., Armin-Hoiland, J., Tong, Y.W., Wang, C.-H., 2017b. Techno-economic and greenhouse gas savings assessment of decentralized biomass gasification for electrifying the rural areas of Indonesia. Appl. Energy 208, 495–510.

Zhang, Q., Zhou, D., Fang, X., 2014. Analysis on the policies of biomass power generation in China. Renew. Sustain. Energy Rev. 32, 926–935.

Zhang, S., Dong, Q., Zhang, L., Xiong, Y., 2015. High quality syngas production from microwave pyrolysis of rice husk with char-supported metallic catalysts. Bioresour. Technol. 191, 17–23.

Zheng, J., Stewart, C.E., Cotrufo, M.F., 2012. Biochar and nitrogen fertilizer alters soil nitrogen dynamics and greenhouse gas fluxes from two temperate soils. J. Environ. Qual. 41 (5), 1361–1370.

Zimmerman, A.R., Gao, B., Ahn, M.-Y., 2011. Positive and negative carbon mineralization priming effects among a variety of biochar-amended soils. Soil Biol. Biochem. 43 (6), 1169–1179.

CHAPTER

21

Redox-Mediated Biochar-Contaminant Interactions in Soil

Yilu Xu[1], *Yubo Yan*[2], *Nadeeka L. Obadamudalige*[1], *Yong Sik Ok*[3], *Nanthi Bolan*[1] *and Qiao Li*[4]

[1]Global Centre for Environmental Remediation (GCER), Faculty of Science, The University of Newcastle, Callaghan, NSW, Australia [2]School of Chemistry and Chemical Engineering, Huaiyin Normal University, Jiangsu, P.R. China [3]Korea Biochar Research Center, O-Jeong Eco-Resilience Institute (OJERI) & Division of Environmental Science and Ecological Engineering, Korea University, Seoul, Republic of Korea [4]School of Environmental and Biological Engineering, Nanjing University of Science and Technology, Nanjing, P.R. China

21.1 REDOX CHARACTERISTICS OF BIOCHAR

Although, the high content of carbon in biochar enables it as a strong reducing agent, it can take part in both oxidation and reduction reactions (redox reactions) due to its ability to donate, accept, or transfer electrons in their surrounding environments (Lehmann and Joseph, 2015; Yuan et al., 2017) (Table 21.1). The reducing power of a biochar is frequently quantified by both the maximum amount of electrons that it can donate (electron donating capacity, EDC) or accept (electron accepting capacity, EAC). The EDC and EAC values of biochar vary with the nature of the feedstock sources and the pyrolysis process conditions (Klupfel et al., 2014; Prevoteau et al., 2016). The redox moieties of biochar are derived from the transformation products generated by the pyrolysis of mainly lignin and cellulose (Prevoteau et al., 2016) with oxygen-containing functional groups forming most of this "redox pool." The electron donating moieties (i.e., reducers) from the biochars are mostly phenolic species, while the compounds accepting electrons (oxidants) are mostly quinones and poly condensed aromatic structures. The extent of aromaticity is also very important

Biochar from Biomass and Waste
DOI: https://doi.org/10.1016/B978-0-12-811729-3.00021-2

TABLE 21.1 Redox-Mediated Biochar-Contaminant Interaction

Biochar Source	Contaminant	Observation	Reference
Chicken manure	Cr and As	Biochar addition reduced Cr(VI) to Cr(III) and As(V) to As(III), and increased Cr(III) adoption yet decreased As(III) adoption. Consequently reduced bioavailability of Cr(III) and increased bioavailability of As(III)	Choppla et al. (2016)
Macadamia (*Macadamia tetraphylla*) nut shell	Nitrogen	Biochar addition reduced denitrification and/or complete denitrification, and NH_4^+ concentration for nitrification. Consequently reduced N_2 loss due to decreased N_2O and NH_3 emission	He et al. (2018), Mandal et al. (2016), Sanchez-Monedero et al. (2018), Thangarajan et al. (2018)
Wheat straw	Fe	Stimulate microbial reduction of the Fe(III) oxyhydroxide mineral ferrihydrite (15 mM) by *Shewanella oneidensis* MR-1, and changed mineral product of ferrihydrite reduction from magnetite (Fe_3O_4) to siderite ($FeCO_3$)	Kappler et al. (2014), Xu et al. (2016), Yuan et al. (2017)
		With the condition of higher particle concentration (5 and 10 g L^{-1}) and presence of soil microorganisms	
Cornstalk/Rice hull	Trichloroethylene (TCE)	Biochar contributed to the sorption and degradation with the presence of nanoscale zero-valent iron (NZVI)	Dong et al. (2017), Yan et al. (2015), Yuan et al. (2017)
Poultry litter/wastewater biosolids/ Fisher activated carbon/rice straw and polymers	Herbicides and explosives	Stimulated chemical reduction of nitro herbicides and explosives by shuttling electrons between reductants (e.g., dithiothreitol) and organic contaminants (i.e., pendimethalin, trifluralin, 2,4-dinitrotoluene and hexahydro-1,3,5-trinitro-1,3,5-triazine (RDX)	Oh et al. (2013), Kemper et al. (2008), Oh and Seo (2016), Xu et al. (2013)
		RDX was rapidly degraded by co-occurrence of sulfide sand black carbons. The reaction products are nitrite and formaldehyde rather than the potentially toxic nitrosated by-products	
Sucrose	Nitrobenzene	N-doped carbon materials function can as efficient catalysts to reduce nitrobenzene to easily degraded products (i.e., aniline) under aquatic system	Liu et al. (2017)
Oak wood	DDX	Biochar in the presence of sulfide can foster the degradation of 1,1-trichloro-2,2-di(4-chlorophenyl)ethane and 1,1-dichloro-2,2-bis (4-chlorophenyl)ethane (DDD) and 1,1-dichloro-2,2-bis(4-chlorophenyl)ethylene	Ding and Xu (2016)
Pine needles/wheat straw/maize straw/*Spartina alterniflora* biomass	Polychlorinated biphenyls (PCBs)/ methylene blue	The persistent free radicals of biochar catalyzed H_2O_2 decomposition producing hydroxyl radicals (•OH) that facilitate efficient degradation of PCB and methylene blue	Fang et al. (2014), Fang et al. (2015), Xu et al. (2014)
Rape straw/rice straw	Pentachlorophenol (PCP)	Accelerate the biodegradation of PCP by enhancing the extracellular electron transfer of microorganisms	Tong et al. (2014), Yu et al. (2015)

with respect to the redox interaction of the biochar with contaminants, since it increases its electronic conductivity. At higher pyrolysis temperature, the formation of graphitic structures enhances the conductivity of the carbon matrix, and also increases the porosity of the biochar (Keiluweit et al., 2010), thereby enhancing the accessibility of the redox moieties by contaminant species.

21.2 REDOX-MEDIATED BIOCHAR-CONTAMINANT INTERACTIONS IN SOIL

Biochar application has been shown to influence contaminants' reactions and bioavailability in soil (Fig. 21.1). The effect of biochar on soil contaminants is mediated through sorption, immobilization, precipitation, and redox reactions. In this section, biochar-induced redox reactions of contaminants are discussed.

21.3 HEAVY METAL(LOID) CONTAMINANTS

Heavy metal(loid)s reach terrestrial and aquatic environments through both pedogenic and anthropogenic processes. Most of the heavy metal(loid)s occur in nature, the major source of which is weathering of soil parent materials, including igneous and sedimentary rocks. For example, the majority of As is derived from geogenic origin. Anthropogenic activities, linked to industrial activities; the management of municipal and industrial waste materials; and the application of composts, manures, and fertilizers to soil are some of the major sources of metal(loid) input to the environment (Bolan et al., 2014). While As input through anthropogenic source is important, the widespread contamination of As in soil and groundwater resources in in many countries, including Bangladesh, India, China,

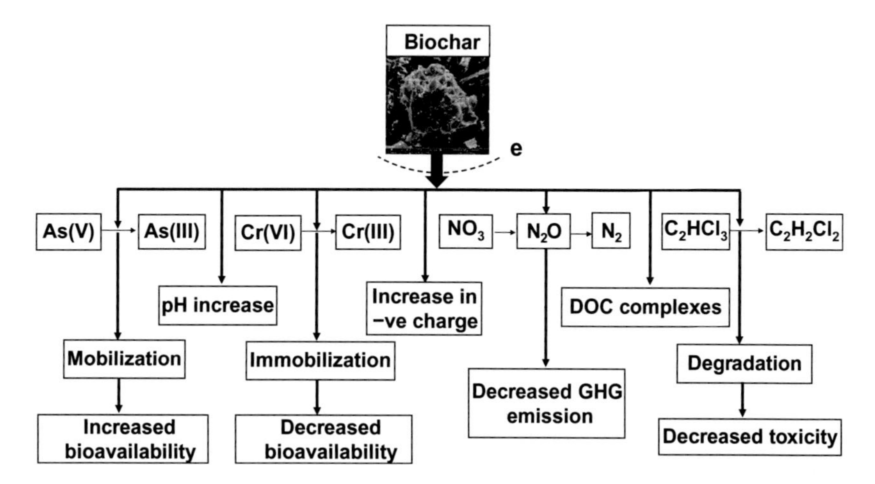

FIGURE 21.1 Biochar-induced reduction of arsenate [As(V)], chromate [Cr(VI)], nitrate (NO_3), and TCE (C_2HCl_3). *DOC*, Dissolved organic carbon.

and Mexico, derived from the release of As from sedimentary rocks (Mahimairaja et al., 2005). Both biosolids and phosphate fertilizer applications are the main source of both toxic and beneficial metal(loid) inputs including Cd, Zn, Cu, and Mb in many countries (Bolan et al., 2014).

21.3.1 Redox Reactions of Metal(loid)s

The most common metal(loid)s that are subject to biotic and abiotic redox reactions include As, Cr, Hg, and Se. In general, metal(loid)s are less soluble in their higher oxidation state, while in the case of metalloids, the solubility and mobility is controlled by both the oxidation state and the ionic form (cation vs anion) (Ross, 1994). The redox reactions are grouped into two categories, assimilatory and dissimilatory. In assimilatory reactions, metal(loid) substrates are involved in the metabolic activity of the organism by serving as terminal electron acceptor; whereas in the case of dissimilatory reactions, the metal (loid) substrates have no functional role, and represent fortuitous reductions coupled to enzymatic oxidations of some other component in the metal(loid) substrates.

The biogeochemistry of both abiotic and biotic redox reactions of metal(loid)s has not been adequately characterized. Metal(loid) transformation is linked to the cytochrome-mediated system (Rosen and Silver, 1987), and metal(loid)-active enzymes are involved in their reduction process. For example, the Se reducing organisms such as *Thauera selenatis* reduce Se(IV) to Se(II) using a selenate reductase (Recha and Macy, 1992), and the subsequent reduction of Se(II) to Se^{o} is catalyzed by nitrite reductase. Arsenic in soils and sediments can be oxidized to As(V) by bacteria. However, under aerobic conditions, As is present as $H_2AsO_4^-$ in acidic environments (e.g., acid soils) and as $HAsO_4^{2-}$ in alkaline environments (e.g., calcareous soils). Under reduced conditions, arsenite [As(III)], which is much more toxic and mobile than As(V), is the major As species, but elemental arsenic (As^{o}) and arsine (H_2As) can also be formed. The reduction of Cr(VI) to Cr(III) can enhance the mobilization and bioavailability of Cr. While oxidation of Cr(III) to Cr(VI) is mostly mediated abiotically through oxidizing agents such as mangenous oxide, reduction of Cr (VI) to Cr(III) is induced by both abiotic and biotic processes (Choppala et al., 2013). In the presence of a ready source of electrons (e.g., ferrous iron [Fe(II)]), Cr(VI), can readily be reduced to Cr(III). Selenium is reduced under both aerobic and anaerobic environments. Selenite is readily reduced to the elemental state by both abiotic process in the presence of chemical reductants and biotic process by reductase enzyme system. Mercury is subject mostly to biological reduction process in soils and sediments. Microorganisms play a major role in the reduction of reactive Hg(II) to toxic methylmercury and to nonreactive Hg^{o} that may be lost through volatilization.

21.3.2 Biochar-Induced Reduction of Arsenic and Chromium

It is reported that redox processes of biochar are important in the immobilization of inorganic contaminants. Recently, Choppala et al. (2016) examined the differential effect of biochar on reduction-mediated mobility and bioavailability of As(V) and Cr(VI). Overall, their study demonstrated that the addition of biochar facilitated the reduction of As(V) to

As(III) and subsequent mobilization and bioavailability of As (Eq. 21.1). However, the biochar-induced reduction of Cr(VI) to Cr(III) was effective in inducing immobilization and decreasing bioavailability of Cr(III) (Eq. 21.2).

$$AsO_4^{3-} + H^+ + 2e^- \rightarrow AsO_3^{3-} + H_2O \quad (21.1)$$

$$Cr_2O_7^{2-} + 14H^+ + 6e^- \rightarrow 2Cr^{3+} + 7H_2O\ (\log K = 4.512) \quad (21.2)$$

Biochar supplies a source of electrons, thereby enabling the reduction of As(V) and Cr (VI) in soils. For example, Dong et al. (2014) revealed that dissolved organic matter (DOM) in biochar acted as both electron donor and acceptor, reducing Cr(VI). The reduction of aqueous toxic Cr(VI) into less toxic Cr(III) by carboxylic and hydroxyl groups on biochar surface has been well studied. This reduction can be facilitated by p-electrons donated from the disordered polycyclic aromatic hydrocarbons sheets (Wang et al., 2010). However, it has been highlighted that pH of the media can significantly impact on the redox reactions (Dixit and Hering, 2003). The reduction reactions of these metal(loid)s generally increase pH and consequent negative charges in both soil and carbon amendments. While the increase in surface negative charge promotes the retention of positively charged Cr(III) species, it decreases the retention of negatively charged As(III) species due to anion exclusion. Although with the immobilization techniques are effective in decreasing bioavailability of contaminants, the total concentration of contaminants in soils is unlikely to change. The biochar-induced reduction of Cr(VI) to Cr(III) facilitates the immobilization of Cr in soil, whereas the reduction of As(V) to As(III) may promote As removal through phytoremediation techniques.

21.4 NUTRIENT CONTAMINANTS

Nutrients such as nitrogen and phosphorus can become contaminants when they reach surface and groundwater sources through soil erosion and leaching. Nitrogen application to soil can also lead to the emission of greenhouse gases, such as nitrous oxide.

21.4.1 Redox Reactions of Nitrogen

Nitrogen undergoes oxidation (nitrification) and reduction (denitrification) reactions in soil. The redox reactions of N in soil are mediated mainly by soil microorganisms (biotic transformation). Nitrification refers to the microbial oxidation of NH_4^+ to NO_3^-, known as nitrification (Eq. 21.3). NH_4^+ ions are released either indirectly from the ammonification reaction of organic matter and organic forms of N fertilizers or directly from the solubilization of ammonium fertilizers. The nitrification reaction includes two stages in which the NH_4^+ ions are first converted into NO_3^-.

$$2NH_4^+ + 4O_2 \rightarrow 2NO_3^- + 4H^+ + 2H_2O \quad (21.3)$$

Nitrate formed through the further oxidation of NO_3^- or added through nitrate compounds is subject to a number of processes, including plant uptake, leaching losses, immobilization, and denitrification. Under reduced conditions, some microorganisms derive their oxygen demand from NO_3^-, resulting in the reduction of NO_3^-, which proceeds in a series of steps, releasing NO_2^-, nitric oxide (NO), nitrous oxide (N_2O), and N_2 gas (Eq. 21.4). Denitrification can lead both the loss of a valuable plant nutrient and the release of N_2O (greenhouse gas).

$$NO_3^- => NO_2 => NO => N_2O => N_2 \qquad (21.4)$$

Denitrification is the last step in the N cycle, where both the biologically and chemically fixed N is returned to the atmospheric pool of inert N_2, especially in the presence of abundant carbon sources (Eq. 21.5). Denitrification is induced by respiratory denitrifiers that obtain energy by linking N-oxide reduction process to electron transport phosphorylation reaction (Tiedje, 1988).

$$5(CH_2O) + 4NO_3^- + 4H^+ \rightarrow 2N_2 + 5CO_2 + 7H_2O \qquad (21.5)$$

21.4.2 Biochar-Induced Reduction of Nitrogen

Numerous studies have shown that biochar has potential to impact N cycling through modulation of primary N transformation processes (Taghizadeh-Toosi et al., 2012; Cayuela et al., 2014; Zhou et al., 2016). Several laboratory and field studies have reported that biochar exerts a substantial control on denitrification and could mitigate from 50% to 90% of N_2O soil emission (Cayuela et al., 2014; Ameloot et al., 2016). Cayuela et al. (2013) carried out laboratory incubation experiments using a range of soils and different biochars produced at 500°C. They reported biotic N_2O production and a consistent decrease in the N_2O/N_2 ratio after biochar application. They suggested that biochar promotes the final step of denitrification (i.e., the conversion of N_2O to N_2) by acting as an "electron shuttle" that enables the allocation of electrons to denitrifying microorganisms. Overall, the role of biochar in redox-regulated nitrogen cycling could be explained by the following mechanisms: (1) biochar enhances soil aeration by reducing bulk density and increasing porosity and thereby depresses the nitrification potential (Alburquerque et al., 2015; Gul et al., 2015); (2) biochar could overexpress the nosZ gene of denitrifiers (Cayuela et al., 2013; Harter et al., 2014), increasing the synthesis of nitrous oxide reductase responsible for catalyzing the reduction of N_2O to N_2; and (3) it also elevates soil pH, increasing N_2O reductase activity that promotes N_2 formation and higher N_2/N_2O ratios (Cayuela et al., 2014; Gul et al., 2015).

21.5 ORGANIC CONTAMINANTS

Biochar contain both organic carbon and inorganic ash components. The high organic carbon content in biochars favors the partition organic contaminants and their subsequent reduction, thereby influencing their mobility and bioavailability in soil.

21.5.1 Redox Reactions of Organic Contaminants

Reductive dehalogenation process is important in managing a persistent class of contaminants. It has often been shown that anaerobic processes involving these compounds can result in dehalogenated compounds that are considered less toxic, less bioaccumulative, and more prone to subsequent degradation processes. Both aromatic (e.g., dioxin) and nonaromatic (e.g., alkyl benzene) organic contaminants are subject to these dehalogenation processes. Halogenated organic compounds that are relatively oxidized due to the presence of highly electronegative halogen component, are more prone to reduction. Thus with increased halogenation of organic compounds, reduction is more possible than oxidation, leading to reductive mineralization (Vogel et al., 1987). Reduction occurs when the electrode potential of soil or groundwater system is less than that of the organic contaminant present in the system. The electrode potential refers to the redox status of the system, indicating potential for the supply of electrons to a reducible compound. The anaerobic biological processes that promote the reductive dehalogenation of organic compounds may play a vital role in the bioremediation of terrestrial (i.e., soil) and aquatic (i.e., groundwater) polluted with these contaminants. The microbial decomposition readily degradable organic matter, such as DOM in these environments, may lead to anaerobic condition due to consumption of oxygen, thereby promoting the dehalogenation process.

21.5.2 Biochar-Induced Reduction of Trichlroethane

Conventionally, organic contaminants absorbed to biochar have been considered to be chemically and biologically inert. However, numerous recent studies disclosed the ability of biochar to catalyze the chemical and biological transformation of organic contaminants, such as hydrocarbons, explosives, pesticides, antibiotics, and solvents, leading to the degradation of these contaminants. For example, Oh et al. (2013) reported that biochar stimulated chemical reduction of nitro herbicides and the explosives by shuttling electron between reductants (i.e., dithiothreitol) and organic contaminants. Ding and Xu (2016) showed that biochar in the presence of sulfide can foster the degradation of persistent organic pollutants including 1,1-trichloro-2,2-di(4-chlorophenyl)ethane and 1,1-dichloro-2,2-bis(4-chlorophenyl)-ethane. Besides, because of the complex compositions of biochar, Oh et al. (2013) mentioned that redox-active metal(loid)s (i.e., Fe, Cu, and Mn) of biochar could also involve in the enhanced reduction of nitro herbicides. It was revealed that $-COOH$ and $-OH$ groups in biochar surface acted as an electron-transfer mediator, to enhance the formation of SO_4^{2-} accelerating the degradation of trichloroethane (TCE) in water.

Similar to the role of biochar in chemical transformation of contaminants, biochar-mediated extracellular electron transfer could also promote the biotransformation of organic contaminants. For example, Tong et al. (2014) and Yu et al. (2015) demonstrated that biochar can promote the microbial degradation of pentachlorophenol, which was ascribed to extracellular electron transfer resulting from the biochar-induced growth and assimilation of exoelectrogens in a mixed culture. TCE is used as an industrial solvent and degreaser, and is a common ingredient in many household products like paints, adhesives, and spot removers. Biochar application has been shown to decrease TCE concentration in soil, and the decrease in TCE concentrations may be attributed to a number of biochar-induced processes, including adsorption, volatilization and reduction (Eq. 21.6).

$$C_2HCl_3 + 3H^+ + 6e^- \rightarrow C_2H_4 + 3Cl^- \ (\log K = 2.10) \quad (21.6)$$

21.6 REDOX REACTIONS IN RELATION TO MOBILITY AND BIOAVAILABILITY

The mobility and bioavailability of contaminants in soil are controlled by their speciation and interaction with soil components (Fig. 21.2). Redox reactions play an important role in the remediation of contaminated soil, sediment, and water. Biochar can be used to manipulate the redox reactions of contaminants, thereby facilitating the remediation of contaminated soil and water sources. Biochar-induced redox reactions can lead to alternations in the toxicity, water solubility, and/or the mobility and bioavailability of the contaminant (Alexander, 1999). An increase in solubility and subsequent mobility can be manipulated to remediate insoluble contaminants in soil because the released product is transformed from the solid phase into the solution phase. Conversely, a decrease in contaminant solubility can be exploited to eliminate the contaminants from soil and groundwater sources through immobilization. In some cases, gaseous contaminant products can be eliminated through volatilization. Thus biochar-induced metal(loid) reduction processes can be utilized for both natural and monitored engineered bioremediation of polluted environments.

Because of the lower mobility of Cr(III) than Cr(VI), the reduction process is likely to result in the immobilization of Cr, thereby mitigating the transport and off-site contamination. For example, reduction of Cr(VI) to Cr(III), and subsequent precipitation of Cr (III) ion, is the most common method for treating wastewater sources that are rich in Cr (VI), such as timber treatment industrial effluents (James, 2001). Using biochar as an energy and C source, direct reduction of Cr(VI) to Cr(III) can be accomplished. Indirect reduction is achieved by sulfate loading on biochar, resulting in the release of H_2S that not only reduces Cr(VI) to Cr(III), but also results in the precipitation of Cr as Cr_2S_3 (Vainshtein et al., 2003).

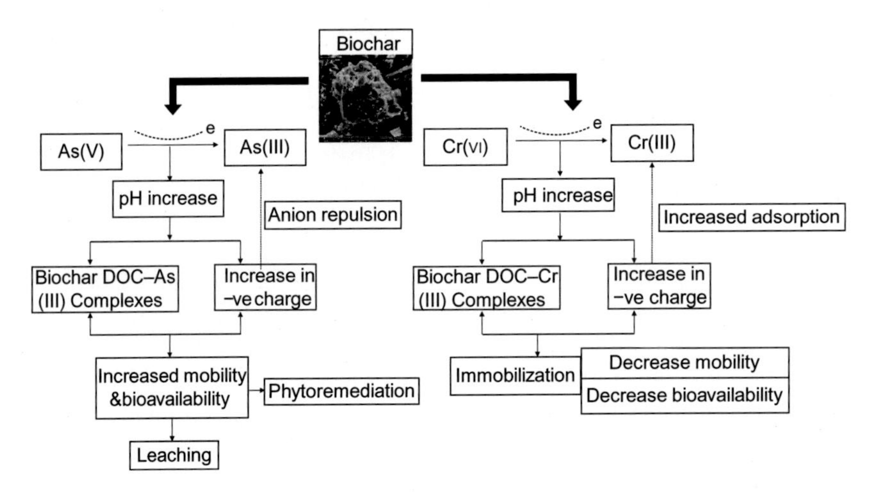

FIGURE 21.2 Biochar-induced reduction of arsenate [As(V)] and chromate [Cr(VI)] in relation to mobility and bioavailability. *DOC*, Dissolved organic carbon.

In the case of Se, the magnitude of Se(VI) reduction is controlled by the supply of C sources, and the reduced insoluble elemental Se can be separated from contaminated water sources. Similarly, methylation reactions result in the formation of gaseous metal (loid) species, which can readily be lost through volatilization (Banuelos and Lin, 2007; Thompson-Eagle and Frankenberger, 1992). Biochar application promotes both the reduction and subsequent methylation of Se(VI), thereby resulting in the removal of Se from both soil and groundwater sources.

21.7 CONCLUSIONS

Biochars application provides electrons, thereby facilitating the reduction of heavy metal(loid)s, such as As(V) and Cr(VI); nutrient elements, such as nitrogen; and organic compounds, such as TCE. Biochar-induced reduction reactions can be manipulated to manage the remediation of contaminated soil. For example, the addition of biochar is effective in promoting the mitigation of Cr toxicity by reducing Cr(VI) to Cr(III) and subsequent immobilization of Cr species. In contrast, the biochar-induced conversion process of As(V) to As(III) species increases mobility of As, thereby enabling the phytoremediation of As in soil. The biochar-promoted reduction of nitrate nitrogen results in the release of nitrous oxide emission. Biochar addition enhances the reduction of TCE, thereby decreasing its bioavailability and toxicity.

References

Alburquerque, J.A., Sanchez-Monedero, M.A., Roig, A., Cayuela, M.L., 2015. High concentrations of polycyclic aromatic hydrocarbons (naphthalene, phenanthrene and pyrene) failed to explain biochar's capacity to reduce soil nitrous oxide emissions. Environ. Pollut. 196, 72–77.

Alexander, M., 1999. Biodegradation and Bioremediation, second ed. Academic Press, San Diego, CA.

Ameloot, N., Maenhout, P., De Neve, S., Sleutel, S., 2016. Biochar-induced N_2O emission reductions after field incorporation in a loam soil. Geoderma 267, 10–16.

Banuelos, G.S., Lin, Z.Q., 2007. Acceleration of selenium volatilization in seleniferous agricultural drainage sediments amended with methionine and casein. Environ. Pollut. 150, 306–312.

Bolan, N., Kunhikrishnan, A., Thangarajan, R., Kumpiene, J., Park, J., Makino, T., et al., 2014. Remediation of heavy metal(loid)s contaminated soils—to mobilize or to immobilize? J. Hazard. Mater. 266, 141–166.

Cayuela, M.L., Sanchez-Monedero, M.A., Roig, A., Hanley, K., Enders, A., Lehmann, J., 2013. Biochar and denitrification in soils: when, how much and why does biochar reduce $N_{(2)}O$ emissions?. Sci. Rep. 3, 1732.

Cayuela, M.L., van Zwieten, L., Singh, B.P., Jeffery, S., Roig, A., Sánchez-Monedero, M.A., 2014. Biochar's role in mitigating soil nitrous oxide emissions: a review and meta-analysis. Agric. Ecosyst. Environ. 191, 5–16.

Choppala, G., Bolan, N., Park, J.H., 2013. Chromium contamination and its risk management in complex environmental settings, Adv. Agron., vol. 120. pp. 129–172.

Choppala, G., Bolan, N., Kunhikrishnan, A., Bush, R., 2016. Differential effect of biochar upon reduction-induced mobility and bioavailability of arsenate and chromate. Chemosphere 144, 374–381.

Ding, K., Xu, W., 2016. Black carbon facilitated dechlorination of DDT and its metabolites by sulfide. Environ. Sci. Technol. 50, 12976–12983.

Dixit, S., Hering, J.G., 2003. Comparison of arsenic(V) and arsenic(III) sorption onto iron oxide minerals: implications for arsenic mobility. Environ. Sci. Technol. 37, 4182–4189.

Dong, H., Zhang, C., Hou, K., Cheng, Y., Deng, J., Jiang, Z., et al., 2017. Removal of trichloroethylene by biochar supported nanoscale zero-valent iron in aqueous solution. Sep. Purif. Technol. 188, 188–196.

Dong, X., Ma, L.Q., Gress, J., Harris, W., Li, Y., 2014. Enhanced Cr(VI) reduction and As(III) oxidation in ice phase: important role of dissolved organic matter from biochar. J. Hazard. Mater. 267, 62–70.

Fang, G., Gao, J., Liu, C., Dionysiou, D.D., Wang, Y., Zhou, D., 2014. Key role of persistent free radicals in hydrogen peroxide activation by biochar: implications to organic contaminant degradation. Environ. Sci. Technol. 48, 1902–1910.

Fang, G., Liu, C., Gao, J., Dionysiou, D.D., Zhou, D., 2015. Manipulation of persistent free radicals in biochar to activate persulfate for contaminant degradation. Environ. Sci. Technol. 49, 5645–5653.

Gul, S., Whalen, J.K., Thomas, B.W., Sachdeva, V., Deng, H., 2015. Physico-chemical properties and microbial responses in biochar-amended soils: mechanisms and future directions. Agric. Ecosyst. Environ. 206, 46–59.

Harter, J., Krause, H.M., Schuettler, S., Ruser, R., Fromme, M., Scholten, T., et al., 2014. Linking N_2O emissions from biochar-amended soil to the structure and function of the N-cycling microbial community. ISME J. 8, 660–674.

He, T., Liu, D., Yuan, J., Luo, J., Lindsey, S., Bolan, N.S., et al., 2018. Effects of application of inhibitors and biochar to fertilizer on gaseous nitrogen emissions from an intensively managed wheat field. Sci. Total Environ. 628–629, 121–130.

James, B.R., 2001. Remediation-by-reduction strategies for chromate-contaminated soils. Environ. Geochem. Health 23, 175–179.

Kappler, A., Wuestner, M.L., Ruecker, A., Harter, J., Halama, M., Behrens, S., 2014. Biochar as an electron shuttle between bacteria and Fe(III) minerals. Environ. Sci. Technol. Lett. 1, 339–344.

Keiluweit, M., Nico, P.S., Johnson, M.G., Kleber, M., 2010. Dynamic molecular structure of plant biomass-derived black carbon (biochar). Environ. Sci. Technol. 44, 1247–1253.

Kemper, J.M., Ammar, E., Mitch, W.A., 2008. Abiotic degradation of hexahydro-l,3,5-trinitro-1,3,5-triazine in the presence of hydrogen sulfide and black carbon. Environ. Sci. Technol. 42, 2118–2123.

Klupfel, L., Keiluweit, M., Kleber, M., Sander, M., 2014. Redox properties of plant biomass-derived black carbon (biochar). Environ. Sci. Technol. 48, 5601–5611.

Lehmann, J., Joseph, S., 2015. Biochar for Environmental Management: Science, Technology and Implementation, second ed. Routledge, London and New York.

Liu, N., Ding, L., Li, H., Jia, M., Zhang, W., An, N., et al., 2017. N-doped nanoporouscarbon as efficient catalyst for nitrobenzene reduction in sulfide-containingaqueous solutions. J. Colloid Interface Sci. 490, 677–684.

Mahimairaja, S., Bolan, N.S., Adriano, D.C., Robinson, B., 2005. Arsenic contamination and its risk management in complex environmental settings. Adv. Agron. 86, 1–82.

Mandal, S., Sarkar, B., Bolan, N.S., Novak, J., Ok, Y.S., van Zwieten, L., et al., 2016. Designing advanced biochar products for maximizing greenhouse gas mitigation potential. Environ. Sci. Technol. 46, 1367–1401.

Oh, S.Y., Seo, Y.D., 2016. Polymer/biomass-derived biochar for use as a sorbent andelectron transfer mediator in environmental applications. Bioresour. Technol. 218, 77–83.

Oh, S.Y., Son, J.G., Chiu, P.C., 2013. Biochar-mediated reductive transformation of nitro herbicides and explosives. Environ. Toxicol. Chem. 32, 501–508.

Prevoteau, A., Ronsse, F., Cid, I., Boeckx, P., Rabaey, K., 2016. The electron donating capacity of biochar is dramatically underestimated. Sci. Rep. 6, 32870.

Rech, S.A., Macy, J.M., 1992. The terminal reductases for selenate and nitrate respiration in *Thauera selenatis* are two distinct enzymes. J. Bacteriol. 174, 7316–7320.

Rosen, B.P., Silver, S., 1987. Ion Transport in Prokaryotes, first ed. Academic Press, San Diego, CA.

Ross, S.M., 1994. Sources and forms of potentially toxic metals in soil–plant systems. In: Ross, S.M. (Ed.), Toxic Metals in Soil-Plant Systems. Wiley, Chichester, pp. 4–25.

Sanchez-Monedero, M.A., Cayuela, M.L., Roig, A., Jindo, K., Mondini, C., Bolan, N.S., 2018. Role of biochar as an additive in organic waste composting. Bioresour. Technol. 247, 1155–1164.

Taghizadeh-Toosi, A., Clough, T.J., Sherlock, R.R., Condron, L.M., 2012. A wood based low-temperature biochar captures NH_3-N generated from ruminant urine-N, retaining its bioavailability. Plant Soil 353, 73–84.

Thangarajan, R., Bolan, N.S., Kunhikrishnan, A., Wijesekara, H., Xu, Y., Tsang, D.C.W., et al., 2018. The potential value of biochar in the mitigation of gaseous emission of nitrogen. Sci. Total Environ. 612, 257–268.

Thompson-Eagle, E.T., Frankenberger, W.T., 1992. Bioremediation of soils contaminated with selenium. In: Lal, R., Stewart, B.A. (Eds.), Advances in Soil Science. Springer, New York, pp. 261–309.

Tiedje, J.M., 1988. Ecology of denitrification and dissimilatory nitrate reduction to ammonium. In: Zehnder, A.J.B. (Ed.), Biology of Anaerobic Microorganisms. John Wiley & Sons, New York, pp. 179–244.
Tong, H., Hu, M., Li, F.B., Liu, C.S., Chen, M.J., 2014. Biochar enhances the microbial and chemical transformation of pentachlorophenol in paddy soil. Soil Biol. Biochem. 70, 142–150.
Vainshtein, M., Kuschk, P., Mattusch, J., Vatsourina, A., Wiessner, A., 2003. Model experiments on the microbial removal of chromium from contaminated groundwater. Water Res. 37, 1401–1405.
Vogel, T.M., Criddle, C.S., McCarty, P.L., 1987. ES critical reviews: transformations of halogenated aliphatic compounds. Environ. Sci. Technol. 21, 722–736.
Wang, X.S., Chen, L.F., Li, F.Y., Chen, K.L., Wan, W.Y., Tang, Y.J., 2010. Removal of Cr(VI) with wheat-residue derived black carbon: reaction mechanism and adsorption performance. J. Hazard. Mater. 175, 816–822.
Xu, S., Adhikari, D., Huang, R., Zhang, H., Tang, Y., Roden, E., et al., 2016. Biochar-facilitated microbial reduction of hematite. Environ. Sci. Technol. 50, 2389–2395.
Xu, W., Pignatello, J.J., Mitch, W.A., 2013. Role of black carbon electrical conductivityin mediating hexahydro-1,3,5-trinitro-1,3,5-triazine (RDX) transformation oncarbon surfaces by sulfides. Environ. Sci. Technol. 47, 7129–7136.
Xu, Y., Lou, Z., Yi, P., Chen, J., Ma, X., Wang, Y., et al., 2014. Improving abiotic reducing ability of hydrothermal biochar by low temperatureoxidation under air. Bioresour. Technol. 172, 212–218.
Yan, J., Han, L., Gao, W., Xue, S., Chen, M., 2015. Biochar supported nanoscale zerovalent iron composite used as persulfate activator for removing trichloroethylene. Bioresour. Technol. 175, 269–274.
Yu, L., Yuan, Y., Tang, J., Wang, Y., Zhou, S., 2015. Biochar as an electron shuttle for reductive dechlorination of pentachlorophenol by *Geobacter sulfurreducens*. Sci. Rep. 5, 16221.
Yuan, Y., Bolan, N., Prévoteau, A., Vithanage, M., Biswas, J.K., Ok, Y.S., et al., 2017. Applications of biochar in redox-mediated reactions. Bioresour. Technol. 246, 271–281.
Zhou, G.W., Yang, X.R., Li, H., Marshall, C.W., Zheng, B.X., Yan, Y., et al., 2016. Electron shuttles enhance anaerobic ammonium oxidation coupled to iron(III) reduction. Environ. Sci. Technol. 50, 9298–9307.

PART IV

FUTURE PROSPECTS

CHAPTER

22

Future Biochar Research Directions*

J.M. Novak[1], *E. Moore*[2], *K.A. Spokas*[3], *K. Hall*[4] *and A. Williams*[5]

[1]U.S. Department of Agriculture, Agricultural Research Service, Coastal Plains Research Center, Florence, SC, United States [2]Department of Environmental Science, School of Earth and the Environment, Rowan University, Glassboro, NJ, United States [3]US Department of Agriculture, Agricultural Research Service, Soil Water Management Research Unit, St. Paul, MN, United States [4]Department of Soil, Water, and Climate, University of Minnesota, St. Paul, MN, United States [5]Davidson Laboratory, Stevens Institute of Technology, Hoboken, NJ, United States

22.1 INTRODUCTION

Biochar can be a material for improving soil properties while also enhancing plant growth for higher crop yields. Unfortunately, it is recognized that there is not a universal biochar for restoration of all soil deficiencies. Thus a more effective paradigm may be to design the biochar to have selective properties tailored for existing or emerging problems. The designer biochar technology was described in Novak and Busscher (2012) and was validated using sandy soils with low fertility, organic carbon (OC) contents, and poor water hydraulics (Novak et al., 2014, 2016; Sigua et al., 2016b). This is also emphasized in the recent meta-analyses of biochar results, indicating that they are most promising in fertility-poor, coarse-textured soils (Jeffery et al., 2011; Crane-Droesch et al., 2013). However, these problems also exist in coastal area sand dunes where plant growth is limited by similar poor plant growth conditions. Therefore one of the topics in this chapter is to develop designer biochar that is tailored for restoration of sand dunes. This topic is timely considering the recent attention to hurricane destruction of dunes at the New Jersey shore (Barone et al., 2014), and after two recent

* USDA is an equal opportunity provider and employer.

Biochar from Biomass and Waste
DOI: https://doi.org/10.1016/B978-0-12-811729-3.00022-4

hurricanes in South Carolina (Hurricane Irma in 2016 and Hurricane Matthew in 2017; https://www.usgs.gov/news/hurricanes).

Biochar utilization should not be limited to solely soil applications. Biochar utilization in the electronics sector is growing since it has recently been reported to act as a supercapacitor and battery (Gao et al., 2017). It is regarded as being the most environmental friendly energy storage technology because the starting biochar matrix is produced from renewable sources (Gu et al., 2015; Gao et al., 2017). Much research is still underway further to develop/optimize the electrical capabilities of biochars through modifying their morphology, porosity, and surface chemistry. This suggests that the designer biochar concept can also be applied to the electronic sector. The last section of this chapter will discuss the historical perspectives of charcoal (biochar) as an energy/conductor source and then review potential pathways at the molecular scale for its electrical performance.

22.2 INTENSIFICATION OF FARMING AND PROJECTED SOIL DEGRADATION

Meeting the projected global food demand in the next 30 years will cause considerable stress on current agronomic practices and soil quality conditions (Tilman et al., 2011; Lal, 2015). Some reports suggest that meeting the global food demand will require agriculture crop yields to increase by 60%–110% above current crop production levels (FAO, 2009; Ray et al., 2013). On the other hand, clearing more land for agricultural production is not a sustainable practice (Godfray et al., 2010; Phalan et al., 2011), so intensification (e.g., higher seeding rates, more crop growth cycles per annum, more tillage, etc.) of crop production on existing land is occurring. Increasing farming intensification is projected to advance soil quality degradation of existing farm land through loss of OC by tillage, depletion of soil nutrients, loss of soil aggregation, and by increases in soil erosion rates and acidification (Lal, 2009, 2010). For continual crop production, degraded soils will eventually need to have their fertility and physical properties restored (Lal, 2015). Biochar usage as a soil amendment is one proposed mechanism to cope with these potential soil degradation issues (Atkinson et al., 2010; Spokas et al., 2012; Biederman and Harpole, 2013).

22.3 PYROLYSIS AND BIOCHAR PROPERTIES

Biochar is a solid material produced during the thermal pyrolysis of organic feedstocks. Many different types of organic-based feedstocks can be used for biochar production, including agricultural wastes products (e.g., manure, crop residues), industrial byproducts (e.g., wood debris, flooring) and municipal wastes (e.g., cardboard, biosolids). These feedstocks are subject to thermal processing at temperatures ranging between 300°C and 800°C that result in volatile compound release and carbonization of structural components into biochar. Since carbonization is a destructive process, plant/wood structural compounds (i.e., cellulose, lignin) are converted into an assortment of condensed aromatic material, nonaromatic ring-type substances, tars, and inorganic elements (e.g., Ca, K, P) in their ash fraction (Antal and Grønli, 2003; Cantrell et al., 2012; Ippolito et al., 2015).

Carbonization causes the releases of gasses that peel apart the condensed polyaromatic sheets creating fissures, pores, and void spaces that increase their high surface area. At lower pyrolysis temperatures (<350°C), biochars possess more functional groups, such as oxygen-rich C = O, C = OH, and COOH groups. These oxygen-rich functional groups have unpaired electrons that allow biochar to participate in redox reactions (Klupfel et al., 2014) and for the biochar electrically to rebalance itself by attracting positively charged inorganic species (i.e., Fe, Al, Mn) (Joseph et al., 2010). On the other hand, pyrolysis at temperatures (>400°C) causes transformation of carbonaceous feedstocks into polycondensed aromatic compounds. The structural features permit biochars to serve as electron shuttle pathways between Fe-minerals and microbes (Kappler et al., 2014). Because of these characteristics, biochars can increase soil OC contents (Laird et al., 2009), electrostatically bind cations to functional groups (Pignatello et al., 2015), participate in redox reactions (Joseph et al., 2010), supply critical plant nutrients (Ippolito et al., 2015), and influence other soil chemical reactions (Beesley and Marmiroli, 2011; Li et al., 2017). Unfortunately, a key principal determined through a meta-analysis of the biochar literature has shown that not all biochars have abilities to produce crop yields improvements, in spite, of all these beneficial properties (Jeffery et al., 2011; Spokas et al., 2012). However, a clear dependency on soil type is evident (Jeffery et al., 2011). A more efficient usage of biochars could be to create a biochar with physical and chemical properties that can remedy specific soil deficiencies.

22.3.1 Designing Biochars for Specific Purposes

Previous analysis of several hundred biochar research articles reported dissimilar responses of biochar at improving soils quality characteristics and, at times, having minimal or no impact at increasing crop yields (Jeffery et al., 2011; Spokas et al., 2012; Crane-Droesch et al., 2013). In response to these reported mixed results, Novak et al. (2008) developed the concept of *Designer Biochar* where the biochar is manufactured with chemical and physical properties that are matched to improve specific soil limitations. Several reports demonstrated that biochars can be produced with properties that expressly improve soil fertility (Novak et al., 2009, 2014; Ippolito et al., 2016), increase soil moisture retention (Novak et al., 2012, 2014; Ippolito et al., 2016) and reduce soil physical limitations (Busscher et al., 2010).

The above reported examples of designer biochar paradigm were tested under controlled laboratory and greenhouse conditions, so it can be argued that more vetting is needed under actual field conditions. These field tests using designer biochar could be directed to soil quality problems in nonagronomic landscapes, such as marine/seashore settings.

22.4 BIOCHAR USAGE IN SALINE SOILS AND SAND DUNE RESTORATION

Designer biochars can be further vetted through evaluation as an amendment for improving plant cover on sand dunes. Sandy beaches and dunes account for roughly one-third of the world's ice-free coastlines (Hardisty, 1994), including densely populated coastal cities. Most beaches have vegetated sand ridges called dunes, built up by deposition of beach sand blown inland or transported by surf during storms (Maun, 2009). The presence of plants stabilizes the dune by trapping the blowing sand, allowing for accumulation and dune enlargement. Some native dune plant species have adapted physiology (e.g., fast growing roots and stems) to deal with increasing sand deposits (Maun, 2009). Water catchment in coastal dunes is also linked to vegetation and dune stabilization (Van der Meulen, 1982). Thus plants have a vital role in dune establishment and their longevity. For coastal beach protection, it is important to have vegetated sand dunes to minimize sand loss and movement via wave action from storms. Additionally, overlying vegetation must be maintained, and protected from damage by human traffic to maximize the benefits of coastal dunes (Wootton et al., 2016).

Coastal dunes provide many valuable ecosystem services, including coastal protection, erosion control, water catchment and purification, raw materials, maintenance of wildlife, carbon sequestration, tourism, recreation, and education (Barbier et al., 2011). Coastal protection is extremely important in the current state of climate change, sea level rise, and extreme weather events, leading to the loss of coastal landscape buffers to natural disasters. During Hurricane Sandy, the New Jersey coastline with engineered dunes had greater protection of landward structures from storm surges and wave action, whereas communities that did not have protective dunes or only possessed low narrow dunes suffered the greatest damage (Barone et al., 2014). As shown, dunes with vegetation are more stable and resilient (Fig. 22.1).

After the Indian Ocean Tsunami, Liu et al. (2005) reported that the loss of 150 lives occurred at a single resort located behind an area where a foredune was removed to improve the beach and ocean views. Vegetation root structure is crucial for stabilizing and strengthening dunes to control erosion and maintain coastal protection, which has significant security and economic value (Huang et al., 2007).

Saltwater intrusion is another destructive consequence of sea level rise and extreme weather, which is damaging coastal agricultural lands and threatening global food security (NIBIO, 2017). Estimates indicate that approximately 30% of irrigated croplands are

(A)

(B)

FIGURE 22.1 (A) Unstable coastal sand dune in Long Beach Island, NJ, after Hurricane Sandy. (B) Stable restored dune with planted vegetation in Long Beach Island, NJ, after Hurricane Sandy. Source: *Dr. Amy Williams at Stevens Institute of Technology, Newark, NJ. (Permission was granted for their use.)*

adversely affected by soil salinization (Chaves et al., 2009). Biochar application to coastal dunes can be used to mitigate the impacts of saltwater intrusion onto coastal agricultural lands. Research has shown that the greatest positive impacts of soil biochar additions, including increased water retention (Arthur and Ahmed, 2017), are observed in nutrient poor, acidic, coarse-textured soils (Jeffery et al., 2011). Other studies have shown that biochar application increases plant growth, biomass, yield, photosynthesis, nutrient uptake, and modified gas exchange, while minimizing Na^+ uptake under drought or salt stress (Ali et al., 2017). Specifically focusing on individual types of plants, biochar additions have been shown to alleviate salt-induced mortality in two herbaceous plant species exposed to otherwise fatal saline conditions (Thomas et al., 2013), and mitigate salinity stress in potato cultivation (Akhtar et al., 2015). Biochar also enhances sand cultivated growth of the cyanobacteria *Microcoleus vaginatus*, as well as increasing fixed sand weight in biological soil crusts (Meng and Yuan, 2014), indicating that biochar can stimulate a diverse range of photosynthesizers that provide multiple mechanisms of dune stabilization.

Dune vegetation growth also contributes to carbon sequestration, which varies depending on the type of vegetation, sedimentation, and coastal geomorphology (Barbier et al., 2011). Enhancing carbon sequestration in coastal systems is a useful potential strategy for reducing greenhouse gas concentrations and abate climate change impacts (i.e., sea level rise). If biochar is produced with a high enough temperature, then the buried biochar itself represents sequestered carbon because of its resistance to oxidation (Glaser et al., 2002; Hamer et al., 2004). Burial of organic matter on coastal systems has had major consequences on the evolution of the geosphere and atmosphere. The evolution of oxygenic photosynthesis was accompanied by enhanced carbon burial, leading to net accumulation of oxygen in the atmosphere (Holland, 2006). Today in the Anthropocene, the burial of biochar in coastal dunes can be part of a larger effort to remove carbon from the atmosphere (Kammann et al., 2017).

Biochars can be designed for coastal sand dunes to improve their fertility and enhance the positive benefits of healthy dune vegetation (Fig. 22.2). Various treatments, including slow release fertilizers, compost/seaweed, water absorbing gels, and mycorrhizae fungi are recommended strategies for promoting vegetation growth/health in coastal dune restoration projects (Van der Meulen, 1982). Different biochar feedstocks and their inherent physical properties can be used to address different soil quality issues that have been addressed by other techniques (Mandal et al., 2016). This allows biochar amendments to further enhance plant growth cover and provide additional benefits. Switchgrass biochar with supplemented nitrogen has been shown to be suitable in highly weathered coastal plain sandy soils (Sigua et al., 2016a). Designer biochar pellets that are produced from appropriate feedstock and impregnated with inorganic fertilizers can improve OC and nutrient availability, while sequestering salts (Na or Cl) to optimize plant growth and dune structural integrity. Many benefits, including enhanced vegetation, coastal protection, prevention of saltwater intrusion, and carbon sequestration, make designer biochar amendments an attractive strategy for coastal dune restoration.

22.5 ELECTRICAL PROPERTIES OF CHARCOAL/BIOCHAR

Charcoal production has a long and rich history (Brown et al., 2015). Early production used either brick kilns or buried feedstocks materials to achieve carbonization, and the

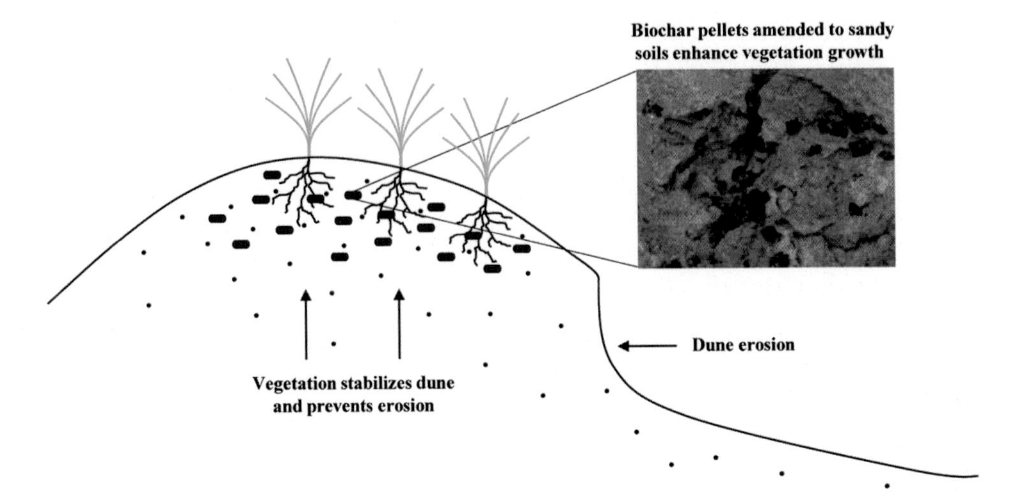

FIGURE 22.2 Pictogram of a coastal dune with vegetation replenishment to prevent dune erosion. Biochar (dark pellets) are amended to highly weathered coastal plain sandy soils to enhance vegetation growth, dune structural integrity, coastal protection, prevention of saltwater intrusion, and carbon sequestration. Source: *Eli Moore and Kate Hall.*

resulting material had a variety of functions (e.g., fuel, black pigment). Noting the distinction of this carbon-rich product from coal, Cross and Bevan (1882) proposed the term of *pseudocarbons* for these elemental forms of carbon produced by the pyrolysis of carbonaceous feedstocks; however, this label did not stick. "Amorphous carbon" is another term used to describe these same materials and is often used to distinguish charcoal from highly-ordered graphitic carbon. Unfortunately, even today, terminology relating to carbon materials remains convoluted (e.g., organic C does not equal biochar C). As defined earlier in this book, "biochar" is a relatively new term describing charcoal that is intended for application to soil; however, the unique electrical properties of biochar (charcoal) are also opening the door to new non-soil applications for this material (see Chacón et al., 2017; Cheng et al., 2017). Because this section draws heavily on historic research into the electrical conductivity properties of these carbonaceous materials, the term charcoal will primarily be used to avoid confusion among the forms of thermally altered carbon as a function of their purpose (e.g., energy, carbon sequestration, or soil application). The concepts and structural discussions, however, apply to biochar and all thermally altered carbons.

The first investigations into the electrical properties of charcoal date back to the late 1800s. In 1846 Faraday (1846) was one of the first to discover charcoal's diamagnetic behavior (repelled by either an N or S pole of a magnet) resulting from its unique electron configuration. By 1898 Cellier (1898) had found the ratio of thermal and electrical conductivities for various carbonized materials were significantly different, which was contrary to the universal relationship known for metal conductors. As these early investigations were underway, the developed world was growing increasingly dependent on coal for energy production. However, at the end of the 19th century, the overall efficiency of burning coal was only around 3% and needed to be improved. In 1894 Ostwald (1894), suggested an innovative solution of a coal battery that would generate electricity from the reaction of carbon with oxygen and would be more efficient than the combustion of coal and subsequent conversion of steam into electricity. A short time later, a 1.5 kW battery was designed, transforming coal C and oxygen to electricity (Jacques, 1896). However, after 6 months, this battery failed due to the formation of carbonate species in the electrolyte, thereby altering the chemical reactions and reducing the output of electricity (Borchers, 1894). This was the first use of pyrolysis C (biochar) as a power source.

Charcoal's ability to conduct electricity is a result of the resonance (sp^2 hybridization orbitals) of aromatic carbon and the delocalized electron clouds that result and surround these materials. Recall that electron delocalization lowers the potential energy of the system, and thus achieves a more chemically stable C than the original feedstock material (Morrison and Boyd, 1979). Additionally, the decentralized electrons provide charge separation in space, thereby creating a negatively charged volume envelope that surrounds the charcoal particle in space (e.g., Tokita et al., 2006; Fig. 22.3). This decentralized electron cloud can act as an electron superhighway, like in transition metals, which explains the exceptional conductivity of these materials. The electrostatics associated with this delocalized cloud are also key in accounting for the measured cation exchange capacity of charcoals (calculated through Gouy-Chapman theory of electrical double layers; Corapcioglu and Huang, 1987). The static electricity arising from the negatively charged surfaces is easily observed on biochar particles during handling and weighing.

Research on single cell wall carbon nanotubes has found that the highest interaction potentials are near the center of pores and not the external surface area (e.g., surface sorption; Kaneko et al., 2012). This interaction potential results from the conductive nature of the carbonized material and is not due to the individual chemical surface moieties. Therefore entry into charcoal pores is not solely driven by diffusional transport, since the electrostatics of the carbon pores contribute an electrostatic attraction force as well (Kaneko et al., 2012; Ohba et al., 2012), thereby overcoming the time that would typically be required for pore-diffusion process to occur in liquids and gases. However, once a compound enters the small nanopores of charcoal, the electron orbitals of the sorbed molecule

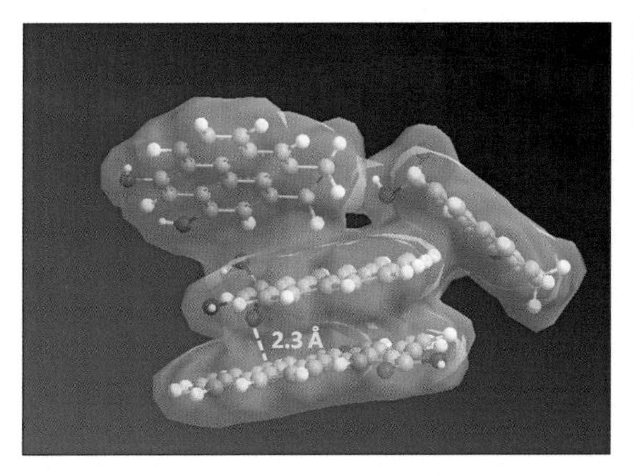

FIGURE 22.3 Electron cloud overlap and potential conductivity throughout a biochar sample. Image created using CHEM3D 15.0. Source: *Kate Hall.*

can interact (as a function of the orbital geometry, hybridization, etc.) with the decentralized aromatic sp^2 orbital, resulting in electrostatic expansion forces that exert the pressure needed to swell the material. This reasoning has also been linked to why crystals grow more slowly on amorphous carbonized materials, since different charged dissolved species would be impacted differently (Franklin, 1951).

While many mechanical properties of charcoal (e.g., density, flammability) correlate well with production temperature, the electrical conductivity of charcoal lacks a clear relationship with production conditions. The general trend is an increase in conductivity with pyrolysis temperature; however, many anomalies in this trend have been observed, particularly at mid-range temperatures (~500°C), where unexplainable jumps in conductivity have occurred. Additionally, there are five orders of magnitude increase in charcoal electrical conductivity with increasing pyrolysis temperatures from 650°C to 1050°C. More importantly, the observed increase in charcoal conductivity does not correlate with changes in the X-ray diffraction spectra, Fourier transformed infrared spectra, or total elemental analysis data (Mochidzuki et al., 2003). Therefore there is little chemical evidence to aid in identifying the mechanisms responsible for alterations in charcoal electrical properties.

As far back at 100 years, researchers have suggested that charcoal production variability contributes to research obstacles in understanding charcoal conductivity mechanisms (Mixter, 1893; Bancroft, 1920). For example, the elemental composition of the solid product has been observed to vary between 70% and 90% in the same batch, with C contents increasing to the center of charcoal produced in a static kiln. In addition to pyrolysis conditions, differences in post-production handling (e.g., water cooling, oxygen exposure, storage conditions; Puri, 1970; Mattson and Mark, 1971) and actions such as grinding (Laskowski, 1987) can further alter the charcoal surface charge and control whether it is positive or negative. It has also been observed that contamination of the charcoal surface with inorganic or organic materials also influences its bulk electrical properties (Celik and Somasundaran, 1980), redox potentials (Joseph et al., 2010), and the strength of the electric double layer (Siffert and Hamieh, 1989).

Biochars ability to act as a battery and as a capacitor has been linked to its ability to distribute electrons among organic structures. Generally, heteroatoms with higher number of valence electrons (e.g., P, N) result in a hybridized orbital structure more stable and possess a higher resonance energy than if the C atom was not substituted (Islas et al., 2007). Such changes to the decentralized electron cloud has been speculated to explain the increased conductivity and decreased resistivity without changes in the bulk or surface chemistry of the material. Because, the electrical properties arise from this interaction of aromatic orbital domains within the charcoal biochar aggregates, the differences in production temperature relate to how complete the restructuring of the C bond hybridization was during pyrolysis. Therefore as the aromaticity of the charcoal increases with increasing temperatures, conductivity is also expected to increase. However, once the degree of aromaticity reaches a certain point (at H/C ratio of <0.1), the electrical conductivity no longer increases and becomes constant (Eley, 1948). Lower temperature non-homogeneous carbonized materials, with partial aromatic clusters (substituted with higher valance elements) seem to be key to the electrical properties of charcoal. To design biochars for improved electrical conductivity, a key element in their production would be choosing the appropriate pyrolysis temperature regime for the feedstock(s).

Free radicals can be supported by this electron cloud surrounding the aromatic domains within the charcoal and their presence may help to explain the imperfect correlation between temperature and conductivity. Free radicals are commonly included in mechanisms during the carbonization phase of biomass pyrolysis (Jakab et al., 2001). The presence of free radicals following coal pyrolysis have been confirmed (Petrakis and Grandy, 1981) as well as in the original biomass sources (Evans and Milne, 1987). Some have even linked the chemical and physical reactivity of charcoals with the presence or absence of these free radicals (Bourke et al., 2007). Additionally, there have been studies that have demonstrated that even mechanical action (e.g., grinding) can form these free radicals and they do not necessarily need to originate from pyrolysis (Urbański, 1967).

Electrochemical capacitors (ECs), which also can be referred to as supercapacitors, electrical double-layer capacitors, or power capacitors, have recently gained a significant amount of attention in the electrical engineering field (Zhang et al., 2009). ECs exhibit 20 to 200 times larger capacitance per unit volume than the conventional capacitors (Chen et al., 2006). This increase in power density is one of the most attractive properties, giving rise to the focus of research on microporous carbons for their potential use in the capacitors (Jianwen et al., 2004). The amount of filler required to achieve high dielectric properties is lower for carbonized materials than ceramics (Jianwen et al., 2004), thereby opening the potential end use of charcoal as an electrode in the EC devices (i.e., Genovese and Lian, 2017; Wang et al., 2017). These are solely a few select examples, as more knowledge is gained on the unique electrical and magnetic properties of charcoal, more tantalizing applications will be conceivable.

22.6 CONCLUSIONS

Biochar utilization in the agronomic sector is at a global scale. Thousands of scientific articles as well as many text books have expounded on its potential to improve degraded soils, reduce nutrient leaching, and impact gas emissions from soils. Biochar's multifunctional properties continue to be discovered and its application into the environmental and electronic sectors continues to grow. In the future, climate variability is expected to make growing crops more difficult and increase storm intensity/activity that will further stress food production. Additionally, more storm activity, as has been recently shown on the Eastern seaboard of the United States, will likely cause more stress to coastal areas through ocean surges and flooding of communities. It is offered that biochars can be designed to improve soil plant relationships in sand dunes. A better plant growth environment, whereby plant cover is increased, should contribute to their improved stability from wind and water erosion. Sand dunes act as buffers, so the benefit to coastal areas would be less flooding of communities, wetlands, and saltwater intrusion into coastal groundwater sources. This is a new scientific area for improving dune stability using biochars that are designed to match existing or perceived future problems. Meanwhile, the electronic sector continues to capitalize on biochars as energy sources and in conduits for energy transfer. Improvement in biochars' electronic capabilities through activation with chemicals or inorganic material is a form of designing biochar. Thus we offer that there is a need to continue expansion of the designer biochar technology into other sectors because of it multifunctional properties.

References

Akhtar, S.S., Anderson, M.N., Liu, F., 2015. Biochar mitigates salinity stress in potato. J. Agron. Crop Sci. 201 (5), 368–378.

Ali, S., Rizwan, M., Qayyum, M.F., Ok, Y.S., Ibrahim, M., Riaz, M., et al., 2017. Biochar soil amendment on alleviation of drought and salt stress in plants: a critical review. Environ. Sci. Pollut. Res. 24 (14), 12700–12712.

Antal, M.J., Grønli, M., 2003. The art, science, and technology of charcoal production. Ind. Eng. Chem. Res. 42, 1619–1640.

Arthur, E., Ahmed, F., 2017. Rice straw biochar affects water retention and soil air movement in a sand-textured tropical soil. Arch. Agron. Soil Sci. 63 (14), 2035–2047.

Atkinson, C., Fitzgerald, J., Hipps, N., 2010. Potential mechanism for achieving agricultural benefits from biochar application to temperate soils: a review. Plant Soil 337, 1–18.

Bancroft, W.D., 1920. Charcoal before the War. I. J. Phys. Chem. 24 (2), 127–146.

Barbier, E.B., Hacker, S.D., Kennedy, C., Kock, E.W., Stier, A.C., Silliman, B.R., 2011. The value of estuarine and coastal ecosystem services. Ecol. Monogr. 81 (2), 169–193.

Barone, D.A., McKenna, K.K., Farrell, S.C., 2014. Hurricane Sandy: beach-dune performance at New Jersey beach profile sites. Shore Beach 82 (4), 13–23.

Beesley, L., Marmiroli, M., 2011. The immobilization and retention of soluble arsenic, cadmium, and zinc by biochar. Environ. Pollut. 159, 474–480.

Biederman, L.A., Harpole, W.S., 2013. Biochar and its effects on plant productivity and nutrient cycling: a meta-analysis. GCB Bioenergy 5, 202–214.

Borchers, W., 1894. Direct production of electricity from coal and combustible gasses. Electr. Rev. (Lond.) 35, 887–890.

Bourke, J.M., Manley-Harris, C., Fushimi, K., Dowaki, T., Nunoura, T., Antal, M.J., 2007. Do all carbonized charcoals have the same chemical structure? 2. A model of the chemical structure of carbonized charcoal. Ind. Eng. Chem. Res. 46 (18), 5954–5967.

Brown, R., del Campo, B., Boateng, A.A., Garcia-Perez, M., Masek, O., 2015. Fundamentals of biochar production. In: Lehmann, J., Joseph, S. (Eds.), Biochar for Environmental Management: Science, Technology and Implementation, second ed. Earthscan, Routledge Publ., London, UK, pp. 39–61.

Busscher, W.J., Novak, J.M., Evans, D.E., Watts, D.W., Niandou, M.A.S., Ahmedna, M., 2010. Influence of pecan biochar on physical properties of a Norfolk loamy sand. Soil Sci. 175, 10–14.

Cantrell, K.B., Hunt, P.G., Uchimiya, M., Novak, J.M., Ro, K.S., 2012. Impact of pyrolysis temperature and manure source on physicochemical characteristics of biochar. Bioresour. Technol. 107, 419–428.

Celik, M.S., Somasundaran, P., 1980. Effect of pretreatments on flotation and electrokinetic properties of coal. Colloids Surf. 1 (1), 121–124.

Cellier, L., 1898. On the conductivity of carbon for heat and electricity. Lond. Edinb. Dublin Philos. Mag. J. Sci. 45 (272), 124.

Chacón, F.J., Cayuela, M.L., Roig, A., Sánchez-Monedero, M.A., 2017. Understanding, measuring and tuning the electrochemical properties of biochar for environmental applications. Rev. Environ. Sci. Biotechol. 16 (4), 695–715.

Chaves, M.M., Flexas, J., Pinheiro, C., 2009. Photosynthesis under drought and salt stress; regulation mechanism from whole plant to cell. Ann. Bot. 103, 551–560.

Chen, C., Zhao, D., Wang, X., 2006. Influence of addition of tantalum oxide on electrochemical capacitor performance of molybdenum nitride. Mater. Chem. Phys. 97, 156–161.

Cheng, B.H., Zeng, R.J., Jiang, H., 2017. Recent developments of post-modification of biochar for electrochemical energy storage. Bioresour. Technol. 246, 224–233.

Corapcioglu, M.O., Huang, C.P., 1987. The surface acidity and characterization of some commercial activated carbons. Carbon 25 (4), 569–578.

Crane-Droesch, A., Abiven, S., Jeffery, S., Torn, M.S., 2013. Heterogeneous global crop yield response to biochar: a meta-regression analysis. Environ. Res. Lett. 8, 044049.

Cross, C.F., Bevan, E.J., 1882. XXXVIII. On pseudo-carbons. Philos. Mag. Ser. 5 13 (82), 325–329.

Eley, D.D., 1948. Phthalocyanines as semiconductors. Nature 162, 819.

Evans, R.J., Milne, T.A., 1987. Molecular characterization of the pyrolysis of biomass. Energy Fuels 1 (2), 123–137.

FAO, 2009. Global Agriculture Towards 2050. FAO, Rome, Italy.

Faraday, M., 1846. On new magnetic actions, and on the magnetic condition of all matter. Philos. Trans. R. Soc. Lond. 136, 21–62.

Franklin, R.E., 1951. Crystallite growth in graphitizing and non-graphitizing carbons. Proc. R. Soc. Lond. A 209, 196–218.

Gao, Z., Zhang, Y., Song, N., Li, X., 2017. Biomass-derived renewable carbon materials for electrochemical energy storage. Mater. Res. Lett. 5 (2), 69–88.

Genovese, M., Lian, K., 2017. Polyoxometalate modified pine cone biochar carbon for supercapacitor electrodes. J. Mater. Chem. A 5, 3939–3947.

Glaser, B., Lehmann, J., Zech, W., 2002. Ameliorating physical and chemical properties of highly weathered soils in the tropics with charcoal—a review. Biol. Fertil. Soils 35, 219–230.

Godfray, H.C.J., Beddington, J.R., Crute, I.R., Haddad, L., Lawrence, D., 2010. Food security: the challenge of feeding 9 billion people. Science 327, 812–818.

Gu, X., Wang, Y., Lai, C., Qiu, J., Li, S., Hou, Y., et al., 2015. Microporus bamboo biochar for lithium-sulfur battery. Nano Res. 8 (1), 129–139.

Hamer, U., Marschner, B., Brodowski, S., Amelung, W., 2004. Interactive priming of black carbon, and glucose mineralization. Org. Geochem. 35, 823–830.

Hardisty, J., 1994. Beach and nearshore sediment transport. In: Pye, K. (Ed.), Sediment Transport and Depositional Processes. Blackwell Publ, London, UK, pp. 216–255.

Holland, H.D., 2006. The oxygenation of the atmosphere and oceans. Philos. Trans. R. Soc. B: Biol. Sci. 361, 903–915.

Huang, J.C., Poor, P.J., Zhao, M.Q., 2007. Economic valuation of beach erosion control. Mar. Resour. Econ. 22 (3), 221–238.

Ippolito, J.A., Spokas, K.A., Novak, J.M., Lentz, R.D., Cantrell, K.B., 2015. Biochar elemental composition and factors influencing nutrient retention. In: Lehmann, J., Joseph, S. (Eds.), Biochar for Environmental Management: Science, Technology and Implementation, second ed. Earthscan, Routledge Publ., London, UK, pp. 139–164.

Ippolito, J.A., Ducey, T.F., Cantrell, K.B., Novak, J.M., Lentz, R.D., 2016. Designer, acidic biochar influences calcareous soil characteristics. 2016. Chemosphere 142, 184–191.

Islas, R., Chamorro, E., Robles, J., Heine, T., Santos, J.C., Merino, G., 2007. Borazine: to be or not to be aromatic. Struct. Chem. 18 (6), 833–839.

Jacques, W.W., 1896. Electricity directly from coal. Harper's New Monthly Mag .

Jakab, E., Blazso, M., Faix, O., 2001. Thermal decomposition of mixtures of vinyl polymers and lignocellulosic materials. J. Anal. Appl. Pyrolysis 58, 49–62.

Jeffery, S., Verheijen, F.A., van der Velde, M., Bastos, A.C., 2011. A quantitative review of the effects of biochar application to soils on crop productivity using meta analysis. Agric. Ecosyst. Environ. 144, 175–187.

Jianwen, X., Wong, M., Wong, C.P., 2004. Super high dielectric constant carbon black-filled polymer composites as integral capacitor dielectrics. In: Proceedings of the 54th Electronic Components and Technology Conference (IEEE Cat. No.04CH37546), vol. 531, pp. 536–541.

Joseph, S.D., Camps-Arbestain, M., Lin, Y., Munroe, P., Chia, C.H., Hook, J., et al., 2010. An investigation into the reactions of biochar in soils. Aust. J. Soil Res. 48, 501s–515.

Kammann, C., Ippolito, J., Hagemann, N., Borchard, N., Cayuela, M., Estavillo, J., et al., 2017. Biochar as a tool to reduce the agricultural greenhouse-gas burden-known, unknowns, and future research needs. J. Environ. Eng. Landsc. Manage. 25 (02), 114–139.

Kaneko, K., Itoh, T., Fujimori, T., 2012. Collective interactions of molecules with an interfacial solid. Chem. Lett. 41 (5), 466–475.

Kappler, A., Wuestner, M.L., Ruecker, A., Harter, J., Halama, M., Behrens, S., 2014. Biochar as an electron shuttle between bacteria and Fe(III) minerals. Environ. Sci. Tech. 1, 339–344.

Klupfel, L., Keiluweit, M., Kleber, M., Sander, M., 2014. Redox properties of plant biomass-derived black carbon (biochar). Environ. Sci. Technol. 48, 5601–5611.

Laird, D.A., Brown, R.C., Amonette, J.E., Lehmann, J., 2009. Review of the pyrolysis platform for coproducing bio-oil and biochar. Biofuels Bioprod. Bioref. 3, 547–562.
Lal, R., 2009. Soils and food sufficiency. A review. Agr. Sustain. Dev. 29, 113–133.
Lal, R., 2010. Managing soils for a warming earth in a food-insecure and energy-starved world. J. Plant Nutr. Soil Sci. 173, 4–15.
Lal, R., 2015. Restoring soil quality to mitigate soil degradation. Sustainability 7, 5875–5895.
Laskowski, J.S., 1987. Coal electrokinetics: the origin of charge at coal/water interface. Prepr. Pap., Am. Chem. Soc., Div. Fuel Chem.; United States, 32(CONF-870410-).
Li, H., Dong, X., da Silva, E.B., de Oliveira, L.M., Chen, Y., Ma, L.Q., 2017. Mechanisms of metal sorption by biochars: biochar characteristics and modifications. Chemosphere 178, 466–478.
Liu, P.L.F., Lynett, P., Fernando, H., Jaffe, B.E., Fritz, H., Higman, B., et al., 2005. Observations by the International Tsunami Survey Team in Sri Lanka. Science 308, 1595.
Mandal, S., Sarkar, B., Bolan, N., Novak, J.M., Ok, Y.S., Van Zwieten, L., et al., 2016. Designing biochar products for maximizing greenhouse gas mitigation potential. Crit. Rev. Environ. Sci. Technol. 46 (17), 1367–1401.
Mattson, J.S., Mark Jr., H.B., 1971. Activated Carbon: Surface Chemistry and Adsorption from Solution. Marcel Dekker, New York, p. 30.
Maun, M.A., 2009. The Biology of Coastal Sand Dunes. Oxford University Press Inc., New York, NY.
Meng, X., Yuan, W., 2014. Can biochar couple with algae to deal with desertification? J. Sustain. Bioenergy Syst. 4, 194–198.
Mixter, W.G., 1893. On the deportment of charcoal with the halogens, nitrogen, sulphur, and oxygen. Am. J. Sci. Ser. 3 45, 363–379.
Mochidzuki, K., Soutric, F., Tadokoro, K., Antal Jr., M.J., Tóth, M., et al., 2003. Electrical and physical properties of carbonized charcoals. Ind. Eng. Chem. Res. 42 (21), 5140–5151.
Morrison, R.T., Boyd, R.N., 1979. Organic Chemistry, third ed. Allyn and Bacon Inc. Publ., Boston, MA.
NIBIO, 2017. Norwegian Institute of Bioeconomy Research. Food security threatened by sea-level rise. Sci. Daily . Available from <http://www.sciencedaily.com/releases.2017/01/170118082423.htm>.
Novak, J., Sigua, G., Watts, D., Cantrell, K., Shumaker, P., Szogi, A.A., et al., 2016. Biochars impact on water infiltration and water quality through a compacted subsoil layer. Chemosphere 42, 160–167.
Novak, J.M., Busscher, W.J., 2012. Selection and use of designer biochars to improve characteristics of southeastern USA coastal plain soils. In: Lee, J.W. (Ed.), Advanced Biofuels and Bioproducts, vol. 1. Springer, New York, pp. 69–97.
Novak, J.M., Busscher, W.J., Watts, D.W., Laird, D.A., Niandou, M.A.S., Ahmedna, M.A., 2008. Influence of pecan-derived biochar on chemical properties of a Norfolk loamy sand soil. American Society of Agronomy-Crop Science of America-Soil Science Society of America Annual Meeting, Houston, 5–9 October 2009. Available from <http://www.biochar-international.org/>.
Novak, J.M., Busscher, W.J., Laird, D.L., Ahmedna, M., Watts, D.W., Niandou, M.A.S., 2009. Impact of biochar on fertility of a southeastern coastal plain soil. Soil Sci. 174, 105–112.
Novak, J.M., Cantrell, K.B., Watts, D.W., Busscher, W.J., Johnson, M.G., 2014. Designing relevant biochars as soil amendments using lignocellulosic-based and manure-based feedstocks. J. Soils Sediments 14, 330–343.
Ohba, T., Kanoh, H., Kaneko, K., 2012. Facilitation of water penetration through zero-dimensional gates on rolled-up graphene by cluster–chain–cluster transformations. J. Phys. Chem. C 116 (22), 12339–12345.
Ostwald, W., 1894. Zeitschrift fur Elektrotechnik und Elektrochemie. July 15, vol. 1 (4). pp. 122–125.
Petrakis, L., Grandy, D.W., 1981. Free radicals in coals and coal conversion. 3. Investigation of the free radicals of selected macerals upon pyrolysis. Fuel 60 (2), 115–119.
Phalan, B., Balmford, A., Green, R.E., Scharlemann, J.P.W., 2011. Minimizing the harm to biodiversity of producing more food globally. Food Policy 36, S62–S71.
Pignatello, J.J., Uchimiya, M., Abiven, S., Schmidt, M.W.I., 2015. Evolution of biochar properties in soil. In: Lehmann, J., Joseph, S. (Eds.), Biochar for Environmental Management: Science, Technology and Implementation, second ed. Earthscan, Routledge Publ., London, UK, pp. 195–233.
Puri, B.R., 1970. In: Walker Jr, P.L. (Ed.), Chemistry and Physics of Carbon, vol. 6. Marcel Dekker, New York, p. 191.
Ray, D., Mueller, N.D., West, P.C., Foley, J.A., 2013. Yield trends are insufficient to double global crop production by 2050. PLoS ONE 8 (6), e66428.

Siffert, B., Hamieh, T., 1989. Effect of mineral impurities on the charge and surface potential of coal: application to obtaining concentrated suspensions of coal in water. Colloids Surf. 35, 27–40.

Sigua, G.C., Novak, J.M., Watts, D.W., Szogi, A.A., Shumaker, P.D., 2016a. Impact of switchgrass biochars with supplemental nitrogen on carbon-nitrogen mineralization in highly weathered Coastal plain Ultisols. Chemosphere. 145, 135–141.

Sigua, G.C., Novak, J.M., Watts, D.W., 2016b. Ameliorating soil chemical properties of a hard setting subsoil layer in Coastal Plain USA with different designer biochars. Chemosphere. 142, 168–175.

Spokas, K.A., Cantrell, K.B., Novak, J.M., Archer, D.W., Ippolito, J.A., Collins, H.P., et al., 2012. Biochar: a synthesis of its agronomic impact beyond carbon sequestration. J. Environ. Qual. 41, 973–989.

Thomas, S.C., Frye, S., Gale, N., Garmon, N., Launchbury, R., Machado, N., et al., 2013. Biochar mitigates effects of salt additions on two herbaceous plant species. J. Agron. Crop Sci. 201 (5), 62–68.

Tilman, D., Blazer, C., Hill, J., Befort, B.L., 2011. Global food demand and the sustainable intensification of agriculture. Proc. Natl. Acad. Sci. U.S.A. 108, 20260–20264.

Tokita, S., Sugiyama, T., Noguchi, F., Fujii, H., Hidehiko, K., Kobayashi, H., 2006. An attempt to construct an iso-surface having symmetry elements. J. Comput. Chem. Jpn. 5 (3), 159–164.

Urbański, T., 1967. Formation of solid free radicals by mechanical action. Nature 216 (5115), 577–578.

Van der Meulen, F., 1982. Vegetation changes and water catchment in a Dutch Coastal Dune area. Biol. Conserv. 24, 305–316.

Wang, Y., Zhang, Y., Pei, L., Ying, D., Xu,, X., Zhao, L., et al., 2017. Converting Ni-loaded biochars into supercapacitors: implication on the reuse of exhausted carbonaceous sorbents. Sci. Rep. 7, Article number: 41523.

Wootton, K., Miller, J., Miller, C., Peek, M., Williams, A., Rowe, P., 2016. New Jersey Sea Grant Consortium Dune Manual. Retrieved from <njseagrant.org/dunemanual>, November 30, 2017.

Zhang, Y., Feng, H., Wu, X., Wang, L., Zhang, A., Xia, T., et al., 2009. Progress of electrochemical capacitor electrode materials: a review. Int. J. Hydrogen Energy 34, 4889–4899.

Index

Note: Page numbers followed by "*f*" and "*t*" refer to figures and tables, respectively.

C

T

n the United States
nasters